AF566451

MTP International Review of Science

Aromatic Compounds

MTP International Review of Science

Publisher's Note

The MTP International Review of Science is an important new venture in scientific publishing, which we present in association with MTP Medical and Technical Publishing Co. Ltd. and University Park Press, Baltimore. The basic concept of the Review is to provide regular authoritative reviews of entire disciplines. We are starting with chemistry because the problems of literature survey are probably more acute in this subject than in any other. As a matter of policy, the authorship of the MTP Review of Chemistry is international and distinguished; the subject coverage is extensive, systematic and critical; and most important of all, new issues of the Review will be published every two years.

In the MTP Review of Chemistry (Series One), Inorganic, Physical and Organic Chemistry are comprehensively reviewed in 33 text volumes and 3 index volumes, details of which are shown opposite. In general, the reviews cover the period 1967 to 1971. In 1974, it is planned to issue the MTP Review of Chemistry (Series Two), consisting of a similar set of volumes covering the period 1971 to 1973. Series Three is planned for 1976, and so on.

The MTP Review of Chemistry has been conceived within a carefully organised editorial framework. The over-all plan was drawn up, and the volume editors were appointed, by three consultant editors. In turn, each volume editor planned the coverage of his field and appointed authors to write on subjects which were within the area of their own research experience. No geographical restriction was imposed. Hence, the 300 or so contributions to the MTP Review of Chemistry come from many countries of the world and provide an authoritative account of progress in chemistry.

To facilitate rapid production, individual volumes do not have an index. Instead, each chapter has been prefaced with a detailed list of contents, and an index to the 10 volumes of the MTP Review of Organic Chemistry (Series One) will appear, as a separate volume, after publication of the final volume. Similar arrangements will apply to the MTP Review of subsequent series.

Butterworth & Co. (Publishers) Ltd.

Organic Chemistry
Series One

Consultant Editor
D. H. Hey, F.R.S.
Department of Chemistry
King's College, University of London

Volume titles and Editors

1 STRUCTURE DETERMINATION IN ORGANIC CHEMISTRY
Professor W. D. Ollis, F.R.S., *University of Sheffield*

2 ALIPHATIC COMPOUNDS
Professor N. B. Chapman, *Hull University*

3 AROMATIC COMPOUNDS
Professor H. Zollinger, *Swiss Federal Institute of Technology, Zurich*

4 HETEROCYCLIC COMPOUNDS
Dr. K. Schofield, *University of Exeter*

5 ALICYCLIC COMPOUNDS
Professor W. Parker, *University of Stirling*

6 AMINO ACIDS, PEPTIDES AND RELATED COMPOUNDS
Professor D. H. Hey, F.R.S. and Dr. D. I. John, *King's College, University of London*

7 CARBOHYDRATES
Professor G. O. Aspinall, *Trent University, Ontario*

8 STEROIDS
Dr. W. F. Johns, *G. D. Searle & Co., Chicago*

9 ALKALOIDS
Professor K. F. Wiesner, F.R.S., *University of New Brunswick*

10 FREE RADICAL REACTIONS
Professor W. A. Waters, F.R.S., *University of Oxford*

INDEX VOLUME

Physical Chemistry
Series One
Consultant Editor
A. D. Buckingham
Department of Chemistry
University of Cambridge

Volume titles and Editors

1 **THEORETICAL CHEMISTRY**
Professor W Byers Brown, *University of Manchester*

2 **MOLECULAR STRUCTURE AND PROPERTIES**
Professor G. Allen, *University of Manchester*

3 **SPECTROSCOPY**
Dr. D. A. Ramsay, F.R.S.C., *National Research Council of Canada*

4 **MAGNETIC RESONANCE**
Professor C. A. McDowell, F.R.S.C., *University of British Columbia*

5 **MASS SPECTROMETRY**
Professor A. Maccoll, *University College, University of London*

6 **ELECTROCHEMISTRY**
Professor J. O'M Bockris, *University of Pennsylvania*

7 **SURFACE CHEMISTRY AND COLLOIDS**
Professor M. Kerker, *Clarkson College of Technology, New York*

8 **MACROMOLECULAR SCIENCE**
Professor C. E. H. Bawn, F.R.S., *University of Liverpool*

9 **CHEMICAL KINETICS**
Professor J. C. Polanyi, F.R.S., *University of Toronto*

10 **THERMOCHEMISTRY AND THERMODYNAMICS**
Dr. H. A. Skinner, *University of Manchester*

11 **CHEMICAL CRYSTALLOGRAPHY**
Professor J. Monteath Robertson, F.R.S., *University of Glasgow*

12 **ANALYTICAL CHEMISTRY —PART 1**
Professor T. S. West, *Imperial College, University of London*

13 **ANALYTICAL CHEMISTRY — PART 2**
Professor T. S. West, *Imperial College, University of London*

INDEX VOLUME

Inorganic Chemistry
Series One
Consultant Editor
H. J. Emeléus, F.R.S.
Department of Chemistry
University of Cambridge

Volume titles and Editors

1 **MAIN GROUP ELEMENTS—HYDROGEN AND GROUPS I–IV**
Professor M. F. Lappert, *University of Sussex*

2 **MAIN GROUP ELEMENTS—GROUPS V AND VI**
Professor C. C. Addison, F.R.S. and Dr. D. B. Sowerby, *University of Nottingham*

3 **MAIN GROUP ELEMENTS—GROUP VII AND NOBLE GASES**
Professor Viktor Gutmann, *Technical University of Vienna*

4 **ORGANOMETALLIC DERIVATIVES OF THE MAIN GROUP ELEMENTS**
Dr. B. J. Aylett, *Westfield College, University of London*

5 **TRANSITION METALS—PART 1**
Professor D. W. A. Sharp, *University of Glasgow*

6 **TRANSITION METALS—PART 2**
Dr. M. J. Mays, *University of Cambridge*

7 **LANTHANIDES AND ACTINIDES**
Professor K. W. Bagnall, *University of Manchester*

8 **RADIOCHEMISTRY**
Dr. A. G. Maddock, *University of Cambridge*

9 **REACTION MECHANISMS IN INORGANIC CHEMISTRY**
Professor M. L. Tobe, *University College, University of London*

10 **SOLID STATE CHEMISTRY**
Dr. L. E. J. Roberts, *Atomic Energy Research Establishment, Harwell*

INDEX VOLUME

Organic Chemistry
Series One

Consultant Editor
D. H. Hey, F.R.S.

MTP International Review of Science

Volume 3

Aromatic Compounds

Edited by **H. Zollinger**
Swiss Federal Institute of Technology

Butterworths · London
University Park Press · Baltimore

THE BUTTERWORTH GROUP

ENGLAND
Butterworth & Co (Publishers) Ltd
London: 88 Kingsway, WC2B 6AB

AUSTRALIA
Butterworths Pty Ltd
Sydney: 586 Pacific Highway 2067
Melbourne: 343 Little Collins Street, 3000
Brisbane: 240 Queen Street, 4000

NEW ZEALAND
Butterworths of New Zealand Ltd
Wellington: 26–28 Waring Taylor Street, 1

SOUTH AFRICA
Butterworth & Co (South Africa) (Pty) Ltd
Durban: 152–154 Gale Street

ISBN 0 408 70277 X

UNIVERSITY PARK PRESS

U.S.A. and CANADA
University Park Press
Chamber of Commerce Building
Baltimore, Maryland, 21202

Library of Congress Cataloging in Publication Data

Zollinger, Heinrich, 1919–
Aromatic compounds.

(Organic chemistry, series one, v. 3) (MTP international review of science)
Includes bibliographical references.
1. Aromatic compounds. I. Title.
QD251.2.074 vol. 3 [QD331] 547'.008s [547'.6]
ISBN 0–8391–1031–6 72–11632

First published 1973 and © 1973
MTP MEDICAL AND TECHNICAL PUBLISHING CO. LTD.
St. Leonard's House,
St. Leonard's Gate,
Lancaster, Lancs.
and
BUTTERWORTH & CO. (PUBLISHERS) LTD.

Filmset by Photoprint Plates Ltd., Rayleigh, Essex
Printed in England by Redwood Press Ltd., Trowbridge, Wilts
and bound by R. J. Acford Ltd., Chichester, Sussex

Consultant Editor's Note

The subject of Organic Chemistry is in a rapidly changing state. At the one extreme it is becoming more and more closely involved with biology and living processes and at the other it is deriving a new impetus from the extending implications of modern theoretical developments. At the same time the study of the subject at the practical level is being subjected to the introduction of new techniques and advancements in instrumentation at an unprecedented level. One consequence of these changes is an enormous increase in the rate of accumulation of new knowledge. The need for authoritative documentation at regular intervals on a world-wide basis is therefore self-evident.

The ten volumes in Organic Chemistry in this First Series of biennial reviews in the MTP International Review of Science attempt to place on record the published achievements of the years 1970 and 1971 together with some earlier material found desirable to assist the initiation of the new venture. In order to do this on an international basis Volume Editors and Authors have been drawn from many parts of the world.

There are many alternative ways in which the subject of Organic Chemistry can be subdivided into areas for more or less self-contained reviews. No single system can avoid some overlapping and many such systems can leave gaps unfilled. In the present series the subject matter in eight volumes is defined mainly on a structural basis on conventional lines. In addition, one volume has been specially devoted to methods of structure determination, which include developments in new techniques and instrumental methods. A further separate volume has been devoted to Free Radical Reactions, which is justified by the rapidly expanding interest in this field. If there prove to be any major omissions it is hoped that these can be remedied in the Second Series.

It is my pleasure to thank the Volume Editors who have made the publication of these volumes possible.

London D. H. Hey

Preface

Following the principles on which the International Review of Science series are based, authors of this Volume treat various aspects of aromatic chemistry. For some 80 years after Kekulé postulated the structure of benzene, aromatic chemistry was the chemistry of benzene, naphthalene and the higher polycyclic hydrocarbons derived from benzene, including heteroaromatic compounds like pyridine. In the 1930s, Hückel's $4n+2$ rule and the discovery of azulene derivatives by Plattner were landmarks which started an expansion of aromatic chemistry in the theoretical aspects as well as the discovery of new classes of compounds. In the 1950s, the more exact elucidation of reaction mechanisms—particularly those of aromatic substitutions of all types—made progress.

The contributions of this volume present a description of the important problems being discussed at the present time: the intriguing problem of appropriate definitions of aromaticity, mechanistic results, a discussion of the large number of new aromatic compounds in the series of non-alternant hydrocarbons, and the annulenes.

It was planned to include three chapters on technologically important aromatic compounds. Unfortunately, it was possible to realise that plan only for the two chapters on aromatic intermediates and on organic dyestuffs and optical brighteners. We hope that the second series will also contain a chapter on aromatic polymers.

Aromatic polymers are of interest because of their importance in temperature resistant materials used in space research, for fire-resistant fibres[1] and similar purposes. The chemistry of new aromatic polymers is treated in a series of papers, e.g. References 2–6.

References

1. Preston, J. (1969). *Applied Polymer Symposia,* **9,** 213
2. Noren, G. K. and Stille, J. K. (1971). *Macromol. Reviews,* **5,** 385
3. Preston, J., Dewinter, W. F. and Black, W. B. (1969). *J. Polymer Sci. A–1,* **7,** 283
4. Bingham, A. and Ellis, B. (1969). *J. Polymer Sci. A–1,* **7,** 3229
5. Alvino, W. M. and Frost, L. W. (1971). *J. Polymer Sci. A–1,* **9,** 2209
6. Montando, G., Finocchiaro, P. and Caccamese, S. (1971). *J. Polymer Sci. A–1,* **9,** 3627

Zurich H. Zollinger

Contents

1
Electrophilic Reactions

R. TAYLOR
University of Sussex

1.1 INTRODUCTION

This review of the period January 1970–September 1971, inclusive, covers both the title reactions and those which produce carbonium ions at the α-positions in side chains. Papers describing substituent effects are also included because of their relevance to electrophilic substitution.

More new electrophilic substitutions have been reported, the arguments for[1] and against[2] the involvement of π-complex rate-determining transition states have been re-emphasised, and reviews have described the effects of the COOH(R)[3] and amide[3a] substituents, substitution in cyclopentadienide anions[4], sulphonation[5], and nitration[6], the problems of nitrosation and encounter control in the latter being stressed. Progress has been made towards quantitative correlations of *ortho*-substituent effects, and the importance of electronic rather than steric effects in determining isomer ratios in electrophilic substitution re-emphasised[7] (cf. Ref. 8). Notable in substituent effect theory is the continuing flow of evidence for hyper-conjugation which is now so impressive in total that even if the proofs coming from electron spin resonance studies were neglected, general acceptance of this phenomenon now seems inevitable.

1.2 THEORIES AND EVALUATIONS OF SUBSTITUENT EFFECTS

Determinations of σ and σ^+-values continue and extend the range available for use in the Yukawa–Tsuno analysis[9] of electrophilic substitution (Table 1.1). Values of σ were obtained either from ionisation of benzoic acids, side chain brominations, or from ^{19}F n.m.r. measurements, the latter being less reliable in view of recent observations (see below); σ^+-values were obtained either from t-cumyl chloride solvolyses or from pyrolysis of l-arylethyl acetates. Notable features are as follows:

The phosphorus-containing substituents have σ_p and σ_m values which are more positive than the σ_I values, consequently they must be conjugatively electron-withdrawing[11] (cf. Ref. 26). The halogeno-phosphorus and halogeno-iodo substituent effects are parallel, indicating a common electronic origin, and the greater electron-withdrawal by the chloro-relative to the fluoro-containing substituents is very surprising, as is the small electron withdrawal by the charged nitrogen and sulphur substituents. Since $\sigma^+_{p\text{-Me}} = -0.311$, the vinyl group appears to have a $-I$ effect (cf. the ethynyl substituent[27]) and the low σ_p value for the pivaloyl substituent suggests steric hindrance to co-planarity of the carbonyl group with the ring[21]. The difference between σ^+_p and σ_p for the silicon-containing substituents indicates hyperconjugative electron release from the carbon–silicon bond[19], and the smaller activation by the *p*-SH relative to the *p*-SMe substituent contrasts to the results for the carbon and oxygen analogues and shows that the hyperconjugative order of electron release does not arise from steric hindrance to solvation[22].

Other work has shown the quantitative similarities of the NF_2 and NO_2 substituents[28], and a large range of σ-values have been obtained for NHX, OX and N=X substituents; in the former two categories X produces a similar attenuation effect[29]. σ_I-Values for the 1-*ortho*- and 1-*meta*-carboranyl

Table 1.1 Substituent constants determined 1970–1971

Substituent	σ_I	σ_R	σ_m	σ_p	σ_m^+	σ_p^+	*Ref.*
$PO{\cdot}Ph_2$	0.27(*p*), 0.26(*m*)*		0.42	0.68			10, 11
$PS{\cdot}Ph_2$	0.23(*p*), 0.14(*m*)*		0.38	0.63			10, 11
$PO{\cdot}(OEt)_2$	0.32(*p*), 0.34(*m*)*						10, 11
$PO{\cdot}Bu_2$	0.34(*p*), 0.38(*m*)*						10
PPh_2			0.12	0.26			11
$P(OEt)_2(C_6H_4F)_2$	0.15	0.06					12
$P(C_6H_4F)_2$	0.45	0.12					12
PF_4	0.45	0.35					13
PF_2	0.39						13
PCl_2	0.44						13
IF_4	1.00	0.15					14
IF_2	0.86	−0.03					14
ICl_2	1.00	0.03					14
$I(OAc)_2$	0.82	0.06					14
$I(OCOCF_3)_2$	1.22	0.12					14
OCH_3	0.25	−0.43					15
$OCHF_2$	0.45	−0.24					15
OCF_3	0.55	−0.18					15
SCH_3	0.13	−0.16					15
SCH_2F	0.27	−0.07					15
$SCHF_2$	0.33	0.06					15
SCF_3	0.44	0.06					15
SO_2CH_3	0.48	0.16					15
SO_2CH_2F	0.55	0.22					15
SO_2CHF_2	0.59	0.28					15
SO_2CF_3	0.73	0.31					15
Me_2S^+			0.06	0.13			16
Me_3N^+			0.06	0.13			16
SO_2^-			−0.02	−0.05			17
SO_3^-			+0.05	+0.09			18
CH_2SiMe_3				−0.29		−0.54	19
$CH(SiMe_3)_2$				−0.33		−0.62	19
$C(SiMe_3)_3$				−0.27		−0.52	19
$CH_2{=}CH{-}CH_2{-}$					−0.023	−0.221	20
$Me_2{=}CH{-}CH_2{-}$						−0.262	20
$CH_2{=}CH{-}CMe_2{-}$						−0.212	20
$CH_2{=}CH{-}CH_2{-}CH_2{-}$						−0.262	20
HCO				0.45			21
MeCO				0.48			21
$MeCH_2CO$				0.48			21
Me_2CHCO				0.47			21
Me_3CO				0.32			21
PhCO				0.46			21
SH						−0.365	22
SPh					0.145		23
CH_2Ph						−0.27	23
OPh						−0.53	23
CF_3					0.565		24
$SiMe_2F$				0.17			24a
$SiMeF_2$				0.23			24a
SiF_3				0.011			24a
$SiMe_2Cl$			0.066	0.022			24a
$SiMeCl_2$			0.089	0.079			24a
$SiCl_3$			0.190	0.198			24a

Table 1.1—*continued*

Substituent	σ_I	σ_R	σ_m	σ_p	σ_m^+	σ_p^+	*Ref.*
$SiMe_2Br$				0.096			24a
$SiMeBr_2$				0.294			24a
$SiBr_3$				0.406			24a
$SiMe_2OMe$			0.041	−0.02			24a
$SiMe(OMe)_2$			0.037	0.007			24a
$Si(OMe)_3$			0.091	0.134			24a
$CNCl_2C_6H_3$						1.22	24b

*The values derived from the *meta*- and *para*-compounds should of course be the same; they differ somewhat from earlier reports[25].

substituents are 0.47 and 0.27 respectively; a 2-methyl and a 2-phenyl substituent change the former value to 0.45 and 0.51 respectively but a 7-methyl substituent has no effect. Since $\sigma_p = 0.13$ and $\sigma_m = 0.10$ for the *ortho*-substituent[30], it must be capable of conjugative electron release.

Small rings transmit electronic effects in the order: cyclopropane ~ oxirane > aziridine, and since the former is comparable in this respect to a double bond[31], it must be able to transmit these effects conjugatively. The order of the first two rings confirms an earlier report[32] in which, however, the cyclopropyl ring was only 27% as efficient as a double bond. The alkaline solvolyses of $Ar_nMe_{3-n}SiH$ and $(ArCH_2)_nMe_{3-n}SiH$ compounds gives the inductive transmission coefficient as 0.33[33] (cf. the average value of 0.36[34]).

In the acid cleavage of $Ar(C{\equiv}C)_nGeEt_3$ compounds there is an excellent correlation of rates for $n = 0$ with those for $n = 1$, 2, or 3, and ρ varies from −4.8 to −0.87. In this system where the positive charge in the transition state is so easily delocalised into the ring, the main effect of substituents must be to stabilise the ring charge[35]. Likewise base cleavages of $Ar(C{\equiv}C)_nMEt_3$ compounds (M = Si or Ge, n = 1–3) correlate with σ-values, indicating that separation of substituent effects into field and resonance components is unsatisfactory[36]; in particular the Dewar–Grisdale[37] method predicts the reactivities less well than do σ-values.

Lewis and Robinson have argued[38] that a single paper on the ^{13}C n.m.r. shifts of methyl substituted compounds[39] is insufficient evidence to propose abandoning the concept (supported by tens of thousands of chemical observations) that alkyl groups are electron supplying, especially since the origins of chemical shifts are less than clearly understood. Another ^{13}C n.m.r. study confirms the greater importance of field or inductive effects at *ortho*-positions[40].

The danger of using ^{19}F n.m.r. shifts to determine substituent effects has been highlighted by the data for $(Me_3Si)_xCH_{3-x}C_6H_4F$ compounds. These indicate (contrary to all chemical observations) that $(Me_3Si)_3C$ has a $-I$ effect, and through-space interactions are probably responsible for the anomaly. The shifts also predict the wrong activating order for some substituents in acid cleavage of $ArSiMe_3$ compounds which yields the σ_m^+-values: Me_3SiCH_2(−0.145); $(Me_3Si)_2CH$(−0.160); $(Me_3Si)_3C$(−0.097)[41]. Dewar has proposed that the effects of substituents on ^{19}F shifts and chemical properties are not related since an improved (FMMF) treatment of substituent effects is satisfactory with regard to the latter but not the former[42].

Through-space interactions with fluorine, structure (1), have been proposed following an ^{19}F n.m.r. analysis of CH_2X, CHX_2 and CX_3 substituents where X = F, Cl, Br, OMe, SCF_3 and CN. Only for the latter is there a linear additive substituent effect, the others showing a saturation effect at disubstitution; for the cyano substituents these interactions are not possible[43]. This is interpreted as supporting the proposal that *m*-orientation of the CF_3

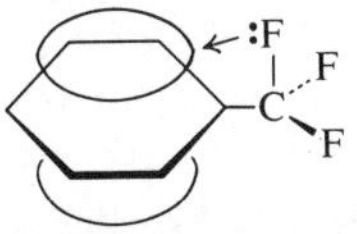

(1) 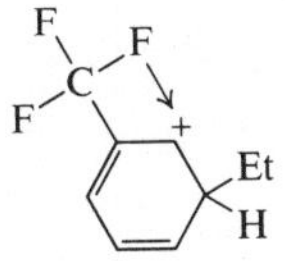 (2) 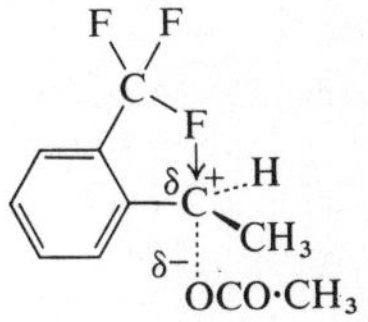 (3)

substituent does not arise from C—F hyperconjugation, but rather from back donation of *p*-electrons from fluorine, structure (2), thereby stabilising the transition state for *meta*-substitution[44]. However, this should make σ^+-values for *m*-CF_3 (and also *p*-CF_3) more positive than σ-values, contrary to fact[45].

A clearer example of interaction between fluorine and a carbonium centre comes from pyrolysis of 1-(*o*-trifluoromethyl)ethyl acetate, the rate of which is unexpectedly high[24]. The proposed interaction, structure (3), is sterically much more favourable than those above, and correspondingly more likely. In this reaction, the *ortho*-halogens deactivate more relative to *meta* substituents than they do in electrophilic substitution, and this is general for *all* reactions producing carbonium ions at the α-position in the side chain; the direct field effect is considered responsible[24]. This probably accounts too for the abnormality of *ortho* oxygen- and sulphur-containing substituents in the pyrolysis and all reactions producing side-chain carbonium ions[23]. The pyrolysis has also been used[46] to show that the non-additivity of fluorine substituent effects in protode-silylation and -stannylation of pentafluorobenzene[47] does not arise primarily from solvent effects or steric acceleration; enhanced conjugative electron release[47] and a reduced inductive effect[46] are probably both contributory factors; non-additivity applies in the 3,5-dichloro-4-cyanophenyl substituent[24b].

The charge distribution in the benzenonium ion itself[48], structure (4), is qualitatively similar to that in the pentamethylbenzenonium ion[49], structure (5). It does not seem to be generally appreciated that with these models for

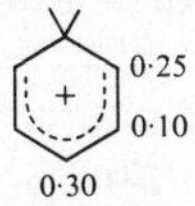

(4) 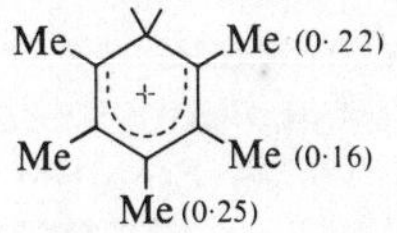 (5) 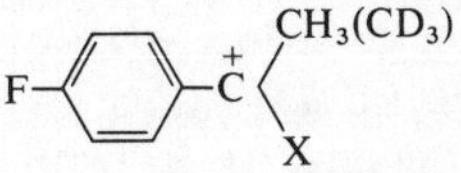 (6)

the transition state of electrophilic substitution it is not necessary to invoke C—F hyperconjugation to explain the effect of the CF_3 substituent anymore than it is necessary to use C—H hyperconjugation to explain the effect of the CH_3 substituent; the model correctly predicts the reactivity and orientation pattern in each case (for fuller discussion see Ref. 50).

^{19}F N.M.R. shifts in methyl-*p*-fluorophenylcarbonium ions and deuterium analogues, structure (6) correlate with the electron supplying effect but not the size of X, showing that hyperconjugation is an electronic rather than a steric phenomenon[51]. Further evidence for hyperconjugation comes from charge transfer studies of $ArCH_2SnR_3$ compounds. The *p*-CH_2SnMe_3 and *p*-CH_2SnPh_3 substituents are very strongly electron supplying ($\sigma^+ = -0.93$ and -0.78, respectively) yet when these are constrained such that co-planarity of the aryl *p*-orbitals and the carbon–metal bond is prevented, electron supply is negligible[52]. A study of the ionisation potentials of the π-systems of a series of silyl- and germyl-substituted indanes, also shows that hyperconjugation is responsible for the strong activation by the CH_2—MR_3 substituent[53]. Measurements of dipole moments and Kerr constants of *p*-$XC_6H_4Bu^t$ compounds are interpreted in terms of hyperconjugation[54] which is also the factor responsible for the Baker–Nathan order in heats of formation of alkylbenzenonium ions in magic acid[55].

The solvolysis of benzyl tosylates has been used as a model for electrophilic substitution and yield polycyclic σ^+-values[56]: naphthalene, -0.345 (1), -0.193 (2); phenanthrene, -0.207 (2), -0.298 (3), -0.356 (9); pyrene, -0.128 (2), -0.342 (4); fluoranthene, -0.483 (3), -0.455 (8); anthracene, -0.389 (2); and biphenylene, -0.13 (2). The latter value is of doubtful validity (cf. Refs. 57, 58 and 87) and the values for the 2-napthalene, 2- and 3-phenanthrene and 2-anthracene positions disagree markedly with those (-0.135, -0.119, -0.196, and -0.295 respectively) given in the literature[59]. The reactivities correlated with CNDO/2 calculations, but not with SCF-π methods.

Cycloalkyl rings in the side chain in the solvolyses of substituted carbinyl chlorides cause ρ-factor variations of up to 20%[60], a correlation between isomer distribution and acidity in electrophilic substitution has been claimed[61], and the methoxy substituent effect is modified by trifluoroacetic acid[62] confirming earlier reports[63]. Ion cyclotron resonance spectroscopy appears to provide a gas-phase analogue of the first step in electrophilic substitution, and reactions analogous to nitration and Hoesch acylation have been elucidated[64].

1.3 THEORETICAL CALCULATIONS OF ELECTROPHILIC REACTIVITY

An MO calculation indicates that aromatic carbon atoms adjacent to fused strained rings have increased electronegativity[65]. However, this treatment seems to have no advantage over that of Streitwieser *et al.* as both fail to account for the large number of accurate quantitative data (see Ref. 58 and references therein). Simple MO theory can predict the aromaticity of a molecule and its tendency to undergo substitution rather than addition[66].

Reactivities of polycyclics have been calculated in terms of electrostatic interaction potentials[67], and all valence CNDO/2 calculations are claimed to correctly predict the effects of substituents[68], though the predicted deactivation *para* to fluorine, contrasts to its activation in many reactions. These calculations predict, contrary to observation that methyl is an electron acceptor in the ground and *meta* transition state[69]; the method is nevertheless more successful than the Hückel and SCF methods[70]. Semi-empirical relationships for electrophilic reactivity have been derived[71]. Other CNDO calculations accurately predict reactivities in mercuration, bromination and hydrogen exchange[72], and similar calculations have been performed for the benzenonium ion[73].

A field and charge-transfer theory has been developed for the quantitative correlation of substituent effects[74]. This theory predicts that the additivity principle should fail[75], as do empirical calculations[76].

1.4 HYDROGEN EXCHANGE

1.4.1 Acid catalysed

Further partial rate factors for detritiation in anhydrous trifluoroacetic acid at 70 °C have been obtained and for fluoranthene these are[77]: 284 (1), 70.2 (2), 8870 (3), 248 (7) and 4380 (8). Values for substituted fluoranthenes were determined also[78].

Rates (10^7k, s^{-1}, at 30 °C) for some naphthalene derivatives are[79]: 1,8-dimethylnaphthalene, 380 (2), 5.6 (3), 704 (4); acenaphthene, 5140 (3), 9.48 (4), 13 800 (5); perinaphthane, 944 (4), 4.2 (5), 2740 (6). The results for the former are in excellent agreement with the additivity principle[80] (cf. Ref. 79), consequently reduced steric hindrance is not responsible for the enhanced reactivity of acenaphthene in hydrogen exchange (nor in the other reactions for which this explanation has been proposed); the high reactivity probably derives from the increased sp^2 character of the carbons in the strained substituent. The slightly enhanced reactivity of perinaphthane (which is not strained) has a parallel in the greater reactivity of tetralin over *o*-xylene, and in each of these molecules with a cyclic substituent, the most stable confirmation provides the maximum opportunity for hyperconjugation and hence enhanced reactivity[80].

Detritiation of *o*-xylene, indane, and tetralin (each with a hydroxy or methoxy substituent in the β-aromatic position) in trifluoroacetic acid at 20 °C shows similar variations in $\alpha:\beta$-ratio to that found for the parent hydrocarbon[81], and which have previously been interpreted in terms of change in bond strain between ground and transition states[82]. Differences in the ratio in the present work are difficult to interpret, however, because of the ease with which these substituents hydrogen bond in this medium[62,63], and the bond strength differs according to the reactivity of the aromatic[83].

Detritiation rates for polycyclics in trifluoroacetic acid (with 2 vol % of carbon tetrachloride to improve solubility) or a standard[84] aqueous perchloric acid–trifluoroacetic medium, when taken along with other data[57,77], yield the reactivity sequence: 3-perylene > 1-pyrene > 2-biphenylene > 6-

chrysene > 3-fluoranthene > 9-phenanthrene > 4-pyrene > 1-triphenylene > 1-naphthalene > 1-phenanthrene > 4-phenanthrene > 3-phenanthrene > 1-fluoranthene > 2-phenanthrene > 2-triphenylene > 2-naphthalene > 1-biphenylene > 2-pyrene > 2-fluoranthene[85]. Correlation of these rates *v.* σ^+-values[56] appeared to yield a curve, but it should be noted that use of other literature σ^+-values[59] gives a much more satisfactory line. The rates were satisfactorily correlated by either the HMO-ω or CNDO/2 methods, except for the strained molecules, e.g. biphenylene, but localisation energies derived from simple HMO theory were poor indices of reactivity[86].

Relative rates of deuteration of methylthio- and 1,1,'-dimethylthioferrocene by deuterotrifluoroacetic acid were: Ferr, 100; 1-MeS-Ferr, 35; 1,1'-$(MeS)_2$-Ferr, 20, which indicates that here the methylthio substituent is electron *withdrawing*[87]. This further demonstrates the very variable electron response of this substituent (cf. Ref. 22) and in the present case it appears that the very high reactivity of the aromatic results in there being no net electron release from the substituent in the transition state. Although sulphur-containing substituents do not normally hydrogen bond in trifluoroacetic acid (t.f.a), this may occur in these molecules which have a very high electron availability. Relative rates for other substituted ferrocenes (ferrocene = 1.0) are shown in structures (7)–(9), and demonstrate reduced substituent effects consistent with high aromatic reactivity.

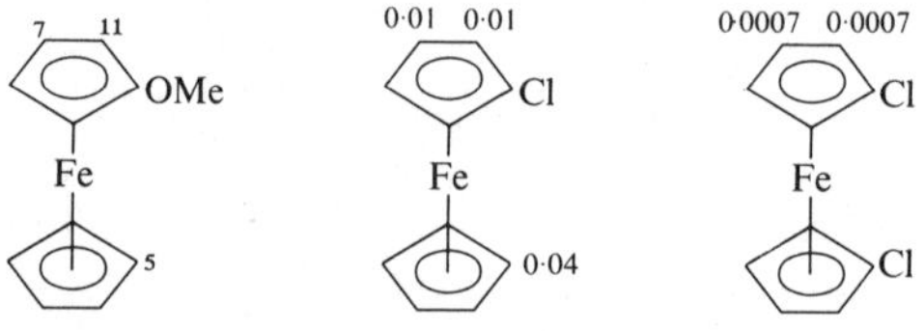

7 11 OMe Fe 5

(7)

0·01 0·01 Cl Fe 0·04

(8)

0·0007 0·0007 Cl Fe Cl

(9)

The value of f_o^{SMe} for detritiation in t.f.a. is 21 500 and taken along with other data shows that $\log f_o$: $\log f_p$ is very consistent for a range of ArXY compounds (X = CH_2, O, or S; Y = Me, H or Ph), consequently the high demand for resonance stabilisation of the transition state for hydrogen exchange masks the differential inductive effects at the *ortho-* and *para-*positions[23]; this contrasts with protodesilylation where wide variations in the ratio occur.

A somewhat unexpected solvent effect on reactivity in dedeuteration of pentamethylbenzene, [2-^{2}H]-1,3-dimethoxybenzene, and [3,4-$^2H_2$]-2,5-diethylthiophene in t.f.a. namely: n-$C_6H_{14} \approx CCl_4 \approx$ liq. $SO_2 \gg TMSO_2 \gg$ DMSO, was attributed to various solvent–t.f.a. and solvent–aromatic interactions[88]. Association between t.f.a. and the last three of the following co-solvents caused reduced rates of dedeuteration of [3,4-$^2H_2$]-2,5-di-t-butylthiophene in n-C_6H_{14}, CCl_4, HOAc, THF, and 1,4-dioxane[89]. Dedeuteration of pentamethyl- and pentamethoxy-benzene in BF_3–EtOH mixtures increases with increase in BF_3 concentration, and is 2000 times faster in BF_3–AcOH mixtures of equal BF_3 concentration; the methoxy compound is 40 000 times as reactive as the methyl compound[90]. The temperature dependence of ρ in hydrogen exchange of anilines has also been investigated[91].

The detritiation of [2-^{3}H]-1,3,5-trimethoxybenzene in eight additional acids shows that the overall value of the Brönsted coefficient α should be 0.56 (cf. 0.52 found previously[92]). However, the catalytic effect of the acid depends on the catalyst type, and Brönsted correlations of like classes of catalysts have slopes varying from 0.56 to 0.71 which indicates that the extent of proton transfer in the transition state is a function of the catalyst type which is not surprising since the extent of transfer has already been shown to depend on the reactivity of the aromatic[93].

Hydrogen exchange of the 2,4,6-trihydroxybenzenonium ion is not a concerted process, and proceeds via initial deprotonation, the rates of protonation of the base intermediate agreeing with rates of exchange extrapolated from lower acidities[94]. Deuteration of phenalone by aqueous deuteroacetic acid occurs at the 2-position contrary to theoretical prediction, and probably due to intramolecular exchange brought about by a water molecule coordinated at the 3-position[95].

Deuteration of aryltricarbonylchromium compounds in various acids[96] has led to the claim that this is seven orders of magnitude more reactive than benzene. However, this value is based on the incorrect assumptions that there is a linear dependence of rate upon acidity function, and that media of identical H_0 values produce the same exchange rates (cf. Ref. 97); this error has been made before[98]. Arenedicarbonyl(triphenylphosphine)chromium compounds undergo exchange much more readily than the above, and 1,3,5-trimethyl substitution gives only a threefold rate increase (cf. Ref. 97) doubtless due to the high reactivity of the aromatic. Furthermore, a methoxy group actually deactivated which is consistent with enhanced hydrogen bonding arising from the high electron availability in these molecules. The $-CH_2Fe(CO)_2C_5H_5$ substituent is more activating than methoxy in deuteration[99].

The dedeuteration of naphthalene in the first excited singlet state takes place more rapidly at the β- than at the α-position in $\sim$65 wt % sulphuric acid, the more normal ($\alpha > \beta$) reactivity being found at lower acidities[100].

Re-examination of the isotope effects in hydrogen exchange has shown that for benzene a constant $k_T : k_D$ value (*c.* 0.62) applies in the 80–87 wt % sulphuric acid range[101], and the curvature of log rate *v.* $-H_0$ plots[102] was confirmed. For anisole (which gives an *o*:*p*-value of 0.51 cf. values of 0.33, 0.42 and 0.55 reported earlier[103]) the kinetic isotope effect decreased with increasing acidity. Toluene, *m*- and *p*-xylene, and bromobenzene also showed isotope effects which neither increased with increasing acidity nor decreased with decreasing reactivity, contrary to earlier predictions[104]

1.4.2 Base catalysed

Substituents in aryltricarbonylchromium compounds produce very little variation in rate, but the rates are very much greater than for normal aryl exchange[105], showing that the $Cr(CO)_3$ group is capable of strong electron withdrawal. The PMe_2 substituent, like the PPh_2 substituent is both activating and gives the reactivity order $m > p > o$, the former substituent being less activating[106]. This parallels previous results for NMe_2 and NPh_2 except that for these latter the order was $o > m > p$, and this is consistent with steric hindrance to substitution *ortho* to the bulkier phosphorus-containing substituent, since base-catalysed exchange is more hindered than acid-catalysed exchange.

Exchange in nitrobenzene (which occurs *meta*) is faster in the presence of a cyclopropyl compared to an isopropyl substituent, and if the former is *ortho* to nitro, side-chain substitution also occurs which is interpreted in terms of coordination between the substituents[107]. Exchange in other nitroaromatics has confirmed the usual pattern of high *ortho* reactivity (arising from the dominance of inductive effects in the reaction) combined with small substituent effects, and rate enhancement by electron-withdrawing groups[108].

The aryl hydrogens of [2,2]-paracyclophane exchange 0.45 times slower than those in benzene, consequently there is no detectable effect arising from the possibility of increased *s*-character of the strained substituent carbons[109].

Exchange of the C—H bonds of *o*- and *m*-carboranes with NH_2^-/NH_3 and $EtO^-/EtOH$ media has a higher activation energy in the latter, and higher for the *m*-isomer (which is the less reactive). Substituents in the carboranes have very little effect on the rate and a relative reactivity series is *o*-carborane > indane > fluorene > *m*-carborane > *p*-carborane > triphenylmethane > diphenylmethane[110].

1.5 METALLATION

1.5.1 Lithiation

Rates of lithiation, potassiation, and caesiation of *o*-, *m*- and *p*-carboranes, with alkyl and phenyl substituent in the former two isomers, and also of 9,10-dichloro-*o*-carborane have been determined in terms of equilibrium acidities. The reactivity order is $o \gg m > p$, methyl substitution produces little effect, and the phenyl, and dichloro substituents produce small and large rate increases respectively[110]. Lithiation of 2-fluoronaphthalene gives 54% 1-, and 46% 3-substitution[205].

1.5.2 Thalliation

With thallium(III) salts in trifluoroacetic acid–acetic acid (the latter reduces the rate) toluene is 7 times as reactive as benzene, gives 10% *ortho*-, 3% *meta*-, and 87% *para*-substitution, and a primary isotope effect $k_H:k_D$ of 3.6. The activation energy for reaction of toluene decreases, and entropy of activation becomes markedly more negative as the acetic acid content of the medium is increased. Thalliation by thallium triacetate in the presence of perchloric and sulphuric acids is complex (as for mercuration) and follows from establishment of equilibria between the following electrophiles, the reactivity order of which is $TlX_3 > TlOAcX_2 > Tl(OAc)_2X > Tl(OAc)_3$. The reaction is also more selective towards *p*-substitution at higher acid concentrations[111].

Similar results have been obtained by Henry[112] who showed the reactivities of anisole, benzoic acid, and benzene to follow the normal electrophilic reactivity order to occur *c*. 200–400 times slower than mercuration, to be free of polysubstitution (cf. mercuration) and to give diarylthallium compounds under certain conditions.

1.5.3 Plumbylation

The *meta* isomer of the dimethoxybenzenes is most reactive towards lead tetra-acetate, providing further evidence that plumbylation is an electrophilic substitution[113].

1.6 REPLACEMENT OF A SUBSTITUENT BY HYDROGEN

The rate-determining step for protodecarboxylation of 2,4-dihydroxybenzoic acid varies according to acid strength, since the solvent isotope effect k_{H_2O}/k_{D_2O} decreases from 2.0 in 1 N HCl to 1.3 in 5 N HCl and then remains constant up to 8 N HCl. Changes in mechanism with acidity have been proposed previously for other decarboxylations[114]. Protodecarboxylation of nitrobenzoic acids in glycerol or aniline provides evidence for the incursion of both S_E1 and S_E2 mechanisms[115].

Acid-catalysed protodestannylation has been used to evaluate the electronic effects of the *p*-[$(Me_3Si)_xMe_{3-x}Si$] groups which produce the relative rates: 0.93($x = 0$), 1.31($x = 1$), 1.67($x = 2$), and 1.76($x = 3$), attributable to inductive and hyperconjugative release from the Si—Si bond. Relative rates of cleavage of the carbon–silicon bond (protodesilylation) were: 1.0($x = 0$), 2.1($x = 1$), 0.35($x = 1$) and 0.056($x = 3$) indicating competition between increasing electron release from the substituent (which aids cleavage) and steric hindrance to cleavage[116]. Protodestannylation in an aqueous ethanol medium increases as the water content changes from 0.6 to 6.0 M, and then remains constant up to 17 M; there is a parallel change in ρ which decreases and then remains constant, and competing solvation of the proton and transition state are believed responsible[117].

Base-catalysed protodestannylation does not involve free carbanions since different rates obtain in media containing D_2O and H_2O; the carbanion would not distinguish between these[118]. The abnormal substituent effects in base cleavage of arylsilanes in aqueous dimethyl sulphoxide[119], of arylstannanes, and in base-catalysed hydrogen exchange have been rationalised in terms of a concerted transition state in which transfer of the proton to the partially carbanionic aromatic is partially rate-determining. Substituents which have mesomeric electron release particularly favour this second step of the reaction, thereby producing abnormal rates[120].

Protodeauration, a new electrophilic substitution (equation 1.1) occurs when phenylauritriphenylphosphine reacts with alcoholic hydrogen chloride[121], and protodebrominations of *N*-alkylbromoanilines[121a], *p*-bromo-phenols[121b], and 3-bromo-4-hydroxy-3,5,2′,6′-tetramethyldiphenyl ether[121c] have been reported.

$$C_6H_5AuP(C_6H_5)_3 \xrightarrow{HCl} C_6H_6 + ClAuP(C_6H_5)_3 \quad (1.1)$$

1.7 REACTIONS INVOLVING GROUP 4 ELECTROPHILES

1.7.1 Alkylation

The boron trifluoride-catalysed ethylation of benzene with [2-^{14}C]-ethyl fluoride produced label scrambling when the reaction was carried out in non-polar solvents (but not when in nitromethane) indicating formation of the free carbonium ion under the former conditions[122]. By contrast, alkylation of benzene with [1,1-$^{2}H_2$]-1-chloropropane (with or without 1,2,4-trichlorobenzene as solvent) was accompanied by only minor label scrambling[123], but these are not necessarily conflicting results since a methyl shift or two hydride shifts are required for scrambling in an n-propyl carbonium ion, but only one hydride shift in an ethyl carbonium ion.

In the benzylation of toluene with substituent benzyl chlorides, the product isomer distribution tends towards statistical as the reactivity of the benzyl halide is increased, and the toluene : benzene reactivity ratio decreases. The selectivity relationship is not obeyed, and the results are claimed to show variation of the nature of the transition state from a σ-complex to a π-complex[124]. Variations of ρ-factor according to electrophile reactivity have of course been observed previously in hydrogen exchange, but in that reaction the minimum toluene : benzene ratio is much larger than in benzylation. Possibly the benzylation results follow from inadequate mixing prior to reaction, this being a factor in producing previous anomalies of this kind. The benzylation of benzene and alkylbenzenes at low temperatures has also been reported[125].

Tritylation of *o*- and *m*-aminophenols occurs *para* to the amino group and not at all with *p*-aminophenol, a result which doubtless has a steric explanation. The corresponding thiophenols were all substituted at sulphur as frequently happens in other electrophilic substitutions of thiophenols[126].

The acid-catalysed cycloalkylation of cymenes with alkenes involves initial hydride abstraction from the isopropyl group by a carbonium ion formed initially in the reaction[127]. Hydride ion abstraction is also involved in transalkylation reactions[128]. Contrary to earlier reports, alkylations with cyclopropanes and n-alcohols show similar product distributions indicating a common (carbonium ion) intermediate[129]. Alkylations of toluene[130] and phenols[131] with isobutylene are both first order in the latter. There have been reports of the alkylation of phenol by styrenes[132], of toluene by propylene in sulphuric acid[133], of chlorobenzene by propylene[134], of benzene by lactones in the presence of aluminium tribromide[135] and by alkenones[135a], and of cyclialkylation of primary phenylalkanols to indanes[135b]. The alkylation of polycyclics by alkenes in sulphuric acid is accompanied by sulphonation which is however the minor reaction[136].

1.7.2 Chloromethylation

Chloromethylation of diphenylmethane is first order in aromatic and disubstitution predominates after extended reaction times[137], this being a common feature of the reaction[114]. Activation energies for mono- and disubstitution of diphenylethane in carbon tetrachloride are 8.75 and 4.9 kcal mol^{-1} respectively[138], consequently disubstitution must occur in the monosubstituted ring. The di(chloromethyl)ation of monochloromethyl-*m*- (and *p*-)xylenes is second order[139], and the activity of catalysts in the chloromethylation of mesitylene is: $ZnCl_2 > SnCl_4 > GeCl_4 >> SbCl_3$ [140]; the accelerating effect of the former catalyst in acetic acid is believed to be due to the formation of $H_2^{2+}[ZnCl_2(OCOCH_3)_2]^{2-}$ which raises the acidity and hence the production of the $^+CH_2OH$ ions[141]. With hydrogen chloride as catalyst, ρ was -3.55 for reaction of substituted benzenes[142], and this value is rather lower than found by previous workers[114]. The mechanism for chloromethylation with $MeOCH_2Cl$ has been investigated[143].

1.7.3 Acylation

There have been a large number of reports of acylation and much of the work confirms the much greater steric hindrance to acetylation than to benzoylation, and also that dual mechanisms are possible, i.e. reaction upon the free acylium ion RCO^+ (the rate of which is reduced by electron supply in R) and nucleophilic displacement by the aromatic upon the polarised acyl halide-catalyst complex (the rate of which is increased by electron supply in R)[114]. Such a change in mechanism is probably responsible for the anomalous reactivity sequence in acylation of benzene and mesitylene with aliphatic and aromatic acyl chlorides[144], and also the *lower* reactivity of chloroacetyl chloride than acetyl or benzoyl chlorides towards reactive aromatics[145]; this latter result is consistent with the displacement mechanism which should be favoured by more reactive aromatics.

The marked difference in steric requirements of acetylation and benzoylation is nicely illustrated by the reactivity of phenanthrene which is: $9>1>4>3>2$ (hydrogen exchange[146]), $9>1>3>2>4$ (nitration[147]), $9>3>1>2>4$ (benzoylation[148]), and $3>9>1>2>4$ (acetylation[148]), the steric requirements of the reactions increasing along this series; the order in the hindrance-free hydrogen exchange is that theoretically predicted. Likewise, whereas benzoylation of 2,6-dimethylnaphthalene goes mainly in the 1-position (as predicted from hydrogen exchange[149]), acetylation goes mainly in the 4-position[150]. Also, acetylation of diphenyl ether and diphenyl sulphide occurs at least 99.6% in the *para*-position in each case[151]. Steric hindrance is probably the factor which prevents diacylation of mesitylene by mesitoyl chloride (in contrast to reactions with other acyl chlorides) and also prevents trisubstitution generally[152].

Other reports of acylation of polycyclics concern benzoylation and fumaroylation of triptycene[153] which (like acetylation[154]) goes only in the 2-position as predicted by strain theory[155], and the acetylation of naphthalene in nitrobenzene in which the $CH_3\overset{+}{C}O{\cdot}\overset{-}{A}lCl_4$ complex can eliminate hydrogen chloride to give a complex ($CH_2{=}C{\cdot}Cl{\cdot}OAlCl_2$) in a reaction of 6 kcal mol^{-1} activation energy. This acetylates naphthalene in a first-order reaction with an activation energy (45.0 kcal mol^{-1}) much higher than for the usual mechanism (13.7 kcal mol^{-1})[156]. The acetylation of anthracene in nitrobenzene has also been studied[157]. An i.r. study of the $CH_3COCl{\cdot}AlCl_3$ addition complex indicates the existence of two ionic structures, and one dative covalent structure namely $MeC(Cl){\cdots}O \rightarrow AlCl_3$ [158].

Re-examination of the acetylation of alkylbenzenes in carbon disulphide shows that disproportionation of the aromatic only occurs for secondary and tertiary alkyl aromatics, provided that there is no delay in adding the acyl chloride to the catalyst–aromatic mixture[159]. Variation in the order of mixing of the reagents, and the amount of catalyst can give a reversal of the normal reactivity order in acylation of benzenes and thiophenes[160], and a contributory factor is complexing of the aromatic with the catalyst. The antimony pentachloride-catalysed benzoylation of mesitylene involves rate-determining reaction of the acylating agent with a 2:1 complex of aromatic and catalyst[161] (cf Ref. 162). Complexing of aluminium chloride with the solvent was postulated as the cause of failure to obtain a Hammett correlation in aroylation of benzene in nitrobenzene, whereas in di- or tetra-chloroethane, correlations with ρ-factors of $+1.3$ and $+1.4$ respectively were obtained[163]. This work appears to be unique in so far as σ^+-values for the substituents in the aroyl chloride were required for the correlations which gave *positive* ρ factors, and this is consistent with conjugation between the positive acylium ionic centre and the substituent, and when the latter is electron-supplying, the reactivity of this electrophile is decreased. Reaction of 3-chloropropionyl chloride with halobenzenes requires a ρ-factor of -9.5 [164].

The benzoylation of phenols in the presence of a strong tertiary amine is general-base catalysed, but in the presence of weaker bases this catalysis only applies to phenols with electron-withdrawing substituents[165]. The *ortho*:*para* ratio in benzoylation of phenol can be varied from 0.03 to 0.75 the highest value applying to the Fries reaction[166], and some of the ratios obtained in this work differ from previous reports. Dewar has shown that variable ratios can be obtained in the Fries reaction depending upon the ratio of ester to catalyst. Comparison of these results with benzoylation of phenol by benzoyl chloride led to the conclusion that two mechanisms operate for the Fries reaction. One is a slow, exclusively *ortho*-orientated process which is probably an intramolecular migration, the other is a faster *ortho*, *para*-orientated process, and which is accelerated by excess catalyst, giving a lower overall *ortho*:*para* ratio. The intramolecular rearrangement may occur via the species in structure (10), but not that in structure (11),

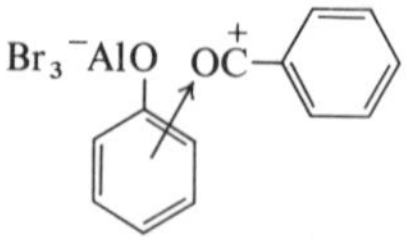

Br_3^-AlO $O\overset{+}{C}$–C_6H_5

(10)

Br_3^-AlO CO

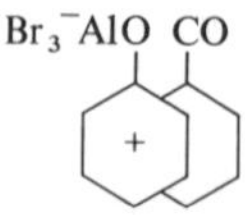

(11)

since the latter would not permit rearrangement of 4-biphenylyl biphenyl-4-carboxylate to give exclusive *ortho* substitution as observed[167]. In the Fries rearrangement of 2-t-butylphenyl propionate, $SbCl_5$ and $ReCl_5$ caused *di*acylation of the ring, and of a range of chlorides, those of silicon and phosphorus do not catalyse the rearrangement at all[168].

Mixtures of products commensurate with the stabilities of the intermediate carbonium ions result from aluminium chloride-catalysed reaction of n-alkenoic acids with benzene[169]. Acylation of *N*-methylaniline is slowed by the addition of salts which operate by decreasing the dissociation of the anhydride, and their effectiveness follows the order: $K^+ > Na^+ > NH_4^+$ [170]. In acetylation of *m*-ferrocenophan-l-ones (structure (12)), the α'/α and β/α ratios decrease markedly as n increased from $2 \rightarrow 4$, and by comparison with acetylation of ferrocenophanes of similar bridge lengths, it appears that

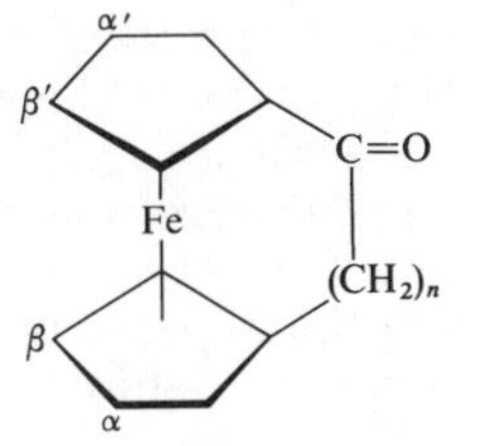

α′ β′ C=O Fe (CH₂)ₙ β α

$n = 2, 3, \text{or } 4$

(12)

the bridge carbonyl exerts a stereospecific deactiving effect in the ring to which it is not attached[171].

Two novel methods of measuring acylation rates have been described. One requires measurement of the adiabatic temperature rise, the other requires addition of excess anisole or mesitylene (which reacts rapidly with the remaining acyl chloride), and this unreacted aromatic is then estimated via a bromate–bromide titration[172]. The acylation of anisole by acetic anhydride catalysed by iodine[173], and by acyl chlorides in perchloric acid[174], have been examined, and the intermediates in *C*-acetylation by acetic anhydride in sulphuric acid are mixed anhydrides of the type $AcOSO_2CH_2COOH$ [175]. Phenanthrene is unlikely to be a by-product of the reaction of 2,3-dibromo-3-phenylpropionic acid with benzene[176], contrary to an earlier report, and no isotope effect is found in the non-catalytic reaction of benzoyl chloride with *p*-chloroaniline in toluene, but the presence of a catalyst gives an isotope effect[177] which is rather surprising. Surprising also, in view of results described above is the observation[178] that increasing the electronegativity of R in RCOOR′ *increases* the aluminium chloride acylation of benzene, as did increased branching in R′. The origin of the isotope effect in the carboxylic acid-catalysed benzoylation of *p*-chlorotoluene has been discussed[178a].

A report of cycloacylation shows that isomeric 4-[2- and 3-bromo(or chloro)-4-methoxy-5-methylphenyl]butanoic acids give three isomeric bromo-(or chloro-)tetralones, the relative yield of which were rationalised in terms of the differing steric and electronic requirements of bromine and chlorine[179].

1.8 REACTIONS INVOLVING GROUP 5 ELECTROPHILES

1.8.1 Diazonium coupling

A large number of diazonium coupling reactions have been examined in terms of the Hammett equation. For coupling of aryldiazonium ions with acetanilide, ρ was $+3.1$, but for the most activating substituents, departure from linearity was observed, and probably arises from the fact that these rates were only three times less than the calculated encounter rate[180]; for coupling with *p*-cresol, *p*-methoxyphenol, and aniline the corresponding ρ-factors were 4.27, 4.09 [181], and approximately 4 [182], this latter being slightly greater for the reaction in water compared to ethanol. Onset of encounter control may account for the departure from linearity in coupling of aryldiazonium ions with naphth-2-ol-6-sulphonic acid ($\rho = 3.2$), or non-additivity of substituent effects in the polysubstituted diazo ions may be responsible[183]. The chloride-ion catalysed diazotisation of *p*-toluidine may be nearly encounter controlled[183a]. For coupling of aryldiazonium ions with 1-naphtholate anion, ρ was 4.15, and 3.94 for 2-substitution in the *less reactive* naphth-1-ol-4-sulphonic acid[184]; this seeming anomaly stems from the general lower sensitivity of *ortho*-substitution to substituents in diazonium ions. Naphth-1-ol and its anion differ in reactivity by 7–9 orders of magnitudes depending upon the diazonium compound used[184]. The claim that diazotised sulphanilic acid and *p*-nitroaniline react much faster with *p*-methoxy phenol than predicted[181] probably has no significance since it assumes that the Hammett equation can be applied to *ortho*-substitution.

A linear σ-complex–hydrogen–base transition state is not a necessary requirement for base catalysis since coupling of the sterically similar 8-phenylnaphth-2-ol and 8-(2′-pyridyl)-naphth-2-ol (both at the 1-position) gave a kinetic isotope effect ($k_H = k_D = 2.7$) and (external) base catalysis for the former compound, but not the latter in which the pyridine ring clearly acts as the base[185].

The coupling of 2-diazophenol-4-sulphonic acid with naphth-1-ol-2-sulphonic acid is novel since it is the resonance stabilised diazonium phenolate ion which reacts, and the reaction is general-base catalysed[186]. *p*-Nitrobenzenediazonium ions and naphth-2-ol-6,8-disulphonic acid anions rapidly form a complex with a charge-transfer structure (as do some other aromatic–diazo ion combinations), and these are proposed as intermediates in the overall substitution pathway[187]. The formation of aggregates of diazonium ions is postulated to account for the reaction-order of 1.65 in coupling of 4,4′-bisdiazobiphenyl with the naphth-2-ol-6,8-disulphonic acid triple anion[188]. Urea catalyses the reaction due to disaggregation of the (less reactive) oligomers, base catalysis, and a dielectric effect[189].

Reaction of *p*-methylbenzenediazonium ions with *N*,*N*-dimethylaniline in tetramethylene sulphone, acetonitrile, water and nitromethane are similar in rate and unaffected by the gegen(an)ion, consequently there can be little solvent reorganisation in the transition state. These results, which contrast markedly to nitration, are consistent with the weakness of the diazonium ion so that it is poorly solvated and interaction in ion pairs is weak, and this is exemplified by the reduction in rate of only 26% on forming a charge-transfer complex[190].

1.8.2 Nitrosation

Substantial primary isotope effects (1.26–4.5) have been observed in nitrosation of a range of aromatic substrates, so that decomposition of the Wheland-type intermediate would seem to be rate-determining. The most reactive substrates reacted by a diffusion controlled process so that the small isotope effects (*c.* 1.3) here arise from the viscosity difference D_2O and H_2O (1.25). The isotope effect (believed to be general in nitrosation) is consistent with the high stability of the nitrosating species, i.e. k_{-1}/k_2 in the normal 2-step A-S_E2 process is large, so no conclusions as to the nature of the electrophile can be made from kinetic studies since its reaction with the aromatic takes place in a rapid pre-equilibrium[191]. Nevertheless, certain alternatives can be shown to be unlikely and from analysis of the acidity dependence of the nitrosation rate of phenol and anisole in perchloric acid, and from other data, it seems probable that the nitrosonium ion is the reactive species above 5M acid[192].

1.8.3 Nitration

In nitration of 4-phenylpyridine and 4-benzylpyridine (which both undergo reaction as the conjugate acids), the deactivation by the positively charged substituent is consistent with a field effect (calculated assuming a unit charge placed at the centroid of the pyridine ring). The *meta*-positions were also more deactivated than by the corresponding —Cl, —CH_2NO_2, and —CH_2CN substituents, indicating the operation of field rather than inductive effects[193].

A field effect also clearly accounts for the 200-fold greater reactivity of the compound shown in structure (13) over that in structure (14)[194], and evidence for this effect also comes from studying nitration of compounds of the

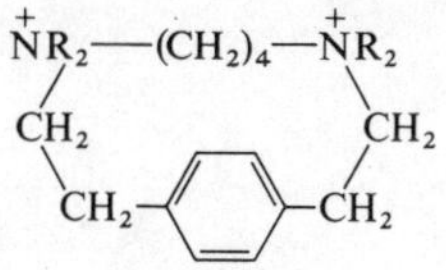

$\overset{+}{N}R_2$—$(CH_2)_4$—$\overset{+}{N}R_2$, CH_2, CH_2, CH_2, CH_2

(13)

$H_3\overset{+}{N}$–CH_2, CH_2–$\overset{+}{N}H_3$, CH_2, CH_2

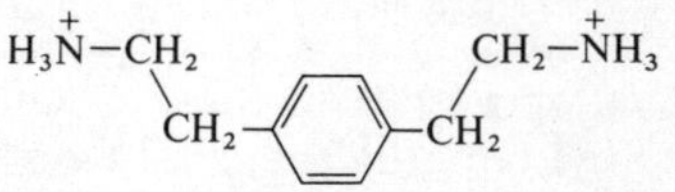

(14)

types ArX and $ArCH_2X$. When X is charged, the electronic transmission factor is about twice that when X is uncharged[195]. New partial rate factors evaluated in this work are as follows: benzotrichloride (f_o, 3.5×10^{-5}; f_m, 1.8×10^{-4}; f_p, 1.0×10^{-4}) and β,β,β-trichloroethylbenzene (f_o, 9.6×10^{-2};

f_m, 2.7×10^{-1}; f_p, 7.3×10^{-1}), and for the former there is very little discrimination between deactivation of the *meta* and *para*-positions ($\log f_p$; $\log f_m$ = 1.07), a situation analogous to that produced by the NMe_3^+ substituent ($\log f_p : \log f_m = 1.08$, Table 1.2) for which the possibility of hyperconjugation does not apply. Consequently, carbon–chlorine hyperconjugation is not significant. For the CF_3 substituent $\log f_p : \log f_m = 1.31$ so this discrimination suggests that C—F hyperconjugation is important, and is estimated as contributing 17% to the total free energy of deactivation of the *para*-position.

Further support for the value of *c.* 0.5 for the transmission coefficient [defined[195] as $(\log f_m^{CH_2X} - \log f_m^{CH_3})/\log f_m^X$] for positive poles comes from nitration of compounds with sulphur and selenium poles[197] (Table 1.2), the coefficients being 0.44 and 0.59, respectively. The very large change in rate

Table 1.2 Partial rate factors for nitration of ArX

X	$10^8 f_o$	$10^8 f_m$	$10^8 f_p$
SMe_2^+	0.044	1.10	0.147
$CH_2SMe_2^+$	38 400	93 800	210 000
$SeMe_2^+$	0.719	25.3	3.37
$CH_2SeMe_2^+$	56 400	35 000	415 000
NMe_3^+		9.05	2.24
$CH_2NMe_3^+$		20 300	3980

at the *para*-positions for these poles compared to the nitrogen pole, strongly indicates the operation of $-M$ effects from the former. The deactivating effect of Ph_4P^+ is reported, not surprisingly, to be greater than that of $PhMe_3P^+$[198].

Partial rate factors for nitration by nitric acid in acetic acid of substituted naphthalenes, structure (15), taken along with literature data, give a reasonable correlation with PPP localisation energies[199]. However, the factors are

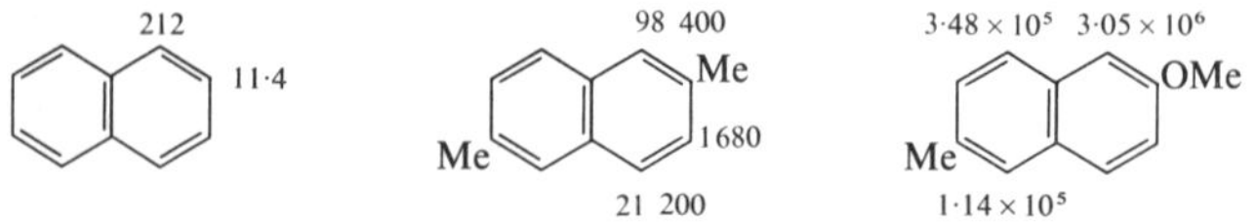

212 11·4

98 400 Me 1680 Me 21 200

3·48 × 10⁵ 3·05 × 10⁶ OMe Me 1·14 × 10⁵

(15)

very much smaller than those calculated by the additivity principle from monomethyl naphthalene data[200] and it is therefore relevant that Schofield *et al.* have argued that partial rate factors $>10^3$ for nitric acid in acetic anhydride are likely to be affected by encounter control[201]. Nitrations in media of similar viscosity which produce factors in excess of this, are likely to involve different electrophiles[202]. Dewar on the other hand obtained a partial rate factor of $>2 \times 10^6$ for 8-substitution of 10-methyl-9,10 borazarophenanthrene and has argued that encounter control does not operate below factors of this magnitude[203]. The $\alpha:\beta$-isomer ratio for naphthalene under other conditions varied thus: 8($AcONO_2$–CCl_4, 40 °C), 10.3 (94% HNO_3–AcOH, reflux), 14 (94% HNO_3–CCl_4, 40°C), and 18.5 (HNO_3–H_2SO_4, 5–50 °C), and increased with increasing polarity for some of the solvents; added metal oxides, nitrates, and copper acetate had no effect on the ratio[204]. Nitric acid in acetic anhydride, and nitronium borofluoride gave different results to nitric acid in nitromethane, sulphuric acid, and acetic acid, in nitration of fluoronaphthalenes[205]. For 1-fluoronaphthalene the reactivity order is 4>5>8>2>7 for the former media and 4>5>8>7>2 for the latter (cf. 4>2>8>5 predicted from hydrogen exchange[149]), and for 2-fluoronaphthalene, 8>5>6>1>4 for the former media and 8>1>5>6>4 for the latter).

The reactivity of toluene towards nitration by acetyl nitrate in acetic anhydride has been re-examined by kinetic and competition methods. The former gave a reactivity relative to benzene of 50±6, and the latter 38±5, the difference being ascribed to probable side reactions for the latter. Since the value is higher than for nitrating media in which the nitronium ion is firmly established as the electrophile, a simple nitronium ion mechanism is unlikely to be applicable to nitration in acetic anhydride[206]. Nitration of a range of alkylbenzenes $PhCH_2R$ by nitric acid–acetic anhydride in nitromethane shows that R has little effect on the partial rate factors which are *c.* 35 (*o*), 3.5 (*m*) and 65 (*p*)[207]. Nitration of nitrophenylpropanes confirms the enhanced electron release from the cyclopropyl group[208].

Nitration of dibenzyl by nitronium borofluoride in sulpholane gives dinitration even when the aromatic is in fivefold excess. Since mixing control

Table 1.3 Partial rate factors for nitration of ArX

X	f_o	f_m	f_p
F	0.045	0.0021	0.58
Cl	0.067	0.0018	0.246
Br	0.077	0.0016	0.200
I	0.17	0.0049	0.400

is therefore dominant under these conditions, nitration of aromatics of similar reactivity to dibenzyl will tend to give apparent relative reactivities near to unity, regardless of the true reactivities of the substrates toward nitronium ion. Reduction of the nitronium salt concentration causes a larger reduction of dinitration than does changing the bibenzyl:nitronium salt ratio, and this also indicates mixing control[209].

A study of nitration of halogenobenzenes in sulphuric and perchloric acids at various temperatures has provided partial rate factors (Table 1.3) and thermodynamic parameters[210]. Iodobenzene behaved normally here, and its previously observed[211] erroneous behaviour in nitration by acetyl nitrate in nitromethane was shown not to arise from nitrodeiodination which could have been responsible[212]. The order of *para*-halogen reactivities parallels that in hydrogen exchange[213], but deactivation by the *ortho*- and to a lesser extent the *meta*-fluoro substituent is greater than predicted from that reaction, and this most likely reflects the differing balance of electrostatic/inductive and mesomeric effects in the two reactions. For the polychlorobenzenes, a very significant breakdown of additivity occurred, especially for the tetra- and penta-chloro compounds. Indeed, the reactivity of the latter is such that the *para*-chloro substituent *activates* when substituted in 1,2,4,5-tetrachlorobenzene (as it does also in nitrobenzene). The reason for this breakdown in not yet fully clear, but it happens also with pentafluorobenzene in other reactions[46, 47].

The effect of the *meta* halogens in nitration, by nitric acid in acetic anhydride, of 4-halogenoanisoles (at the 2-position) is given by the values: f_m^{F}, 0.119; f_m^{Cl}, 0.069; f_m^{Br}, 0.077, and comparison with these in Table 1.3 shows the differences expected on reactivity-selectivity grounds; relative rates of nitrodehalogenation were also obtained (see below)[214].

Nitration by nitric acid in carbon tetrachloride has received attention. Nitration of mesitylene and *p*-xylene is sixth-order in nitric acid and zeroth-order in aromatic, this latter being true also of benzene and toluene[215]. As with other reactions in carbon tetrachloride[114], a negative temperature coefficient is obtained, the rate decreasing 2000-fold for an increase of 46 °C. Addition of trifluoroacetic acid lowers the rate and the order in nitric acid to about four, suggesting that the transition state complex is highly solvated by nitric acid molecules, some of which are replaced by trifluoroacetic acid molecules. Attention has been drawn to the importance in these media of water which, not surprisingly, increases the rate[216], but in this work the rate was found to be dependent upon aromatic concentration for the nitration of benzene, and this was true also for toluene at low concentration. The nitronium ion was concluded to be the electrophile and this reacts close to the rate of encounter for more reactive substrates, e.g. *o*-xylene and mesitylene. The pre-equilibrium step for the formation of the nitronium ion was envisaged as in equation (1.2) giving an order in nitric acid of five. Both sets of work are in agreement in that for example the zeroth-order rate at

$$\begin{array}{ccccc} & & NO_2^+ & & H_3O^+ \\ & & \vdots & & \vdots \\ 5HNO_3 & \rightleftharpoons & NO_3^- \cdot HNO_3 & + & NO_3^- \cdot HNO_3 \end{array} \qquad (1.2)$$

25 °C for nitration of mesitylene is *c.* 4×10^{-8} mol l^{-1} s^{-1} in 15×10^{-2} M nitric acid (cf. an earlier report[217]).

The nitration of diphenyl sulphoxide in 83–100 wt % sulphuric acid takes place on both neutral molecule and conjugate acid, and the extent of *meta* substitution also increases with acidity, thereby showing increasing reaction on the conjugate acid in that direction[218]. Nitration of benzaldehyde by nitric acid in trifluoroacetic acid at 25 °C or in sulphuric acid at 75 °C shows greater selectivity under the former condition as is also found for nitration of benzonitrile and nitrobenzene[219]. The nitration of biphenyl and the nitrobiphenyls by nitric acid in acetic anhydride has been re-examined[220] (cf. Reference 221). Though there are quantitative differences in these studies, they agree in so far as the reactivity and *ortho*:*para* ratio decreases along the series: biphenyl > 3-nitro- > 4-nitro- > 2-nitro-biphenyl.

Nitration of benzene with nitric acid under electrolytic conditions at −40 °C gave a red coloration attributed to the cation, structure (16), the reaction product being nitrobenzene, but with nitric acid–sulphuric acid only *m*-dinitrobenzene was obtained[222]. The reaction between aniline and fuming nitric acid is first order in each and yields aniline nitrate, *p*-nitrosoaniline, nitroanilines, 2,4-dinitroaniline and nitrosobenzene; ammonium metavanadate increased the rate fourfold[223]. The mechanism of the nitration of 1,2,3,10*b*-tetrahydrofluoranthene has been described[224], and there have

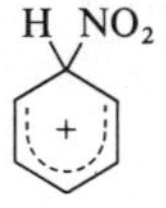

H NO_2

+

(16)

been reports of the nitration of the sodium salt of anthraquinone-1-sulphonic acid[225], and of 4-(*p*-toluenesulphonamide)acetanilide and its methyl derivative with nitric acid, and nitrates, in acetic acid[226].

Nitration, nitrosation, bromination, and Vilsmeier formylation all occur in the cyclopentadienyl ring of triphenylphosphonium cyclopentadienylide showing the greater electrophilic reactivity of cyclopentadiene compared to that of benzene[227]. The 2-position is the preferred position of substitution whereas calculations predict this to be the 3-position. However, 2-substitution involves favourable linear conjugation between phosphorus and the cyclopentadiene ring in the transition state, whereas 3-substitution requires cross-conjugation[228]. The nitration of 8-cyanoheptafulvene occurs readily in the side chain (and this is true also of bromination, formylation, and trifluoroacetylation) thereby providing evidence for the aromatic character of the molecule[229].

Finally, the chemical and engineering aspects of aromatic nitration have been renewed with particular emphasis on kinetics, mechanism, agitation effects and safety factors[230].

1.9 REACTIONS INVOLVING GROUP 6 ELECTROPHILES

1.9.1 Hydroxylation

The low temperature reaction of hydrogen peroxide with aromatics in the presence of hydrogen fluoride gives phenols and the proposal that OH^+ is the electrophile; however, the high *ortho*:*para* ratio suggests that free radicals may be involved[231].

1.9.2 Alkoxylation

Reaction with t-butylhydroperoxide–aluminium chloride gave the relative rates: *m*-xylene (137); *p*-xylene (63.5); *o*-xylene (62.1); toluene (13.8); benzene (1.0), and an isomer distribution for toluene of 50% *ortho*, 8% *meta*, and

36% *para*. The electrophile is believed to be $Bu^t\overset{+}{O}$ giving $ArOBu^t$ which is cleaved by aluminium chloride and subsequently hydrolysed[232].

1.9.3 Acetoxylation

Isolation of a diene intermediate in acetoxylation of *o*-xylene shows this reaction to proceed via an addition-elimination mechanism[233].

1.9.4 Sulphonoxylation

Nitrosulphonoxylation of alkylbenzenes gives partial rate factors typical of an electrophilic substitution and the Baker–Nathan reactivity order at the *para* (but not *meta*) positions. *Ortho*-nitrosulphonoxylation was less selective than *meta*-, consistent with the electronic properties of these substituents[234] and the reaction of bis(*m*-nitrobenzenesulphonyl)peroxide with toluene also gave the normal substitution pattern and was not accelerated by radical promoters[235].

1.9.5 Sulphonation

The *ortho*:*para* ratio in sulphonation of toluene can vary from 0.1 to 0.5 according to the reaction conditions, and the extent of *ortho* substitution tends to decrease as the reactivity of the sulphonating species increases[236]. Sulphonation of polymethylsulphonic acids shows modification of the reactivity consistent with buttressing effects[237], and steric hindrance probably accounts for the reduced rate increase per additional methyl group in a re-examination[114] of the sulphonation of toluene, *m*-xylene, and mesitylene, though protonation effects were proposed[238]. Naphthalene sulphonation kinetics have been re-examined[114], and compared with those for thianaphthalene and 2-methylnaphthalene[239]. Sulphonation of 4-hydrazobenzene involves attack by $H_3SO_4^+$ (<94% H_2SO_4), $H_2S_2O_7$ (94–100% H_2SO_4) and then via partly rate-determining proton removal (>100% H_2SO_4)[240].

1.9.6 Sulphonylation

Aluminium chloride-catalysed *p*-toluene-sulphonylation in dichloromethane gives the third-order rate law (equation 1.3), a kinetic isotope effect $k_H:k_D$ = 2.8 (toluene) and 1.98 (chlorobenzene) and f_p values of 52.5 (Me), 0.224 (Cl)

$$\text{Rate} = k_3[\text{ArSO}_2\text{Cl}]\,[\text{AlCl}_3]\,[\text{ArH}] \qquad (1.3)$$

and 0.164 (Br). Spectroscopic evidence indicated the electrophile to be the $AlCl_3$–TsCl complex and not the free ion. The smaller partial rate factors and lack of isotope effect noted by Jensen and Brown[241] (using benzenesulphonyl chloride) was attributed to the differing reactivities of the electrophile in the two reactions[242], but the small structural difference seems inadequate for this explanation; we should also expect chlorobenzene to have the larger isotope effect. Additional f_p^{Me} values were 31.4 (in $MeNO_2$) and 38.3 (in $PhNO_2$), the electrophile here being an ion pair; the low value of f_p^{OMe} (150) in nitromethane may arise from formation of an anisole–aluminium chloride complex[243].

1.10 REACTIONS INVOLVING GROUP 7 ELECTROPHILES-HALOGENATION

Halogenation of alkylbenzenes has received further attention. The wide variation in effect of Friedel–Crafts catalysts on selectivity in chlorination of toluene is indicated by the f_p^{Me} values of 7.3 ($AlCl_3$) and 590 (I_2), and notably the *ortho*:*para* ratio can be varied from 1.66 to 3.36 [244]. Partial rate factors for molecular bromination pass through a maximum in 93.3 wt % acid (cf. the rate maximum observed in hydrogen exchange in 96 wt % acid[245]) and the very high value of f_m^{t-Bu} (119) is attributed to the importance of ground state solvation of t-butylbenzene in this acid[246]. This argument also accounts for the high f_m^{t-Bu} values previously obtained in chlorination[247] and hydrogen exchange[248], and for the reversal of the *p*-alkyl activation order on carrying out reactions in trifluoroacetic acid. Partial rate factors for bromination of toluene in liquid SO_2 are f_o 3700; f_m, 12, and f_p, 57 000, the largest recorded, and fluorobenzene gives only *para*-substitution[249].

Chlorination of *m*- and *p*-xylene, naphthalene and triphenylene in 96% aqueous dioxan is catalysed by both acid and silver perchlorate and rates, which correlate with acidity function in 0.2–0.6 M acid, indicate $ClOH_2^+$ and $ClAgCl^+$ to be the electrophiles[250]. Iodination by iodine in nitric acid–sulphuric acid gives $\rho = -4.50$ and proceeds via interaction of NO_2^+ and I_2 to give I^+ [251]; iodine in oleum is also a useful iodinating reagent[251a]. Bromination of polymethylbenzenes in water gives $\rho = -10.7$ but this involves use of σ^+-values for *ortho*-substituents which may not be valid[252].

Molecular bromination of methylnaphthalenes shows that steric hindrance to this reaction is less than commonly supposed (see also Reference 8), since the additivity principle holds very well[253]. Further, the reactivity order for 2-methylnaphthalene was 1>8>4>5>6>3>7 (cf. 1>8>4>6>5>3>7 in hindrance-free hydrogen exchange[149]) and previously undetected bromodimethylnaphthalenes were isolated. π-Electron densities correlated the reaction rates better than localisation energies.

Iodination by iodine–peracetic acid shows promise as a preparative method since it is faster than traditional iodination[254], and the increased steric hindrance may give improved preparative selectivity over other halogenations. Chlorination of aromatics by titanium tetrachloride in peroxyacids is also promising in view of the total absence of *ortho*-substitution[255]. By contrast, chlorination of phenols by cupric chloride in toluene or aqueous solution gave very high *ortho*:*para* ratios and the process (probably involving the electrophile $CuCl_4^{2-}$) is continuous in the presence of excess HCl [256]. The palladium(II)-catalysed chlorination of azobenzene is also stereospecific due to formation of an intermediate complex which favours subsequent *ortho*-substitution[257]. Wide variation in the *ortho*:*para* ratio with change in medium dielectric has been reported for chlorination of anisole[258].

A number of reports have concerned the kinetic order and effectiveness of catalysts in halogenation of phenols and ethers[259]. Bases accelerate iodination of phenols and imidazole (which give isotope effects) according to their nucleophilicities, but with dimethylaminobenzenesulphonic acid (which gives no isotope effect) the acceleration is proportional to basicity, and the formation of an alternative more reactive iodinating species through pre-equilibrium between iodine and base was proposed[260].

An unusual reactivity order namely PhSMe > PhOMe for bromination in aqueous acetic acid is dramatically reversed in the anhydrous medium, due apparently to the effect of water on the stability of the PhSMe·Br_2 complex[261], though without rates relative to benzene this conclusion is difficult to evaluate. The bromination of phenols with cyclic substituents in the 3,4-positions shows substitution patterns consistent with explanations of the Mills–Nixon effect or, for the compounds with 7- and 8-membered substituents, with hindrance to 8-substitution[262]. Steric hindrance to both the electrophile and co-planarity of the methoxy substituent with the benzene ring accounts for non-additivity in chlorination of polymethylanisoles with sulphuryl chloride, and which gives the partial rate factors, f_o^{OMe}, 660, f_p^{OMe}. 4800 [263]. Values of f_m^{Cl} of 0.0019 and 0.0026 were derived, assuming the additivity principle, from chlorination of 2-chloroacetanilide[264], and these compare with literature values of 0.0021–0.030; the lower values seem likely to be most correct. Non-co-planarity effects are proposed to account for the bromination of 4-hydroxy-3,5,2′,6′-tetramethyldiphenyl ether in the 3-position (and not the 4′-position as previously believed)[264a]. The kinetics of chlorination of 1-aminoanthraquinones[265], of phenoxyacetic acid and derivatives[266], and bromination of amines[267] have been described.

The selectivity in catalysed chlorination of benzene and chlorobenzene is inversely proportional to the catalytic activity as expected, and changing the dielectric of the medium produces a similar correlation[268]. Bromination of 4-nitrobiphenyl by bromine on a silica surface is first order in both reagents, giving a normal *ortho*:*para* ratio and the electrophile is probably SiOH·Br_2 [269]. Bromination of 2,9- and 3,9-dimethylanthracene is an electrophilic substitution in the absence of light and oxygen[269a].

Halogenation of silanes has received particular attention. Dichlorophenyl-(trichloromethyl)silane is as expected less reactive than dichloromethylphenylsilane and could be brominated in the presence of catalysts[270]; no cleavage of the C—Si bond occurred in contrast to results for phenyltrichloro- and other silanes[271]. Substitution generally occurred *meta* to $SiCl_3$ in chlorination of chloro phenyltrichlorosilanes, though for the *p*-chloro compound, migration of the silyl substituent occurs[272]. Catalysed chlorination of trichloro-1-naphthylsilane goes as expected in the α-position of the unsubstituted ring, and without detectable C—Si cleavage[273]. Orientation in chlorination of phenylfluorosilanes $PhSiMe_{3-n}F_n$ changes from weak *ortho*, *para* (with activation) to strongly *meta* (with deactivation) as n increases from $0 \rightarrow 3$ (with ferric chloride as catalyst) but iodine catalysis gives anomalous results[274]; C—Si cleavage is least significant when $n = 3$. Hyperconjugation is proposed to account for the *ortho*, *para*-orientation (except when $n = 3$) in chlorination of the corresponding benzyl compounds $PhCH_2SiMe_{3-n}F_n$[275].

Two interesting reports of the direct fluorination of the aromatic ring have appeared. A 6% solution of fluorine and benzene (0.7:1 mole ratio) in acetonitrile at −35 °C gave mono- and *ortho*-, *meta*- and *para*-difluorobenzenes in 60:4:1:5 relative ratio. Isomer distributions for toluene (19% *ortho*, 4% *meta*, 15% *para*), nitrobenzene (12% *ortho*, 70% *meta*, 8% *para*), and an $\alpha:\beta$ ratio of 3.0 for naphthalene all indicate an electrophilic substitution. Higher concentrations of reagents gave polymeric by-products and attempted fluorinations of bromo- and chloro-benzenes and methylbenzoate were unsuccessful[276]. A typical electrophilic substitution pattern was obtained in fluorination of benzene, fluorobenzene, benzotrifluoride and nitrobenzene with xenon difluoride in the presence of hydrogen fluoride catalyst, and the xenon can be quantitatively recovered; the electrophile may be a radical cation[277].

Chlorination of carbaphosphaboranes proceeds with decreasing ease along the series *ortho* > *meta* > *para* (as for the carboranes), and the reactivity relative to the corresponding carboranes is greater for the *ortho*, similar for the *meta*, and less for the *para*[278]. The rearrangement of *N*-chloroanilines to chloroanilines has been shown to proceed via formation of the nitrenium ion (equation 1.4), since solvolysis of these compounds in ethanol correlates

$$PhN(R)Cl \xrightleftharpoons{\text{slow}} PhN^{+}R\ Cl^{-} \longleftrightarrow {}^{+}C_6H_5{=}NR\ Cl^{-}\ \text{etc.} \longrightarrow ClC_6H_4NHR \qquad (1.4)$$

with σ^+-values ($\rho = -6.35$), consequently ionisation to the nitrenium ion is the slow step of the reaction; the mechanism of the subsequent step awaits resolution. The higher ρ-factor than is usual in side-chain solvolyses reflects the greater ability of divalent nitrogen to accommodate a positive charge, compared to trivalent carbon[279].

1.11 REPLACEMENT OF A SUBSTITUENT X BY A SUBSTITUENT Y

Some new electrophilic substitutions have been reported, namely halodeauration, mercurideauration, and acyldeauration. For most of these reactions (with phenylauritriphenylphosphines) the reaction is implied by the isolation of the substituted auritriphenylphosphine, though for cleavage by mercury, the arylmercurial was isolated as well[121].

Cleavage of halogens *para* to methoxy by nitric acid in acetic anhydride gives the relative rates of 0.18(I), 0.079(Br), 0.061(Cl) and 1.0(H). This has been interpreted as indicating the effect of the halogens on the rate of attack on the ring carbon to which they are attached[214]. However, breakage of the carbon-halogen bond may be partially rate-determining and the above reactivity order parallels the expected ease of $C^{\delta-}—X^{\delta+}$ polarisation of the bond to carbon. Indeed the relative leaving abilities in electrophilic substitution are reported[280] as: $Cl^+ \sim NO_2^+ \sim R^+ < Br^+ < D^+ \sim ArN_2^+ \sim SO_3 \sim RCO^+ < NO^+ \sim H^+ \sim I^+ \sim$ Hg. A reactivity order which is surprising and contrary both to the above and earlier work[281] comes from chlorodealkylation of 2,6,4-$(Bu^t)_2RC_6H_2OH$ compounds, the relative rates of 4-substitution being 240(R = H), 92(R = Me) and 72(R = Bu^t)[282].

The preparative importance of nitrosodestannylation has been stressed. The weakness of the C—Sn bond facilitates ready cleavage by nitrosyl chloride in methylene chloride at low temperatures to give 25–80% of the nitrosoaromatics which are otherwise difficult to obtain[283].

An S_E1 mechanism has been proposed for iododemercuration of diarylmercurials containing strong electron-withdrawing substituents, since reaction of these is virtually instantaneous. The unsatisfactory kinetics in this work were attributed to initial formation of ArHgX which should react faster by the S_E1 mechanism than would Ar_2Hg[284]. The S_E1 mechanism was also proposed for mercuridemercuration of pentafluorophenyl mercuric bromide by labelled mercuric bromide in dimethyl sulphoxide and benzene. In this reaction, as in the above, nucleophilic catalysts are presumed to coordinate on to mercury and the complex then ionises in the rate-determining step (or, alternatively, the coordination on to mercury is synchronous with the ionisation[285]). The kinetics here are also unsatisfactory, and the conclusions must be treated with caution since previous claims for an S_E1 process have been shown to be eroneous[114].

References

1. Olah, G. A. (1971). *Accounts Chem. Res.,* **4,** 240
2. Ridd, J. H. (1971). *Accounts Chem. Res.,* **4,** 248
3. Kohnstam, G. and Williams, D. H. L. (1969). *Chem. Carboxylic Acid Esters,* 765

3a. Shafer, J. A. (1970). *Chem. Amides,* 685

4. Rybinskaya, M. I. and Korneva, L. M. (1971). *Usp. Khim.,* **40,** 444
5. Hiyarna, H. (1969). *Kagaku Kogyo,* **20,** pp. 313-317 and 417-423
6. Hoggett, J. G., Moodie, R. B., Penton, J. R. and Schofield, K. (1971). *Nitration and Aromatic Reactivity.* (Cambridge University Press); Albright, L. F., Hanson, C. (1969). *Loss. Prev.,* **3,** 26
7. Charton, M. (1971). *J. Org. Chem.,* **36,** 882
8. Norman, R. O. C. and Taylor, R. (1965). *Electrophilic Substitution in Benzenoid Compounds,* pp. 147-148 (Amsterdam: Elsevier)
9. Yukawa, Y. and Tsumo, Y. (1959). *Bull. Chem. Soc. Japan,* **32,** 971
10. Tsvetkov, E. N., Malevannaya, R. A., Lobanov, D. I., Osipenko, N. G. and Kabachnik, M. I. (1969). *J. Gen. Chem. USSR,* **39,** 2369
11. Tsvetkov, E. N., Makhamatkhanov, M. N., Lobanov, D. I. and Kabachnik, M. I. (1970). *J. Gen. Chem. USSR,* **40,** 2376
12. Chang, B. C., Denney, D. Z. and Denney, D. D. (1971). *J. Org. Chem.,* **36,** 998
13. Rakshys, J. W., Taft, R. W. and Sheppard, W. A. (1968). *J. Amer. Chem. Soc.,* **90,** 5236
14. Lyalin, V. V., Syrova, G. P., Orda, V. V., Alekseeva, V. V. and Yagupolskii, L. M. (1970). *J. Org. Chem. USSR,* **6,** 1433

15. Syrova, G. P., Sedova, L. N., Gandelsman, L. Z., Alekseeva, L. A. and Yagupolskii, L. M. (1970). *J. Org. Chem. USSR*, **6,** 2293
16. Hojo, M., Utaka, M. and Yoshida, M. (1971). *Tetrahedron,* **27,** 4255
17. Lindberg, B. J. (1970). *Acta Chem. Scand.,* **24,** 2852
18. McDaniel, D. H. and Brown, H. C. (1958). *J. Org. Chem.,* **23,** 420
19. Cook, M. A., Eaborn, C. and Walton, D. R. M. (1970). *J. Org. Chem.,* **24,** 293
20. Jones, L. B. and Foster, J. P. (1970). *J. Org. Chem.,* **35,** 1777
21. Bowden, K. and Shaw, M. J. (1971). *J. Chem. Soc. B,* 161
22. Bailey, F. P. and Taylor, R. (1971). *J. Chem. Soc. B,* 1446
23. Taylor, R. (1971). *J. Chem. Soc. B,* 1450
24. Taylor, R. (1971). *J. Chem. Soc. B,* 622
24a. Mares, F., Plzak, Z., Hetflejs, J. and Chvalovsky, V. (1971). *Chem. Commun.,* **36,** 2957
24b. Gassman, P. G. and Fentiman, A. F. (1970). *Tetrahedron Lett.,* 1021
25. Tsvetkov, E. N., Lobanov, D. I., Izosenkova, L. A. and Kabachnik, M. I. (1969). *J. Gen. Chem. USSR,* **39,** 2126
26. Bott, R. W., Dowden, B. F. and Eaborn, C. (1965). *J. Chem. Soc.,* 6306
27. Eaborn, C., Thompson, A. R. and Walton, D. R. M. (1969). *J. Chem. Soc. B,* 859
28. Baum, K. (1970). *J. Org. Chem.,* **35,** 1203
29. Exner, O. and Lakomy, J. (1970). *Collect. Czech. Chem. Commun.,* **35,** 1371
30. Zakharkin, L. I., Kalinin, V. N. and Snyakin, A. P. (1970). *J. Gen. Chem.,* **40,** 2411
31. Turner, A. N., Lutz, R. E., McFarlane, N. S. and Boykin, D. W. (1971). *J. Org. Chem.,* **36,** 1107
32. MacDonald, C. J. and Schaefer, T. (1970). *Can. J. Chem.,* **48,** 1033
33. Schutt, G. and Hansen, P. (1970). *Z. Anorg. Allg. Chem.,* **374,** 94
34. March, J. (1968). *Advanced Organic Chemistry,* 242. (New York: McGraw-Hill); Taft, R. W. and Lewis, I. C. (1958). *J. Amer. Chem. Soc.,* **80,** 2436
35. Eaborn, C., Eastmond, R. and Walton, D. R. M. (1970). *J. Chem. Soc. B,* 752
36. Eaborn, C., Eastmond, R. and Walton, D. R M. (1971). *J. Chem. Soc. B,* 127
37. Dewar, M. J. S. and Grisdale, P. J. (1962). *J. Amer. Chem. Soc.,* **84,** 3548
38. Lewis, P. M. E. and Robinson, R. (1970). *Tetrahedron Lett.,* 2783
39. Jackman, L. M. and Kelley, D. P. (1970). *J. Chem. Soc. B,* 102
40. Hess, R. E. (1971). *J. Org. Chem.,* **36,** 2201
41. Bassindale, A. R., Eaborn, C. and Walton, D. R. M. (1970). *J. Org. Chem.,* **21,** 91
42. Dewar, M. J. S., Golden, R. and Harris, J. M. (1971). *J. Amer. Chem. Soc.,* **93,** 4187
43. Sheppard, W. A. (1971). *Tetrahedron,* **27,** 945
44. Sheppard, W. A. (1965). *J. Amer. Chem. Soc.,* **87,** 2410
45. Holz, D. (1971). *Chem. Rev.,* **71,** 139; *Progress in Physical Organic Chemistry,* **8,** 1. (New York: Wiley)
46. Taylor, R. (1971). *J. Chem. Soc. B,* 255
47. Eaborn, C., Treverton, J. A. and Walton, D. R. M. (1967). *J. Org. Chem.,* **9,** 259
48. Olah, G. A., Schlosberg, R. H., Kelly, D. P. and Mateson, G. H. (1970). *J. Amer. Chem. Soc.,* **92,** 2546
49. Maclean, C. L. and Mackor, E. L. (1961). *Molec. Phys.,* **4,** 241; Colpa, J. P., Maclean, C. L. and Mackor, E. L. (1963). *Tetrahedron,* **19** (Suppl. 2), 65
50. Reference 8, pp. 306–310
51. Timberlake, J. W., Thompson, J. A. and Taft, R. W. (1971). *J. Amer. Chem. Soc.,* **93,** 274
52. Hanstein, W., Berwin, H. J. and Traylor, T. G. (1970). *J. Amer. Chem. Soc.,* **92,** 7476
53. Pitt, C. G. (1970). *J. Org. Chem.,* **23,** C35
54. Aroney, M. J., Calderbank, K. E., Le Fevre, R. J. W. and Pierens, R. K. (1970). *J. Chem. Soc. B,* 1120
55. Arnett, E. M. and Larsen, J. W. (1969). *J. Amer. Chem. Soc.,* **91,** 1438
56. Streitweiser, A., Hammond, H. A., Jagow, R. H., Williams, R. M., Jesaitis, R. G., Chang, C. J. and Wolf, R. (1970). *J. Amer. Chem. Soc.,* **92,** 5141
57. Blatchly, J. M. and Taylor, R. (1964). *J. Chem. Soc.,* 4641; (1968), 1402
58. Taylor, R. (1971). *J. Chem. Soc. B,* 536
59. Brown, H. C. and Inukai, T. (1961). *J. Amer. Chem. Soc.,* **83,** 4825
60. Tanida, H. and Tsushima, T. (1970). *J. Amer. Chem. Soc.,* **92,** 3397
61. Allan, Z. J. (1970). *Tetrahedron Lett.,* 2425
62. Levy, G. C., Nelson, G. L. and Cargioli, J. D. (1971). *Chem. Commun.,* 506

63. Baker, R., Bott, R. W., Eaborn, C. and Greasley, P. M. (1964). *J. Chem. Soc.*, **627**; Taft, R. W., Price, E., Fox, I. R., Lewis, I. C., Anderson, K. K. and Davis, G. T. (1963). *J. J. Amer. Chem. Soc.*, **85**, 3146; Peterson, P. E. (1966). *J. Org. Chem.*, **31**, 439
64. Benezra, S. A., Hoffman, M. K. and Bursey, M. M. (1970). *J. Amer. Chem. Soc.*, **92**, 7501
65. Rieke, R. D. (1971). *J. Org. Chem.*, **36**, 227
66. Dixon, W. T. (1970). *J. Chem. Soc. B*, 612
67. Bancroft, K. C. C. and Howe, G. R. (1970). *Tetrahedron Lett.*, 2035
68. Howe, G. R. (1970). *Chem. Commun.*, 868
69. Howe, G. R. (1971). *J. Chem. Soc. B*, 984
70. Howe, G. R. (1971). *J. Chem. Soc. B*, 981
71. Bancroft, K. C. C. and Howe, G. R. (1971). *J. Chem. Soc. B*, 1221
72. Isaacs, N. S. and Cvitas, D. (1971). *Tetrahedron*, **27**, 4139
73. Jakubetz, W. and Schuster, P. (1971). *Angew. Chem.*, **10**, 497
74. Godfrey, M. (1971). *J. Chem. Soc. B*, 1534
75. Godfrey, M. (1971). *J. Chem. Soc. B*, 1545
76. Taylor, R., unpublished work
77. Bancroft, K. C. C. and Howe, G. R. (1970). *J. Chem. Soc. B*, 1541
78. Bancroft, K. C. C. and Howe, G. R. (1971). *J. Chem. Soc. B*, 400
79. Opie, M. C. A., Wright, G. J. and Vaughan, J. (1971). *Aust. J. Chem.*, **24**, 1205
80. Ansell, H. V. and Taylor, R. (1971). *Tetrahedron Lett.*, 4915
81. Selander, H. and Nilsson, J. L. G. (1971). *Acta Chem. Scand.*, **25**, 1182
82. Vaughan, J. and Wright, G. J. (1968). *J. Org. Chem.*, **33**, 2580
83. Eaborn, C. and Spillett, R. E., personal communication
84. Eaborn, C. and Taylor, R. (1961). *J. Chem. Soc.*, 247
85. Streitwieser, A., Lewis, A., Schwager, I., Fish, R. W. and Labana, S. (1970). *J. Amer. Chem. Soc.*, **92**, 6525
86. Streitwieser, A., Mowery, P. C., Jesaitis, R. G. and Lewis, A. (1970). *J. Amer. Chem. Soc.*, **92**, 6529
87. Kursanov, D. N., Setkina, V. N., Fedin, E. I., Nefedova, M. N. and Khatami, A. I. (1970). *Proc. Acad. Sci. USSR*, **192**, 347
88. Serebryanskaya, A. I., Amanchayan, T. S. and Shatenshtein, A. I. (1970). *J. Gen. Chem. USSR*, **40**, 858
89. Zvyagintseva, E. N., Gorbenko-Germanov, D. S. and Shatenshtein, A. I. (1970). *J. Gen. Chem. USSR*, **40**, 1142
90. Sannikov, A. P., Utyanskaya, E. Z. and Shatenshtein, A. I. (1970). *J. Gen. Chem. USSR*, **40**, 1619
91. Biherwolf, T. E., Linder, R. E. and Ling, A. C. (1970). *J. Chem. Soc. B*, 1673
92. Kresge, A. J. and Chiang, Y. (1959). *J. Amer. Chem. Soc.*, **81**, 5509; (1961). **83**, 2877
93. Kresge, A. J., Slae, S. and Taylor, D. W. (1970). *J. Amer. Chem. Soc.*, **92**, 6309
94. Kresge, A. J., Chiang, Y. and Shapiro, S. A. (1971). *Can. J. Chem.*, **49**, 2777
95. El-Anani, A. A., Greig, C. C. and Johnson, C. D. (1970). *Chem. Commun.*, 1024
96. Kursanov, D. N., Setkina, V. N., Baranetskaya, N. K., Zdomovich, V. I. and Anisimov, K. N. (1970). *Proc. Acad. Sci. USSR*, **190**, 127
97. Reference 8, Chapter 8
98. Bykova, E. V., Setkina, V. N. and Kursanov, D. N. (1968). *Proc. Acad. Sci. USSR*, **178**, 31
99. Anderson, S. N., Ballard, D. H. and Johnson, M. D. (1971). *Chem. Commun.*, 779
100. Stevens, C. G. (1971). *Diss. Abs. B*, **31**, 5299
101. Olsson, S. and Russell, M. (1970). *Arkiv Kemi*, **31**, 439, 455; Olsson, S. (1970). *Arkiv Kemi*, **32**, 89, 105
102. Eaborn, C. and Taylor, R. (1960). *J. Chem. Soc.*, 3301
103. Satchell, D. P. N. (1956). *J. Chem. Soc.*, 3911; Lauer, W. M and Day, J. T. (1955). *J. Amer. Chem. Soc.*, **77**, 1904
104. Melander, L. and Olsson, S. (1956). *Acta Chem. Scand.*, **10**, 879
105. Setkina, V. N., Baranetskaya, N. K., Anisimov, K. N. and Kursanov, D. N. (1970). *Bull. Acad. Sci. USSR*, 429
106. Yakovlova, E. A., Isaeva, G. G., Makhamatkhanov, M. M., Zhidkova, A. M., Tsvetknov, E. N., Kabachnik, M. I. and Shatenshtein, A. I. (1970). *J. Gen. Chem. USSR*, **40**, 1614
107. Shabarov, Yu. S. and Kazbulatova, N. A. (1970). *Vestn. Mosk. Univ. Khim.*, **11**, 239; *Chem. Abs.*, **73**, 65884t; Kazbulatova, N. A., Isaeva, G. G., Yakovleva, E. A., Shabarov, Yu. S. and Shatenshtein, A. I. (1970). *Vestn. Mosk. Univ. Khim.*, **11**, 746; *Chem. Abs.*, **74**, 87074b

108. Guthrie, R. D. and Wesley, D. P. (1970). *J. Amer. Chem. Soc.*, **92,** 4057
109. Streitwieser, A. and Ziegler, G. R. (1971). *Tetrehedron Lett.*, 415
110. Yakovleva, E. A., Isaeva, G. G., Kalinin, V. N., Zakharkin, L. I. and Shatenshtein, A. I. (1970). *J. Gen. Chem. USSR,* **40,** 2658; and with Petrov, E. S., Yakushin, F. S. and Vukmirovich, Z. (1970). *J. Org. Chem.*, **23,** 313. These reports are essentially identical
111. Briody, J. M. and Moore, R. A. (1970). *Chem. Ind.*, 803
112. Henry, P. M. (1970). *J. Org. Chem.*, **35,** 3083
113. Norman, R. O. C. and Thomas, C. B. (1970). *J. Chem. Soc. B,* 421
114. For leading references and detailed analyses, see Taylor, R. (1972). *Comprehensive Chemical Kinetics, Vol. 13.* (Amsterdam: Elsevier)
115. Manikyam, D. S. and Sundaram, E. V. (1970). *Bull. Chem. Soc. Japan,* **43,** 3651
116. Cook, M. A., Eaborn, C. and Walton, D. R. M. (1970). *J. Org. Chem.*, **23,** 85, cf. Benkeser, R. A., Hickner, R. A. and Hoke, D. I. (1958). *J. Amer. Chem. Soc.*, **80,** 2279
117. Eaborn, C., Thompson, A. R. and Walton, D. R. M. (1971). *J. Org. Chem.*, **29,** 257
118. Alexander, R., Eaborn, C. and Taylor, T. G. (1970). *J. Org. Chem.*, **21,** 65
119. Cretney, J. and Wright, G. J. (1971). *J. Org. Chem.*, **28,** 49
120. Bassindale, A. R., Eaborn, C., Taylor, R., Thompson, A. R., Walton, D. R. M., Cretney, J. and Wright, G. J. (1971). *J. Chem. Soc. B,* 115
121. Perevalova, E. G., Baukova, T. V., Goryunov, E. I. and Grandberg, K. I. (1970). *Bull. Acad. Sci. USSR,* 2031
121a. Effenberger, F. and Menzel, P. (1971). *Angew. Chem.*, **10,** 493
121b. O'Bara, E. J., Balsley, R. B. and Starer, I. (1970). *J. Org. Chem.*, **35,** 16
121c. Hamilton, S. B. and Blanchard, H. S. (1970). *J. Org. Chem.*, **35,** 3348
122. Natsubori, A. and Nakane, R. (1970). *J. Org. Chem.*, **35,** 3372
123. Lee, C. C. and Woodcock, D. J. (1970). *Can. J. Chem.*, **48,** 858
124. Olah, G. A., Tashiro, M. and Kobayashi, S. (1970). *J. Amer. Chem. Soc.*, **92,** 6369
125. Finocchiaro, P. (1971). *Tetrahedron,* **27,** 581
126. Eberhardt, M. K. and Chuchani, G. (1970). *Tetrahedron,* **26,** 955
127. Boone, D. E., Eisenbraun, E. J., Flanagan, P. W. and Grigsby, R. D. (1971). *J. Org. Chem.*, **36,** 2042
128. Khalaf, A. A. and Roberts, R. M. (1970). *J. Org. Chem.*, **35,** 3717
129. Stolyarov, B. V., Isidorov, V. A. and Ioffe, B. V. (1970). *Proc. Acad. Sci. USSR,* **191,** 208
130. Shah, A. K. and Sharma, M. M. (1969). *Indian Chem. Eng.*, **11,** 125; *Chem. Abs.*, **73,** 98186v
131. Gehlawat, J. K. and Sharma, M. M. (1970). *J. Appl. Chem.*, **20,** 93
132. Kurashev, M. V., Paushkin, Y. M., Mamedaliev, G. M. and Sementsova, L. G. (1970). *Neftekhimiya,* **10,** 840; *Chem. Abs.*, **74,** 63728s
133. Feizkhanov, F. A., Fattakhova, R. S. and Churkin, Yu. V. (1970). *Tr. Nauch. Inst. Neftekhim. Proizvod.*, **2,** 22; *Chem. Abs.*, **74,** 87111m
134. Plotkina, N. I., Gein, N. V. and Plyusnin. (1971). *Zh. Prikl. Khim.*, **44,** 1099
135. Waples, D. W. and Brauman, J. I. (1971). *Chem. Commun.*, 1075
135a. Ansell, M. F. and Mahmud, S. A. (1971). *Tetrahedron Lett.*, 4129
135b. Khalaf, A. A. and Roberts, R. M. (1971). *J. Org. Chem.*, **36,** 1040
136. Pakter, M. K. and Gorokhova, Z. Ya. (1971). *Zh. Prikl. Khim.*, **44,** 1116
137. Kantsova, L. V. and Yukelson, I. I. (1970). *Zh. Prikl. Khim.*, **43,** 2570
138. Kantsova, L. V. and Yukelson, I. I. (1970). *Kinet. Katal.*, **11,** 1334
139. Mironov, G. S., Farberov, M. I. and Budnii, I. V. (1970). *Kinet. Katal.*, **11,** 1349
140. Lyushin, M. M., Mekhtiev, S. D. and Guseinova, S. N. (1971). *Neftekhimiya,* **11,** 42; *Chem. Abs.*, **74,** 111264m
141. Lysuhin, M. M., Mekhtiev, S. D. and Guseinova, S. N. (1970). *J. Org. Chem. USSR,* **6,** 1445
142. Mironov, G. S., Budnii, I. V., Farberov, M. I. and Shein, V. D. (1970). *J. Org. Chem. USSR,* **6,** 1231
143. Svetlov, A. K. and Demenkova, T. N. (1970). *Ref. Zh. Khim.*, abstract 11B 1565; *Chem. Abs.*, **74,** 140488t
144. Gore, P. H., Hoskins, J. A. and Thorburn, S. (1970). *J. Chem. Soc. B,* 1343
145. Belenkii, L. I., Yakubov, A. P. and Goldfarb, Ya. L. (1970). *J. Org. Chem. USSR,* **6,** 2531
146. Bancroft, K. C. C., Bott, R. W. and Eaborn, C. (1964). *Chem. Ind.*, 1951
147. Dewar, M. J. S. and Warford, E. W. T. (1956). *J. Chem. Soc.*, 3570; *Chem. Ind.*, 98
148. Gore, P. H., Thadani, C. K., Thorburn, S. and Yusuf, M. (1971). *J. Chem. Soc. C,* 2329
149. Eaborn, C., Golborn, P., Spillett, R. E. and Taylor, R. (1968). *J. Chem. Soc. B,* 1112

150. Gore, P. H. and Yusuf, M. (1971). *J. Chem. Soc. C*, 2586
151. Finocchiaro, P. (1969). *Boll. Sedute. Acad. Gioenia. Sci. Natur. Catania.*, **9**, 573; *Chem. Abs.*, **72**, 66115j
152. Gore, P. H. and Hoskins, J. A. (1970). *J. Chem. Soc. C*, 517
153. Skvarchenko, V. R., Brunovlenskaya, I. I., Novikov, A. M. and Levina, R. Ya. (1970). *J. Org. Chem. USSR*, **6**, 1514
154. Paget, C. J. and Burger, A. (1965). *J. Org. Chem.*, **30**, 1329
155. Taylor, R., Wright, G. J. and Homes, A. J. (1967). *J. Chem. Soc. B*, 780
156. Anous, M. M. T., Bassilios, H. F., Salem, A. Y. and Arab, M. M. (1970). *Z. Phys. Chem.*, **71**, 1
157. Anous, M. M. T., Bassilios, H. F., Salem, A. Y. and El-Hares, M. (1970). *Ann. Chim.*, **60**, 605
158. Cassimatis, D., Bonnin, J. P. and Theophanides, T. (1970). *Can. J. Chem.*, **48**, 3860
159. Olafsson, P. G. (1970). *J. Org. Chem.*, **35**, 4257
160. Belenkii, L. I., Yakubov, A. P. and Goldfarb, Ya. L. *J. Org. Chem. USSR*, **6**, 2537
161. Corriu, R. and Coste, C. (1969). *Bull. Soc. Chim. Fr.*, **9**, 3272
162. Jensen, F. R., Marino, G. and Brown, H. C. (1958). *J. Amer. Chem. Soc.*, **80**, 3039
163. Hashimoto, I., Nojiri, T. and Ogata, Y. (1970). *Tetrahedron*, **26**, 4603
164. Virtanen, P. O. I., Malo, H. and Routsalainen. H. (1970). *Suom. Kemistilehti B*, **43**, 512; *Chem. Abs.*, **74**, 63686b
165. Vinogradova, S. V., Vasnev, V. A., Korshak, V. V., Vasilev, A. V. and Perfilov, Yu. I. (1970). *Bull. Acad. Sci. USSR*, 2018
166. Gore, P. H., Smith, G. H. and Thorburn, S. (1971). *J. Chem. Soc. C*, 650
167. Dewar, M. J. and Hart, L. S. (1970). *Tetrahedron*, **26**, 973, 1001
168. Martin, R., Betoux, J. M. and Coton, G. (1970). *C. R. Acad. Sci. Ser. C*, **271**, 760
169. Ansell, M. F. and Whitfield, G. F. (1971). *J. Chem. Soc. C*, 1098
170. Szponar, Z. (1970). *Rocz. Chem.*, **44**, 185; *Chem. Abs.*, **73**, 55348g
171. Lentzner, H. L. and Watts, W. E. (1970). *Chem. Commun.*, 906
172. Pettiford, L. R. (1971). *J. Chem. Soc. B*, 1088
173. Nerisson, P. and Wojtkowiak, B. (1970). *C. R. Acad. Sci. Ser. C*, **271**, 1499
174. Corriu, R. and Dabosi, G. (1971). *Bull. Soc. Chim. Fr.*, 1670
175. Casadervall, A. and Commeyras, A. (1970). *Bull. Soc. Chim. Fr.*, 1850, 1856
176. Barnes, R. A. and Neto, A. C. (1970). *J. Org. Chem.*, **35**, 4259
177. Litvinenko, L. M., Semenyuk, G. V. and Oleinik, N. M. (1970). *J. Org. Chem. USSR*, **6**, 2552
178. Traven, V. F. and Stephanov, B. I. (1971). *Zh. Org. Khim.*, **7**, 511
178a. Litvinkenko, L. M., Semenyak, G. U. and Oleinik, V. M. (1970). *J. Org. Chem. USSR*, **6**, 2552
179. Chatterjee, A. and Banerjee, D. (1971). *J. Chem. Soc*, 1859
180. Machacek, V., Panchartek, J., Sterba, V. and Vecera, M. (1970). *Collect. Czech. Chem. Commun.*, **35**, 844
181. Dobas, I., Panchartek, J., Sterba, V. and Vecera, M. (1970). *Collect. Czech. Chem. Commun.*, **35**, 1288
182. Beranek, V. and Vecera, M. (1970). *Collect. Czech. Chem. Commun.*, **35**, 3402
183. Kavalek, J., Panchartek, J. and Sterba, V. (1970). *Collect. Czech. Chem. Commun.*, **35**, 3470
183a. Aboul-Seoud, A. and Ahmed, M. (1970). *Bull. Soc. Chim. Belges*, **79**, 53
184. Kropacova, H., Panchartek, J., Sterba, V. and Valter, K. (1970). *Collect. Czech. Chem. Commun.*, **35**, 3287
185. Snyckers, F. and Zollinger, H. (1970). *Tetrahedron Lett.*, 2759; *Helv. Chim. Acta*, **53**, 1294
186. Jermini, C., Koller, S. and Zollinger, H. (1970). *Helv. Chim. Acta*, **53**, 72
187. Koller, S. and Zollinger, H. (1970). *Helv. Chim. Acta*, **53**, 78
188. Gloor, B. and Zollinger, H. (1971). *Helv. Chim. Acta*, **54**, 553
189. Gloor, B. and Zollinger, H. (1971). *Helv. Chim. Acta*, **54**, 563
190. Penton, J. R. and Zollinger, H. (1971). *Helv. Chim. Acta*, **54**, 573
191. Challis, B. C., Higgins, R. J. and Lawson, A. J. (1970). *Chem. Commun.*, 1223
192. Challis, B. C. and Lawson, A. J. (1971). *J. Chem. Soc. B*, 770
193. de Sarlo, F. and Ridd, J. H. (1971). *J. Chem. Soc. B*, 712
194. Mossa, G., Ricci, A. and Ridd, J. H. (1971). *Chem. Commun.*, 332

195. de Sarlo, F., Grynkiewicz, G., Ricci, A. and Ridd, J. H. (1971). *J. Chem. Soc. B,* 719
196. Grynkiewicz, G. and Ridd, J. M. (1971). *J. Chem. Soc. B,* 716
197. Gillow, H. M., de Shazo, M. and van Cleave, W. C. (1971). *J. Org. Chem.,* **36,** 1745
198. Modro, T. A. and Pickos, A. (1970). *Bull. Acad. Pol. Sci. Ser. Sci. Chim.,* **18,** 347; *Chem. Abs.,* **73,** 130429q
199. Clark, D. T. and Fairweather, D. J. (1969). *Tetrahedron,* **25,** 5525
200. Alcorn, P. G. E. and Wells, P. R. (1965). *Aust. J. Chem.,* **18,** 1377, 1391
201. Hoggett, J. G., Moodie, R. B. and Schofield, K. (1969). *Chem. Commun.,* 695
202. Reference 6, p. 29
203. Dewar, M. J. S. and Logan, R. H. (1968). *J. Amer. Chem. Soc.,* **90,** 1924
204. Sekiguchi, S., Kunii, K., Maeno, M. and Matsui, K. (1970). *Kogyu Kagaku Zasshi,* **73,** 339; *Chem. Abs.,* **73,** 13807a
205. Bechner, J. P. (1970). *Diss. Abs. B,* **31,** 577
206. Hartshorn, S. R., Moodie, R. B. and Schofield, K. (1971). *J. Chem. Soc. B,* 1256
207. Nias, J. M. A. and Wepster, B. M. (1971). *Rec. Trav. Chim.,* **90,** 1081
208. Shabarov, Yu. S., Mochalov, S. S. and Khryaschevskaya, O. M. (1970). *J. Org. Chem. USSR,* **6,** 2446
209. Christy, P. F., Ridd, J. H. and Stears, N. D. (1970). *J. Chem. Soc. B,* 797
210. Coombes, R. G., Grout, D. H. G., Hoggett, J. G., Moodie, R. B. and Schofield, K. (1970). *J. Chem. Soc. B,* 347
211. Roberts, J. D., Sanford, J. K., Sixma, F. L. J., Cerfontain, H. and Zagt, R. (1954). *J. J. Amer. Chem. Soc.,* **76,** 4525
212. Reference 8, p. 74
213. Eaborn, C. and Taylor, R. (1961). *J. Chem. Soc.,* 2388
214. Perrin, C. L. and Skinner, G. A. (1971). *J. Amer. Chem. Soc.,* **93,** 3389
215. Bonner, T. G., Hancock, R. A., Rolle, F. R. and Yousif, G. (1970). *J. Chem. Soc. B,* 314
216. Coombes, R. G. (1969). *J. Chem. Soc. B,* 1256
217. Bonner, T. G., Hancock, R. A. and Rolle, F. R. (1968). *Tetrahedron Lett.,* 1665
218. Marziano, N. C., Maccarone, E. and Passerini, R. C. (1971). *J. Chem. Soc. B,* 745
219. Östman, B. and Lindgren, I. (1970). *Acta Chem. Scand.,* **24,** 1105
220. Sharnin, G. P., Moisak, I. E., Gryazin, E. E. and Sofronova, T. N. (1969). *Inst. Vyssh. Vcheb. Zaved. Khim. Tekhnol.,* **12,** 907; *Chem. Abs.,* **72,** 21186s
221. Taylor, R. (1966). *J. Chem. Soc. B,* 727; *Tetrahedron Lett.,* 6093; Simamura, O. and Mizuno, Y. (1958). *J. Chem. Soc.,* 3875
222. Benkovskii, V. G., Bogoslovskaya, T. M., Naurozov,. M. K. and Serikov, Zh. (1970). *Tr. Inst. Khim. Nefti. Prir. Solei. Akad. Nauk. Kaz.,* **1,** 78; *Chem. Abs.,* **74,** 3183a
223. Kishore, K. and Upadhyaya, S. N. (1970). *J. Ind. Chem. Soc.,* **47,** 727
224. Crombie, D. A., Dean, R. E. and Shaw, S. A. (1970). *Chem. Ind.,* 1469
225. Adamek, M., Remes, M. and Hruda, J. (1970). *Chem. Prum.,* **20,** 500; *Chem. Abs.,* **74,** 111273p
226. Burminstrov, S. I., Mysak, D. D. and Vishnevskaya, A. I. (1970). *J. Org. Chem. USSR,* **6,** 1273
227. Yoshida, Z., Yoneda, S., Murata, Y. and Hashimoto, H. (1971). *Tetrahedron Lett.,* 1523
228. Lloyd, D. and Singer, M. I. C. (1971). *Chem. Ind.,* 786
229. Oda, M. and Kitahava, Y. (1971). *Bull. Chem. Soc. Japan,* **44,** 296
230. Albright, L. F. and Hanson, C. (1969). *Loss. Prev.,* **3,** 26
231. Vesely, J. A. and Schmerling, L. (1970). *J. Org. Chem.,* **35,** 4028
232. Hashimoto, J. and Koike, W. (1970). *Bull. Chem. Soc. Japan,* **43,** 293
233. Blackstock, D. J., Fischer, A., Richards, K. E., Vaughan, J. and Wright, G. J. (1970). *Chem. Commun.,* 641
234. Dannley, R. L., Gagen, J. E. and Stewart, O. J. (1970). *J. Org. Chem.,* **35,** 3076
235. Levi, E. M., Kovacic, P. and Gormish, J. F. (1970). *Tetrahedron,* **26,** 4537
236. Koeberg-Telder, A. and Cerfontain, H. (1971). *Rec. Trav. Chim.,* **90,** 193
237. Prinsen, A. J., Koeberg-Telder, A. and Cerfontain, H. (1970). *Tetrahedron,* **26,** 1953
238. Spryskov, A. A. and Yakovleva, Z. A. (1970). *Izv. Vyssh. Ucheb. Zared. Khim. Tekhnol.,* **13,** 1139; *Chem. Abs.,* **74,** 3187e
239. Pakter, M. and Gorokhova, Z. Ya. (1971). *Zh. Prikl. Khim.,* **44,** 589
240. Buncel, E., Strachan, W. M. J. and Cerfontain, H. (1971). *Can. J. Chem.,* **49,** 152
241. Jensen, F. R. and Brown, H. C. (1958). *J. Amer. Chem. Soc.,* **80,** 4038, 4042, 4044, 4046

242. Kobayashi, M., Minato, H. and Kohara, Y. (1970). *Bull. Chem. Soc. Japan,* **43,** 234
243. Kohara, Y., Kobayashi, M. and Minato, H. (1970). *Bull. Chem. Soc. Japan,* **43,** 520
244. Lebedev, N. N. and Baltadzhi, I. I. (1969). *J. Org. Chem. USSR,* **5,** 1604
245. Eaborn, C., Jackson, P. M. and Taylor, R. (1966). *J. Chem. Soc. B,* 613
246. Stock, L. M. and Wasielewski, M. R. (1971). *J. Org. Chem.,* **36,** 1002
247. Himoe, A. and Stock, L. M. (1969). *J. Amer. Chem. Soc.,* **91,** 1452
248. Eaborn, C. and Taylor, R. (1959). *Chem. Ind.,* 949; (1961). *J. Chem. Soc.,* 247
249. Causelier, J. P. (1971). *Bull. Soc. Chim. Fr.,* 1785
250. de la Mare, P. B. D. and Main, L. (1971). *J. Chem. Soc., B,* 90
251. Sedov, A. M. and Novikov, A. N. (1971). *Zh. Org. Khim,* **7,** 517
251a. Arotsky, J., Butler, R. and Darby, A. C. (1970). *J. Chem. Soc. C,* 1480
252. Rothenberg, F., Alcais, P. and Dubois, J. (1971). *Bull. Soc. Chim. Fr.,* 592
253. Kim, J. B., Chen, C., Krieger, J. K., Judd, K. R., Simpson, C. C. and Berliner, E. (1970). *J. Amer. Chem. Soc.,* **92,** 910
254. Ogata, Y. and Urasaki, I. (1970). *J. Chem. Soc. C,* 1689
255. Grossert, J. S. and Chip, G. K. (1970). *Tetrahedron Lett.,* 2611
256. Crucker, H. P. and Walser, R. (1970). *J. Chem. Soc. C,* 1982
257. Fahey, D. R. (1971). *J. Organometal. Chem.,* **27,** 283
258. Seguchi, K., Asano, T., Sera, A. and Goto, R. (1970). *Bull. Chem. Soc. Japan,* **43,** 3318
259. Sridhar, N., Rajaram, J. and Kuriacose, J. C. (1971). *Curr. Sci.,* **40,** 10; Kothandaraman, G., Rajaram, J. and Kuriacoise, J. C. (1970). *Proc. Indian Acad. Sci.,* **72,** 144 Gnanapragasam, N. S. and Joseph, N. (1971). *Curr. Sci.,* **40,** 83; Ganesan, R. and Kalachandra, S. (1970). *Z. Phys. Chem.,* **72,** 269; Rajaram, J. and Kuriacose, J. C. (1970). *Indian J. Chem.,* **8,** 145; Rao, D. C., Rajaram, J. and Kuriacose, J. C. (1971). *Chem. Commun.,* 754
260. Schutte, L. and Havinga, E. (1970). *Tetrahedron,* **26,** 2297
261. Ahmed, S. and Wardell, J. L. (1971). *Tetrahedron Lett.,* 3089
262. Nilsson, J. L. G., Selander, H., Sievertsson, H., Skanberg, I. and Svensson, K. (1971). *Acta Chem. Scand.,* **25,** 94
263. Bolton, R. (1970). *J. Chem. Soc. B,* 1770
264. el Dusouqui, O. M. H., Hassan, M. and Ibrahim, B. (1970). *J. Chem. Soc. B,* 926
264a. Hamilton, S. B. and Blanchard, H. S. (1970). *J. Org. Chem.,* **35,** 3348
265. Gerasimenko, Yu. E., Tkachenko, S. S. and Shteinberg, Ya. B. (1969). *Reaktz. Sposobnost Org. Soedin.,* **6,** 1102; *Chem. Abs.,* **73,** 65718s
266. Zubarev, S. V., Zakharenkova, V. I., Igoshev, A. D., Sobolev, A. S. and Kaluzhskaya, I. N. (1970). *Zh. Prikl. Khim.,* **43,** 2686
267. Uzan, R. and Dubois, J. E. (1971). *Bull. Soc. Chim. Fr.,* 598
268. Lebedev, N. N. and Baltadzhi, I. I. (1969). *Izv. Vyssh. Ucheh. Zaved. Khim. Technol.,* **12,** 1522; *Chem. Abs.,* **72,** 110515r
269. Rosen, M. J. and Gandler, J. (1971). *J. Phys. Chem.,* **75,** 887
269a. Flood, J., Mosnaim, A. D. and Nonhebel, D. C. (1970). *Chem. Commun.,* 761
270. Motsarev, G. V. (1970). *J. Gen. Chem. USSR,* **40,** 1263
271. Bott, R. W., Eaborn, C. and Jackson, P. M. (1967). *J. Organometal. Chem.,* **7,** 79
272. Chernyshev, E. A., Charskaya, B. A. and Mikheev, E. P. (1970). *Proc. Acad. Sci. USSR,* **194,** 777
273. Motsarev, G. V., Inshakova, V. T., Kolbasov, V. I., Dzhagatspanyan, R. V. and Snegova, A. D. (1971). *J. Gen. Chem. USSR,* **41,** 110
274. Lepeska, B., Bazant, V. and Chvalovsky, V. (1970). *J. Organometal. Chem.,* **23,** 41
275. Vcelak, J. and Chvalovsky, V. (1970). *J. Organometal. Chem.,* **23,** 47
276. Grakauskas, V. (1970). *J. Org. Chem.,* **35,** 723
277. Shaw, M. J., Hyman, H. H. and Filler, R. (1970). *J. Amer. Chem. Soc.,* **92,** 6498
278. Zakharkin, L. I. and Kyskin, V. I. (1970). *J. Gen. Chem. USSR,* **40,** 2221
279. Gassman, P. G. and Campbell, G. A. (1971). *J. Amer. Chem. Soc.,* **93,** 2567
280. Perrin, C. L. (1971). *J. Org. Chem.,* **36,** 420
281. Reference 8, pp. 259–260
282. Baciocchi, E. and Bocca, P. L. (1969). *Ric. Sci.,* **39,** 68
283. Bartlett, E. M., Eaborn, C. and Walton, D. R. M. (1970). *J. Chem. Soc. C,* 1717
284. Beletskaya, I. P., Savinykh, L. V. and Reutov, O. A. (1971). *J. Organometal. Chem.,* **26,** 13
285. Beletskaya, I. P., Zakharycheva, I. I. and Reutov, O. A. (1970). *Proc. Acad. Sci. USSR,* **195,** 861

2
Mechanisms and Reactivity in Aromatic Nucleophilic Substitution Reactions

C. F. BERNASCONI
University of California, Santa Cruz

2.1 INTRODUCTION

With the publication of Bunnett and Zahler's[1] landmark review in 1951 the subject of nucleophilic aromatic substitution acquired the status of a well defined field of organic chemistry. At that time only two mechanisms were known.

(a) The addition–elimination (AE, S_NAr) mechanism for reactions involving substrates activated by electron-withdrawing substituents, S, as in equation (2.1). This mechanism was for the first time strongly advocated by

$$\text{ArX(S)} + \ddot{Y}^{(-)} \underset{k_{-1}}{\overset{k_1}{\rightleftharpoons}} \text{(1)} \xrightarrow{k_2} \text{ArY(S)} + X^- \quad (2.1)$$

Bunnett and Zahler[1] as an attractive alternative to an aromatic analogue of the S_N2-mechanism promoted mainly by Chapman *et al.*[2,3].

(b) The unimolecular or S_N1-mechanism for the thermal decomposition of diazonium ions in aqueous solution, is usually represented as in equation

$$C_6H_5N_2^+ \xrightarrow[\text{slow}]{-N_2} C_6H_5^+ \xrightarrow[\text{fast}]{H_2O} C_6H_5OH \quad (2.2)$$

(2.2). Recently, some doubts as to its existence have been raised[4] though later partially retracted[5].

The field of nucleophilic aromatic substitutions has seen many developments in the past two decades; the most notable has undoubtedly been the discovery of the elimination–addition (EA, benzyne, aryne) mechanism. It was first firmly established by Roberts *et al.*[6,7] who found that upon treatment with KNH_2 in liquid ammonia [1-^{14}C]chlorobenzene yielded, after allowance for a small kinetic isotope effect, equal parts of [1-^{14}C]aniline and [2-^{14}C]aniline. A symmetrical intermediate, 'benzyne', (2) was thus indicated and the mechanism represented as follows.

$$C_6H_5{}^*Cl \xrightarrow{KNH_2} \text{(2)} \xrightarrow[NH_3]{KNH_2} C_6H_5{}^*NH_2 + C_6H_5NH_2{}^* \quad (2.3)$$

The discovery of benzyne or dehydrobenzene has triggered the development of a whole new field of organic chemistry which reaches far beyond the scope of nucleophilic aromatic substitution. In fact, arynes can be generated under a variety of non-basic conditions and the thus formed intermediate readily reacts with a large number of reagents in cyclo-additions, Diels–Alder additions and others; benzyne chemistry has been most comprehensively reviewed in a monograph by Hoffmann[8]. There are also some more recent reviews[9–11].

Photonucleophilic aromatic substitutions, another new type of mechanism was discovered in 1956 [12]; very recently substitution mechanisms involving free radicals have been proposed[13–17].

As is testified by a large number of recent reviews[18–23], a great deal of work has been done in elucidating structures and reactivities of Meisenheimer or Jackson–Meisenheimer Complexes[24,25] which are models for the intermediate in the S_NAr mechanism.

In this article I will concentrate on topics dealing with the addition–elimination mechanism which have not yet received much attention in earlier reviews[18,26–34] and on some important new findings in the last 3 years. Due to space limitation and since excellent reviews are already available, the benzyne[8–11] and S_N1 mechanism[30] will not be treated here. It is my belief that such a biased review reflecting the interests of the author will be more stimulating.

2.2 THE ADDITION-ELIMINATION MECHANISM (S_NAr)

2.2.1 Amines as nucleophiles

2.2.1.1 Base catalysis

(a) *Rate laws and their information content* – Kinetic studies involving primary and secondary amines as nucleophiles played a central role not only in firmly establishing the two- or multi-step nature of the mechanism but in answering questions regarding relative rates of intermediate complex formation and decomposition. This is most easily seen with reference to reaction (2.4), which is representative for the most frequently studied type of aromatic substrates.

$$\text{1-X-2,4-}(NO_2)_2C_6H_3 + R^1R^2NH \underset{k_{-1}}{\overset{k_1}{\rightleftharpoons}} \underset{(3)}{\text{X},\overset{+}{N}HR^1R^2\text{-2,4-}(NO_2)_2C_6H_3^-} \xrightarrow[k_3^{B_i}(B_i)]{k_2} \text{1-}NR^1R^2\text{-2,4-}(NO_2)_2C_6H_3 \quad (2.4)$$

In contrast to the situation where the nucleophile is an anion (equation (2.1)), the intermediate (3) can proceed to products either spontaneously (k_2) or through general base catalysis ($k_3{}^{B_i}$). The bases B_i are typically lyate ions, tertiary amines or others which may be specifically added to the reaction mixture or it is frequently the nucleophile itself. Thus by adding or omitting bases in the reaction mixture the experimenter may sometimes exert considerable control over the relative rates of product formation *v.* reversion to reactants of the intermediate and thus on the overall rate-controlling steps. This is expressed quantitatively by the rate expression (2.5), which

$$\frac{\text{rate}}{[\text{ArX}][R^1R^2NH]} = k_A = \frac{k_1k_2 + k_1\sum_i k_3^{B_i}[B_i]}{k_{-1} + k_2 + \sum_i k_3^{B_i}[B_i]} \qquad (2.5)$$

is obtained by applying the Bodenstein[35] or commonly called 'steady-state'[36] approximation to reaction (2.4).

Since in a study of base catalysis the concentration of all bases except one is generally held constant, equation (2.5) may be rewritten as (2.6) where k_2' represents the true uncatalysed rate coefficient, k_2, plus the sum of all the

$$k_A = \frac{k_1 k_2' + k_1 k_3{}^{B_j}[B_j]}{k_{-1} + k_2' + k_3{}^{B_j}[B_j]} \tag{2.6}$$

$k_3{}^{B_i}[B_i]$ terms for which i is $\neq j$, i.e. for which the base concentration is held constant. Provided that k_2' is smaller than k_{-1}, equation (2.6) in principle describes a dependence of k_A on $[B_j]$ which is linear at low concentrations but becomes curved and finally levels off as the base concentration is increased. This is because at low concentrations $k_3{}^{B_j}[B_j] \ll k_{-1} + k_2'$ and equation (2.6) becomes equation (2.7) which describes a straight line*, whereas

$$k_A = \frac{k_1 k_2'}{k_{-1} + k_2'} + \frac{k_1 k_3^{B_j}}{k_{-1} + k_2'}[B_j] \tag{2.7}$$

at very high concentrations $k_3^{B_j}[B_j] \gg k_{-1} + k_2'$ and equation (2.6) simplifies to equation (2.8), i.e. the overall reaction is no longer sensitive to the base catalysis.

$$k_A = k_1 \tag{2.8}$$

Only a comparatively small number of aminolysis reactions have shown such a characteristic curved plot[37–52, 71]. In the majority of reactions no base catalysis is observed[28, 39, 40, 42, 44, 48, 49, 53–63] or where there is base catalysis, k_A depends linearly on $[B_j]$ over the entire concentration range[37, 42, 43, 59–61, 63a, 64–70].

In terms of the addition-elimination mechanism, absence of base catalysis can mean that $k_2 \gg k_{-1}$, or that one has reached the region where levelling off takes place ($\sum_i k_3{}^{B_i}[B_i] \gg k_{-1}$) since both these conditions simplify equation (2.5) to (2.8). In principle, one could easily distinguish between the two alternatives by choosing all base concentrations equal to zero. But since the concentration of the nucleophile cannot be zero, from a strictly mathematical point of view the absence of base catalysis only warrants the conclusion that $k_2 + k_3{}^{R^1R^2NH}[R^1R^2NH] \gg k_{-1}$ with the implied possibility that $k_2 \ll k_{-1}$ but $k_3{}^{R^1R^2NH}[R^1R^2NH] \gg k_{-1}$ even at a low amine concentration. For a typical low amine concentration of 10^{-3} M such interpretation would call for $k_3{}^{R^1R^2NH}/k_{-1} \geqslant 10^4$ and $k_3{}^{R^1R^2NH}/k_2 \geqslant 10^6$. As will become apparent from the discussion of Table 2.1 below, this is inconsistent with the large body of experimental data; the few reactions where $k_3{}^{R^1R^2NH}/k_2$ may be as high as 10^5 are in fact those where $k_3{}^{R^1R^2NH}[R^1R^2NH]$ is much smaller rather than much larger than k_{-1} making base catalysis by the amine very visible throughout the entire concentration range. Thus we conclude that when no catalysis is observed at low amine concentration the inference that $k_2 \gg k_{-1}$ is fully justified.

Absence of base catalysis, even when $k_2 \ll k_{-1}$, could in principle be

*In practical situations very often $k_2' \ll k_{-1}$ so that the denominators in equation (2.7) are simply k_{-1}; in other cases k_2' and k_{-1} are quite similar in magnitude (see Table 2.1) which makes the plots of k_A *v.*$[B_j]$ curved throughout the entire concentration range.

explained if the rate coefficients $k_3^{B_i}$ are small compared to k_2, so that even at high base concentration $k_2 \gg \Sigma k_3^{B_i}[B_i]$ and thus the base-catalysed terms cannot be detected experimentally since equation (2.5) simply becomes

$$k_A = \frac{k_1 k_2}{k_{-1}} \tag{2.9}$$

There is plenty of evidence however (Table 2.1) which suggests that whenever $k_2 \ll k_{-1}$ base catalysis is in fact very visible due to $k_3^{B_i} \gg k_2$. Only when a very weak base is used or when large steric effects come into play, a $k_3^{B_i}$ coefficient may become vanishingly small as in the case of the reaction of 2,4-dinitrofluorobenzene with piperidine in benzene which is not catalysed by the bulky triethylamine[59]. Use of another, sterically less hindered base can however clarify such situations; in the reaction under discussion strong base catalysis by triethylenediamine was indeed observed in a later study designed to clarify this point[66, 71].

From this discussion the conclusion emerges that for all practical purposes absence of base catalysis—preferably tested with several different bases—means that $k_2 \gg k_{-1}$.

The linear response to base concentration characteristic for the majority of base-catalysed reactions is rationalised by assuming $k_3^{B_j}[B_j] \ll k_{-1}$, as well as $k_2' \ll k_{-1}$, which converts equation (2.6) into

$$k_A = \frac{k_1 k_2'}{k_{-1}} + \frac{k_1 k_3^{B_j}[B_j]}{k_{-1}} \tag{2.10}$$

It is apparent that a considerable body of information can be obtained from the way a reaction responds to the addition of base. When the response is curvilinear and the dependence of k_A on all bases including the nucleophile is determined, k_1 and the ratios k_2/k_{-1}, $k_3^{B_i}/k_{-1}$ and $k_3^{B_i}/k_2$ can usually be calculated on the basis of equation (2.6) according to procedures described[37–45]. When the response is linear, the inference that $k_3^{B_i}/k_{-1} \ll 1$ and $k_2/k_{-1} \ll 1$ can be made. From the slopes and intercepts of the plots of k_A *v.* the concentration of the various bases the ratio $k_3^{B_i}/k_2$ can be determined on the basis of equation (2.10). When no base catalysis is observed, k_1 which coincides with k_A can be obtained and the inference that $k_2/k_{-1} \gg 1$ can be made.

(b) *Small accelerations: base catalysis or medium effect*—It should be noted at this point that a number of reactions, particularly involving activated chlorobenzenes in non-polar solvents have been found to undergo mild accelerations with increasing amine concentration[28, 47, 72, 74–77], or by addition of various other substances[28, 73]. The second-order rate constant usually obeys equation (2.11) and typical k_i''/k' ratios vary between 0.5 and 5. Though

$$k_A = k' + k_i''[B_i] \tag{2.11}$$

interpreted as base catalysis by some authors[28, 72] with the implication that $k_{-1} \gg k_2$ and thus $k_i''/k' = k_3^{B_i}/k_2$, such an interpretation has been criticised by Bunnett[62] mainly on the grounds that change to a stronger base did not bring about a more efficient catalysis. According to Bunnett[62], reactions for which the ratio $k_i''/k' \geqslant 50$, the inference that it is genuine base catalysis is

Table 2.1 S_NAr reactions of amines

	Solvent	Ar-X	X(pK_a)	*Amine* (pK_a)	*Temp/* °C	*Base catal.*	$k_1/M^{-1}s^{-1}$	k_2/k_{-1}	$(k_3^B/k_{-1})/M^{-1}$	$(k_3^B/k_2)/M^{-1}$	*Ref.*
1	H_2O	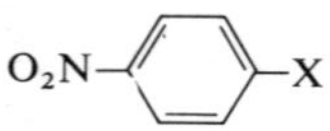	$OP(OMe)O_2^-$ (~7)	Piperidine (11.06)	39	no	7.8×10^{-6} [a]	≫1			78
2	H_2O		OPO_3^{2-} (12.3)	$MeNH_2$ (10.6)	39	no	3.7×10^{-7}	≫1			39
3	H_2O	O_2N–X	OPO_3^{2-} (12.3)	Me_2NH (10.6)	39	HO^- Me_2NH	1.03×10^{-5}	0.12[b]	82 HO^- 2.3 Me_2NH	683 HO^- 19.2 Me_2NH	39
4	H_2O		OPO_3^{2-} (12.3)	Piperidine (11.06)	39	HO^- [c]	7.3×10^{-6}	0.091 [b]	34 HO^-	374 HO^-	39
5	H_2O		OPO_3^{2-} (12.3)	Morpholine (8.36)	39	HO^-	d	<1 [d]			39
6	H_2O	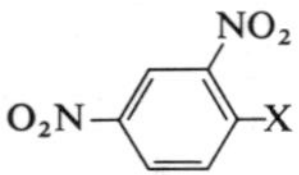 NO_2	F (~3)	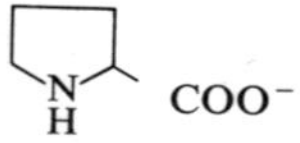 N-H ring COO^- (11.2)	10	no	2.64	≫1			49
7	10%Dioxone –90% H_2O	O_2N–X	OPh (10)	Bu^nNH_2 (10.59)	29.4	no	1.48×10^{-3}	≫1			51
8	60% Dioxane –40% H_2O	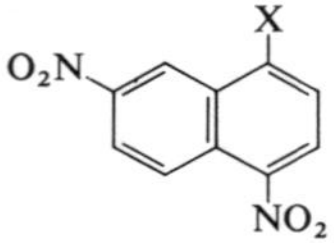 O_2N, X, NO_2	OPh (c. 16.7)	Bu^nNH_2 (10.59)	25	Bu^nNH_2		<1[s]			70
9	10% Dioxane –90% H_2O	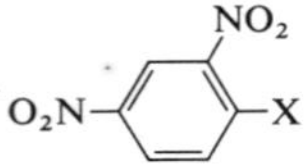	F (~3)	Bu^tNH_2 (10.8)	20	no	5.9×10^{-3} [e,f]	≫1			42
10	10% Dioxane –90% H_2O	NO_2, O_2N–X	OPh (10)	$PhCH_2NH_2$ (9.37)	29.4	no	8.9×10^{-4}	≫1			44
11	10% Dioxane –90% H_2O		OPh (10)	Pyrrolidine (11.27)	29.4	no	4.57×10^{-2}	≫ 1			49

Table 2.1—*continued*

	Solvent	Ar-X	X(pK_a)	*Amine* (pK_a)	*Temp*/°C	*Base catal.*	$k_1/M^{-1}s^{-1}$	k_2/k_{-1}	$(k_3^B/k_{-1})/M^{-1}$	$(k_3^B/k_2)/M^{-1}$	*Ref.*
12	10% Dioxane -90% H_2O	$2,4-(O_2N)_2C_6H_3-X$	OPh (10)	$C_4H_7N(H)COO^-$ (11.2)	29.4	HO^{-} [g]	2.6×10^{-2}	0.22	5.36 HO^-	24.4 HO^-	49
13	10% Dioxane -90% H_2O	$2,4-(O_2N)_2C_6H_3-X$	OMe (*c.* 16.7)	Pyrrolidine (11.27)	29.4	no		$\gg 1$			38a
14	10% Dioxane -90% H_2O	$2,4-(O_2N)_2C_6H_3-X$	$O-C_6H_3-2,4-(NO_2)_2$ (4.09)	Piperidine (11.06)	29.4	no	0.80	$\gg 1$			40
15	10% Dioxane -90% H_2O	$2,4-(O_2N)_2C_6H_3-X$	$O-C_6H_4-4-NO_2$ (7.14)	Piperidine (11.06)	29.4	HO^{-} [c]	0.15	≈ 2	≈ 50 HO^-	≈ 25 HO^-	40, 41
16	10% Dioxane -90% H_2O	$2,4-(O_2N)_2C_6H_3-X$	SPh (6.52)	Piperidine (11.06)	29.4	HO^{-} [c]	8.4×10^{-4}	0.15	6.55 HO^-	43.6 HO^-	40, 44
17	10% Dioxane -90% H_2O	$2,4-(O_2N)_2C_6H_3-X$	OPh (10)	Piperidine (11.06)	29.4	HO^- Pip	3.10×10^{-2}	0.069	121 HO^- 7.86 Pip	1750 HO^- 112 Pip	40, 44
18	10% Dioxane -90% H_2O	$4-O_2N-2-NC-C_6H_3-X$	OPh (10)	Piperidine (11.06)	29.4	HO^- Pip	8.2×10^{-4}	0.035	63 HO^- 3.22 Pip	1800 HO^- 92 Pip	51
19	10% Dioxane 90% H_2O	$2,4-(O_2N)_2C_6H_3-X$	OCH_3 (*c.* 16.7)	Piperidine (11.06)	29.4	HO^- Pip	3.77×10^{-3}	$\approx 10^{-4}$	25.7 HO^- 0.15 Pip	$\approx 2.5 \times 10^5$ HO^- ≈ 1500 Pip	40, 44
20	10% Dioxane -90% H_2O	$2,4-(O_2N)_2C_6H_3-X$	OPh (10)	$PhCH_2NHCH_3$ (9.58) (MBA)	29.4	HO^- MBA	6.4×10^{-3}	0.005	17.2 HO^- 0.96 MBA	3440 HO^- 192 MBA	44

Table 2.1—*continued*

	Solvent	Ar-X	X(pK_a)	Amine (pK_a)	Temp/ °C	Base catal.	$k_1/M^{-1}\,s^{-1}$	k_2/k_{-1}	$(k_3^B/k_{-1})/M^{-1}$	$(k_3^B/k_2)/M^{-1}$	Ref.
21	10% Dioxane –90% H_2O	2,4-dinitrophenyl-X (O_2N, NO_2)	OPh (10)	Morpholine (8.36)	29.4	HO^- Mor	3.33×10^{-3}	0.0015	11.9 HO^- 0.13 Mor	7900 HO^- 86.5 Mor	45
22	50% Dioxane –50% H_2O		Cl (*c.* 3)	Bu^nNH_2 (10.59)	24.8	no [h]	2.2×10^{-3}	$\gg1$ [h]			28
23	50% Dioxane –50% H_2O		Cl (*c.* 3)	Piperidine (11.06)	0	no	1.4×10^{-2}	$\gg1$			53
24	60% Dioxane –40% H_2O		F (*c.* 3)	$PhNH_2$ (4.60)	0	no [i]	6.35×10^{-3}	$\gg1$ [i]			62
25	60% Dioxane –40% H_2O		F (*c.* 3)	$PhNHCH_3$ (4.84)	0	HO^- [c]	2.0×10^{-4}	0.071	48.8 HO^-	683 HO^-	37
26	60% Dioxane –40% H_2O	2,4-dinitrophenyl-X (O_2N, NO_2)	O–C_6H_4–NO_2 (7.14)	Piperidine (11.06)	29.4	HO^- [c]	0.265	0.60	18.9 HO^-	31.5 HO^-	50
27	60% Dioxane –40% H_2O		OPh (10.0)	Piperidine (11.06)	29.4	HO^- Pip	3.5×10^{-2}	0.0121	12.2 HO^- 0.73 Pip	984 HO^- 61 Pip	38, 50
28	60% Dioxane –40% H_2O		O–C_6H_4–OCH_3 (10.2)	Piperidine (11.06)	29.4	HO^- Pip	2.46×10^{-2}	0.0112	15.8 HO^- 0.81 Pip	1170 HO^- 62 Pip	50
29	60% Dioxane –40% H_2O		O–$C_6H_2(CH_3)_3$ (10.4)	Piperidine (11.06)	29.4	HO^- Pip	1.43×10^{-2}	0.0114	13.5 HO^- 0.73 Pip	1056 HO^- 64 Pip	50
30	Methanol		F (*c.* 3)	Piperidine (11.06)	25	no	6.46	$\gg1$			63
31	Methanol		OCH_3 (~16.7)	Piperidine (11.06)	67.9	CH_3O^- Pip [j]		$\ll1$ [k]		'∞' CH_3O^- [k]	63a
32	Methanol		F (*c.* 3)	$PhNH_2$ (4.60)	29.4	no	1.26×10^{-2}	$\gg1$			62

Table 2.1—*continued*

	Solvent	Ar-X	X(pK_a)	*Amine* (pK_a)	*Temp/* °C	*Base catal.*	k_1/M^{-1} s^{-1}	k_2/k_{-1}	$(k_3^B/k_{-1})/M^{-1}$	$(k_3^B/k_2)/M^{-1}$	*Ref.*
33	Ethanol	2,4-dinitrophenyl-X (O_2N–C_6H_3(NO_2)–X)	F (*c.* 3)	$PhNH_2$ (4.60)	50	no	1.68×10^{-2}	$\gg 1$			56
34	Ethanol	2,4-dinitrophenyl-X (O_2N–C_6H_3(NO_2)–X)	F (*c.* 3)	PhNHMe (4.84)	46	AcO^{-1}		$\ll 1$		150 AcO^{-1}	37
35	DMSO	O_2N–C_6H_4–X	F (*c.* 3)	Piperidine (11.06)	50	no	2.74×10^{-2}	$\gg 1$			57
36	DMSO	2,4-dinitrophenyl-X (O_2N–C_6H_3(NO_2)–X)	F (*c.* 3)	Piperidine (11.06)	25	no	630	$\gg 1$			117
37	DMSO	2,4-dinitrophenyl-X (O_2N–C_6H_3(NO_2)–X)	cyclohexyl–O (*c.* 17)	Piperidine (11.06)	69.8	no	7.4×10^{-6}	$\gg 1$			112
38	DMSO	2,4-dinitrophenyl-X (O_2N–C_6H_3(NO_2)–X)	F (*c.* 3)	trans-2,6-Dimethylpiperidine	25	no	1.1×10^{-2}	$\gg 1$			117
39	DMSO	O_2N–C_6H_4–X	F (*c.* 3)	$PhNH_2$ (4.60)	25	no	9.8×10^{-8}	$\gg 1$			58
40	Acetone	2,4-dinitrophenyl-X (O_2N–C_6H_3(NO_2)–X)	F (*c.* 3)	Piperidine (11.06)	−30	Pip	37.6	0.79	1260 Pip	1610 Pip	47
41	Acetone	2,4-dinitrophenyl-X (O_2N–C_6H_3(NO_2)–X)	Cl (*c.* −3)	Piperidine (11.06)	30	Pip [m]	0.73 [m]	0.3 [m]	120 Pip [m]	400 Pip [m]	47
42	Acetone	2,4-dinitrophenyl-X (O_2N–C_6H_3(NO_2)–X)	Cl (*c.* −3)	$PhNH_2$ (4.60)	50	$PhNH_2$	6.54×10^{-5}	0.2	9.2 $PhNH_2$	46.5 $PhNH_2$	46, 47
43	Acetone	2,4-dinitrophenyl-X (O_2N–C_6H_3(NO_2)–X)	F (*c.* 3)	$PhNH_2$ (4.60)	50	$PhNH_2$	5.25×10^{-3}	$\ll 1$	0.625 $PhNH_2$	very large $PhNH_2$	46, 47

Table 2.1—*continued*

	Solvent	Ar-X	X(pK_a)	*Amine* (pK_a)	*Temp/* °C	*Base catal.*	k_1/M^{-1} s^{-1}	k_2/k_{-1}	(k_3^B/k_{-1})/M^{-1}	(k_3^B/k_2)/M^{-1}	*Ref.*
44	Ethylacetate	$2,4-(NO_2)_2C_6H_3X$	Cl (*c.* −3)	$PhNH_2$ (4.60)	35	no	8.89×10^{-6}	≫1			54
45	Ethylacetate	$2,4-(NO_2)_2C_6H_3X$	F (*c.* 3)	$PhNH_2$ (4.60)	35	$PhNH_2$	r	≪1 [r]		'∞' $PhNH_2$ [r]	55
46	Benzene	$2,4-(NO_2)_2C_6H_3X$	Cl (*c.* −3)	Bu^nNH_2 (10.59)	25	no	3.9×10^{-4}	≫1			48
47	Benzene	$2,4-(NO_2)_2C_6H_3X$	Cl (*c.* −3)	Bu^tNH_2 (10.8)	25	no	1.62×10^{-6}	≫1			48
48	Benzene	$2,4-(NO_2)_2C_6H_3X$	Cl (*c.* −3)	Piperidine (11.06)	25	no	8.5×10^{-2}	≫1			59
49	Benzene	$2,4-(NO_2)_2C_6H_3X$	F (*c.* 3)	Bu^nNH_2 (10.59)	25	Bu^nNH_2 Pyridine	0.71 [b]	0.24 [b]	50 [b] Bu^nNH_2 43 [b] Py	210 [b] Bu^nNH_2 183 [b] Py	48
50	Benzene	$2,4-(NO_2)_2C_6H_3X$	F (*c.* 3)	Bu^sNH_2 (~10.6)	25	Bu^sNH_2	0.13 [b]	0.17 [b]	10.4 [b] Bu^sNH_2	61.5 [b] Bu^sNH_2	48
51	Benzene	$2,4-(NO_2)_2C_6H_3X$	F (*c.* 3)	Bu^tNH_2 (10.8).	25	Bu^tNH_2	0.01 [b]	0.19 [b]	20.9 [b] Bu^tNH_2	108 [b] Bu^tNH_2	48
52	Benzene	$2,4-(NO_2)_2C_6H_3X$	F (*c.* 3)	$PhCH_2NH_2$ (9.37)	25	BA [n] DABCO [o] Pyridine	~0.1	~0.1	32.7 BA 24.4[p] DABCO[o] 9.3 Pyr	~320 BA ~24.4[p] DABCO[o] ~93 Pyr	43
53	Benzene	$2-NO_2C_6H_4X$	F (*c.* 3)	Piperidine (11.06)	25	no	5.6×10^{-4} [f,q]	≫1			61
54	Benzene	$4-O_2NC_6H_4X$	F (*c.* 3)	Piperidine (11.06)	85	Pip	r	≪1 [r]		'∞' Pip [r]	61
55	Benzene	$2,4-(NO_2)_2C_6H_3X$	F (*c.* 3)	Piperidine (11.06)	25	Pip DABCO [o] Pyridine	$\frac{k_1k_2}{k_{-1}}=0.5$	≪1		1230 Pip 64.5 DABCO 4.3 Pyr.	59, 66

Table 2.1—*continued*

	Solvent	Ar-X	X(pK_a)	*Amine* (pK_a)	*Temp/* °C	*Base catal.*	$k_1/M^{-1}\,s^{-1}$	k_2/k_{-1}	$(k_3^B/k_{-1})/M^{-1}$	$(k_3^B/k_2)/M^{-1}$	*Ref.*
56	Benzene	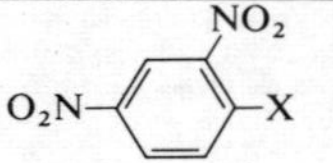 NO_2, O_2N, X	F (c 3)	2-Methyl piperi-dine (MP)	25	MP	$\frac{k_1k_2}{k_{-1}} = 2.28 \times 10^{-3}$	$\ll 1$		36.6 [b] MP	64
57	Benzene	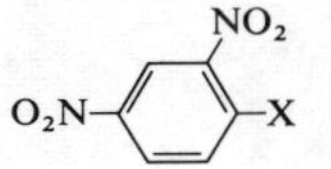 NO_2, O_2N, X	F (c 3)	Trans-2,6-di-methylpiperidine (DMP)	25	DMP	$\frac{k_1k_2}{k_{-1}} \approx 10^{-5}$	$\ll 1$		$\sim$2 DMP	60
58	Benzene	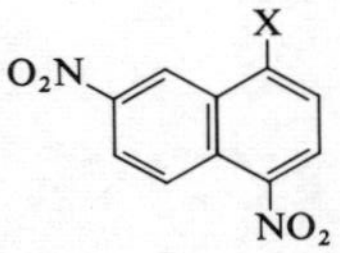 X, O_2N, NO_2	F (c 3)	Piperidine (11.06)	25	Pip	r	$\ll 1$	$\frac{k_1k_3^B}{k_{-1}} = 1.7$ *Pip*	'∞' [r] Pip	69
59	Benzene	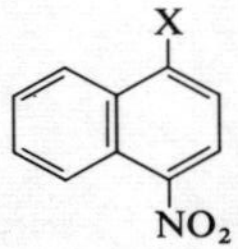 X, NO_2	F (c 3)	Piperidine (11.06)	25	Pip	r	$\ll 1$	$\frac{k_1k_3^B}{k_{-1}} = 2.72 \times 10^{-4}$ *Pip*	'∞' [r] Pip	70
60	Benzene	X, NO_2	F (c 3)	Bu^nNH_2 (10.59)	25	Bu^nNH_2	r	$\ll 1$	$\frac{k_1k_3^B}{k_{-1}} = 1.62 \times 10^{-5}$ Bu^nNH_2	'∞' [r] Bu^nNH_2	70
61	Benzene	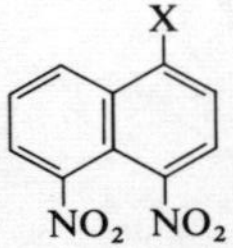 X, NO_2, NO_2	F (c 3)	Bu^nNH_2 (10.59)	25	Bu^nNH_2	r	$\ll 1$	$\frac{k_1k_3^B}{k_{-1}} = 2.28 \times 10^{-4}$ Bu^nNH_2	'∞' [r] Bu^nNH_2	70
62	Benzene	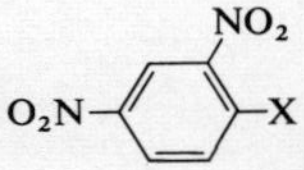 NO_2, O_2N, X	OPh (10.0)	Piperidine (11.06)	25	Piperidine Pyridine	r	$\ll 1$	$\frac{k_1k_3^B}{k_{-1}} = 1.35 \times 10^{-2}$ *Pip*	'∞' Pip [r]	65

Table 2.1—*continued*

	Solvent	Ar-X	X(pK_a)	*Amine* (pK_a)	*Temp*/ °C	*Base catal.*	$k_1 M^{-1} s^{-1}$	k_2/k_{-1}	$(k_3^B/k_2)/M^{-1}$	$(k_3^B/k_2)/M^{-1}$	*Ref.*
63	Benzene	O_2N–⟨ring⟩(–NO_2)–X	⟨H⟩–O (*c.* 17)	Piperidine (11.06)	100	Piperidine	r	$\lll 1$	$\frac{k_1 k_3^B}{k_{-1}} = \sim 8\times10^{-7}$ *Pip*	'∞' Pip [r]	69
64	Benzene		F (*c.* 3)	$PhCH_2NHCH_3$ (MBA) (9.58)	25	MBA DABCO [o] Pyridine	$\frac{k_1 k_2}{k_{-1}}$ 4.5×10^{-3}	$\lll 1$		595 MBA 630 DABCO [o] 16 Pyr	43
65	Benzene		F (*c.* 3)	Morpholine (8.36)	25	Mor DABCO [o] Pyr	$\frac{k_1 k_2}{k_{-1}} = 5.3\times10^{-3}$	$\lll 1$		982 Mor 444 DABCO [o] 58 Pyr	68
66	Benzene	O_2N–⟨ring⟩(–NO_2)–X	Cl (*c.* −3)	*p*-Anisidine (5.3)	25	*p*-An DABCO [o] Pyr	$\frac{k_1 k_2}{k^{-1}} = 8.0\times10^{-7}$	$\lll 1$		7.2 *p*-An 99 DABCO [o] 39 Pyr	67
67	Benzene		F (*c.* 3)	*p*-Anisidine (5.3)	25	*p*-An DABCO [o] Pyr	$\frac{k_1 k_2}{k_{-1}} < 10^{-5}$	$\lll 1$		>22 *p*-An >5500 DABCO [o] >750 Pyr	67
68	$CHCl_3$	O_2N–⟨ring⟩(–NO_2)(–NO_2)–X	Cl (*c.* −3)	Imidazole (6.95)	20	Im.	0.27 [b]	0.22 [b]	21.8 im [b]	99 [b] Im	52

a Nucleophilic attack is slightly catalysed by OH^-.
b Calculated from the original rate data by the reviewer.
c No catalysis by the amine could be detected.
d No elaborate study reported.
e At $[Bu^tNH_2] = 0.005$ M.
f There is a mild increase in the rate constant with increasing amine concentration which is not considered to be base catalysis.
g Amine catalysis was not investigated.
h Despite claims to the contrary by the author; see discussion about mild rate accelerations.
i The rate is mildly accelerated by HO^- which, however, is not considered to be genuine base catalysis.
j Amine catalysis assumed by authors though not investigated.
k No non-catalysed contribution could be detected.
l At $[PhNH_2] = 5\times10^{-3}$ M. (k_1 decreases slightly with increasing $[PhNH_2]$ due to charge-transfer complex formation.)
m There is some doubt about these numbers in the mind of the reviewer, see discussion.
n BA = Benzylamine.
o 1,4-Diaza-[2.2.2]-bicyclo-octane.
p Statistically corrected.
q At [Pip] = 0.0376 M.
r No first-order term in amine concentration could be detected.
s Preliminary results

warranted but is definitely not when $k_i''/k' \leqslant 5$. In the difficult cases where k_i''/k' is somewhere between these limits the interpretation of some recent examples[47, 74, 77] seems mainly to reflect the author's biases. Considering the principles governing the various rate coefficients k_{-1}, k_2 and $k_3^{B_i}$, as discussed in Section 2.2.1.2, should help in making sounder judgements.

Mild accelerations by HO^- have also been interpreted as base catalysis of the addition step (k_1)[39, 78] though without mechanistic rationalisation.

2.2.1.2 *Factors determining incidence of base catalysis*

Some of the factors determining whether a reaction is susceptible to base catalysis or not, i.e. whether $k_2 < k_{-1}$, or $k_2 \gg k_{-1}$ are reasonably well understood, others are more obscure. By summarising data of the kind which can be obtained from the study of the influence of bases on the rate (Table 2.1) it is hoped to clarify some of the points and to stimulate research regarding others. Though Table 2.1 is not an exhaustive list of aminolysis reactions, it includes all cases known to the reviewer where base catalysis was found excluding heterocyclic systems; there are many more examples in the literature pertaining to non-catalysed reactions, but the examples included suffice to illustrate the factors responsible for this behaviour.

(a) *Leaving group*—Among the several factors the influence of the leaving group on k_2/k_{-1} appears the most straightforward. In a variety of nucleophilic substitutions leaving group capability (nucleofugicity[79]) has been recognised to increase with decreasing pK of the conjugate acid of the leaving group[80–82]; the correlations are quite good when leaving groups with the same heteroatom involved in bond breaking are compared. There are a number of examples in Table 2.1 which show this to be the case for the intermediate complex in nucleophilic aromatic substitutions as well. Chloride along with bromide and iodide ions, though no examples of the latter two are included in Table 2.1, are among the best leaving groups studied. Except for a few cases discussed below (entry 66 and 68, but probably not 41) where other factors contribute to a drastic lowering of the k_2/k_{-1} ratio, genuine base catalysis has not been detected with chloride as leaving group, implying $k_2/k_{-1} \gg 1$.

Wherever a direct comparison is possible, fluoride ion is always a much poorer leaving group than chloride ion and consequently these reactions are associated with a smaller k_2/k_{-1} ratio (entries 42 and 43; 44 and 45; 46 and 49; 47 and 51; 48 and 55; 66 and 67).

In this discussion it is implicitly assumed that the changes in the k_2/k_{-1} ratios as a function of the leaving group are for the main part caused by changes in k_2 rather than k_{-1}, since it is k_2 which refers to the bond-breaking process. However, k_{-1} is expected to be affected to some extent as the electrophilicity of the tetrahedral carbon in the intermediate is altered by the inductive effect and the polarisability of the leaving group. Fluorine with its strong inductive effect as reflected in the much higher rates of nucleophilic attack (k_1) on fluoro than on analogous chloro compounds (entries 42 and 43; 46 and 49; 47 and 51; 48 and 55) is a case in point. This inductive effect may have a considerable enhancing effect on k_2/k_{-1} by reducing k_{-1} and thus fluoride may appear as more nucleofugic than it really is.

The most extensive correlation of k_2/k_{-1} ratios with the basicity of the leaving group is seen by comparing entries 14, 15, 17, 19 and entries 26–29. A plot (not shown) of log k_2/k_{-1} *v.* p*K* of XH for the three entries 15, 17 and 19 gives a good straight line with a slope $\beta \approx -0.39$.

Interestingly, the $k_3^{B_i}/k_{-1}$ ratios are much less sensitive to the leaving group p*K* than the k_2/k_{-1} ratios and do not always correlate well with the p*K* value (compare, e.g. entries 15, 17, 19). Another way to look at it is to consider the $k_3^{B_i}/k_2$ ratios which are a measure of how effectively a base can catalyse the reaction. From entries 15, 17, 19, and from entries 26–28 it is apparent that k_3^{HO}/k_2 is much larger for poor leaving groups than it is for good ones, which is probably most easily understood from a reactivity–selectivity point of view.

For extremely poor leaving groups like alkoxide ions and, particularly in non-polar solvents, fluoride ion, the ratio k_3^{B}/k_2 may become so large as to make the k_2 term undetectable (entries 31, 45, 54, 58–63, 67); though fluoride ion is not a particularly poor leaving group in protic solvents – it is, e.g. better than phenoxide ion as seen from entries 6 and 12 – it becomes so in non-polar aprotic solvents due to the high demand for solvation of the fluoride ion[83] which cannot be provided by the solvent.

Another interesting observation is the higher nucleofugicity of *p*-nitrophenoxide ion compared to thiophenoxide (entries 15 and 16) despite the lower p*K* of thiophenol. This is in line with the general finding that sulphur bases are less nucleofugic than oxygen bases of the same p*K* [82].

(b) *Entering amine: basicity and solvent effect* – Let us begin with the assumption that variation in the amine will affect k_2/k_{-1} chiefly through changes in k_{-1}. Entries 17, 20, 21 allow direct comparison of three secondary aliphatic amines of different basicity, entries 49 and 52 of two primary amines. The trend in the k_2/k_{-1} ratios bears out the expected increase with increasing p*K* of the amine.

Since the reversion of the zwitterionic intermediate (3) to reactants leads to a charge neutralisation, k_{-1} is expected to increase strongly in non-polar solvents[84] thus reducing the k_2/k_{-1} ratio unless a compensating effect of larger magnitude operates on k_2. This expectation is borne out as can be seen by comparing entries 15 with 26, 17 with 27, 27 with 62, 40 and 55. A more dramatic manifestation of the same principle is the more frequent observation of base catalysis in non-polar solvents even in reactions with moderately good leaving groups (entries 33 and 45; 44 and 66; 37 and 63).

Even in benzene the chloride ion is, however, still nucleofugic enough as to make $k_2/k_{-1} \gg 1$ for strongly basic amines like piperidine (entry 48). The report of strong piperidine catalysis for the reaction of piperidine with 2,4-dinitrochlorobenzene in the much more polar solvent acetone (entry 41) is therefore very much out of line with other findings so that a re-investigation of this reaction seems warranted.

There are some features about the influence of the amine on k_2/k_{-1} which are less well understood. The most notable is that the k_2/k_{-1} ratio is frequently much lower for secondary compared to primary amines of comparable basicity, the consequence of which is that the reactions with secondary amines are often base catalysed, whereas the reactions of primary amines are not. This can be verified by comparing entries 2 and 3, 10 and 20, 7 and

17, 24 and 25, 33 and 34. Only in solvents of low polarity such as benzene and ethylacetate, where the expulsion of fluoride ion becomes very difficult, do primary amines give rise to ratios $k_2/k_{-1}<1$ (entries 45, 49–52, 60, 61, 67). Indeed when going to the weakly basic *p*-anisidine, it is even possible to see base catalysis with 2,4-dinitrochlorobenzene (entry 66).

(c) *Entering amine: steric effects and influence of the* o-*nitro group*—The smaller k_2/k_{-1} ratios for secondary as compared to primary amines have been attributed to steric compressions in the intermediate complex[42, 62]. On reverting to reactants this steric strain can be partially released since the transition state is looser than the intermediate. This is supposed greatly to enhance k_{-1} and therefore reduce k_2/k_{-1}. As there may be some steric compression in the transition state which is also the transition state for nucleophilic attack, k_1 should at the same time be reduced. Comparison of entries 24 and 25 show *N*-methylaniline is indeed *c.* 30 times less nucleophilic than aniline which is in agreement with this interpretation.

Secondary aliphatic or alicyclic amines on the other hand are better nucleophiles than their primary counterparts as is evident by a comparison of k_1 values (entries 2 and 3, 10 and 20, 7 and 11, 17), but, nevertheless, lead to smaller k_2/k_{-1} ratios with the exception of pyrrolidine (entries 11 and 17, 13 and 19).

In terms of a simple steric effect it appears difficult to explain how both k_1 and k_{-1} should be larger with secondary amines, that the relatively minor change from piperidine to pyrrolidine should presumably decrease k_{-1} by at least 10^6-fold (entries 13 and 19) and that for the reactions 49–51 the k_2/k_{-1} ratios are about the same despite the large variation in bulkiness of the amine. Furthermore, a study of the reaction of 1,3,5-trinitrobenzene with several amines (equation (2.12)) revealed that k_{-1} was substantially

$$\text{1,3,5-}(NO_2)_3C_6H_3 + R^1R^2NH \underset{k_{-1}}{\overset{k_1}{\rightleftharpoons}} \text{[}O_2N, NO_2, NO_2\text{-cyclohexadienide with H and } {}^{+}NHR^1R^2\text{]} \qquad (2.12)$$

the same with the amines $MeNH_2$ [51], Me_2NH [51], Bu^nNH_2 [85], piperidine[85] and pyrrolidine[85] although two *o*-nitro groups would be expected to increase steric compression in the intermediate.

Recent observations in our laboratory[86] suggest the *o*-nitro group may play a certain role in determining the value of k_2/k_{-1}. This is best seen by comparing entries 49 with entries 60 and 61 where k_2/k_{-1} changes from 0.24 in the presence of an *o*-nitro group to $\ll 1$ (estimate $<10^{-4}$ [86]) when the nitro group is replaced by a condensed benzene ring of similar steric requirement.

The effect of an *o*-nitro group on the k_2/k_{-1} ratio can be rationalised in terms of intra-molecular hydrogen bonding in the intermediate as depicted in structures (4) and (5) for a secondary and a primary amine respectively. Independent evidence that such intramolecular hydrogen bonding indeed occurs even in aqueous solution has been obtained from a proton transfer rate study[87]. It suggests a hydrogen bond strength of about 2.3 kcal mol^{-1} in aqueous solution for a 2,4,6-trinitrobenzene derivative. In a less polar solvent and with a smaller number of nitro groups sharing the negative charge this hydrogen bond is likely to be appreciably stronger; in fact full proton transfer to the *o*-nitro group has been suggested to occur in benzene[52].

In the argument to follow, which is equally valid if the proton is completely transferred, we consider separately how k_{-1} and k_2 are influenced by hydrogen bonding. Reversion to reactants (k_{-1}) requires the breaking of the hydrogen bond and thus some extra activation energy regardless of the nature of

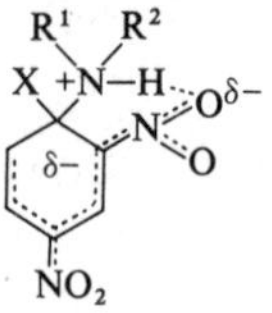

(4) 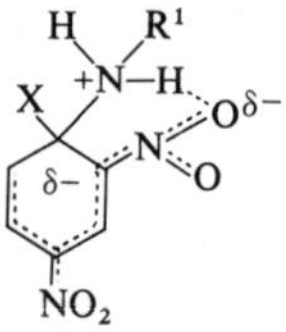 (5)

the amine. The extra activation energy will be reflected in a lowering of k_{-1} which may be about the same for both types of amines. (In case the proton sits on the nitro group the extra activation energy arises from the need to deprotonate the weak nitronic acid rather than the more strongly acidic ammonium ion.)

As will be discussed in Section 2.2.1.4, the mechanism associated with k_2 must involve a transfer of an ammonium proton in (4) or (5) to the leaving group, even in a polar solvent. In (4) there is only one such proton available and it is tied up in a hydrogen bond. To make it available for catalysis of leaving group expulsion the hydrogen bond must be broken which requires extra activation energy and reduces k_2 just as it reduces k_{-1}. In (5) on the other hand, there is a non-hydrogen-bonded proton available and to a first approximation no extra activation energy is required. This is, of course, not exactly true since transferring the free proton to the incipient anion X^- reduces the acidity of the bonded hydrogen and thus reduces the stabilisation through hydrogen bonding, but this in turn is partially compensated by an increase in acidity due to transformation from an aliphatic to an aromatic amine. The net effect should be that k_2, if at all reduced by intramolecular hydrogen bonding in (5), decreases much less than k_2 in (4).

In conclusion, the introduction of an *o*-nitro group decreases k_{-1} to approximately the same extent for both types of amines but decreases k_2 for secondary amines more than for primary amines which results in a smaller k_2/k_{-1} ratio for secondary amines. This effect should be largest in non-polar solvents; the only well documented cases found so far are indeed reactions carried out in benzene solution (compare entry 49 with 60, 61). At press time we have, however, preliminary evidence of base catalysis in the reaction of n-butylamine with 4,7-dinitronaphthyl methyl ether (no *o*-nitro group) in 60% dioxane-40% water (entry 8). This would be the first example where $k_2/k_{-1}<1$ in a reaction of a primary amine in a protic solvent.

The drastic increase in the k_2/k_{-1} ratio on moving the nitro group from the *para* to the *ortho* position for the reaction of the same secondary amine (entries 53 and 54) calls for some comment. It can be rationalised by assuming that, in general, the hydrogen bond decreases k_{-1} more effectively than it does k_2 and that the discrepancy between the effect on k_{-1} and k_2 is magnified when the hydrogen bond is strong. This is likely to be the case in benzene solution and even more so when there is only one nitro group in the molecule carrying most of the negative charge. It will bring about a change from $k_2 \ll k_{-1}$ to $k_2 \gg k_{-1}$ by introducing an *o*-nitro group. The impossibility of detecting the non-base catalysed pathway ($k_3{}^B/k_2 = \infty$) in the absence of an *o*-nitro group in entries 58–61, which contrasts with findings in the presence of such group (entries 55, 56, 64, 65), has been interpreted along similar lines[61] and so has the relatively low k_2/k_{-1} ratio in entry 68, where it has been suggested[52] that the proton sits on the imidazole nitrogen far removed from the *o*-nitro groups thus preventing interaction with the nitro group.

On the other hand, in hydroxylic solvents the hydrogen bond is relatively weak and the differential effect on k_2 and k_{-1} slight, as can be seen by comparing entries 17 and 18, where k_2/k_{-1} is almost the same with or without *o*-nitro group.

One observation which has not yet found a satisfactory explanation is the enormous increase in k_2/k_{-1} when changing from piperidine to pyrrolidine (entries 11 and 17, 13 and 19). It appears, however, from the mentioned study[51, 85] on reaction 12 and even more from a kinetic study of the separately observable formation and decomposition of the intermediate complex involving 2,4-dinitro-1-naphthyl ethyl ether as a substrate in dimethyl sulphoxide[88] that a change in k_2 rather than in k_{-1} must be the main cause.

2.2.1.3 Mechanism of the base catalysed step ($k_3{}^{B_i}$)

Three principal mechanisms have been proposed:

(a) Rate-limiting proton transfer from (3) to the base with subsequent rapid leaving-group removal. This interpretation was suggested in 1958[37] but lost favour in later discussions[18, 38–40, 44, 89], largely on the basis of the work of Eigen[90] who showed that proton transfer to strong bases are very fast and often diffusion controlled. The mechanism has also been directly shown to be untenable for at least one case by the fact that base-catalysed expulsion of phenoxide ion in the reaction of 2,4-dinitrophenyl phenyl ether with piperidine (entry 27) shows an oxygen isotope effect[89] when the ether oxygen is replaced by ^{18}O.

The question whether such a mechanism is possible with very weak bases will, however, have to be raised in connection with the discussion of the 'solvent' catalysed' or 'spontaneous' leaving group removal (k_2) in Section 2.2.1.4.

(b) Concerted breaking of the N—H and C—X bonds of (3) brought about by attack of base on H. This E2-like mechanism was also suggested in 1958 [37], has lost favour in later discussions[38–40, 44], but has been advocated in a recent review[18] and even more recently by others[47, 91] (see also Section 2.2.1.5).

(c) The mechanism which has been favoured by the majority of recent discussions was first proposed in 1960[92] and involves a fast acid-base equilibrium between the initially formed zwitterionic intermediate (3) and its conjugate base (6) followed by a rate-limiting expulsion of X with general acid catalysis by BH. (Equation (2.13a) and (2.13b).)

$$(3) + B_i \xrightleftharpoons{K_3^{B_i}} (6) + B_iH \tag{2.13a}$$

$$(6) + B_iH \xrightarrow{\tilde{k}_3^{B_i}} [(7)] \rightarrow \text{product} \tag{2.13b}$$

Note that in terms of this mechanism the phenomenological rate coefficient $k_3^{B_i}$ is given by

$$k_3^{B_i} = K_3^{B_i}\tilde{k}_3^{B_i} \tag{2.14}$$

For later reference we call it the SB–GA (for specific base–general acid catalysis) mechanism.

Steady-state kinetics does not allow a choice between the E2-type and the SB–GA mechanisms. But recently, Bunnett and Orvik[93] were able to follow the kinetics of intermediate complex formation and decomposition separately in the reaction of 2,4-dinitro-l-naphthyl ethyl ether with n-butyl and t-butylamine in dimethyl sulphoxide. It was shown that equilibrium (2.15) lies strongly on the side of 9, and that the removal of ethoxide ion from 9 is

$$(8) + RNH_2 \rightleftharpoons (9) + R\overset{+}{N}H_3 \tag{2.15}$$

first order with respect to the concentration of RNH_3^+. This is in complete agreement with the SB–GA mechanism but extremely difficult to reconcile with an E2-type mechanism since involvement of RNH_3^+ in the rate limiting transition state assigns a logical function to RNH_3^+ only in the former but not in the latter mechanism[93].

2.2.1.4 Mechanism of the uncatalysed step (k_2)

The large efforts devoted to the elucidation of the mechanism of base catalysis contrasts with the meagre attention paid to the spontaneous, or presumably solvent assisted[30], decomposition of (3) to products. Actually, enough data are available in the literature to allow a reasonable mechanism to be advanced and certain possibilities to be excluded.

In polar protic media it may indeed appear attractive to attribute an important role to the solvent. There are several possibilities.

(a) One might visualise the same type of SB–GA mechanism as for base catalysis, only that the solvent plays the role of B, the lyonium ion the role of BH as, e.g. in equations (2.16a) and (2.16b)[93]

$$(3)+H_2O \overset{K_2}{\rightleftharpoons} (6)+H_3O^+ \qquad (2.16a)$$

$$(6)+H_3O^+ \overset{\tilde{k}_2}{\rightarrow} \text{products} \qquad (2.16b)$$

Like $k_3^{B_i}$ (equation (2.14)), k_2 would be a composite quantity

$$k_2 = K_2\tilde{k}_2 \qquad (2.17)$$

where K_2 is the acid-dissociation constant of (3). Since proton transfer to the solvent cannot be fast (assuming, e.g. a p$K \approx 10$ for (3) the deprotonation rate coefficient can at most be *c.* 1 mol^{-1} s^{-1} since the rate of the reverse reaction cannot be faster than diffusion controlled[90]), proton transfer would mainly occur through catalysis by lyate ions or whatever efficient base is present in the system (amine) as shown in equation (2.18);

$$(3)+HO^- \overset{K_3^{HO}}{\rightleftharpoons} (6)+H_2O \qquad (2.18)$$

the second step is still as in equation (2.16b). This is of no consequence to equation (2.17) since

$$k_2 = K_3^{HO}\tilde{k}_2[H_3O^+][HO^-] = K_3^{HO}K_w\tilde{k}_2 = K_2\tilde{k}_2 \qquad (2.19)$$

where K_w is the self-ionisation constant of water.

As attractive as this mechanism may appear, it can be ruled out on the following grounds. Let us first go back to the SB–GA mechanism for base catalysis (e.g. (2.13a) and (2.13b)) and estimate the order of magnitude for the rate coefficient $\tilde{k}_3^{B_i}$ in step (2.13b) for the specific case where B_i = MeO$^-$, X = MeO in methanol solvent (equation (2.20). The decomposition of complex (11) (equation (2.21) with k_d = 42 s^{-1} [94] can be regarded as a model for reaction (2.20) and hence we assume $\tilde{k}_3^{MeO} \approx k_d$.

(6 a) [CH_3O, NR^1R^2, NO_2, NO_2] $\xrightarrow[CH_3OH]{\tilde{k}^{MeO}}$ [NR^1R^2, NO_2, NO_2] + CH_3O^- (2.20)

(11) [CH_3O, OCH_3, NO_2, NO_2] $\xrightarrow[CH_3OH]{k_d}$ [OCH_3, NO_2, NO_2] + CH_3O^- (2.21)

The argument to follow is not very sensitive to the exact correspondence between k_d and $\tilde{k}_3^{MeO}$ as long as $\tilde{k}_3^{MeO}$ is not several orders of magnitude smaller than k_d; if on the contrary $\tilde{k}_3^{MeO} > k_d$, the argument becomes even more convincing. There is some indication that the latter inequality prevails, since complexes like (11) with two oxygen substituents in the 1-position appear to be particularly stable[95] (reduction of k_d), and the transition state for the decomposition of (6a) may benefit more from the larger resonance stabilisation between the amino group and the benzene ring in the incipient aniline than is possible for the anisole forming step (k_d) since the resonance interaction of the methoxyl group is smaller[96]; this would enhance $\tilde{k}_3^{\mathrm{MeO}}$ over k_d. Thus we conclude conservatively $\tilde{k}_3^{\mathrm{MeO}} \geqslant 42$; in aqueous solution $\tilde{k}_3^{\mathrm{HO}}$ for the same reaction is likely to be very similar[97].

With this estimate of $\tilde{k}_3^{\mathrm{HO}}$ for the poorest leaving group (methoxide ion) we can now indicate a lower limit for k_3^{HO}/k_2 ratios expected on the basis of mechanism of equation (2.16). Since from equation (2.14)

$$k_3^{\mathrm{HO}} = K_3^{\mathrm{HO}}\tilde{k}_3^{\mathrm{HO}} = K_w K_2 \tilde{k}_3^{\mathrm{HO}} = 10^{14} K_2 \tilde{k}_3^{\mathrm{HO}} \tag{2.22}$$

one obtains in combination with equation (2.17)

$$k_3^{\mathrm{HO}}/k_2 = 10^{14}\tilde{k}_3^{\mathrm{HO}}/\tilde{k}_2 \tag{2.23}$$

Allowing for a diffusion-controlled reaction of (6) with H_3O^+, step (2.16b) ($\tilde{k}_2 = 10^{10}\ \mathrm{mol\,l^{-1}\,s^{-1}}$) one obtains from equation (2.23) and from $\tilde{k}_3^{\mathrm{HO}} \geqslant 42$ $k_3^{\mathrm{HO}}/k_2 \geqslant 4 \times 10^5$. Since for poor leaving groups the hydronium-ion catalysed conversion of (6) to products is certainly much slower than diffusion controlled ($\tilde{k}_2 \ll 10^{10}\ \mathrm{mol\,l^{-1}\,s^{-1}}$) whereas for good leaving groups $\tilde{k}_3^{\mathrm{HO}} \gg 42$, a more realistic estimate is $k_3^{\mathrm{HO}}/k_2 \gg 4 \times 10^5$ which would imply that k_2 is too small to allow experimental detection of the non-catalysed pathway under most practical conditions and contradicts experimental findings. Inspection of Table 2.1 shows that with the exception of entry 19 experimental k_3^{HO}/k_2 ratios never even approach the value of 4×10^5 and thus the SB–GA mechanism becomes very unattractive.

(b) Rate determining proton transfer from (3) to the solvent leads to similarly low k_2 values as for the SB–GA mechanism and can be ruled out along similar lines.

(c) An E2-like mechanism is not easily ruled out. However, a much more attractive alternative involving intramolecular acid catalysis of leaving group departure is indicated by the transition state structures (12) and (12a),

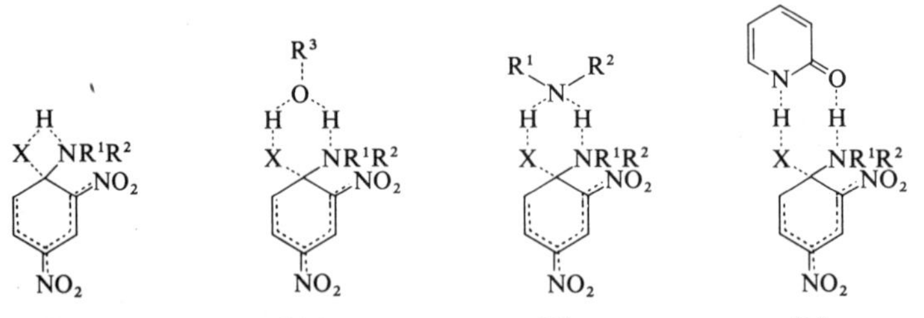

(12) (12a) (13) (14)

briefly mentioned as a possibility by Kirby and Jencks[39]. Whether (12) or (12a) prevails cannot be decided, though in a hydroxylic solvent (12a) or even a variant including several solvent molecules between the ammonium ion and leaving group may be more favourable; a transition state resembling (12) seems best fitted to explain why a non-catalysed pathway can at all be effective in benzene even with fluoride ion as leaving group.

This mechanism easily accounts for the failure to generally observe electrophilic catalysis, which had been one of the strongest arguments against the SB–GA mechanism for *base catalysis*[18, 89]; electrophilic catalysis by a weak acid cannot successfully compete with the intramolecular process which is favoured on entropy grounds whereas strong acids would tie up a large portion of the nucleophile and effectively decrease the rate.

It should be noted that although (12a) bears formal resemblance to (13) suggested for piperidine catalysis[32] in non-polar solvent and to (14) proposed for bifunctional catalysis by α-pyridone[98], there is a difference. (13) and (14) assign an active role to the catalyst, whereas the role of the solvent in (12a) is the one of a passive proton transfer agent, just as in ordinary intermolecular acid–base reactions in hydroxylic solvents[90], the main driving force is provided by the pK difference between ammonium ion and incipient leaving group anion rather than between ammonium ion and the catalyst as in (13) and (14).

2.2.1.5 Kinetic hydrogen isotope effects

The study of kinetic hydrogen isotope effects in aminolysis reactions (N—H *v.* N—D) has not shed as much light on the mechanism of base catalysis as was hoped for. Generally, k_H/k_D is not much different from unity[28, 40, 59, 65, 67, 69, 77, 91, 99], and was never found to be larger than 1.5 [69]. Though there seems to be a pattern in that the magnitude of k_H/k_D is usually larger when a very poor leaving group is involved[65, 69, 91, 99] (strong base catalysis), the significance of these observations is by no means clear, particularly in view of the fact that some reactions which are definitely base catalysed have $k_H/k_D \approx 1$ (e.g. entry 55 in Table 2.1) and others with good leaving groups like bromide have a relatively high k_H/k_D ratio[77].

Although hydrogen isotope effect experiments have often been of great diagnostic value in assigning the rate limiting step where proton transfer is from carbon[100–103], they do not appear equally informative when proton abstraction is from more electronegative atoms. As a matter of fact, there is no general agreement as to when primary isotope effects can be expected in such situations[104, 105]. It has also been pointed out by Bell[106] that when the reaction coordinate involves several other bond-breaking and bond-forming processes besides proton transfer there are indeed many factors which can render an isotope effect very small. Hence arguments favouring any of the mechanisms for base catalysis and in particular the E2-like based on such isotope effects[18, 28, 89, 77, 91] do not seem warranted at this time.

2.2.2 Further aspects of the S_NAr-mechanism in reactions including other nucleophiles

2.2.2.1 Ortho v. para activation

The question under what circumstances an *ortho*-nitro group (or other activating substituent) is more activating than in *para* position is still under active investigation[107–115]. The prevailing view has been that in the absence of other influences a nitro group in *ortho*-position is inherently more activating than in *para*-position[26, 28, 116] presumably due to its stronger inductive effect. This situation is particularly manifest in reactions of 2,4-dihalogenonitrobenzenes[26]. In reactions of *ortho*-halogenonitrobenzenes with alkoxide ions steric interference with the attainment of co-planarity by the nitro group in the transition state was invoked to explain a slightly lower reactivity than that of *para*-halogenonitrobenzenes[26]. The appreciably higher *ortho* activation, notably in non-polar solvents when the nucleophile is a primary or secondary amine was explained by 'built-in solvation' from the *ortho*-nitro group[116], which was visualised either as electrostatic or as a hydrogen-bonding interaction, a choice between the two being unwarranted on the basis of the experiments.

Some other effects which are not well understood may, however, completely overshadow this steric hindrance of resonance in the *o*-nitro group. In the reactions of 1-halogeno-2-nitro- and 1-halogeno-4-nitro-benzenes with alkoxide ions the *ortho*:*para* ratio increases when going from MeO^- to EtO^- to Pr^iO^- [114]; this is in agreement with earlier findings on reactions with sterically hindered amines[30, 117]. The same substrates reacting with Bu^tO^- in Bu^tOH can lead to an *ortho*:*para* ratio as high as 360 [112], although a different laboratory[110] reported much lower ratios. The situation is further complicated by the great influence of the cation. For example, substitution of potassium by benzyltrimethylammonium lowers the *ortho*:*para* ratio from 360 to 26 in the case of the fluoronitrobenzenes[112]; the higher ratio with potassium is attributed[112] to a specific stabilisation of the transition state for *ortho*-nitro-substituted substrates by a potassium bridge between the oxygens of the nucleophile and the nitro group. Similar views on the '*ortho*-effect' of metal cations have been expressed by Shein[107–109].

2.2.2.2 *Early and late transition states*

The transition state for intermediate complex formation (rate determining step) in the reaction of nitrofluorobenzenes with sulphite ion probably resembles reactants rather than the intermediate complex in contrast to methoxydefluorination reactions[114]. This is based on the following observations:

In reactions of 2-nitro-, 4-nitro- and 2,4-dinitrofluorobenzene the methoxide ion is a better nucleophile than the sulphite ion (e.g. $k(MeO^-)/k(SO_3^{2-}) = 11.9$ for 2-nitrofluorobenzene), whereas with 3,5-dinitrofluorobenzene sulphite ion is the better reagent ($k(MeO^-)/k(SO_3^{2-}) = 0.25$). Whereas the activation towards attack by methoxide ion is about the same with a *para*-nitro group as with two *meta*-nitro groups, the reaction of sulphite ion with 3,5-dinitrofluorobenzene is *c.* 150 times faster than with 4-nitrofluorobenzene. In a transition state which is close to the intermediate complex (MeO^-) the *p*-nitro group can almost exert its full mesomeric effect and thus can be about as effective as two *meta*-nitro groups which act mainly through their inductive effect. If the transition state resembles the reactants (SO_3^{2-}), the resonance effect of the *para*-nitro group is much weaker so that two *meta*-nitro groups are more effective in activating the substrate. That the sulphite ion leads to an early transition state is reasonable since it is highly polarisable[118].

Another manifestation of an early transition state when the nucleophile is a sulphur base can be seen in the reactions of picryl halides with 2,4,6-trinitrothiophenolate and picrate ion respectively. Since there is great steric interaction when the two reagents approach each other, only the reaction with the sulphur base (early transition state) occurs, whereas the picrate ion does not react at all[119].

2.2.2.3 *Effects of substituents, leaving group, nucleophile and medium on reactivity*

A considerable amount of effort is being devoted by some Russian workers to the establishment of reactivity orders for a large number of typical S_NAr reactions. These include the study of substituent effects in the aromatic substrate[108, 120, 121] and in particular correlations by the Hammett[122] equation, correlation of nucleophilic reactivities[125] with the Swain–Scott[126] equation, the effect of the solvent[75, 108, 109, 111, 124, 127–133] and of the leaving group[75, 120, 134–136]

2.2.2.4 *Salt effects and micellar catalysis*

Measurements of the activity coefficients of 2,4-dinitrochlorobenzene and aniline in aqueous salt solution showed that the salt effect on the initial state can be more important than on the transition state[137], a fact which had been overlooked in many earlier discussions. For example, KCl, $NaNO_3$ and NaBr increase the rate of reaction between 2,4-dinitrochlorobenzene and aniline practically exclusively by increasing the activity of the initial state, NaCl and Na_2SO_4 enhance the rate by destabilising the initial state more than the transition state and only $Me_4N^+Cl^-$ and $Me_4N^+ClO_4^-$ were found to stabilise considerably the transition state and thus to offset the somewhat stabilising salt effect upon the initial state. In ion–molecule reactions (e.g. 2,4-dinitrochlorobenzene and HO^-) cations of low charge density generally stabilise the transition state[137].

The effect of micelle formation upon the rate of S_NAr reactions has also been investigated[138–141]. Cationic micelles such as formed from cetyltrimethylammonium bromide (CTA) accelerate, anionic micelles of sodium lauryl sulphate (NaLS) retard the reactions of 2,4-dinitrochlorobenzene with anionic nucleophiles[139, 140]. This can be explained by incorporation of the aromatic substrate into the micelle, which depending on its charge type, attracts or repels the anionic nucleophile[139, 140]. On the other hand, incorporation into a non-ionic micelle has no kinetic effect which again can be rationalised with the above electrostatic model. An alternative explanation is that a cationic micelle could decrease the free energy of a large anionic transition state relative to that of the initial state (neutral molecule plus small anion) and an anionic micelle could have the opposite effect[139, 140].

Micellar effects on reactions of a neutral nucleophile such as aniline with

2,4-dinitrohalogenobenzenes[141] which are catalysed by both cationic and non-ionic micelles, and even to some extent by anionic micelles are less well understood, in particular the effective catalysis by non-ionic detergents such as polyethers. A possible explanation is that the incipient ammonium group in the first (rate determining) transition state is hydrogen bonded to the ether oxygen atom of the micelle[141].

The mentioned studies all refer to reactions with nucleophilic attack (k_1 in equation (2.1)) as the rate-determining step. The effect of micelles on the reversion of the intermediate to reactants (k_{-1} in equation (2.1)) was also studied for some highly activated aromatic substrates which form stable σ-complexes or Meisenheimer complexes (see also Section 2.2.3)[142–144]. Cationic micelles and sometimes non-ionic micelles[142, 144] decrease k_{-1}, anionic micelles have no effect; the magnitude of retardation is generally

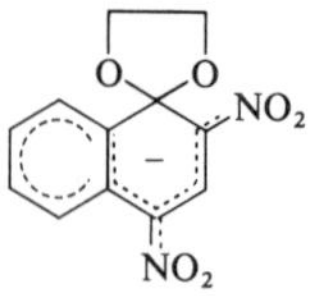

(15)

much smaller than the magnitude of acceleration of k_1, with the notable exception of the spiro complex (15) the decomposition rate of which is enormously retarded by CTA, which was attributed to a transition-state effect[144].

2.2.3 Meisenheimer complexes

The interactions of highly activated aromatic species with bases to form detectable or isolable Meisenheimer complexes (M.C.) have received a great deal of attention in the last few years. Since this field has recently been thoroughly reviewed by several authors[18–23], only some of the most novel developments not yet reviewed were chosen for inclusion in this report.

2.2.3.1 Some novel Meisenheimer complexes

A novel type of M.C. of the general structure (16) with R = H, Me or Bu^n have been isolated by Taylor[145, 146] upon treatment of 1,3,5-trinitrobenzene with tetramethylammonium borohydride (R = H) or with the appropriate tetra-alkyl boron salts (R = Me, Bu^n) respectively. Similar findings are reported by Kaplan and Siedle[147]; these latter authors also report the observa-

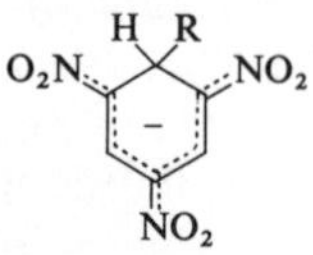

(16) 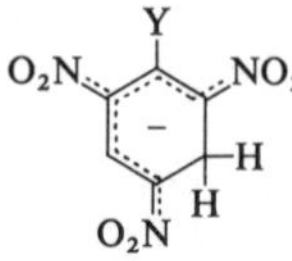 (17) 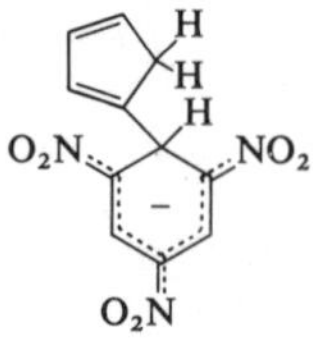 (18) 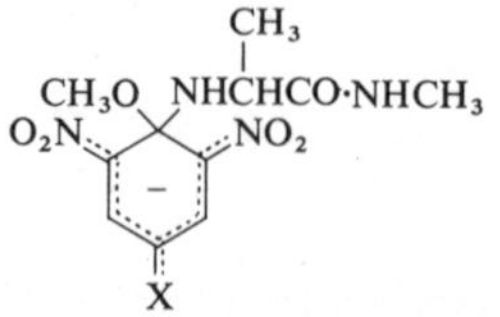 (19)

tion of (17) (Y = Cl, MeO, Me) from reaction of 1-Y-2,4,6-trinitrobenzene with tetramethylammonium octahydrotriborate. σ-Complex formation between 1,3,5-trinitrobenzene and aryl copper compounds[148] or cyclopentadienylcopper[149] and related species[149, 150] have been described among which (18) is a representative example.

Boulton's[151–153] group was able to generate complexes of structure (19) by treating 2,6,-dinitro-4-X-anilino-*N*-methylpropionamides (X = NO_2, CN, $MeSO_2$) with methoxide ion or amines in methanol solution. This is remarkable since mono-alkyl substituted picramides are known to either undergo proton abstraction at the amino nitrogen or to be attacked at the 3-position upon treatment with methoxide ion[154, 155]. Intramolecular hydrogen bonding between the amido-N—H and an *ortho*-nitro group was suggested to provide the extra stabilisation of the 1-complex (19) over the competing products[153].

Treatment of (19) with methanolic HCl produced an immediate spectral change with a subsequent slow decay to regenerate the starting materials[152]; the fast spectral change suggests the formation of the neutral σ-complex with one of the nitro groups protonated[152]. Neutral σ-complexes of similar type have been first described by Clapp *et al.*[156].

2.2.3.2 *1-σ versus 3-σ-complexes*

The phenomenon, first observed by Servis[155, 157] for the interaction of 2,4,6-trinitroanisole with methoxide ion in dimethyl sulphoxide, that many highly activated aromatic ethers undergo a rapid reversible nucleophilic addition to a non-substituted position (20) which is followed by the slower formation of the thermodynamically more stable adduct at the carbon bearing the alkoxy substituent (21) is still widely investigated and has recently been shown to occur in nitro activated benzene[158–164], naphthalene[164–165] and pyridine[166] derivatives bearing a variety of substituents.

Until recently, direct evidence for 3-σ-complexes could only be obtained in dimethyl sulphoxide or dimethyl sulphoxide–methanol mixtures due to their short lifetime in pure methanol. The first indication of a 3-σ-complex, (20),

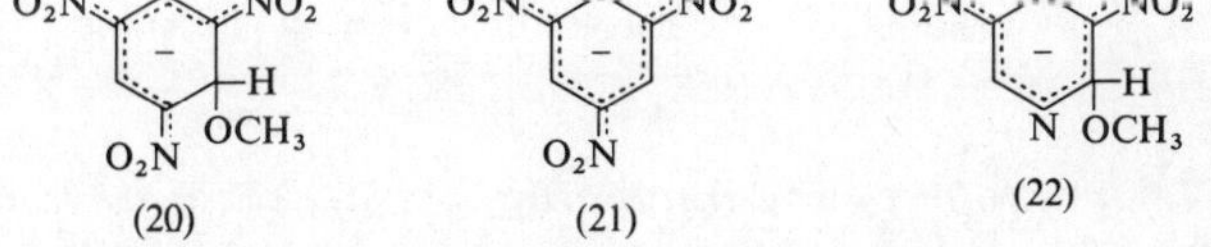

(20) (21) (22)

formation in methanol came from a calorimetric study[158, 167]; more direct structural (u.v./vis. spectra) evidence for (20) has become available through a temperature-jump–stopped-flow study[168, 169] whereas a stopped-flow study provided evidence for the transient formation of (22)[170].

Various explanations, including steric and solvation effects, not all of them very satisfactory[95], have been advanced for the much slower formation of the thermodynamically more stable 1-σ-complex; they have recently been summarised by Crampton[21] and by Strauss[23] and also discussed by Miller *et al.*[171]. As illustrated with the 'classical' example of 2,4,6-trinitroanisole, Bernasconi[95, 168] has suggested that resonance stabilisation[171] of the ground state ((23) is probably the most important resonance form) which is essentially conserved in the 3-σ-complex (24) but is lost in the 1-σ-complex (21) is probably an important factor and can satisfactorily explain why the 3-σ-complex (20) is formed faster despite its lower stability compared to (21). The much smaller discrepancy between kinetic and thermodynamic stability

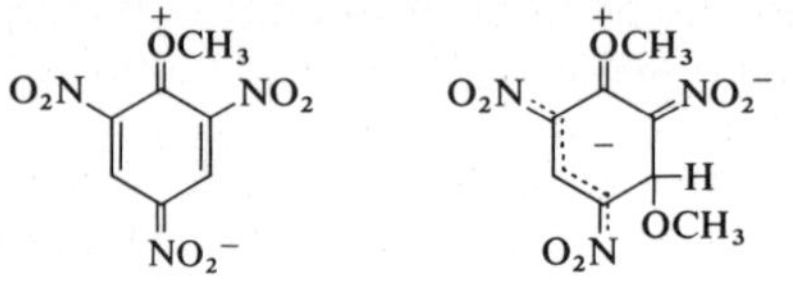

(23) (24)

when the *para* substituent is F, Cl or CF_3 [162, 165], substituents which unlike the nitro group cannot interact so well with the 1-methoxy group, is in agreement with this interpretation.

Bowden and Cook[172] found kinetic evidence for transient formation of a 5-σ-complex between 1-substituted 2,4-dinitrobenzene and hydroxide ion in aqueous dimethyl sulphoxide.

2.2.3.3 *Solvent effects*

The great stabilising effect exerted on Meisenheimer complexes by dipolar aprotic solvents is well known[21, 23, 83, 173]. Calorimetric measurements of heats of transfer from methanol to methanolic dimethyl sulphoxide solutions for 2,4,6-trinitroanisole, sodium methoxide and the 1-σ-complex (21) indicate that the increase in the exothermicity of complex formation in a mixture of dimethyl sulphoxide–methanol (95:5) compared to methanol is to be attributed to both desolvation of the methoxide ion (10.6 kcal mol^{-1}) and to increased stabilisation of the complex (*c.* 7 kcal mol^{-1}) [174].

The stabilising effect of dimethyl sulphoxide on 1-σ- and 3-σ-complexes appears to be about the same since the ratio of equilibrium constants for their formation are essentially independent of solvent composition[161, 162, 166]. This is also true for the ratio of their rates of formation indicating that stabilisation of the respective transition states is also the same [161, 162].

2.2.3.4 *Nucleophilicity and nucleofugicity in protic solvents*

In an attempt to separate the effect of solvent from the inherent nucleophilic and nucleofugic reactivity of alkoxide ions in protic media, the kinetics of σ-complex formation between 1,3,5-trinitrobenzene and HO^-, MeO^- and EtO^- in 'essentially aqueous' solution were measured[175]. The rates of complex formation are somewhat reduced for MeO^- (threefold) and EtO^- (fourfold) compared to the reactions in their respective pure solvents[95] (see reference 176 for similar but more extensive data in various alcohols) as expected on the basis of classic solvent theory[84], whereas the reverse rates are practically unaffected by the solvent change. This is in interesting contrast to the much greater solvent dependence of the complex formation constants between 1,3,5-trinitrobenzene and ethanethiolate ions[177]. From these data [175,177], the relative carbon basicities in an aqueous solution containing 19% ethanol come out to be $HO^-:MeO^-:EtO^-:EtS^- = 1:0.92$ (in 22% MeOH–78% H_2O):23:73 which is quite different, in particular with regard to EtO^-, from the reactivity order 1:0.3:2.2:45 calculated by Crampton and Ghariani[177] for a 'common solvent'.

2.3 RADICAL MECHANISMS

2.3.1 The $S_{RN}1$-mechanism

Recently it has been proposed[13] that the reactions of 5- and 6-iodopseudocumenes with potassium amide in liquid ammonia to form 5- and 6-pseudocumidines proceed through a radical mechanism in competition with a benzyne route. A pure benzyne mechanism, as found for the chloro and bromo analogues should yield the products 6- and 5-pseudocumidine in the 'aryne ratio' of *c.* 1.5 regardless of starting material. The ratios for the iodo compounds were substantially different and in the direction of retention of the original orientation, suggesting the competing mechanism to be non-rearranging. The fact that the presence of radical trapping agents shifted the product ratio toward the 'aryne ratio' was taken as evidence for a radical mechanism which was tentatively sketched as follows:

$$\text{electron donor} + ArI \rightarrow [ArI]^{-\cdot} \tag{2.24}$$

$$[ArI]^{-\cdot} \rightarrow Ar\cdot + I^- \tag{2.25}$$

$$Ar\cdot + NH_2^- \rightarrow [ArNH_2]^{-\cdot} \tag{2.26}$$

$$[ArNH_2]^{-\cdot} + ArI \rightarrow ArNH_2 + [ArI]^{-\cdot} \tag{2.27}$$

$$\text{termination steps} \tag{2.28}$$

This mechanism has been designated '$S_{RN}1$', standing for 'substitution, radical–nucleophilic, unimolecular' (step (2.25)). Subsequent work[14] has shown that the $S_{RN}1$-route can be strongly promoted by addition of alkali metals as electron donors and can force a number of other substitutions to proceed through this mechanism. A case in point are the reactions of *o*-halogenoanisoles with sodium or potassium amide in ammonia which epitomise cine[26] substitution by the aryne mechanism and afford overwhelmingly the *m*-anisidine[7]. With an excess of potassium and potassium amide over the anisole, *o*-anisidine is the exclusive product[14].

2.3.2 Free radicals in S_NAr reactions

Based mainly on e.s.r. and n.m.r. evidence indicating the presence of radical anions generated from nitroaromatics and nucleophiles[16], particularly in dipolar aprotic solvents, Shein[15] proposed that σ-complex ($ArXY^-$) formation may proceed through prior radical anion formation rather than by direct nucleophilic addition. This is illustrated by equations (2.29) and (2.30)

$$ArX + Y^- \rightleftharpoons [ArX]^{-\cdot} + Y\cdot \tag{2.29}$$

$$[ArX]^{-\cdot} + Y\cdot \rightleftharpoons ArXY^- \rightarrow ArY + X^- \tag{2.30}$$

A second mechanism was also proposed to account for data suggesting a radical chain process as described in equations (2.31)–(2.33)

$$ArX + Y^- \rightleftharpoons [ArX]^{-\cdot} + Y\cdot \tag{2.31}$$

$$[ArX]^{-\cdot} + Y^- \rightleftharpoons [ArXY]^{2-\cdot} \tag{2.32}$$

$$[ArXY]^{2-\cdot} + ArX \rightarrow ArXY^- + [ArX]^{-\cdot} \tag{2.33}$$

Although the presence of radicals in many of the investigated systems seems to be beyond doubt, Shein's work does not give compelling evidence that these radicals are indeed located on the reaction coordinate of the substitution reaction or σ-complex formation rather than representing competing side reactions.

Acknowledgement

I am grateful to Professor J. F. Bunnett for numerous suggestions and criticism of the manuscript.

References

1. Bunnett, J. F. and Zahler, R. E. (1951). *Chem. Rev.* **49,** 275
2. Chapman, N. B., Parker, R. E. and Soanes, P. W. (1954). *J. Chem. Soc.,* 2109
3. Cavell, E. A. S. and Chapman, N. B. (1953). *J. Chem. Soc.,* 3392
4. Lewis, E. S., Hartung, L. D. and McKay, R. M. (1969). *J. Amer. Chem. Soc.,* **91,** 419
5. Footnote 7 in Lewis, E. S. and Chalmers, D. J. (1971). *J. Amer. Chem. Soc.,* **93,** 3267
6. Roberts, J. D., Simmons, Jr., H. E., Carlsmith, L. A. and Vaughan, C. W. (1953). *J. Amer. Chem. Soc.,* **75,** 3290
7. Roberts, J. D., Semenov, D. A., Simmons, H. E., Jr. and Carlsmith (1956). *J. Amer. Chem. Soc.,* **78,** 601
8. Hoffmann, R. W. (1967). *Dehydrobenzene and Cycloalkynes.* (New York and London: Academic Press)
9. Gilchrist, R. L. and Rees, C. W. (1969). *Carbenes, Nitrenes and Arynes.* (London: Nelson)
10. Krishnappa, S. (1970). *J. Scientific Ind. Res.,* **29,** 538
11. Heaney, H. (1970). *Fortschr. Chem. Forsch.,* **16,** 35
12. Havinga, E., de Jongh, R. O. and Dorst, W. (1956). *Rec. Trav. Chim.,* **75,** 378
13. Kim, J. K. and Bunnett, J. F. (1970). *J. Amer. Chem. Soc.,* **92,** 7463
14. Kim, J. K. and Bunnett, J. F. (1970). *J. Amer. Chem. Soc.,* **92,** 7464
15. Shein, S. M., Bryukhovetskaya, L. V., Khmelinskaya, A. D., Starichenko, V. F. and Ivanova, T. M. (1969). *Reakts. Sposobnost' Organ. Soedin.,* **6,** 1087
16. Shein, S. M., Brovko', V. V. and Khmelinskaya, A. D. (1970). *J. Org. Chem. USSR,* **6,** 784
17. Starichenko, V. F., Sokolov, V. V. and Shein, S. M. (1971). *Isv. Akad. Nauk USSR, Ser. Khim. Nauk,* 1836
18. Buncel, E., Norris, A. R. and Russell, K. E. (1968). *Quart. Rev. Chem. Soc.,* **22,** 123

19. Foster, R. and Fyfe, C. A. (1966). *Rev. Pure Appl. Chem.*, **16,** 61
20. Buck, P. (1969). *Angew. Chem., Int. Ed. Engl.*, **8,** 120
21. Crampton, M. R. (1969). *Advan. Phys. Org. Chem.*, **7,** 211
22. Hall, T. N. and Poranski, Jr., C. F. (1970). *The Chemistry of the Nitro and Nitroso Groups,* Part 2, 329. (H. Feuer, editor) (New York: Interscience)
23. Strauss, M. J. (1970). *Chem. Rev.*, **70,** 667
24. Meisenheimer, J. (1902). *Liebigs Ann. Chem.*, **323,** 205
25. Jackson, C. J. and Gazzolo, F. H. (1900). *J. Amer. Chem. Soc.*, **23,** 376
26. Bunnett, J. F. (1958). *Quart. Rev. Chem. Soc.*, **12,** 1
27. Sauer, J. and Huisgen, R. (1960). *Angew. Chem.*, **72,** 294
28. Ross, S. D. (1963). *Progr. Phys. Org. Chem.*, **1,** 31
29. Miller, J. (1968). *Aromatic Nucleophilic Substitution.* (New York: Elsevier)
30. Pietra, F. (1969). *Quart. Rev. Chem. Soc.*, **23,** 504
31. de Boer, T. J. and Dirkx, I. P. (1969). *The Chemistry of the Nitro and Nitroso Groups,* Part 1, 487. (H. Feuer, editor) (New York: Interscience)
32. Capon, B. and Rees, C. W. (1964). *Ann. Reports,* **61,** 278
33. Capon, B., Perkins, M. J. and Rees, C. W. (1965). *Organic Reaction Mechanisms,* 133 (New York: Interscience); (1966). ibid, 160; (1967). ibid, 166
34. Alder, R. W. (1968). *Organic Reaction Mechanisms,* 187. (B. Capon and C. W. Rees, editors); (New York: Interscience) (1969). ibid, 215
35. Bodenstein, M. (1913). *Z. Physik. Chem.*, **85,** 329
36. Hammett, L. P. (1970). *Physical Organic Chemistry,* 2nd edn., 78. (New York: McGraw-Hill)
37. Bunnett, J. F. and Randall, J. J. (1958). *J. Amer. Chem. Soc.*, **80,** 6020
38. Bunnett, J. F. and Garst, R. H. (1965). *J. Amer. Chem. Soc.*, **87,** 3879
38a. Bunnett, J. F. and Jackson, L. (1969). Personal communication
39. Kirby, A. J. and Jencks, W. P. (1965). *J. Amer. Chem. Soc.*, **87,** 3217
40. Bunnett, J. F. and Bernasconi, C. F. (1965). *J. Amer. Chem. Soc.*, **87,** 5209
41. Bernasconi, C. F. (1965). *Ph.D. Thesis,* (ETH Zürich)
42. Beale, J. H. (1966). *Ph.D. Thesis,* (Brown University)
43. Bernasconi, C. F. and Zollinger, H. (1967). *Helv. Chim. Acta,* **50,** 3
44. Bernasconi, C. F. (1967). *J. Org. Chem.*, **32,** 2947
45. Bernasconi, C. F. and Schmid, P. (1967). *J. Org. Chem.*, **32,** 2953
46. Banjoko, O., Bevan, C. W. L. and Hirst, J. (1969). *Nigerian J. Sci.*, in the press
47. Bamkole, T. O. and Hirst, J. (1969). *J. Chem. Soc.*, **B,** 848
48. Pietra, F. and Vitali, D. (1968). *J. Chem. Soc. B,* 1200
49. Bunnett, J. F. and Hermann, D. H. (1970). *Biochemistry,* **9,** 816
50. Bunnett, J. F. and Bernasconi, C. F. (1970). *J. Org. Chem.*, **35,** 70
51. Bernasconi, C. F. and Schmid, P. (1969). Unpublished results
52. Minetti, R. and Bruylants, A. (1970). *Bull. Acad. Roy. Belg.*, **56,** 1047
53. Bunnett, J. F. and Pruitt, K. M. (1957). *J. Elisha Mitchell Sci. Soc.*, **73,** 297
54. Ross, S. D. and Kuntz, I. D. (1954). *J. Amer. Chem. Soc.*, **76,** 3000
55. Parker, R. E. (1963). *Advan. Fluorine Chem.*, **3,** 63
56. Hammond, G. S. and Parks, L. R. (1955). *J. Amer. Chem. Soc.*, **77,** 340
57. Suhr, H. (1964). *Chem. Ber.*, **97,** 3277
58. Suhr, H. (1965). *Liebigs. Ann.*, **687,** 175
59. Pietra, F. and Fava, A. (1963). *Tetrahedron Lett.*, 1535
60. Pietra F. and Del Cima, F. (1966). *Tetrahedron Lett.*, 1925
61. Pietra, F. and Del Cima, F. (1967). *Tetrahedron Lett,* 4573
62. Bunnett, J. F. and Garst, R. H. (1965). *J. Amer. Chem. Soc.*, **87,** 3875
63. Bunnett, J. F., Kato, T. and Nudelman, N. S. (1969). *J. Org. Chem.*, **34,** 785
63a. Bunnett, J. F. and Garst, R. H. (1968). *J. Org. Chem.*, **33,** 2320
64. Pietra, F. (1965). *Tetrahedron Lett.*, 745
65. Pietra, F. (1965). *Tetrahedron Lett.*, 2405
66. Bernasconi, C. F. and Zollinger, H. (1966). *Helv. Chim. Acta.*, **49,** 103
67. Bernasconi, C. F. and Zollinger, H, (1966). *Helv. Chim. Acta,* **49,** 2570
68. Becker, G., Bernasconi, C. F. and Zollinger, H. (1966). *Helv. Chim. Acta.*, **49,** 10
69. Pietra, F., Vitali, D. and Frediani, S. (1968). *J. Chem. Soc. B,* 1595
70. Bernasconi, C. F. and de Rossi, R. H. (1971). Unpublished results
71. Bernasconi, C. F. and Zollinger, H. (1965). *Tetrahedron Lett.*, 1083

72. Suhr, H. (1963). *Ber. Bunsenges.*, **67**, 893
73. Bernasconi, C. F., Kaufmann, M. and Zollinger, H. (1966). *Helv. Chim. Acta*, **49**, 2563
74. Beltrame, P., Bonomi, L., Da Re, P. and Simonetta, M. (1967). *Gazz. Chim. Ital.*, **97**, 470
75. Danilova, N. K. and Shein, S. M. (1970). *Reakts. Sposobnost' Organ. Soedin.*, **7**, 476, 718
76. Shein, S. M. and Sutchkova, L. A. (1970). *Reakts. Sposobnost Organ. Soedin*, **7**, 732
77. Lisitsyn, V. N. and Shul'chishin, V. A. (1970). *J. Org. Chem. USSR*, **6**, 313
78. Kirby, A. J. and Younas, M. (1970). *J. Chem. Soc. B*, 1187
79. Mathieu, J., Allais, A. and Valls, J. (1960). *Angew. Chem.*, **72**, 71
80. Thornton, E. R. (1964). *Solvolysis Mechanisms*, 163. (New York: Ronald Press)
81. Kirby, A. J. and Jencks, W. P. (1965). *J. Amer. Chem. Soc.*, **87**, 3209
82. Johnson, S. L. (1967). *Advan. Phys. Org. Chem.*, **5**, 294
83. Parker, A. J. (1962). *Quart. Rev. Chem. Soc.*, **16**, 163
84. Ingold, C. K. (1953). *Structure and Mechanism in Organic Chemistry*, 345. (Ithaca, N.Y.: Cornell University Press)
85. Bernasconi, C. F. (1970). *J. Amer. Chem. Soc.*, **92**, 129
86. Bernasconi, C. F., de Rossi, R. H. and Schmid, P. (1971). Unpublished results
87. Bernasconi, C. F. (1971). *J. Phys. Chem.*, **75**, 3636
88. Bunnett, J. F., Smith, L. A. and Sekiguchi, S. (1971). Personal communication
89. Hart, C. R. and Bourns, A. N. (1966). *Tetrahedron Lett.*, 2995
90. Eigen, M. (1964). *Angew. Chem. Int. Ed.*, **3**, 1
91. Giardi, I., Illuminati, G. and Sleiter, G. (1968). *Tetrahedron Lett.*, 5505
92. Bunnett, J. F. and Davis, G. T. (1960). *J. Amer. Chem. Soc.*, **82**, 665, footnote 27
93. Bunnett, J. F. and Orvik, J. A. (1970). **92**, 2417
94. Bernasconi, C. F. (1968). *J. Amer. Chem. Soc.*, **90**, 4982
95. Bernasconi, C. F. (1970). *J. Amer. Chem. Soc.*, **92**, 4682
96. Boulton, J. J. K., Jewess, P. J. and McFarlane, N. R. (1971). *J. Chem. Soc. B*, 928
97. Bernasconi, C. F. and Bergstrom, R. G. (1971). *J. Org. Chem.*, **36**, 1325
98. Pietra, F. and Vitali, D. (1966). *Tetrahedron Lett.*, 5701
99. Pietra, F. and Vitali, D. (1968). *Chem. Commun.*, 692
100. Saunders, W. H., Jr. (1966). *Survey Progr. Chem.*, **3**, 109
101. Simon, H. and Palm, D. (1966). *Angew. Chem.*, **78**, 993
102. Zollinger, H. (1964). *Advan. Phys. Org.*, **2**, 163
103. Melander, L. (1960). *Isotope Effects on Reaction Rates.* (New York: Ronald Press)
104. Swain, C. G., Kuhn, D. A. and Schowen, R. L. (1965). *J. Amer. Chem. Soc.*, **87**, 1553
105. Bunton, C. A. and Shiner, V. J., Jr. (1961). *J. Amer. Chem. Soc.*, **83**, 3214
106. Bell, R. P. (1965). *Discuss. Faraday Soc.*, **39**, 16
107. Shein, S. M., Litvak, V. V. and Khmetinskaya, A. D. (1968). *J. Org. Chem. USSR*, **4**, 632
108. Shein, S. M. and Danilova, N. K. (1968). *J. Org. Chem. USSR*, **4**, 1875, 1882
109. Shein, S. M. and Evstifeev, A. V. (1969). *J. Org. Chem. USSR*, **5**, 906
110. Shein, S. M. and Evstifeev, A. V. (1969). *Reakts. Sposobnost' Organ. Soedin.*, **6**, 371
111. Danilova, N. K. and Shein, S. M. (1967). *Reakts. Sposobnosts Organ. Soedin.*, **4**, 649
112. Pietra, F., Vitali, D. Del Cima, F. and Cardinali, G. (1970). *J. Chem. Soc.*, 1659
113. Pietra, F. and Del Cima, F. (1968). *Chem. Commun.*, 216
114. Bevan, C. W. L., Foley, A. J., Hirst, J. and Uwamu, W. O. (1970). *J. Chem. Soc. B*, 794
115. Bamkole, T. O. and Hirst, J. (1971). *Chem. Commun.*, 69
116. Bunnett, J. F. and Morath, R. J. (1955). *J. Amer. Chem. Soc.*, **77**, 5051
117. Pietra, F. and Del Cima, F. (1968). *J. Org. Chem.*, **33**, 1411
118. Edwards, J. O. and Pearson, R. G. (1962). *J. Amer. Chem. Soc.*, **84**, 16
119. Sharnin, G. P., Nurgatin, V. V. and Buzykin, B. I. (1968). *J. Org. Chem. USSR*, **4**, 834
120. Vizgart, R. V., Ozdrovskaya, I. M., Nazaretyan, V. P. and Yagupol'skii, L. M. (1968). *J. Org. Chem. USSR*, **4**, 1452
121. Zagulayeva, V. A., Shein, S. M., Shvets, A. I., Mamayev, V. P. and Krivopalov, V. P. (1970). *Reakts. Sposobnost' Organ. Soedin.*, **7**, 1133
122. Hammett, L. P. (1940). *Physical Organic Chemistry*, 184. (New York: McGraw-Hill)
123. Shein, S. M. and Kozorez, L. A. (1966). *Reakts. Sposobnost' Organ. Soedin.*, **3**, 45
124. Shein, S. M. and Suchkova, L. A. (1968). *Reakts. Sposobnost' Organ. Soedin.*, **5**, 310
125. Wisghert, R. W. and Ozdrowska, I. M. (1966). *Reakts. Sposobnost' Organ. Soedin.*, **3**, 16
126. Swain, C. G. and Scott, C. B. (1953). *J. Amer. Chem. Soc.*, **75**, 4231
127. Shein, S. M., Danilova, N. K. and Kuznetsova, N. I. (1970). *Reakts. Sposobnost' Organ. Soedin.*, **7**, 458, 476

128. Ozdrowska, I. M. and Wisghert, R. W. (1966). *Reakts. Sposobnost' Organ. Soedin.*, **3**, 146
129. Mamayev, V. P., Zagulyaeva, O. A. Shein, S. M., Shvets, A. I. and Krivopalov, V. P. (1968). *Reakts. Sposobnost' Organ. Soedin.*, **5**, 824
130. Solodova, K. V. and Shein, S. M. (1970). *J. Org. Chem. USSR*, **6**, 1475
131. Sutschkova, L. A. and Shein, S. M. (1969). *Reakts. Sposobnost' Organ. Soedin.*, **6**, 586, 867
132. Shein, S. M. and Evstifeev, A. V. (1968). *Reakts. Sposobnost' Organ. Soedin.*, **5**, 817
133. Shein, S. M. and Rodionov, P. P. (1970). *Reakts. Sposobnost' Organ. Soedin.*, **7**, 1150
134. Wisghert, R. W. and Ozdrowska, G. M. (1966). *Reakts. Sposobnost' Organ. Soedin.*, **3**, 35
135. Shein, S. M. and Solodova, K. V. (1970). *J. Org. Chem. USSR*, **6**, 1479
136. Shein, S. M., Evstifeev, A. V. and Starichenko, V. F. (1968). *J. Org. Chem. USSR*, **5**, 384
137. Bunton, C. A. and Robinson, L. (1968). *J. Amer. Chem. Soc.*, **90**, 5965
138. Herries, D. G., Bishop, W. and Richards, F. M. (1964). *J. Phys. Chem.*, **68**, 1842
139. Bunton, C. A. and Robinson, L. (1968). *J. Amer. Chem. Soc.*, **90**, 5972
140. Bunton, C. A. and Robinson, L. (1969). *J. Org. Chem.*, **34**, 780
141. Bunton, C. A. and Robinson, L. (1970). *J. Amer. Chem. Soc.*, **92**, 356
142. Fendler, E. J. and Fendler, J. H. (1970). *Chem. Commun.*, 816
143. Casilio, L. M., Fendler, E. J. and Fendler, J. H. (1971). *J. Chem. Soc. B*, 1377
144. Fendler, J. H., Fendler, E. J. and Merrritt, M. V. (1971). *J. Org. Chem.*, **36**, 2172
145. Taylor, R. P. (1970). *Chem. Commun.*, 1463
146. Taylor, R. P. (1970). *J. Org. Chem.*, **35**, 3578
147. Kaplan, L. A. and Siedle, A. R. (1971). *J. Org. Chem.*, **36**, 937
148. Björklund, C., Nilsson, M. and Wennerström, O. (1970). *Acta Chem. Scand.*, **24**, 3599
149. Wahren, R. and Wennerström, O. (1970). *Acta Chem. Scand.*, **24**, 3064
150. Nilsson, M., Ullenius, C. and Wennerström, O. (1971). *Tetrahedron Lett.*, 2713
151. Bergman, E., McFarlane, N. R. and Boulton, J. J. K. (1970). *Chem. Commun.*, 511
152. Boulton, J. J. K. and McFarlane, N. R. (1971). *J. Chem. Soc. B*, 925
153. Boulton, J. J. K., Jewess, P. J. and McFarlane, N. R. (1971). *J. Chem. Soc.*, 928
154. Crampton, M. R. and Gold, V. (1966). *J. Chem. Soc. B*, 893
155. Servis, K. L. (1967). *J. Amer. Chem. Soc.*, **89**, 1508
156. Clapp, L. B., Lacey, H., Beckwith, G. G. Srivastava, R. M. and Muhammad, N. (1968). *J. Org. Chem.*, **33**, 4262
157. Servis, K. L. (1965). *J. Amer. Chem. Soc.*, **87**, 5495
158. Fendler, E. J., Fendler, J. H., Griffin, C. E. and Larsen, J. W. (1970). *J. Org. Chem.*, **35**, 287
159. Terrier, F., Millot, F. and Letellier, P. (1970). *Bull. Soc. Chim. France*, 1743
160. Terrier, F. and Simonnin, M.-P. (1971). *Bull. Soc. Chim. France*, 671
161. Dearing, C., Terrier, F. and Schaal, R. (1970). *Compt. Rend. Acad. Sci. Paris*, **271C**, 349
162. Terrier, F., Dearing, C. and Schaal, R. (1971). *Reaction Transition States.* (London: Gordon and Breach)
163. Terrier, F., Hallé, J.-C., Simonnin, M.-P. and Lecourt, M. J. (1971). *Org. Mag. Res.*, **3**, 361
164. Millot, F. and Terrier, F. (1969). *Bull. Soc. Chim. France*, 2692
165. Millot, F. and Terrier, F. (1971). *Bull. Soc. Chim. France*, 3897
166. Chatrousse, A.-P., Terrier, F. and Schaal, R. (1970). *Compt. Rend. Acad. Sci. Paris*, **271C**, 1477
167. Larsen, J. W., Fendler, J. H. and Fendler, E. J. (1969). *J. Amer. Chem. Soc.*, **91**, 5903
168. Bernasconi, C. F. (1971). *J. Amer. Chem. Soc.*, **93**, 6975
169. Bernasconi, C. F. (1971). *Abstracts IUPAC-Congress, Boston*, 112
170. Terrier, F., Schaal, R. and Chatrousse, A. P. (1971). *Abstracts IUPAC-Congress Boston*, 112
171. Biffin, M. E. C., Miller, J., Moritz, A. G. and Paul, D. P. (1969). *Aust. J. Chem.*, **22**, 2561
172. Bowden, K. and Cook, R. S. (1970). *Tetrahedron Lett.*, 249
173. Parker, A. J. (1969). *Chem. Rev.*, **69**, 1
174. Larsen, J. W., Amin, K. and Fendler, J. H. (1971). *J. Amer. Chem. Soc.*, **93**, 2910
175. Bernasconi, C. F. and Bergstrom, R. G. (1971). *J. Org. Chem.*, **36**, 1325
176. Gan, L. H. and Norris, A. R. (1971). *Canad. J. Chem.*, **49**, 2490
177. Crampton, M. R. and Ghariani, M. El. (1971). *J. Chem. Soc. B*, 1043

3
Aromatic Rearrangements

HENRY J. SHINE
Texas Tech University, Lubbock, Texas

3.1 INTRODUCTION

This review will attempt to cover developments in aromatic rearrangements since the writing of the reviewer's book[1], the manuscript for which was finished in the fall of 1965. Since 1965 the ideas of orbital symmetry and concerted rearrangements have been applied successfully (via the use of molecular-orbital diagrams) to some aromatic rearrangements (e.g. the Claisen). The entirely new technique of chemically-induced dynamic nuclear polarisation (CIDNP) has changed completely the views of some rearrangements which once seemed so right but may now turn out to have been wrong (e.g., the Sommelet). A few new rearrangements have been discovered, while in some rearrangements new data have clarified or changed earlier ideas (e.g., the Fischer–Hepp and nitramine).

Aromatic rearrangements usually fall into two structural classes. In one, a group attached to an aromatic ring migrates to another position on the ring, either intra- or intermolecularly. In the other, a group, which is part of a ring substituent, leaves the substituent and enters the ring. Again, the rearrangement may be either intra- or intermolecular. The two classes are illustrated in general terms in equations (3.1) and (3.2).

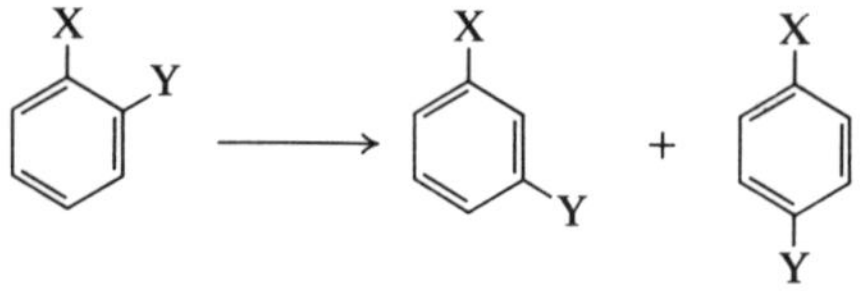

X, Y (ortho) ⟶ X, Y (meta) + X, Y (para) (3.1)

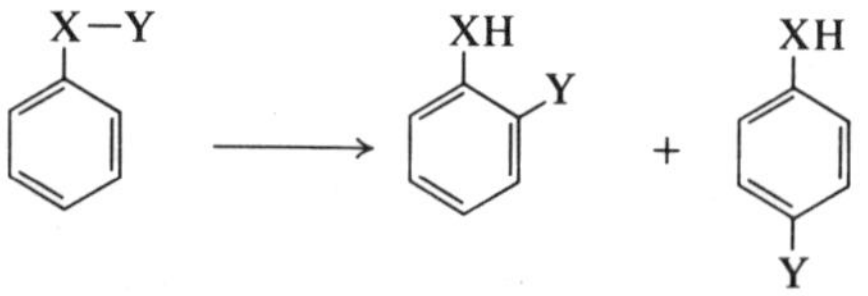

X–Y ⟶ XH, Y (ortho) + XH, Y (para) (3.2)

Much of the research in aromatic rearrangements has been devoted to mechanisms – why and how the rearrangements occur. The rearrangements are most often acid catalysed, but well known base-catalysed, thermal and photo-induced rearrangements occur. Questions of intra- and intermolecularity pervade the field, and where intramolecular rearrangement occurs, particularly, the details of migration are often the source of much speculation and debate. The role (or in some minds, the existence) of π-complexes is still under question. Rules governing concerted rearrangements offer helpful guidance, although one is not certain that they are applicable to aromatic rearrangements in which heteroatoms are involved.

Some rearrangements of the type in equation (3.2) are allowed concerted sigmatropic shifts (e.g. *o*-Claisen), while others which appear to be intramolecular are not allowed (e.g. the rearrangement of alkyl aryl ethers). In such cases one wonders how intramolecularity is achieved, and it is in some of these cases that arguments over π-complexes and other representations, e.g. ion pairs, arise. Examples will be encountered in the pages that follow.

3.2 ALKYL- AND HALOGENOAROMATICS

3.2.1 Alkylaromatics

Strong acids (usually Lewis acids) cause alkyl- and halogenoaromatics to rearrange. Rearrangement may be intra- or intermolecular.

For example, the rearrangement of *o*-xylene (1) to *m*-xylene (2), and of (2) to *p*-xylene (3) involves intramolecular 1,2-shifts of a methyl group. The detail of these 1,2-shifts is controversial. In some quarters it is believed that a π-complex intermediate (5) is formed while in others it is thought that a

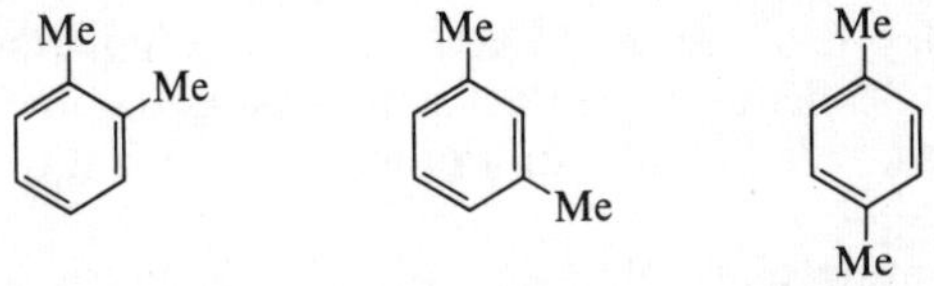

(1) (2) (3)

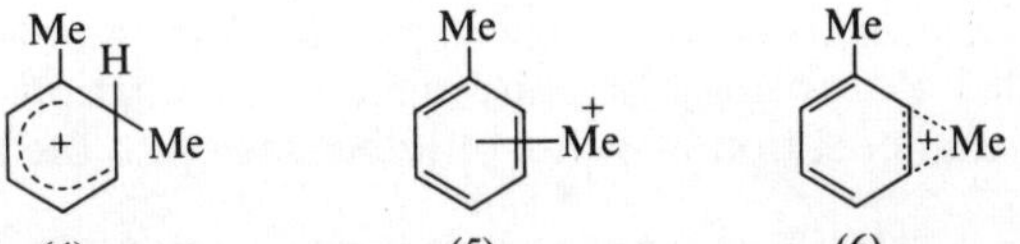

(4) (5) (6)

bridged carbonium ion intermediate (6) is formed. In either case, migration of the methyl group is initiated by protonation (4).

Each of these intermediates represents some degree of simultaneous bonding between the migrating group and the *o*- and *m*-carbon atoms. There is an interesting problem about both representations. The two highest occupied MOs (HOMOs) of benzene are degenerate. An electron-donating group (e.g., alkyl) in the ring lifts the degeneracy in such a way as to make the MO which is labelled (A) in Figure 3.1 higher in energy than the MO

Figure 3.1

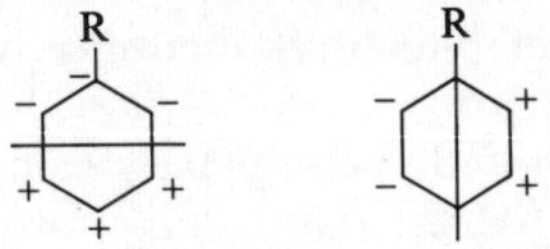

(A) (B)

labelled (B). If we treat the transition state of the rearrangement of *o*-xylene to *m*-xylene as consisting of toluene and methyl cation and consider orbital (A) as the dominant factor, we will see from (A) that concerted migration is ruled out. The node in the MO rules out the π-complex and bridged-ion representation. A loose association between cation and toluene molecule would have to suffice for intramolecular rearrangement of the methyl cation. The problem loses its severity, however, because photoelectron spectroscopy (PS)[2] has shown that the two degenerate HOMOs of benzene (energy 9.24 eV) are not perturbed seriously by a methyl group. The two orbitals, albeit no longer degenerate, are close enough (energy 8.82 eV) not to be resolved by PS[3]. Concerted migrations of a methyl group in *o*-xylene are apparently feasible for this reason.

The foregoing treatment does not account for the fact that branched alkyl groups do not migrate predominantly by 1,2-shifts. Migration of branched groups may occur either intra- or intermolecularly. Two intermolecular routes are available. Thus, *o*-t-butyltoluene will transfer the t-butyl group (after protonation) to solvent toluene. Transfer will occur mostly to the *p*-position following the rules of electrophilic substitution, but location of the t-butyl group in the *m*-position finally predominates if equilibration occurs (p. 69).

If the alkyl group has an α-hydrogen, intermolecular transfer may occur via an alkylation–dealkylation sequence, first proposed by Streitwieser and Reif[4] and adapted recently to ethyltoluene isomerisation over alumina-silica by Csicsery[5].

Although intramolecular migration of substituents is known to go by 1,2-shifts, it is thought in some cases that direct 1,3- and 1,4-intramolecular migration occurs. The groups in these cases usually have d-orbitals (e.g., SO_3H^+, I^+) which permit bonding overlap with a highest-filled aromatic π-orbital. Olah has proposed a new type of 1,3-intramolecular migration, however, called a *bond-to-bond* rearrangement, (equation (3.3)) to account for the rapid isomerisation of *o*-di-t-butylbenzene into the *p*-isomer[6,7].

H Me C—Me Me (7) ⇌ Me—C—Me Me (8) ⇌ H (3.3)

According to this proposal, formation of a *m*-sigma complex does not readily occur. Instead, the migrating t-butyl group goes to the *p*-position via intermediates (7) and (8). It is not clear how this representation (e.g. (7), (8)) differs from Dewar's π-complex representation[9]. An alternative explanation for the rearrangement of *o*-di-t-butylbenzene has been proposed by the reviewer[1].

In connection with π- and σ-complexes, calculations[8] by the CNDO/2 method have indicated that the more stable form of protonated benzene is an edge-type π-complex[1]. Conversion into a σ-complex is calculated to require more energy than isomerisation to an equivalent π-complex (20 *v.*

15 kcal mol^{-1})[8]. This situation is like, and might be thought to provide support for, the *o*-di-t-butyl case above, but, the CNDO/2 calculations are for the gas phase, and comparison with solution-phase rearrangements will be useless until the relative effects of solvation on intermediates are known. For example, the xylenes in solution do not fit the pattern of the gas-phase calculations.

3.2.2 Halogenoaromatics

Among halogenoaromatic molecules migration of fluorine does not occur, migration of chlorine is most likely to be intramolecular, while migration of bromine may be both intra- and intermolecular[1,10,11]. Iodoaromatics in similar circumstances may disproportionate readily[12].

3.2.3 Equilibration

If *o*- or *p*-xylene is treated with an excess of acid, e.g. aluminium chloride, complete conversion into *m*-xylene occurs. The reason for this is that the most stable xylenium ion (9) is formed and remains in solution until work-up removes a proton.

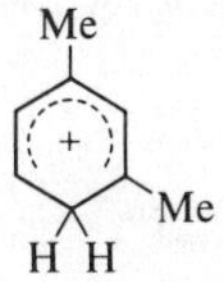

Me

Me

H H

(9)

If, on the other hand, any one of the xylenes is treated with a small amount of acid, equilibration eventually occurs in which all three xylenes are at *thermodynamic* equilibrium, determined by the relative stabilities of the xylenes and not by the relative stabilities of xylenium ions[1,11]. The same holds true for all di- and poly-substituted aromatics. Equilibrium may be reached by intra- or intermolecular migrations. Among equilibrated disubstituted aromatics, the *m*-isomer always predominates. Steric and electronic factors govern the distribution, and these have been discussed elsewhere in fair detail[1,13,14].

3.2.4 Jacobsen and Reverdin rearrangements

This is a particular type of alkylaromatic rearrangement occurring mostly when polyalkyl (methyl or ethyl) aromatics are warmed with sulphuric acid. Durene and isodurene rearrange to prehnitene. In contrast, both durene and prehnitene rearrange completely to isodurene in excess hydrogen fluoride–boron trifluoride. In the latter case the most stable tetramethylbenzenium ion is formed. In the former the most stable tetramethylbenzenesulphonic acid is formed, which on work-up is desulphonated and becomes the hydrocarbon[1].

The mechanism of the Jacobsen rearrangement is not known. Because e.s.r. signals have been detected in Jacobsen rearrangements, there is some thought that the rearrangement may involve radicals, but this is a very uncertain basis for diagnosis. At the moment the consensus is that rearrangement occurs after sulphonation, but it is not known whether sulphonation initiates rearrangement (e.g. with the intermediate (10)) or whether the sulphonic acid rearranges (e.g. via the intermediate (11)). Hart and Janssen have used trichloromethylation[15] instead of sulphonation to deduce that (10) is the more likely precursor to Jacobsen rearrangements.

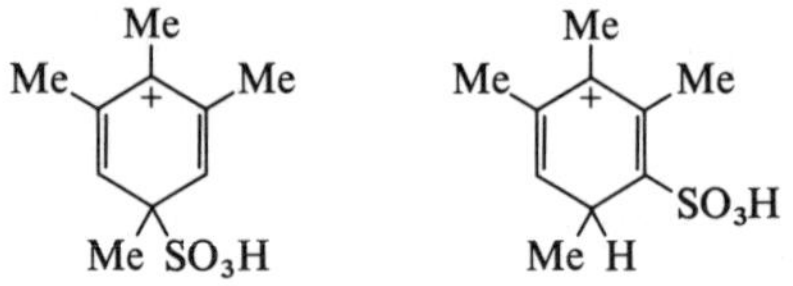

(10) (11)

An analogous situation exists in the Reverdin rearrangement, in which, for example, an electrophile (NO_2^+) attacks the *p*-position of *p*-iodoanisole and leads to 2-iodo-4-nitroanisole[1]. Recently, the extent (*ipso* factors) of attack of NO_2^+ at the *p*-position of anisoles has been measured and amounts to 30.6, 40, 31, and 28% of total attack for *p*-H, *p*-I, *p*-Br and *p*-Cl substituents; but under the conditions used, displacement rather than rearrangement of halogen occurs[16].

3.2.5 Dienone–phenol rearrangement

A substituted dienone may undergo rearrangement and aromatisation in acid solution. Rearrangement is initiated by formation of a benzenium ion, so the dienone–phenol rearrangement has some analogy with the isomerisation of alkylaromatics.

Rearrangement of dienones is usually carried out in acetic anhydride containing a little concentrated sulphuric acid (so-called Thiele conditions) or in hot aqueous sulphuric or hydrochloric acid[1].

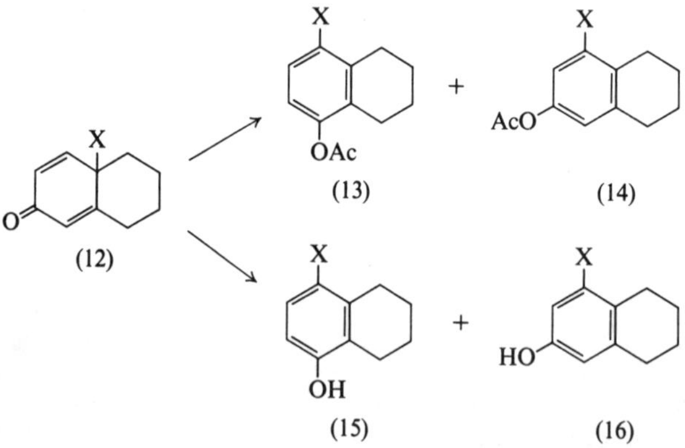

(12) (13) (14) (15) (16)

	(a)	(b)	(c)	(d)
X =	Me	Et	Pr^n	Bu^n

Thiele rearrangement of (12a) gave (13a) and (14a) in the ratio 80:20, while rearrangement in aqueous sulphuric acid gave (15a) and (16a) in the ratio 20:80 [17]. Compounds (13) and (15) are formed from ion (17) (R = H or Ac) by successive 1,2-butyl-type shifts of the saturated ring (ions (18) and (19)). The commonest route to compounds (14) and (16) is a 1,2-alkyl shift in (17), giving

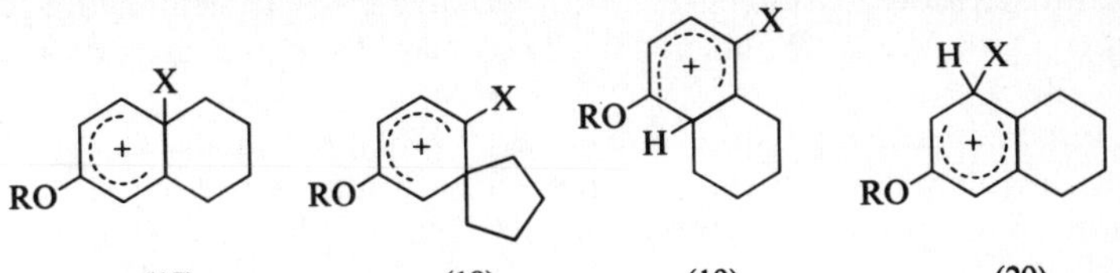

(17) (18) (19) (20)

(20). Deprotonation of (19) and (20) gives the final products. Thus, the medium apparently affects the migratory aptitude of the two groups (methyl and butyl) of (12a), and this has long been a puzzle[1].

New data provide some clarification. First, (12a) has been shown by means of ^{14}C labelling in position 5 to rearrange to (16a) by a second route involving

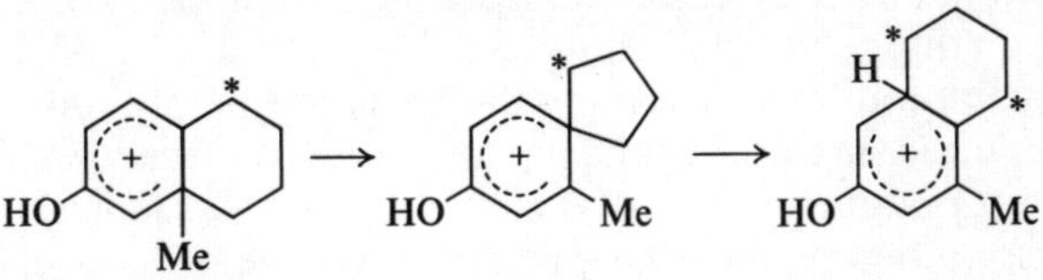

(21a) (22a) (23a)

the ions (21a)–(23a)[18]. Secondly, Thiele rearrangement of (12b) gave only (14b)[19]. Thirdly, systematic investigation of the series (12a–d) at 50 °C has given the results in Table 3.1 [20].

Table 3.1 Composition (%) of products of rearrangement of (12a-d)
(From H. J. Shine and C.E. Schoenig, unpublished results.)

Compound (12)	$Ac_2O–H_2SO_4$ (13)	(14)	20–*N* H_2SO_4 (15)	(16)
(a)	84	16	14	86
(b)	10	90	23	77
(c)	7	93	2	98
(d)	1	99	8	92

The data show that the *m*-type product predominates in all of the tetrahydronaphthols and in all of the acetates except those in which X = Me. Obviously, it is not only the medium which affects product distribution but also the nature of the angular group in (12). It is not surprising when *m*-products ((14) and (16)) predominate. The groups Et, Pr^n and Bu^n will have migratory aptitudes similar to the cyclic system. As soon as ion 20 (X = Me–Bu^n) is formed, it is likely to deprotonate and form product. The second route to a *m*-product (ions (21a)–(23a)) competes with the spiran migrations that would give a *p*-product, and so diminishes the latter again. We do not know, of course, that the second route is followed by any angular group other than Me. The formation of spirans which lead to *p*-products is also surely reversible at the (18) stage, by return to (17) in competition with reaction to (19). In these spirans, furthermore, X suffers steric interference by methylene hydrogens of the spiran (X = Me–Bu^n) and by CH hydrogen of the cyclohexadienyl ring (X = Et–Bu^n).

The most troublesome question is why (13a) predominates over (14a). Perhaps the answer lies in the possible stabilisation by acetoxy-group participation, as in (24a), in which migration (as shown) leads to product. Such

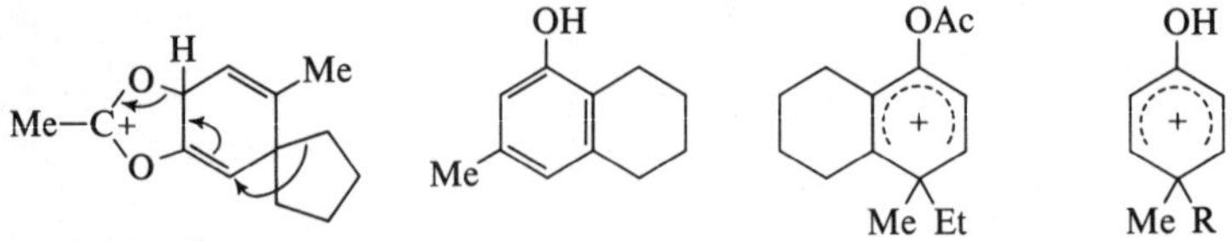

(24a) (25) (26) (27)

stabilisation would then be a deciding factor in competitive alkyl-group migration (i.e. methyl *v.* the cyclic butyl).

Product distribution (at least among the tetrahydronaphthols) is not controlled by the relative stabilities of the *products*, for Dreiding[17] has shown that (25) (which is not obtained ordinarily) is the most stable isomer.

Migratory aptitudes in dienone–phenol rearrangements are of current interest. Thus, Et migrates in preference to Me in (26)[21]. The ion (27) (R = $CHCl_2$) is stable in 90% sulphuric acid while, when R = Me, rearrangement occurs[22].

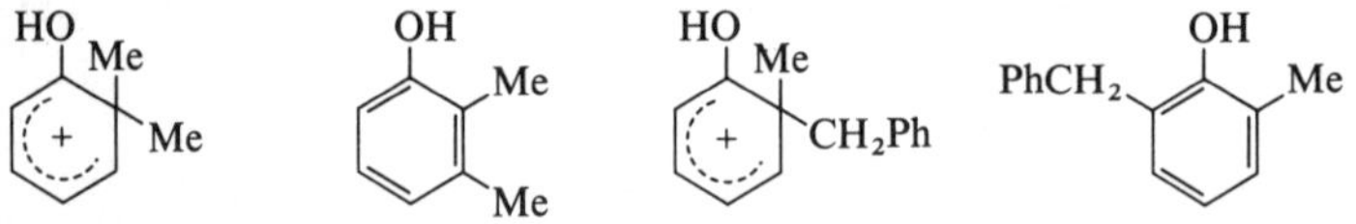

(28) (29) (30) (31)

Whereas (28) gives only (29), (30) gives only (31), a difference which is attributed to π-protonation of the carbonyl group in the first case and n-protonation (followed by a sigmatropic [1,5] shift) in the second[23].

3.3 ETHERS AND ESTERS

3.3.1 Claisen rearrangements

The well-known features of Claisen rearrangements are as follows: allyl phenyl ether (32a) rearranges when heated (neat, in solution, or in the gas phase) to *o*-allylphenol (34a). The rearrangement goes through a dienone intermediate (33a). If the aryl group contains *o*-substituents (32b), prototropy in the intermediate (33b) is impossible. The intermediate may then either revert to (32b) or rearrange to the phenol (36b) after passing through intermediate (35b). If the allyl group contains *o*- and *p*-substituents (32c), neither

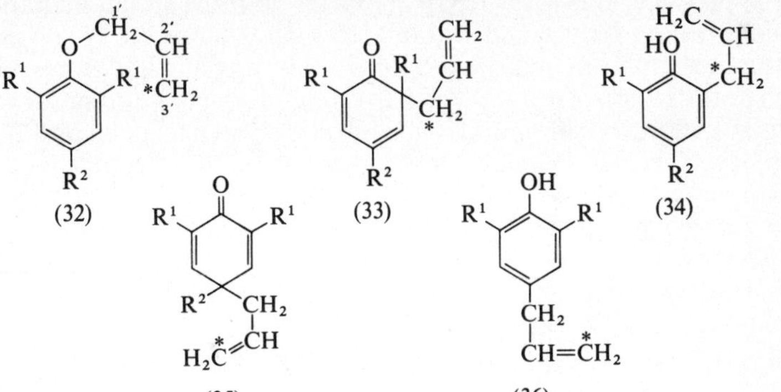

(32) (33) (34)

(35) (36)

(a) $R^1 = R^2 = H$; (b) $R^1 = Me, R^2 = H$;
(c) $R^1 = R^2 = Me$; (d) $R^1 = R^2 = F$

of the dienone intermediates (33c) and (35c) can go on to a phenol and reversion to the ether would occur. Gas-phase rearrangement of pentafluorophenyl allyl ether gave (35d), all other substituents = F) in 32% yield[24].

These successive one-step migrations, which entail bonding at the termini of the allyl group (illustrated by the asterisk) are sigmatropic shifts. The migrations have been treated in these modern terms most clearly by Schmid[25, 26]. In this treatment, the transition state of the rearrangement of, say, allyl phenyl ether is considered to consist of two interacting components, one of which is the phenoxyl radical and the other the allyl radical. Interest is centred on the highest occupied MO (HOMO) of each component. The phenoxyl radical (like the isoelectronic benzyl radical) is regarded as a seven π-MO system, four of which are occupied. The allyl radical is a three π-MO system, two of which are occupied. Each of the occupied MOs is a combination of atomic *p*-orbitals, and these combinations are describable with conventional diagrams (Figure 3.2a). In these diagrams, the signs of the atoms represent the lobes of the *p*-orbitals on one face of the molecule. If we were to look at the other face (see Figure 3.2b), we would write the opposite sign at each atom in Figure 3.2a. The sign distribution for the HOMOs ψ_4 and $\psi_{2'}$ and the rules for sigmatropic rearrangements allow for a concerted, thermally-induced migration of the allyl group from oxygen to an *o*-position in the ring. This is a sigmatropic [3,3] shift. It leads to the intermediate (33a) which undergoes rapid prototropy to give the phenol (34a).

Consideration of analogous HOMO diagrams shows that, in principle, intermediate (33a) can also either revert to allyl phenyl ether or rearrange to intermediate (35a). Each of these possibilities is outweighed by rapid prototropy. It is interesting to note that prototropy cannot be concerted (a sigmatropic [1,3] shift), a fact which was first stressed to the writer by Professor W. C. Herndon. Conversion of the dienone (33a) into the phenol (34a) is intermolecular for rearrangements in solution, and may be a wall reaction for rearrangements in the gas phase.

Consideration of Figure 3.2a shows that, in principle, a direct, concerted [1,5] shift of the allyl group (1′-position) from oxygen to the *p*-position is allowed, but this is stereochemically and energetically too difficult.

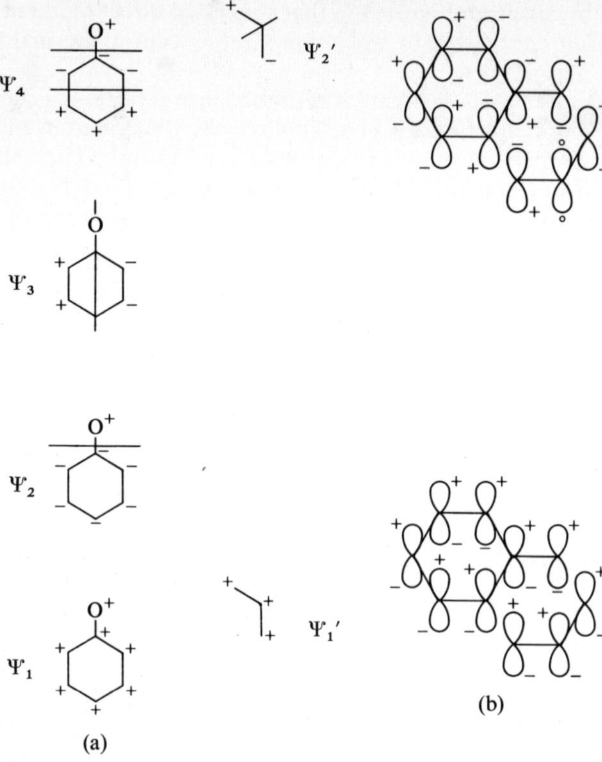

Figure 3.2

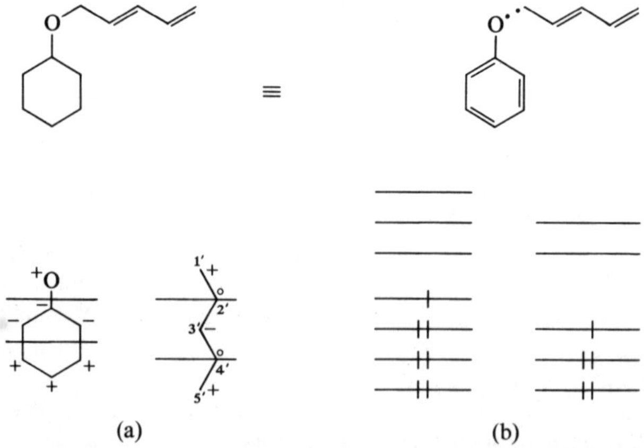

(a) (b)

Figure 3.3

It is also obvious from Figure 3.2a that concerted direct union of the allylic 3′-position from the oxygen to *p*-position cannot occur in thermal rearrangements.

Direct migration of the last kind is permitted in aryl penta-2,4-dienyl ethers, however. The principles, in MO terminology, are illustrated in Figure 3.3 [27]. Two concerted rearrangements are allowed: a [3,3] and a [5,5]. The former would lead to an *o*-Claisen product (37) and the latter directly to a *p*-Claisen

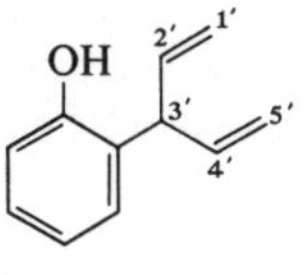

(37)

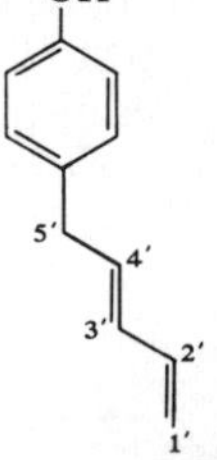

(38)

product (38); both of these occur. Again, in principle, the *p*-Claisen product (38) could be obtained by two consecutive [3,3] shifts. Schmid has shown that, in the main, the two-step rearrangement does not occur and that the direct sigmatropic [5,5] shift overshadows the [3,3] shift.

As yet we have said nothing about the stereochemistry of the Claisen rearrangement. It was clearly deduced long before the rules of orbital symmetry were commonly applied that the best transition state for the Claisen rearrangement is one in which the six atoms directly involved are held in the chair rather than the boat conformation. As a consequence, for example,

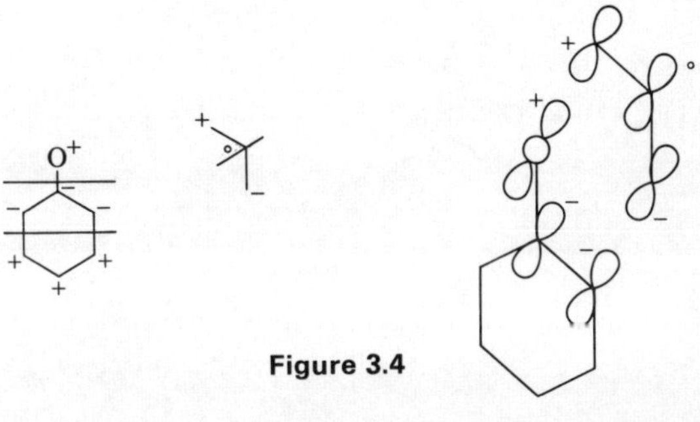

Figure 3.4

trans-allylic ethers rearrange to *trans*-allylic phenols with inversion, while *cis*-allylic ethers rearrange to *trans*-allylic phenols with retention of configuration around the two allylic carbon atoms involved in bond breaking and making. The chair conformation is achieved in each case by the alignment of the allyl group above (or below) and parallel with the aryl ring. In the orbital-symmetry treatment the best transition state is again deduced to be the chair form. Furthermore, the symmetry rules require that bond breaking and bond making occur from one face of each component. This is a requirement for the overlapping of orbital lobes of the same sign, and is illustrated in Figure 3.4. In orbital symmetry terminology the Claisen rearrangement is suprafacial-suprafacial and is written as a [3s,3s] shift. The concerted [5,5] migration in the aryl penta-2,4-dienyl system is a [5s,5s] one.

Stereochemical relationships in rearrangements of a variety of allyl aryl ethers have been discussed recently by Schmid and co-workers[28].

3.3.1.1 *Abnormal Claisen rearrangement*

This, in fact, is well-established as the isomerisation of a γ-substituted allyl group *after* the normal *o*-Claisen phenolic product has been formed. Schmid[29] has made an interesting contribution to understanding the factors which affect the occurrence of an abnormal rearrangement; the data are summarised in equation (3.4). The abnormal product (40) predominates when rearrange-

(39) → (40) + (41) (3.4)

Solvent	(40) %	(41) %
DMF	89	2
DEA	30	72

ment of (39) is carried out in dimethyl formamide (DMF), while the normal *p*-Claisen product (41) predominates in *N*,*N*-diethylaniline (DEA). The normal *o*-Claisen product is not isolated. The reasons are as follows. Ether

(39) → (42) → (43) → (44) → (40) (3.5)

(39) undergoes a normal [3s,3s] shift to give the intermediate (42). If intermolecular prototropy (see p. 73) occurs, the normal product (43) will be formed. Once formed, this undergoes a homosigmatropic [1,5] shift to give the spiran intermediate (44), which in turn undergoes a [1,5] sigmatropic shift to form the so-called abnormal Claisen product (40). Apparently, solvent DMF enhances intermolecular prototropy, and guides the rearrangement

on to (40). In contrast, solvent DEA enhances the second [3s,3s] shift, to end in (41).

3.3.2 Thio-Claisen rearrangement

The sulphur analogues of allylic aryl ethers rearrange to cyclic sulphides as illustrated with allyl phenyl sulphide (45). A thiochroman (46) and a thio-coumaran (47) are formed.

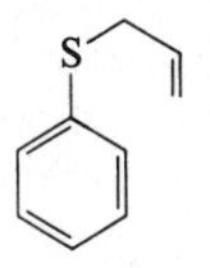

(45) 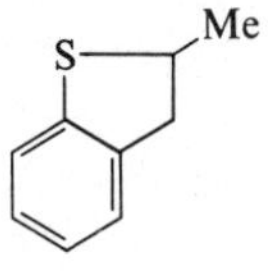 (46) 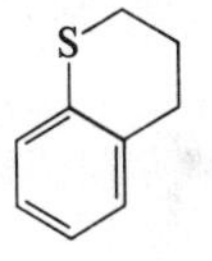 (47) 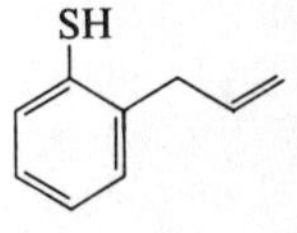 (48)

The rearrangement follows the [3,3]-sigmatropic shift pattern of the *o*-Claisen rearrangement. *o*-Allylthiophenol (48) has been trapped (as the methyl thioether) in the rearrangement of (45)[30]. Corresponding allylquinoline thiols have been trapped (as thiobutyrate esters) in the rearrangement of allyl-3- and allyl-4-quinolyl sulphides[31], while allylthiophene thiol has been trapped (as the thiobenzoate ester) in rearrangement of allyl 2-thienyl sulphide[32]. Makisumi[31] has proposed that the two types of product (e.g. (46) and (47)) arise from the same intermediate (e.g. (48)) but by heterolytic (46) and homolytic (47) involvement of the —SH group.

Thio-Claisen rearrangements may become complicated by competitive 'thioallylic' rearrangements[33, 34] of the allylthio group. Selenoallylic, but not seleno-Claisen, rearrangement occurs, too[35].

3.3.3 Aryl propargyl ethers and sulphides

Cyclisation of aryl propargyl ethers in hot *N*,*N*-diethylaniline has been known for some years. Details of the cyclisation have been elucidated by Schmid[36]. A propargyl ether (49) undergoes a [3,3] sigmatropic rearrange-

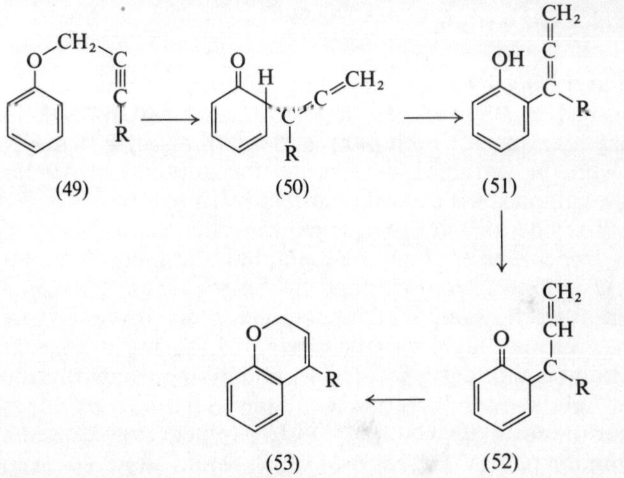

(49) (50) (51)

(3.6)

(53) (52)

ment to an allenyl dienone (50) from which the final product (53) is formed by the steps indicated.

Similar *o*-Claisen-type rearrangements occur in aryl propargyl sulphides[31,37], although the overall course of rearrangement may not be as simple as in the ethers, because concomitant thiopropargyl rearrangement may also occur.[37].

3.3.4 Miscellaneous developments

Very high yields of *o*-Claisen rearrangement products have been obtained by treating allyl aryl ethers with diethylaluminium chloride at room temperature[38]. Reaction is carried out in hexane solution and works extraordinarily well with allyl phenyl ether itself and with chloro substituent in the 2-, 3-, 4-, 2,4-, and 2,4,5- positions. Reaction of allyl 2,6-dichlorophenyl ether gave the *p*-Claisen product in 43 % yield. Use of allyl 2,4,6-trichlorophenyl ether gave 96 % of 2,4,6-trichlorophenol. Neither 2-alloxypyridine nor allyl phenyl sulphide rearranged by this method.

3.3.5 The Fries rearrangement (aryl esters)

This is the rearrangement of an aryl ester to *o*- and *p*-acyl phenols caused by Lewis acids. The questions on whether the rearrangement is intra- or intermolecular are still not answered, and it appears that both routes are encountered[1].

Recently, Dewar and Hart have examined the rearrangement of phenyl benzoate to a mixture of *o*- and *p*-benzoylphenol, and rearrangement of the biphenyl ester of *p*-phenylbenzoic acid to 2-(4′-phenylbenzoyl)hydroxybiphenyl. Two mechanisms are thought to be operating: a π-complex of the edge type and an ion pair[39].

3.3.6 Alkyl aryl ethers

An alkyl aryl ether may rearrange to *o*- and *p*-alkyl phenols in strong acids (sulphuric or Lewis). As in the Fries rearrangement, it appears that alkyl aryl ethers may rearrange by both intra- and intermolecular routes[1], and, in connection with the intramolecular route, the question of whether or not it involves π-complexes is argued continually. In this connection, Spanninger and von Rosenberg[40] have rearranged optically active 2-butyl phenyl ether (54) in chlorobenzene and benzene containing aluminium bromide. By adding Lewis acid to (54) both *o*- (55) and *p*-2-butyl phenol (56) were obtained in which the 2-butyl group had undergone a net inversion (14 and 52 %, respectively). By adding (54) to the Lewis acid, (55) was obtained in which the 2-butyl group had undergone 54 % net retention of configuration. The interpretation[40] of these results is that π-complexes (requiring complete retention of configuration) are ruled out. Instead, two routes are visualised. One of them involves an ion pair (57), collapse of which would allow for formation of (55) with net, but not necessarily complete, retention of configuration. This route would prevail when (54) was not in excess. The other route is a bimolecular one, prevailing in excess of (54) and in which transfer of the 2-butyl group from (58) in an S_N2-like displacement occurs. This route, which is similar to Brown's early proposals for Friedel–Crafts alkylations[1], allows for net

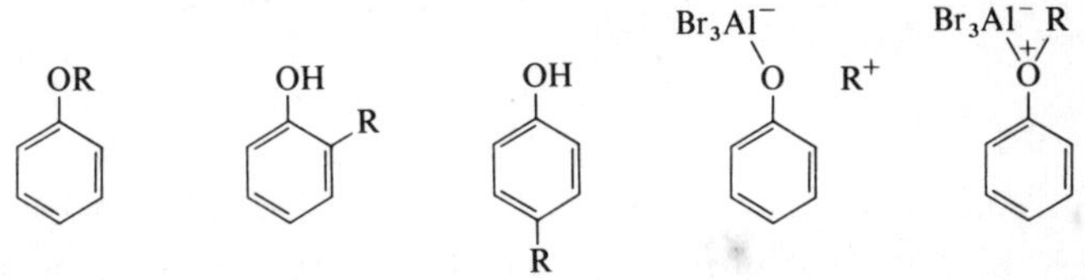

(54) R = 2-butyl (55) (56) (57) (58)

inversion of configuration. Failure to achieve complete optical purity in either case is attributable to concomitant participation of fragments from dissociated ion pairs.

3.4 *N*-SUBSTITUTED AMINOAROMATIC COMPOUNDS

3.4.1 Benzidine rearrangements[41]

The benzidine rearrangement, as commonly understood, is the acid-catalysed rearrangement of an aromatic hydrazo compound. From interest in the acid-catalysed rearrangement, there developed also the recognition of a thermal rearrangement. In due course, two points of view arose in the correlation of the acid-catalysed and thermal rearrangements under a generalised mechanism. One of these, developed by Ingold, Hughes and Banthorpe, is known as the polar-transition-state theory, and regards all benzidine rearrangements as being concerted. In this theory, breaking of the N,N′ bond and making of a new bond (whether N,C or C,C) are parts of a continuous process occurring within a polar transition state. To this theory, the English group fitted all that was known about the products and kinetics of the benzidine rearrangements. The second point of view on a generalised mechanism for benzidine rearrangements holds that an intermediate is interposed between the breaking of the N,N′ bond and the making of a new bond. The most common view, developed by Dewar, is that the intermediate is a π-complex, whose components are orientated in the configuration corresponding with the product of rearrangement.

At the present time, neither of these theories is entirely acceptable, nor is either of them entirely dismissible.

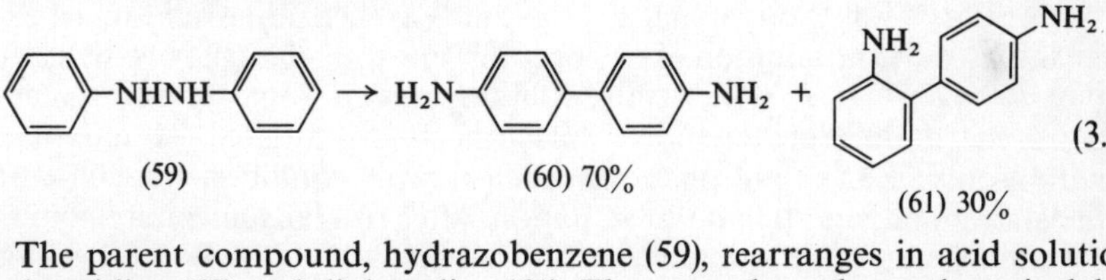

(59) → (60) 70% + (61) 30% (3.7)

The parent compound, hydrazobenzene (59), rearranges in acid solution to benzidine (60) and diphenyline (61). These are the only products isolable in usual circumstances. It is important to note, though, that all of the other possible products (62)–(64) are also formed in particular circumstances.

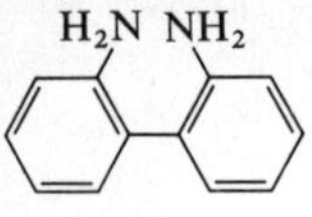

(62) 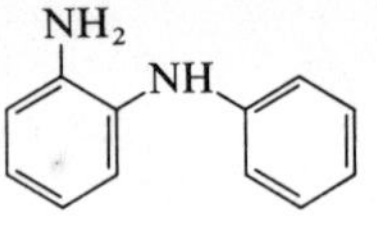 (63) 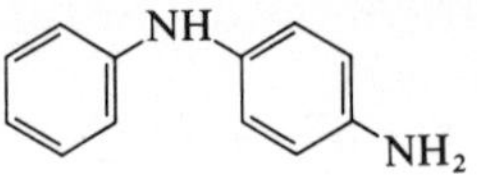 (64)

They are formed in quantities too small for isolation but sufficient to be detectable chromatographically from rearrangements in acidic aqueous ethanol, and are also formed from rearrangement in non-polar solvents.

Products corresponding with (62) are rarely obtained from hydrazobenzenes but are obtained in good yield from rearrangements of 1- and 2-hydrazonaphthalene.

Products corresponding with (61), (63) and (64) are obtainable from rearrangements of hydrazobenzenes which carry appropriate substituents.

According to the polar-transition-state theory, control over product type is exercised in the transition state. In the acid-catalysed rearrangement, heterolysis of the N,N′ bond is initiated by either mono- or di-protonation. During heterolysis and concerted new-bond formation, one half of the molecule becomes positively charged while the other has the character of an aryl amine. Consequently, the ease with which the aryl groups can delocalise either positive charge (one or two units) or a pair of electrons controls the direction of heterolysis. Charge densities and steric effects within the two halves of the polarised transition state determine which new bonds shall be formed.

The polar-transition-state theory treats the thermal benzidine rearrangement similarly, with the view that in polar solvents heterolysis of the N,N′ bond gives a transition state of which one half has positive and the other half negative charge character.

Proponents of the π-complex theory make no extensive attempt to explain product control in terms of selectivity of π-complex formation. π-Complexes bearing zero, one, or two units of positive charge are proposed as corresponding to the thermal rearrangement or to acid-catalysed rearrangements which involve either one or two protons. In contrast with the polar-transition-state theory, the π-complex theory is indefinite and vague on how product control is exercised. Nevertheless, there is reluctance among chemists to discard the theory.

At one time it was thought that all acid-catalysed benzidine rearrangements were initiated by the addition of two protons, one to each of the two hydrazo-group nitrogens. It is now firmly established that some rearrangements require the double protonation while others are initiated by monoprotonation only; that all protonations occur in rapid equilibria which are not rate determining; and that de-protonations after rearrangement are not rate determining either. A series of equations can be written as follows. The rate-determining steps in this series are (3.9) and (3.12), and it is what these steps really represent that is, in some minds, the unanswered question in benzidine rearrangements. For thermal rearrangements, one can write analogous equations ((3.14) and (3.15)) in which the fast step (3.15) represents only the redistribution of hydrogen atoms after the new bonds of the product have been formed.

The polar-transition-state theory is the only theory which attempts to account for mono- and di-protonation. The latest work of Banthorpe[42] is devoted to this aspect. The data are summarised in Table 3.2. The polar-

$$\mathrm{ArNHNHAr + H_3O^+ \rightleftarrows ArN^+H_2NHAr + H_2O} \quad \text{fast} \quad (3.8)$$

$$\mathrm{ArN^+H_2NHAr \rightarrow (product)^+} \quad \text{slow} \quad (3.9)$$

$$\mathrm{(product)^+ + H_2O \rightarrow product + H_3O^+} \quad \text{fast} \quad (3.10)$$

$$\mathrm{ArN^+H_2NHAr + H_3O^+ \rightleftarrows ArN^+H_2N^+H_2Ar + H_2O} \quad \text{fast} \quad (3.11)$$

$$\mathrm{ArN^+H_2N^+H_2Ar \rightarrow (product)^{2+}} \quad \text{slow} \quad (3.12)$$

$$\mathrm{(product)^{2+} + 2H_2O \rightarrow product + 2H_3O^+} \quad \text{fast} \quad (3.13)$$

$$\mathrm{ArNHNHAr \rightarrow initial\ product} \quad \text{slow} \quad (3.14)$$

$$\mathrm{Initial\ product \rightarrow final\ product} \quad \text{fast} \quad (3.15)$$

Table 3.2 Kinetic orders in hydrogen ion and relative rates of reaction of substituted hydrazobenzenes[42]

Substituent(s)	*Order in* H^+	Relative rate*
None	2	1000
2,2′-OMe	1	†
2,2′-F	2	1.6
2,2′-Cl	2	0.022
2,2′-Br	1.2–1.9	0.02†
2,2′-I	1	‡§
4-OMe	1	†
4-NHCOMe	1	†
4-Cl	2	67
4,4′-Cl	2	8.0
4,4′-Br	2	3.3
4,4′-I	2	7.6

*For two-proton processes.
†Pseudo-first-order rate constants for one-proton processes were determined in buffered solutions. 4-OMe was about 20 times faster than 4-NHCOMe.
‡One-proton rate estimated to be smaller than the rate for *o*-hydrazotoluene by a factor of several thousands.
§It is estimated that if a two-proton rate could be measured, it would be *c.* 10^7 times smaller than that of hydrazobenzene.

transition-state theory predicts that a 4-substituent may engender one-proton catalysis if it can delocalise the positive charge easily. The 4-methoxy and 4-acetamido substituents are considered to be doing this (e.g. as in (65)).

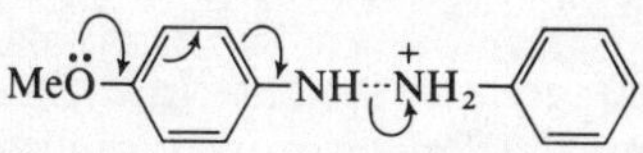

$$\mathrm{Me\ddot{O}{-}C_6H_4{-}NH\cdots\overset{+}{N}H_2{-}C_6H_5}$$

(65)

These substituents, in supplying electrons to the hydrazo group, also increase the basicity of the group and its extent of protonation. Hence, they lead to fast reactions. Substituents in the 2,2′-positions may lead to one-proton catalysis if they can assist in charge delocalisation (e.g. the methoxy group) or inhibit di-protonation (or cation solvation) because of their size (e.g. the iodo group). 2,2′-Substituents will also lead to slower reactions because they inhibit protonation (or cation solvation) either by their size (e.g. the iodo group) or electronic properties (e.g. the fluoro group).

Kurihara and Yoda[43] have also attributed the slow, two-proton rearrangement of hydrazobenzene-2,2′-dicarboxylic acid to the low basicity of the hydrazo compound.

White[44] has shown that *N,N′*-dimethylhydrazobenzene rearranges under one-proton catalysis to all of the possible product types.

3.4.1.1 Disproportionation

During the 1960s, it was recognised that some hydrazobenzenes disproportionated by second-order acid catalysis.

Banthorpe has now shown that even in first-order acid catalysis both disproportionation and rearrangement are governed by the same kinetics[42]. Disproportionation and rearrangement differ in molecularity, however, whether catalysis occurs by the one- or two-proton route. Disproportionation requires participation of two molecules of hydrazo compound while rearrangement involves only one. In disproportionation, therefore, an intermediate must be formed by the rate-determining sequence and then react with a second molecule of hydrazo compound in the product-forming step. The general view is that the intermediate involved in disproportionation cannot be one in which the bonding character of a rearrangement product has been reached for, once that bonding has been reached, rapid proton loss should yield a rearrangement product irreversibly[41]. Ingold at one time proposed, but more recently questioned[45], a quinoid intermediate (66) as the intermediate, which, formed according to polar-transition-state principles, next reacted rapidly with a second hydrazo molecule and underwent cleavage in a redox step. Banthorpe's latest work leads to further questions about the involvement of such quinoid intermediates. For example, disproportionation occurred with four of the five 2,2′-disubstituted hydrazobenzenes in Table 3.2.

$$H_2\overset{+}{N}{=}C_6H_3R{=}C_6H_3R{=}\overset{+}{N}H_2 \; (66) \; + \; R{-}C_6H_4{-}NHNH{-}C_6H_4{-}R \longrightarrow 2\,R{-}C_6H_4{-}NH_2 \; + \; R{-}C_6H_4{-}N{=}N{-}C_6H_4{-}R \qquad (3.16)$$

There is only one rearrangement product in each case, so does that mean that the quinoid intermediate in disproportionation is the same one that precedes rearrangement ((67) and (68))?

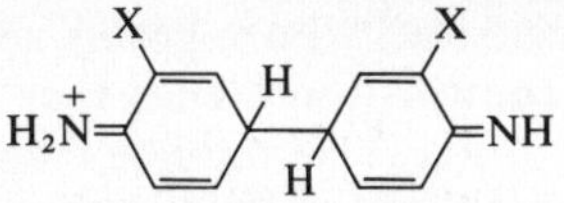

(67) X = OMe, Br 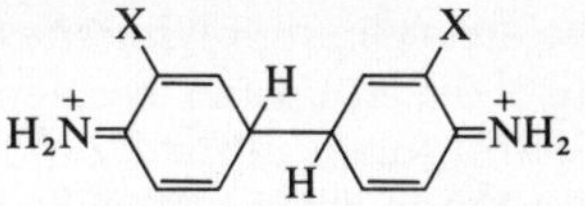 (68) X = F, Cl, Br

If that is so, the product-forming step in disproportionation must be fast enough to compete with proton transfer (leading to the rearrangement product), which is not a reasonable expectation.

Questions of this kind, to which answers are not available, leave the benzidine problem unsettled in spite of the accumulation of so many first-class experimental data.

Banthorpe has concluded from experiments with added polymerisable monomers and added inhibitors of radical reactions that disproportionation does not involve radical intermediates[42]. The reviewer has also failed to detect radicals by e.s.r. in disproportionation reactions. In connection with searching for radicals in benzidine systems, the reviewer has also failed to detect unusual n.m.r. signals (CIDNP) when hydrazobenzene rearranged in the probe of an n.m.r. spectrometer.

3.4.1.2 Concerted rearrangements and Woodward–Hoffmann rules

A curious fact about the benzidine rearrangements is that an application of the Woodward–Hoffmann rules to them is not to be found in the literature. Schmid has pointed out that, in analogy with the rearrangement of aryl pentadienyl ethers, the formation of benzidine from hydrazobenzene can be regarded as a [5,5]-sigmatropic rearrangement[27]. The signs of the two ion-radical HOMOs (Figure 3.5) show that concerted rearrangement to benzidine

Figure 3.5

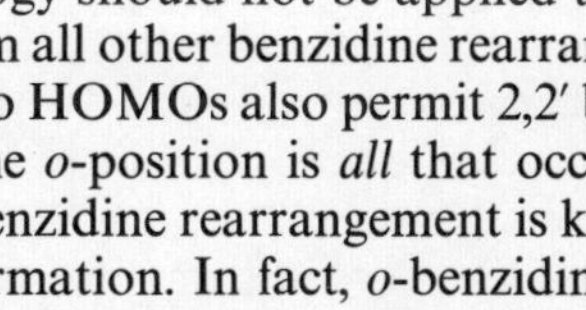

is allowed. But Schmid's analogy should not be applied to benzidine formation as if this were isolated from all other benzidine rearrangements. As in the Claisen rearrangement, the two HOMOs also permit 2,2′ bonding, and it will be recalled that bonding at the *o*-position is *all* that occurs in the simplest Claisen rearrangement. The benzidine rearrangement is known for the *rarity* of examples of *o*-benzidine formation. In fact, *o*-benzidine (62) *is* formed in acid-catalysed rearrangements (but only in small amounts) and this might allow the inclusion of the benzidine rearrangement in the sigmatropic class, but we would have to know why, in that case, the [5,5] so completely overshadowed the [3,3] shift.

The real obstacle to including the benzidine rearrangement in the sigmatropic class is the formation of products, which, according to the HOMOs, are *not* allowed. These are diphenyline (61), which is formed in 30% yield from hydrazobenzene, and *o*-semidine (63), which is obtained in small amounts in particular circumstances.

Most strikingly, also, the product which has brought most of the criticism of the concerted rearrangements of the polar-transition-state theory, *p*-semidine (64), *is* allowed. This rearrangement would be classified as a [1,5]-sigmatropic shift. It is the analogue of the *p*-Claisen rearrangement, but *that* rearrangement occurs in two [3,3]-sigmatropic steps. If we were to treat *p*-semidine formation as a sigmatropic shift, we would rule out direct concerted migration. In strict analogy with the *p*-Claisen rearrangement, the *p*-semidine rearrangement would be preceded by an *o*-benzidine-type step. There is no experimental evidence for that at all.

If it is felt that the acid-catalysed benzidine rearrangements should follow the Woodward–Hoffmann rules, we must conclude that some rearrangements may be concerted while others must go through an intermediate of some sort. At this stage, therefore, it does not seem useful to classify some of the rearrangements as sigmatropic shifts and leave others in limbo.

Possibly, the thermal benzidine rearrangements might be thought to be classifiable as sigmatropic shifts, although again there is no compelling evidence that they should be. The thermal rearrangement of 2,2′-hydrazonaphthalene gives the right products (i.e. for [3,3]- and [5,5]-shifts). On the other hand, the major product from the thermal rearrangement of hydrazobenzene in polar solvents is diphenyline, while in non-polar solvents the major products are the *o*- and *p*-semidines. It will be remembered that *N*-allyl-1-naphthylamine undergoes the so-called amino-Claisen rearrangement, whereas *N*-allylaniline decomposes when heated[1]. Perhaps there is some analogy to be sought between the amino-Claisen and thermal benzidine rearrangements.

No evidence has been found for the formation of radicals when hydrazobenzenes are heated, but Welzel[46] has recently concluded that the best explanation for the formation of *o*- and *p*-semidines from tetraphenylhydrazine is the recombination of thermally-produced diphenylamino radicals.

3.4.2 Analogues of benzidine rearrangements

Diphenyl disulphide was thought some years ago to undergo benzidine-like rearrangement in concentrated sulphuric acid[47]. Reaction with triethyloxonium fluoroborate has now given 2- and 4-(ethylthio)phenyl sulphide. These are analogous to *o*- and *p*-semidine, but unlike semidine formations these compounds are formed intermolecularly[48]. Ethyl phenyl sulphide and 4-methoxyphenyl sulphide were the major products of reaction in anisole solution.

Similar work with diaryl peroxides is not feasible because of the difficulty of preparing and isolating them. Aryl benzenesulphenates (e.g. (69)) rearrange thermally in the absence of light (equation (3.17)). The mechanism of the rearrangement has not been fully explored, but present evidence is thought to indicate initial ion-pair formation from (69) and subsequent intra- and intermolecular paths to the products[49].

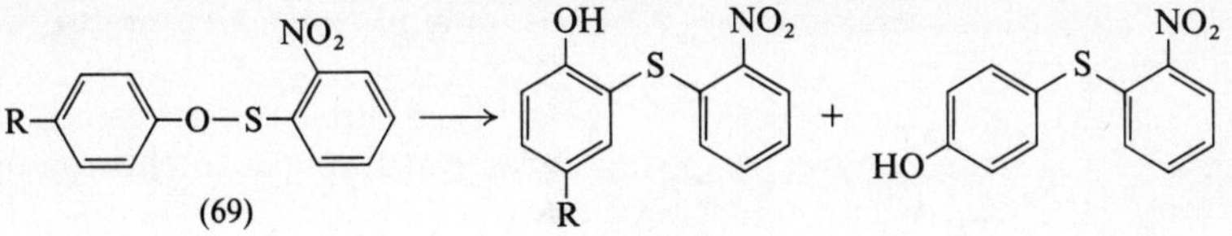

(69) → R = H, 46%; R = Me, 60% + R = H, 31% (3.17)

Similar rearrangements of nitrobenzenesulphenanilides have been observed[50]. By heating 3-nitro- (70) and 2-nitro-benzenesulphenanilide (73) in aniline at 195 °C the *o*- and *p*- rearrangement products (71)(72)–(74)(75) were obtained.

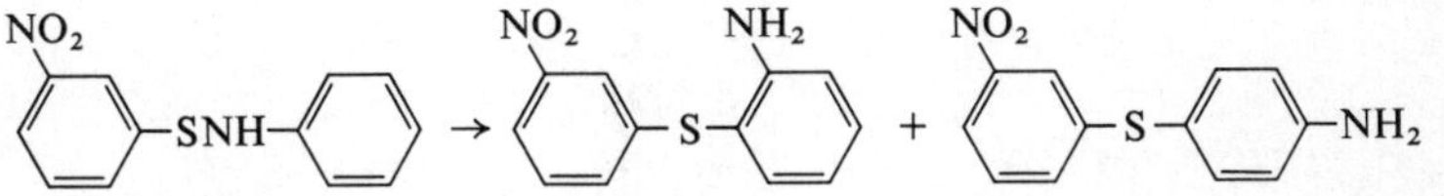

(70) → (71) + (72) (3.18)

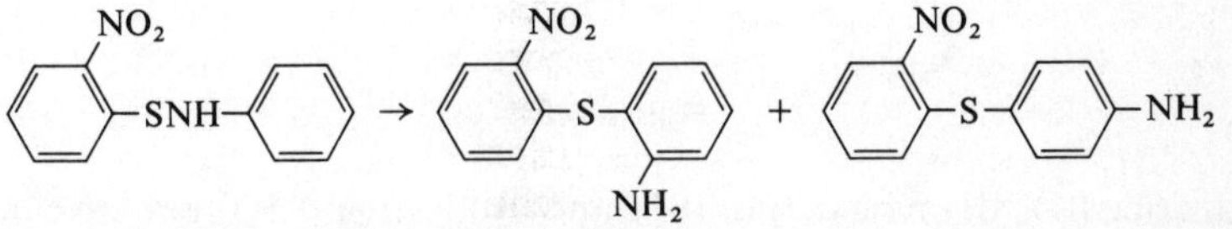

(73) → (74) + (75) (3.19)

Attempts to synthesise *N*,*O*-diarylhydroxylamines (76) have ended in apparent rearrangement products (77)(78)[51,52]. These rearrangements are analogues of the benzidine rearrangement, and, therefore, further exploration will be interesting.

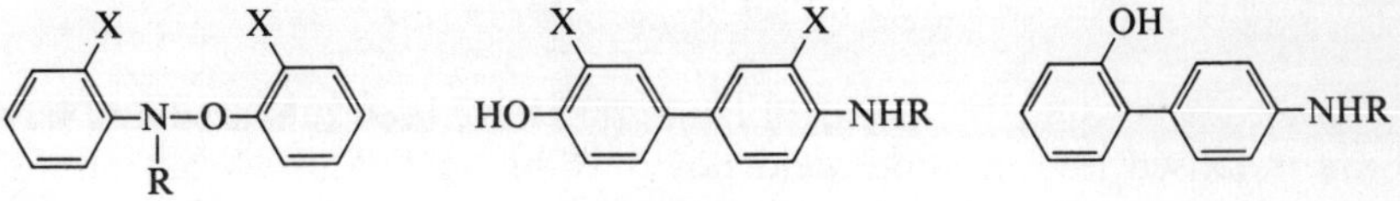

(76) (77a–c) (78a)

(a)[51] R = COMe, X = H
(b)[52] R = $COCH_2Ph$, X = NO_2
(c)[52] R = $COBu^t$, X = NO_2

3.4.3 Fischer indole reaction

Much of the work on this reaction in recent years has been devoted to identifying particular steps and intermediates in the commonly-accepted sequence of reactions[53] (Figure 3.6). Whether or not this sequence is correct in all details is not known, but certain features have now been confirmed. Examples of

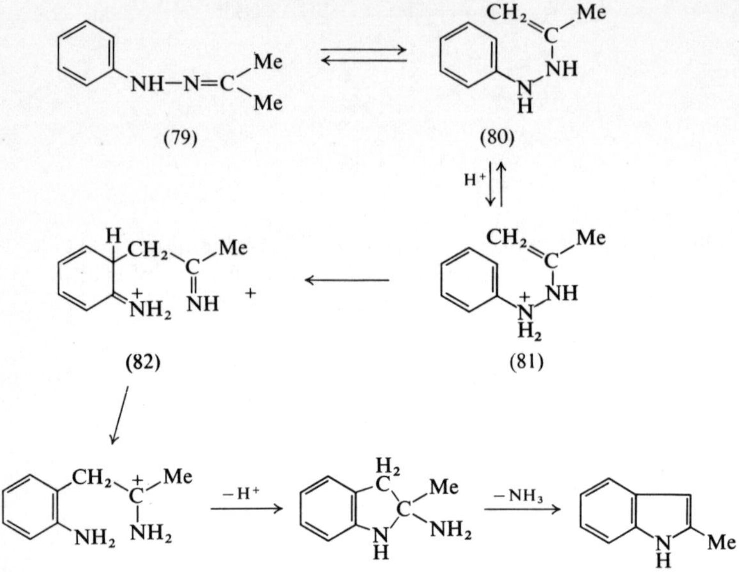

(79) ⇌ (80)

H^+ ⇅

(82) ← (81)

↓

(83) $\xrightarrow{-H^+}$ (84) $\xrightarrow{-NH_3}$ (85)

Figure 3.6

enehydrazine (80), diamine (82) or (83), and cyclisation (84) types have been isolated and/or identified.

Ene-hydrazines (86) and (87) have been isolated[54].

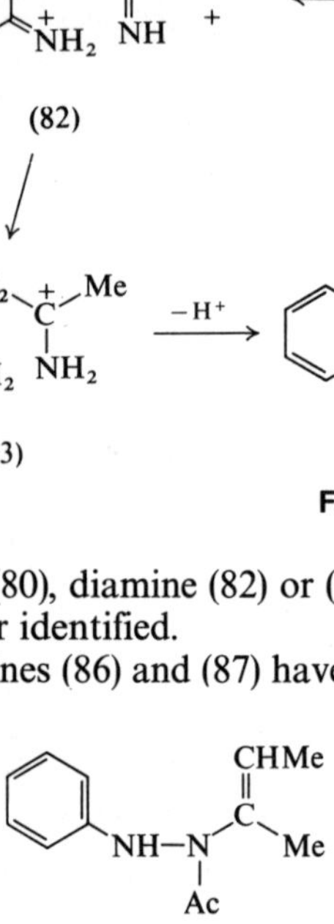

(86) 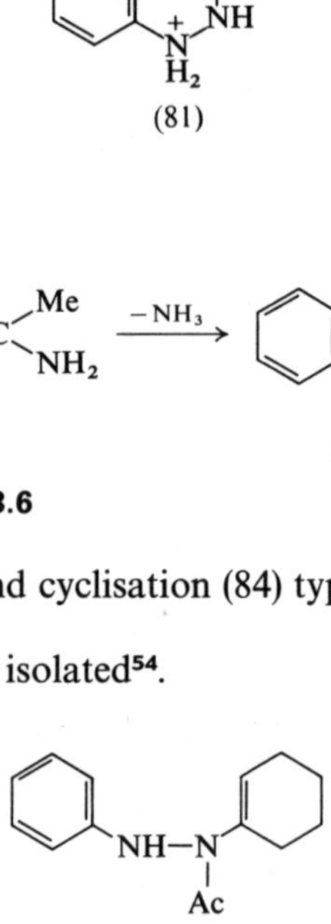 (87)

Several *N,N′*-dimethyl-ene-hydrazines (88) have been prepared and have been indolised (89) in good yield not only by acid catalysis but also by heating[55].

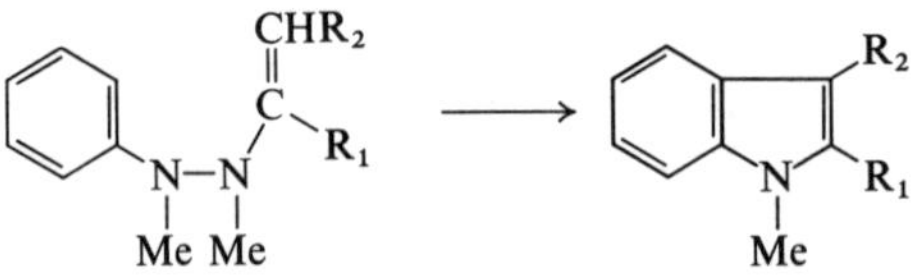

(88) ⟶ (89) (3.20)

Direct evidence for the first time of the equilibrium between hydrazone and ene-hydrazine has been obtained. The n.m.r. spectrum of the *N*-methylphenylhydrazone of ethylacetoacetate (90) showed the presence of both stereoisomers of (90) (a and b) and the two ene-hydrazines (91a) and (91b)[56].

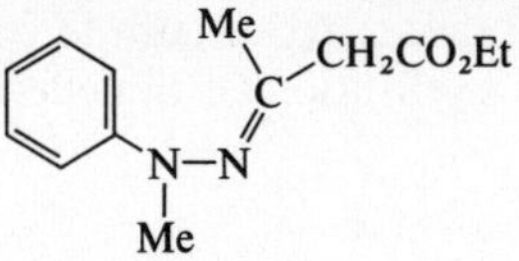

(90a) 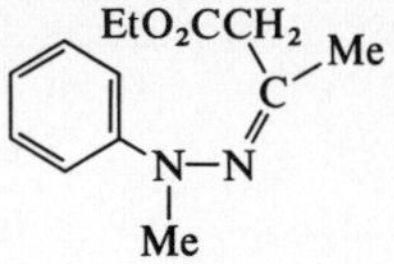 (90b)

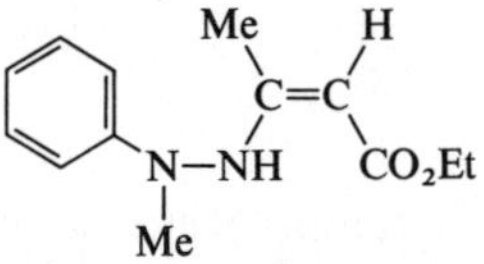

(91a) 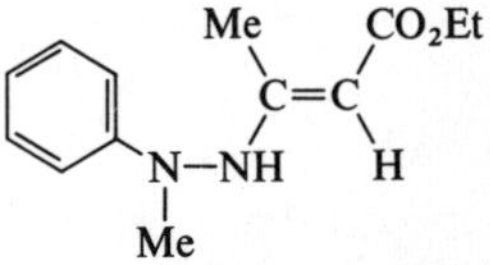 (91b)

As evidence for diamine intermediates, the lactone phenylhydrazone (92) gave the isolable diamine (93) which was indolised (94) with acid[57].

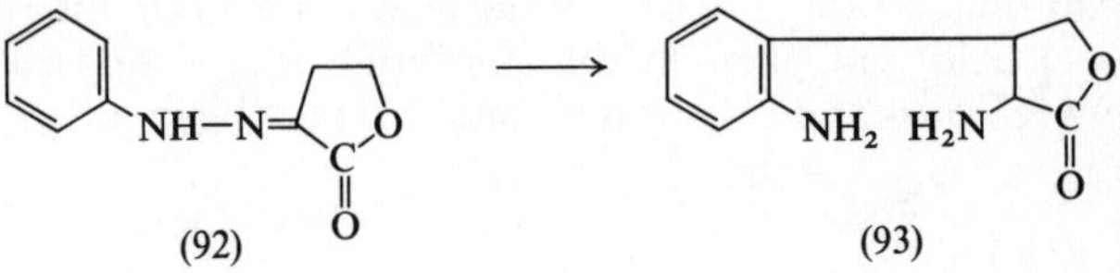

(92) (93) (3.21)

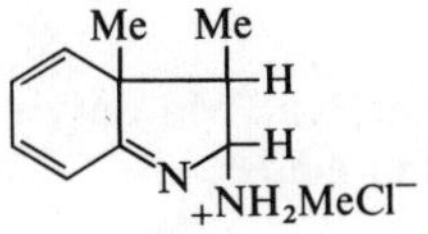

(95) 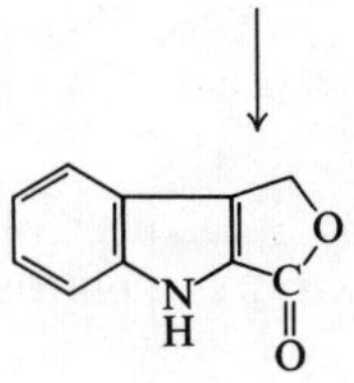 (94)

Cyclic intermediates have been isolated from reactions of *N*,2,6-trimethylphenylhydrazine with carbonyl compounds[58]. Reaction with propanal, for example, gave (95) which was indolised with loss of methylamine and the angular methyl group.

3.4.3.1 Acidity and selectivity in cyclisation

The extent of protonation of Fischer reaction intermediates has, as yet, never been demonstrated. Analogy with the one- and two-proton benzidine rearrangements has been drawn, because one of the key steps in the Fischer reaction is the opening of a hydrazo linkage[1]. Furthermore, the sequence of steps is always written as involving proton transfers, although many Fischer reactions are carried out by heating the phenylhydrazone with a Lewis acid. The effect of acids and acidity has received further attention, and we now find certain results attributed to steric effects of the acid and to competition between mono- and di-protonation.

Lyle and Skarlos[59] have found that the distribution of products (96) or (98) obtained from the phenylhydrazone (97) varies with the acid used. They regard polyphosphoric acid as a proton donor, acetic acid as an acetylating agent, and zinc chloride as a complexing agent, and attribute control over product formation to steric factors in the cyclisation of the ene-hydrazine–acid complex.

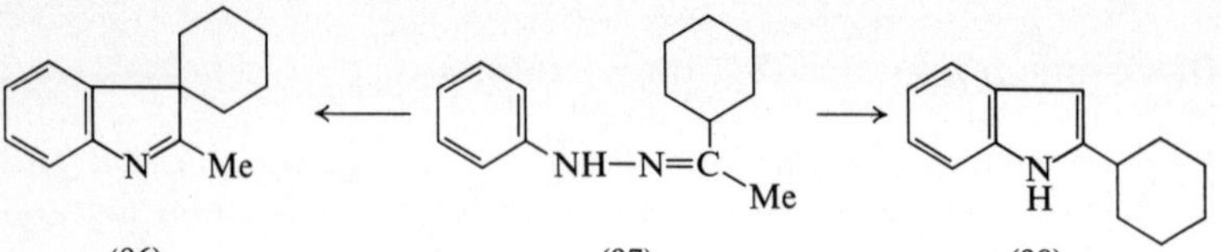

(96) ⟵ (97) ⟶ (98) (3.22)

In contrast, other workers have found that directions of cyclisation (in other phenylhydrazones) vary with acid concentration without changing the acid itself.

For example, the phenylhydrazone of methyl isopropyl ketone in sulphuric acid gave both the indolenine (99) and indole (100) in relative amounts which depended not only on the molarity of the acid but also on the relative concentrations of acid and phenylhydrazone used. It is proposed that competition between one- and two-proton catalysis may be involved[60].

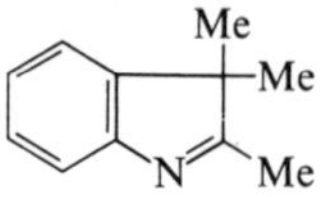

(99) 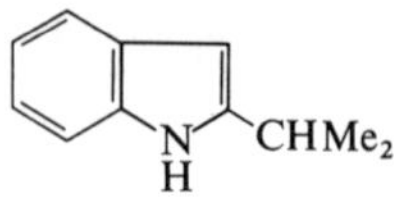 (100)

The same conclusion has been reached by Palmer and MacIntyre from work with phenylhydrazones in phosphoric and other acids[61].

3.4.4 Oxy- and thio-Fischer indole reactions

Whether or not these names are appropriate will be judged by time and the reader. The first name refers to the preparation of benzofurans (e.g. (102)) from *O*-phenyloximes (e.g. (101)). Reactions have been carried out with boron trifluoride etherate in acetic acid at 100 °C [62] and by boiling in a mixture of ethanol and concentrated hydrochloric acid[63].

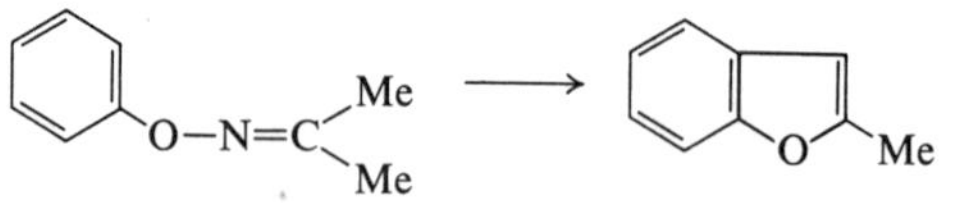

(101) ⟶ (102) (3.23)

Further, oxime (103) has been converted into the iminium salt ((104), cf. (83)), and heating either (103) or (104) in acetic acid-hydrogen chloride gave (105) [64].

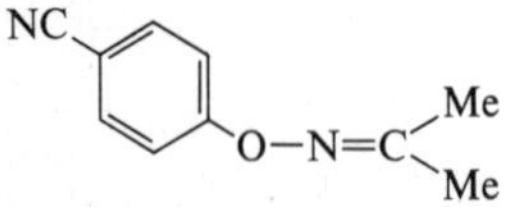

(103) 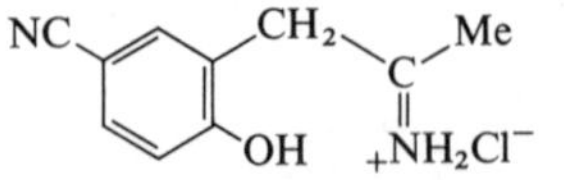 (104) 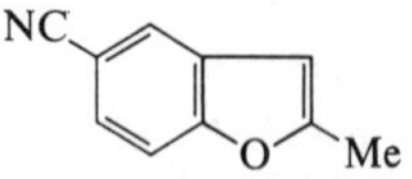 (105)

The second name in the title refers to unsuccessful attempts to prepare benzothiophenes from thio-oximes[63].

3.4.5 Diazoaminoaromatics (1,3-diaryltriazenes)

This acid-catalysed rearrangement is illustrated by equations (3.24) and (3.25). The protonated triazene breaks up into the aryldiazonium ion and arylamine, and these fragments can recombine in two ways.

$$C_6H_5{-}N{=}N{-}\overset{+}{N}H_2{-}C_6H_5 \underset{\text{(ii)}}{\overset{\text{(i)}}{\rightleftharpoons}} C_6H_5{-}N_2^+ + C_6H_5{-}NH_2 \qquad (3.24)$$

$$C_6H_5{-}N_2^+ + C_6H_5{-}NH_2 \xrightarrow{\text{(iii)}} C_6H_5{-}N{=}N{-}C_6H_4{-}NH_2 + H^+ \qquad (3.25)$$

In the first (step (ii)) the triazene is regenerated, and in the second (step (iii)) the aminoazo product is formed[1]. Recent kinetic work by Yamada with *N*-methyltriazenes (106) which cannot tautomerise, has given added support to this generally-accepted sequence[65].

$$X{-}C_6H_4{-}N{=}N{-}N(Me)_2$$

(106)

The work shows the rearrangement to be first-order in acid and first-order in triazene. Substituent effects (X = *p*-methoxy, *p*-chloro-, *p*-methyl, *p*-acetyl and *m*-methyl) indicate that step (i) is rate determining. Recombination at nitrogen (step (ii)) is much faster than at ring carbon (step (iii)), which is as it should be because step (iii) is the way in which triazenes are synthesised.

Equations (3.24) and (3.25) also account for the formation of a new triazene and a 'crossed' amino-azo compound which occurs when rearrangement is carried out in the presence of an added arylamine[1]. But, a question remains about reactions with added arylamine, namely, can the products be formed also by direct, bimolecular reaction between protonated triazene and arylamine? Yamada's kinetic data are not clear enough to rule out some incursion of this pathway.

3.4.6 *N*-Halogenoacylamines (Orton rearrangement)

This rearrangement, exemplified by the conversion of *N*-chloroacetanilide into *o*- and *p*-chloroacetanilide, is recognised as intermolecular[1]. Reaction occurs in hydrochloric acid or, for example, a solution of hydrogen chloride in acetic acid. The products are formed, in fact, by the chlorination of acetanilide, the chlorine having been formed by displacement of positive chlorine from *N*-protonated *N*-chloroacetanilide by chloride ion. The use of hydrogen bromide instead of hydrogen chloride causes bromination, since the halogenating agent bromine chloride is formed.

Orton rearrangements can also be accomplished with the use of carboxylic acids, acidic phenols, and sulphonic acids. These reactions are not as well understood[1]. Recently, toluenesulphonic acid in acetic acid was used[66]. One might wonder if, in reactions such as this, homolysis of the N—Cl bond may not occur, and be responsible eventually for product formation. Homolyses of *N*-chlorodialkylamines in strong acids are well-known (e.g. in the Hofmann–Löffler rearrangement), although in those cases free-radical initiation of some sort (e.g. light) is necessary.

In a related rearrangement it has been found that *N*-t-butyl-*N*-chloroaniline rearranges slowly in methanol to *o*- (52%) and *p*-chloro-*N*-t-butylaniline (14%). A tight ion pair is believed to be involved[67, 68]. The reaction is somewhat similar to the rearrangement of phenylhydroxylamine (Bamberger) in that other nucleophiles enter the ring in some circumstances.

3.4.7 *N*-Nitrosoarylamines (Fischer–Hepp rearrangement)

The Fischer–Hepp rearrangement is best known as the formation of a *p*-nitroso-*N*-alkylaniline (109) from an aromatic secondary nitrosamine (107).

$$\text{RNNO-C}_6\text{H}_5\ (107) + \text{HX} \longrightarrow \text{RNH-C}_6\text{H}_5\ (108) + \text{NOX} \longrightarrow p\text{-ON-C}_6\text{H}_4\text{-NHR}\ (109) + \text{HX} \qquad (3.26)$$

Because the reaction occurs best in reasonably dry solutions of hydrogen chloride, and not very well with other acids, it has been felt that equation (3.26) represents the course of rearrangement (X = Cl)[1, 69]. This pathway has been challenged, and the rearrangement has been claimed to be intramolecular, on the basis of kinetic results with (107) (R = Me) in aqueous hydrochloric acid[70] and its 3-methyl derivative in aqueous sulphuric acid[71]. In each case, both rearrangement and loss of the nitroso group occur, and, after it has completely departed, the nitroso cation is considered not to re-bond at ring carbon.

These results do not necessarily invalidate equation (3.26), which itself is in harmony with the formation of nitrosyl chloride-olefin adducts when reaction has been carried out in the presence of an olefin[1]. It may be that the Fischer–Hepp rearrangement has both intra- and intermolecular pathways, which are solvent-dependent. It would be interesting to know if, furthermore, this rearrangement, like the nitramine, may involve radicals (i.e. an amine cation radical and nitric oxide). None of the proposed mechanisms, however, accounts for the fact that only *p*-nitrosation occurs.

3.4.8 Nitramine rearrangement

This rearrangement is known best by the example shown in equation (3.27). None of the *m*-nitroaniline is formed.

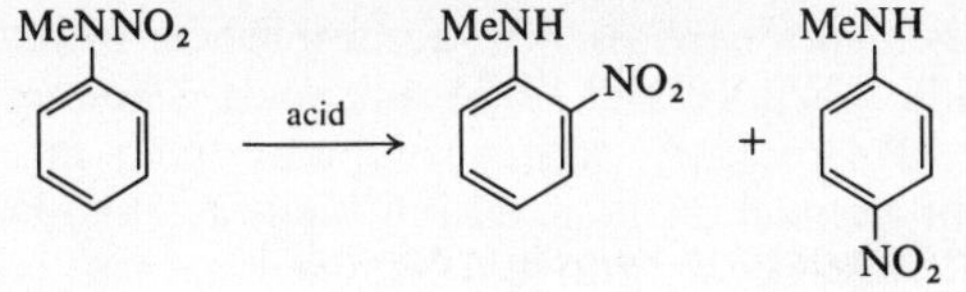

$$\text{PhN(Me)NO}_2 \ (110) \xrightarrow{\text{acid}} o\text{-}O_2NC_6H_4NHMe \ (111) + p\text{-}O_2NC_6H_4NHMe \ (112) \quad (3.27)$$

The mechanism of migration of the nitro group under acid catalysis was a centre of controversy. Dewar advocated the involvement of a π-complex, and Ingold advocated concerted steps, often referred to by others as the cartwheel mechanism. In a series of admirable papers White has shown that not only is neither the π-complex nor cart-wheel mechanism correct, but also the long-held belief that the rearrangement is wholly intramolecular is incorrect too[72].

The rearrangement is first order in nitramine and acid, and is specific-acid catalysed. The N—N bond of the protonated nitramine undergoes homolysis in a rate-determining step, forming an anilinium cation radical and nitrogen dioxide. These are held in a solvent cage in which they may recombine in a way which suits not just unpaired electron density in the cation radical, but also the relative positions of the two radicals within the cage. The pair of radicals may also recombine after leaving the cage, and if they do, the relative amounts of *o*- and *p*-nitro product formed will be different from those of cage recombination. If they do not recombine, the radicals may become reduced to the appropriate arylamine and nitrite ion, or they may be scavenged in some other way.

This sequence of reactions may be summarised schematically in Figure 3.7.

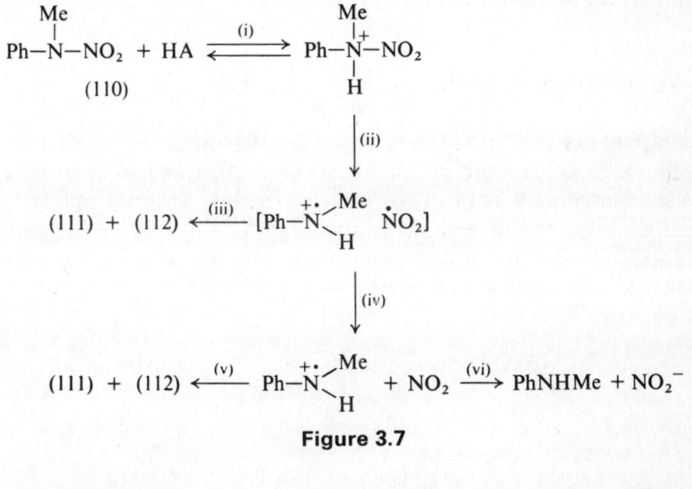

$$\underset{(110)}{\text{Ph—N(Me)—NO}_2} + \text{HA} \xrightleftharpoons{(i)} \text{Ph—}\overset{+}{\text{N}}\text{(Me)(H)—NO}_2$$

$$\downarrow (ii)$$

$$(111) + (112) \xleftarrow{(iii)} [\text{Ph—}\overset{+\bullet}{\text{N}}\text{(Me)H} \quad \dot{\text{N}}\text{O}_2]$$

$$\downarrow (iv)$$

$$(111) + (112) \xleftarrow{(v)} \text{Ph—}\overset{+\bullet}{\text{N}}\text{(Me)H} + \text{NO}_2 \xrightarrow{(vi)} \text{PhNHMe} + \text{NO}_2^-$$

Figure 3.7

The cage (square brackets) represents a variety of orientations of the two radicals. Recombination (step (iii)) and escape (step (iv)) are affected by solvent viscosity, a fact which itself is a critical test of caged-radical processes. White has shown, for example, that as the viscosity was increased from 0.60 cP to 80.5 cP the total yield of 2- and 4-nitro-*N*-methylaniline increased from 76 to 100 %. Radicals which escape the cage have been scavenged with a variety of agents. To be effective they must be reducing agents, such as iodide ion, bisulphite ion, sulphur dioxide, thiourea, and hydroquinone[72].

3.4.8.1 Thermal nitramine rearrangement

Di-*p*-tolynitramine rearranged quantitatively in boiling benzene to 2-nitro-4,4′-dimethyldiphenylamine[73]. In the presence of added tetraphenylhydrazine, 4-nitrodiphenylamine was formed. It was concluded, therefore, that the rearrangement is intermolecular and that nitrogen dioxide was formed and was scavenged by diphenylamino radicals from tetraphenylhydrazine.

In contrast, the decomposition of some phenylnitramines in octan-2-ol at 120 °C gave none of the rearrangement products[74]. The major products are illustrated with (112) (equation (3.28)). The nitramine is thought to rearrange

$$p\text{-}RC_6H_4NHNO_2\ (112) + C_6H_{13}CH(OH)Me \longrightarrow C_6H_5R + C_6H_{13}C(=O)Me + N_2 + H_2O \qquad (3.28)$$

(112) (a) R = Me; (b) R = NO_2

into a hydroxydiazo-oxide which next undergoes decomposition into radicals which lead to the observed products[74]. Apparently the decomposition of nitramines like (112) is highly dependent on structure and solvent, because earlier workers obtained 64% of rearrangement of *p*-nitro-*N*-methylphenylnitramine in tetrachloroethane[1].

3.4.9 Arylsulphamic acids

Phenylsulphamic acid (113) is known to rearrange in acid solution to sulphanilic acid (115). One major question about this has been whether or not the rearrangement is intra- or intermolecular. Another major question has been whether or not orthanilic acid (114) is also formed, as claimed by Bamberger when he first investigated the reaction[1], and if so, if (114) is the source of (115). The isomerisation of (114) to (115) on being heated in sulphuric acid is a known reaction.

$C_6H_5NHSO_3H$ (113) $o\text{-}H_2NC_6H_4SO_3H$ (114) $p\text{-}H_2NC_6H_4SO_3H$ (115)

From recent work in which complete ^{35}S scrambling occurred when (113) (potassium salt) rearranged in dioxane containing ^{35}S-labelled sulphuric acid, Spillane and Scott concluded that the rearrangement is *intermolecular*[75–77]. That is, that protolysis of (113) occurs first and is followed by sulphonation of aniline. But aniline in similar circumstances could not be sulphonated. In contrast, Allan and Vrba find that (113) undergoes rapid ^{35}S exchange (equation 3.29) and conclude (with the aid of other results) that the rearrangement is *intramolecular*[78, 79].

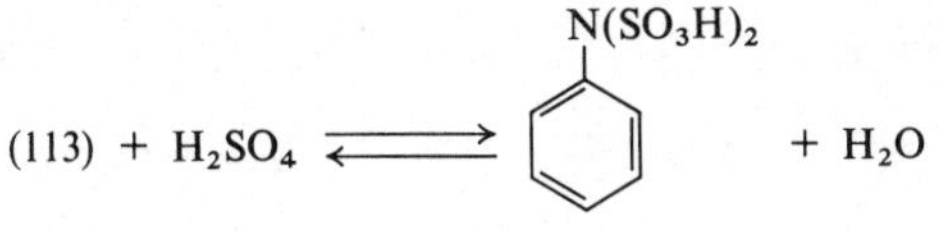

$$\text{(113)} + H_2SO_4 \rightleftharpoons C_6H_5N(SO_3H)_2 \; \text{(116)} + H_2O \qquad (3.29)$$

Interestingly, the rearrangement of 1-naphthylsulphamic acid to the 4-sulphonic acid in dioxane–sulphuric acid is thought by Scott to go concurrently via both intra- and intermolecular routes[80].

Evidently, further clarification of the rearrangement of arylsulphamic acids is needed. Some new results on the rearrangement of arylsulphonic acids may be interesting in this context. Recently, Fischer and Zollinger[80a] showed by ^{35}S tracer technique that 2-naphthol-l-sulphonic acid rearranges into the 2,6-isomer by an intermolecular mechanism in aqueous sulphuric acid, but predominantly in an intramolecular way when working in mixtures of acetic acid with sulphuric acid.

3.4.10 Wallach rearrangement

The Wallach rearrangement is illustrated with azoxybenzene (117). Rearrangement is intermolecular and occurs at a *p*-position.

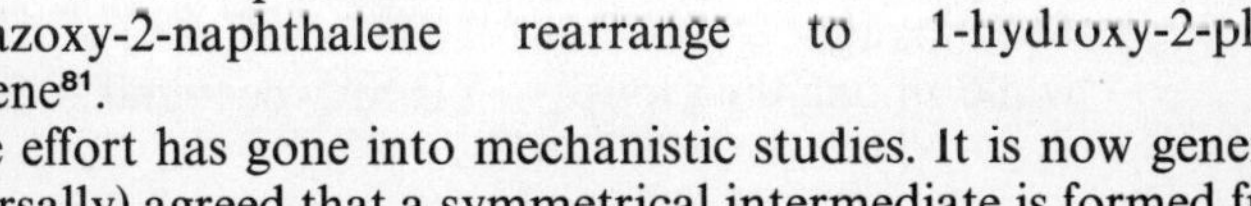

$$C_6H_5\text{–N=N(O)–}C_6H_5 \; \text{(117)} \xrightarrow[H_2O]{H_2SO_4} C_6H_5\text{–N=N–}C_6H_4\text{–OH} \; \text{(118)} \qquad (3.30)$$

If the *p*-positions are blocked, rearrangement may displace a substituent or may occur at an *o*-position[1]. Buncel has now shown that both α- and β-phenylazoxy-2-naphthalene rearrange to 1-hydroxy-2-phenylazonaphthalene[81].

A large effort has gone into mechanistic studies. It is now generally (but not universally) agreed that a symmetrical intermediate is formed from (117) prior to attack by a nucleophile. Much of the work in this area has been done by Buncel, who concludes that the intermediate is the dication (119)[81].

Past interpretations of the formation of (119) have accepted that the two successive protonations of (117), which are required, were fast equilibrium processes. Most recently, Buncel has shown that the rate of rearrangement of (117) is still sensitive to acidity over the range 92.3–99.9% sulphuric acid, and argues, therefore, that the second protonation must be slow and rate-determining. The following picture is presented, in which step (iii) represents the rate-determining protonation.

$$(117)\ ,\ + H^+ \underset{(ii)}{\overset{(i)}{\rightleftharpoons}} C_6H_5{-}N{=}\overset{+}{N}(OH){-}C_6H_5 \xrightarrow[(iii)]{HA} C_6H_5{-}N{=}\overset{+}{N}(OH\cdots H\cdots A){-}C_6H_5$$

$$\xrightarrow[-H_2O,\ -A^-]{(iv)} C_6H_5{-}\overset{+}{N}{\equiv}\overset{+}{N}{-}C_6H_5\ (119) \xrightarrow[Nu^-\ (Nu)]{(v)} Nu{-}C_6H_4{-}N{=}N{-}C_6H_5\ (120)\ + H^+\ (+ 2H^+)$$

Intermediate (119) reacts rapidly with a nucleophile. In 92.3–99.9 % sulphuric acid, for example, (119) is not detectable spectroscopically. In aqueous acids the nucleophile is most likely a water molecule and step (v), therefore, forms the product directly. In very strong acid the nucleophile is bisulphate ion, which in step (v) forms azobenzene-4-(hydrogen sulphate) ((120) Nu=OSO_3H). The latter undergoes a two-proton protolysis, which in very strong acid is so fast that (120) cannot be detected. Solutions of (120) in strong acid rapidly develop the spectrum and characteristics of solutions of (118)[81].

Hendley and Duffey have shown with the use of perdeuterio-(117) that deprotonation of the ring is not part of the rate-determining process[82]. This fits Buncel's mechanism, but the two groups are not in agreement about the rate-determining process. Duffey and Hendley believe that nucleophilic attack occurs on monoprotonated (117) prior to or during the rate-determining step, which is thought to be the scission of the N—O bond[82, 83].

3.5 BASE-CATALYSED REARRANGEMENTS

3.5.1 Smiles rearrangement

The general principle of this intramolecular rearrangement is shown in equation (3.31). The linkage —X— in (121) is often —O—, —$S(O_2)$—, —S(O)—, —S—, while among the groups —YH may be found —OH, —SH, —NH_2, —$CONH_2$, —NHCOR, —SO_2NH_2, —Me, —CH_2R, —CHR_2[1, 84]. The carbon chain joining X and Y may be saturated and unsaturated. In most examples it is part of an aromatic ring. The conversion of (121) into (122) is usually initiated by base. The overall conversion in an aromatic

$$\text{C(-X-}^{*}\text{C}_6\text{H}_5\text{)-C-YH}\ (121) \longrightarrow \text{C(-XH)-C-Y-}^{*}\text{C}_6\text{H}_5\ (122) \qquad (3.31)$$

system is shown with (123) and (125) where it can be seen that nucleophilic displacement of X by Y^- occurs (124). The scope of the rearrangement is

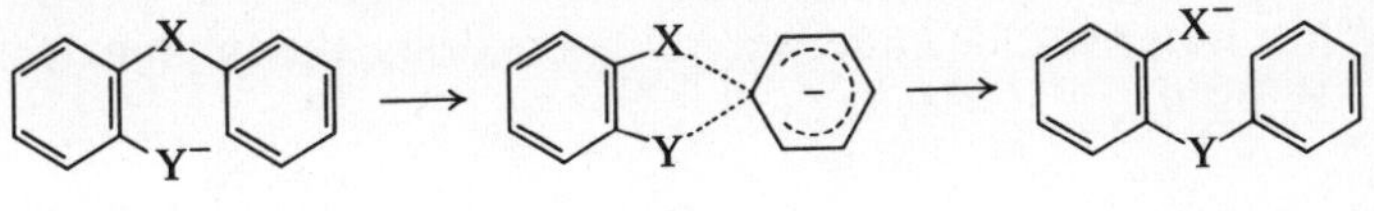

(3.32)

(123) (124) (125)

well known[1,84] and, therefore, will not be discussed.

Drozd[85] and Truce[86,87] have brought out some of the intricacies of the rearrangement of diaryl sulphones. Drozd has found that when a sulphone (126) bearing an *o*-alkyl substituent is treated with n-butyl-lithium in ether, several carbanions may be formed, whose distribution depends on the nature of the substituents in the sulphone. Among these carbanions may be the ring-lithiated compound (127), the open carbanion (128), the cyclic carbanion (129), and the sulphinate (130). The formation of the sulphinate represents, of course, the Smiles rearrangement of (126). Our concern is primarily with

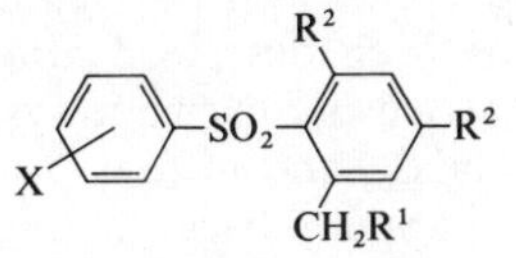

(126) 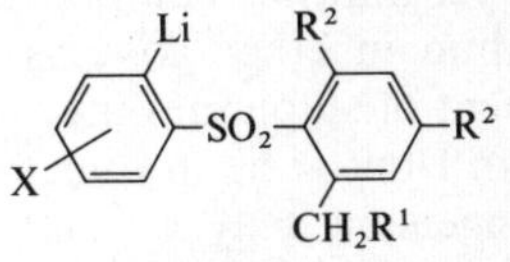 (127)

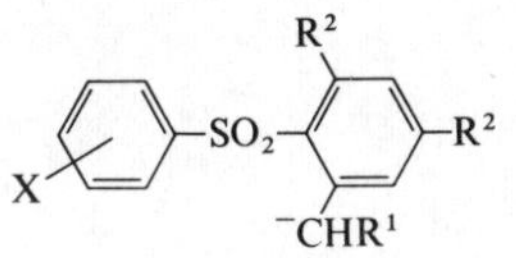

(128) 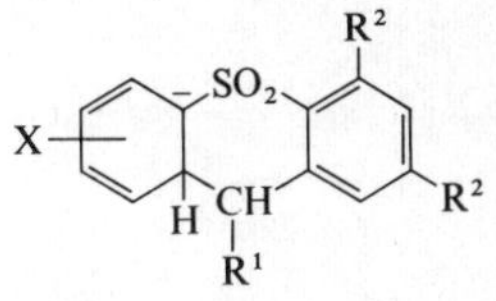 (129)

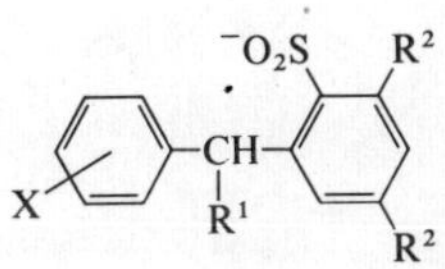

(130)

the behaviour of ions (128) and (129). In some cases (128) and (129) are in an equilibrium which is slowly disturbed by the irreversible formation of (130). This is the situation with phenyl *o*-tolyl sulphone ((126a), $X = R^1 = R^2 = H$). In such cases (130) may be formed by the traditional Smiles route from (128). In many cases the carbanion (129) is believed to be formed irreversibly, and the final slow conversion into the sulphinate (130) is thought to be an intramolecular-rearrangement of (129)[85]. Examples of such cases are the mesityl sulphones (126b–e) in which $R^2 = Me$, $R^1 = H$, and $X = H$, *p*-Cl, *m*-Me and *p*-Me. In other cases the ion (129) is sufficiently stable either to resist conversion into the sulphinate or to allow conversion only very slowly. This behaviour is observed with mesityl 1-naphthyl (126f) and mesityl 2-naphthyl sulphone (126g)[85].

In almost all cases the first product obtainable by treating a rearrangement system with water is the dihydrothioxanthene dioxide corresponding with (129). Thus (131b) is obtainable from (126b) (X = H), and (131g) from (126g). When water is added after several hours delay, the sulphinate which

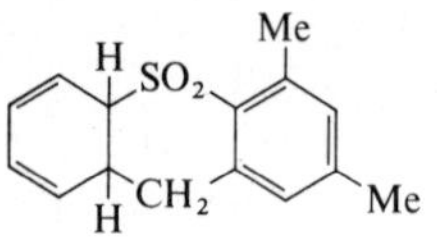

(131b)

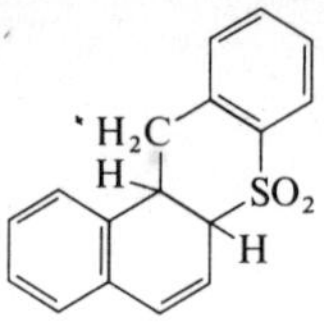

(131g)

will have been formed slowly is the product. There are some cases, too, in which the open ion (128) goes directly to (130) and one of (129) can be found. Examples of this kind are 2-ethylphenyl phenyl sulphone and 2,2′,4,4′-tetramethyldiphenyl sulphone[85].

There is yet another complication. The use of aprotic systems such as n-butyl-lithium in ether governs what has been described above. Smiles rearrangement of sulphones may also be carried out in protic systems, such as potassium t-butoxide in dimethyl sulphoxide, and in that case so-called *cine* rearrangement may occur[85–87]. This happens because the ion (129) becomes protonated and undergoes a β-elimination reaction. An example is given with mesityl 1-naphthyl sulphone ((126f) equation (3.33)).

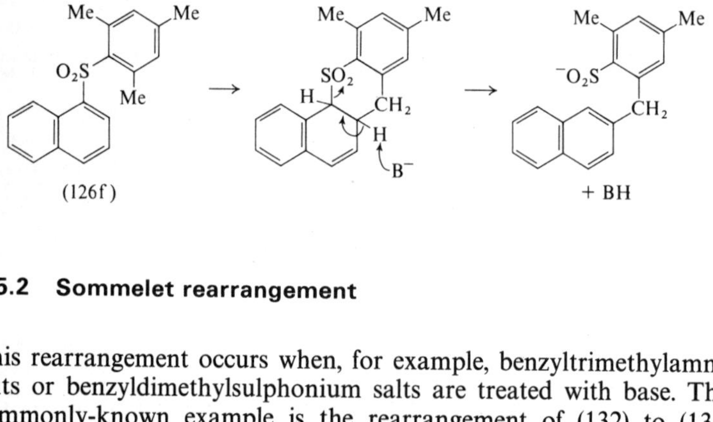

(126f) → → + BH (3.33)

3.5.2 Sommelet rearrangement

This rearrangement occurs when, for example, benzyltrimethylammonium salts or benzyldimethylsulphonium salts are treated with base. The most commonly-known example is the rearrangement of (132) to (133). The rearrangement is usually accompanied by another rearrangement, the Stevens, which does not involve the aromatic ring.

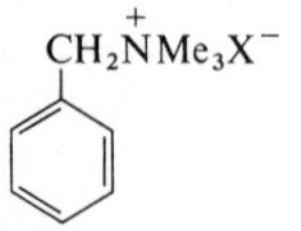

(132) 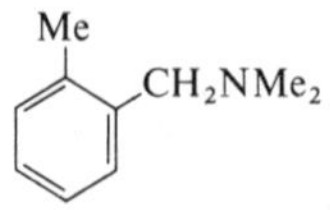 (133) 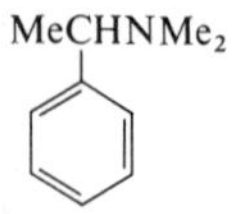 (134) 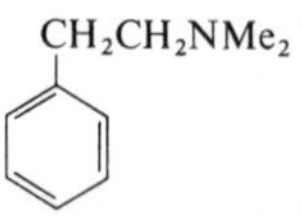 (135)

An example would be the formation of (134) and (135) from (132). The products arise from ylide intermediates.

No example of a *p*-rearrangement was known until recently[88], when ion (136) was found to give 8% of (137) when treated with n-butyl-lithium in hexane.

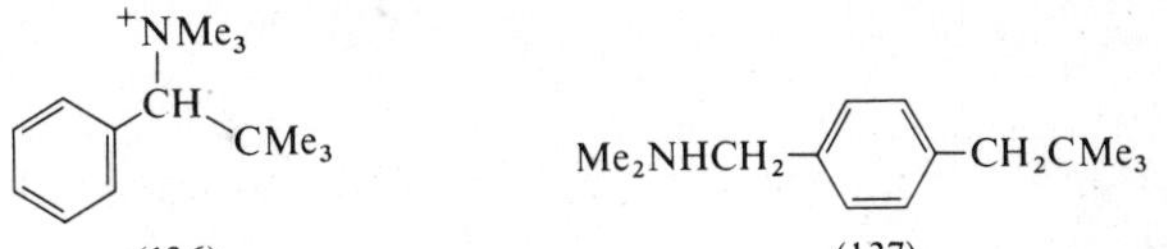

$^{+}NMe_3$ / CH / CMe_3 — (136)

Me_2NHCH_2–C_6H_4–CH_2CMe_3 — (137)

Rearrangement of (132) to (133) involves the intermediate (138). There has never been serious doubt about the formation of (138), but if there were, it should be removed by Pine's report of the formation of (139) (equation (3.34)) which is attributed to the addition of the n-butyl anion to the methylene group[89].

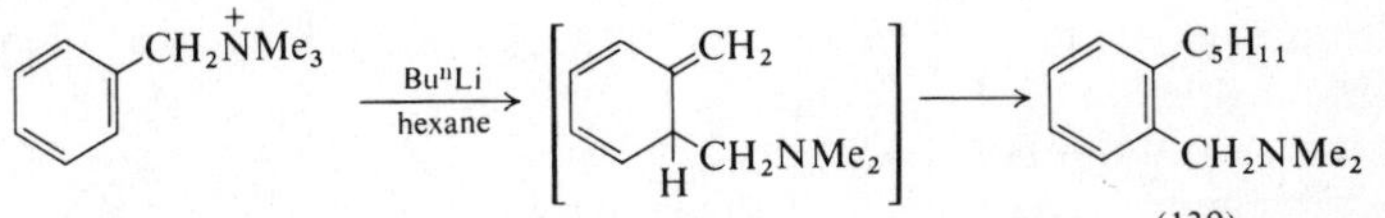

$$C_6H_5CH_2\overset{+}{N}Me_3 \ (132) \xrightarrow[\text{hexane}]{Bu^nLi} [\text{(138)}: =CH_2, \ CH_2NMe_2, \ H] \longrightarrow C_6H_4(C_5H_{11})(CH_2NMe_2) \ (139) \quad (3.34)$$

A few years ago a Stevens rearrangement was regarded by many workers as proceeding through an ion pair, regarded as 'tight' and very short-lived because a migrating optically-active group retained its configuration to a large degree.

But, a great change in thinking about the Stevens rearrangement has occurred since 1969. The discovery of chemically-induced dynamic nuclear polarisation in products of Stevens rearrangements, first by Lepley[90] and subsequently by others[91–95], tells us now that the path of a Stevens rearrangement is not an ionic one, but involves a pair of radicals held in a solvent cage. A more striking change in interpretation of a rearrangement would be difficult to find.

Interpretations of the Sommelet and Stevens rearrangements have always been tied together. Not too long ago, each rearrangement was interpreted as involving a concerted displacement within an ylide. Competition between 3- and 5-membered cyclic transition states was thought to account for the effect of temperature on the division into Stevens and Sommelet pathways[1, 96]. It is recognised now from orbital symmetry considerations that a concerted pathway with retention of configuration in the Stevens rearrangement is forbidden. The concept of concerted migration is replaced with one of a pair of caged radicals in which the migrating group retains its configuration.

What about the Sommelet rearrangement? Concerted rearrangement is *not* forbidden. The situation is like that of the Claisen rearrangement, a seven- and three-electron system (e.g. (140)), and we encounter the interesting dichotomy; when Stevens and Sommelet rearrangements occur together, are they both radical reactions or is the Stevens radical and the Sommelet concerted?

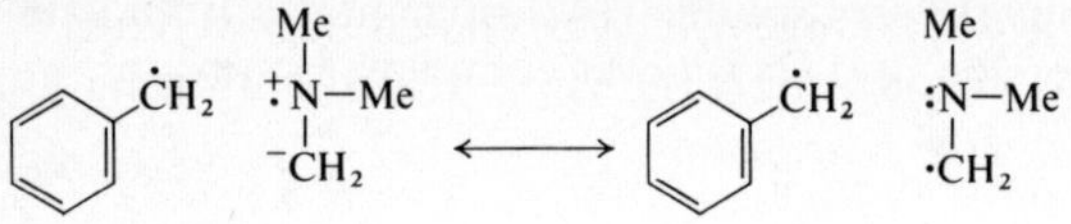

$$\text{(140)}\quad C_6H_5\dot{C}H_2\ \overset{+}{N}(Me)_2\,{}^{-}CH_2 \longleftrightarrow C_6H_5\dot{C}H_2\ :N(Me)_2\,\cdot CH_2 \qquad (3.35)$$

Search for CIDNP in Sommelet rearrangements should answer that question.

3.5.3 Trihalogenobenzenes

The curious rearrangement of 1,2,4-trihalogenobenzenes (e.g. (141)) to the 1,3,5-isomer (142) is an intermolecular disproportionation reaction. The

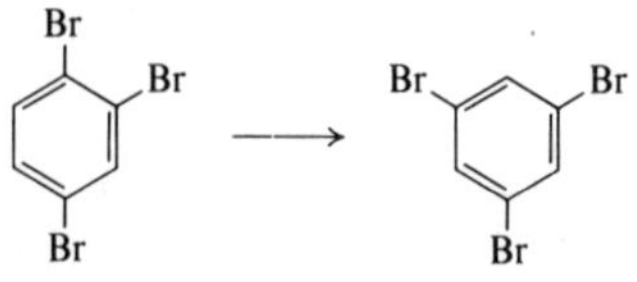

$$\text{(141)} \longrightarrow \text{(142)} \qquad (3.36)$$

reaction is catalysed by strong bases (e.g. amide ion) but best of all by anilide ion. The rearrangement has been called a 'seven-halogen' one by Bunnett and co-workers because they have deduced that a tetrahalogenobenzene (143) plays a key role in it[97–100]. The overall scheme is written as follows*.

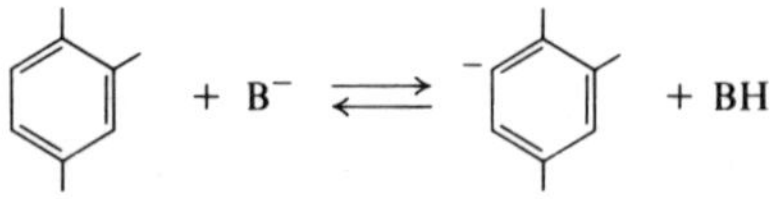

$$\text{(141)} + B^- \rightleftharpoons \text{Ar}^- + BH \qquad (3.37)$$

$$\text{Ar}^- + \text{ArBr}_3 \longrightarrow \text{(143)} + \text{Ar}^- \qquad (3.38)$$

$$\text{Ar}^- + \text{ArBr}_3 \rightleftharpoons \text{Ar}^- + \text{ArBr}_3 \qquad (3.39)$$

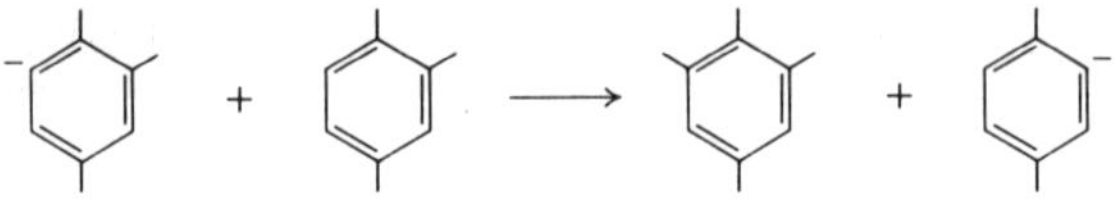

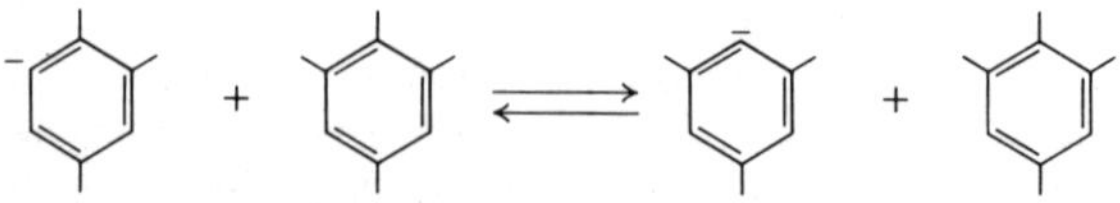

$$\text{Ar}^- + BH \rightleftharpoons \text{(142)} + B^- \qquad (3.40)$$

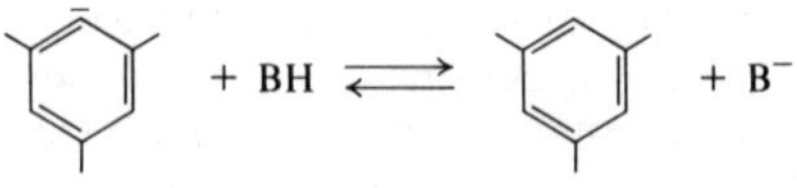

*The ring-substituents in all equations are Br

Equations (3.39) and (3.40) indicate that the rearrangement should be reversible. The formation of (143) in good yield occurs because in the usual reaction conditions the less soluble (142) precipitates from solution (liquid ammonia). Conversion of (141) is never complete, however.

3.5.4 Carboxylates (Henkel and Kolbe–Schmitt reactions)

The disproportionation of potassium benzoate and the rearrangement of dipotassium salts of phthalic and isophthalic acids into the salt of terephthalic acid is an important industrial process (the Henkel process). Extensive work by Ratusky has cleared up most of the questions of mechanism. The usual practice is to heat an acid salt as a melt containing a cadmium(II) salt, and out of the melt the desired terephthalic acid salt crystallises. Ratusky has shown that these reactions are intermolecular and involve decarboxylation-recarboxylation sequences such as illustrated below[101]. Equation (3.41)* illustrates reversible decarboxylation, while equation (3.42) shows how new

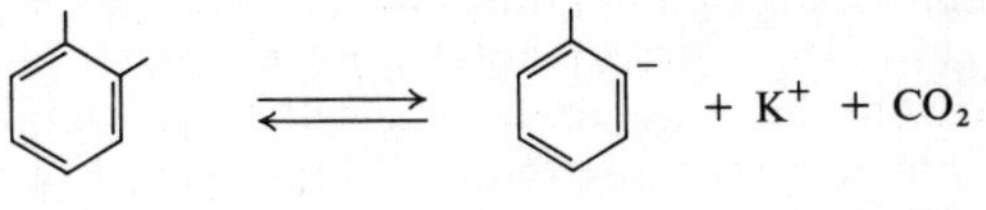

$$\rightleftharpoons \quad + K^+ + CO_2 \qquad (3.41)$$

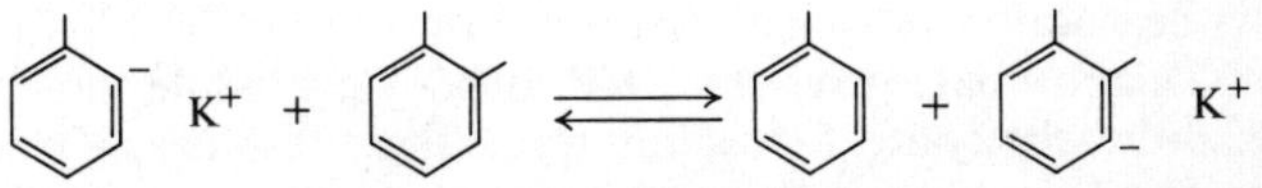

$$K^+ + \quad \rightleftharpoons \quad + \quad K^+ \qquad (3.42)$$

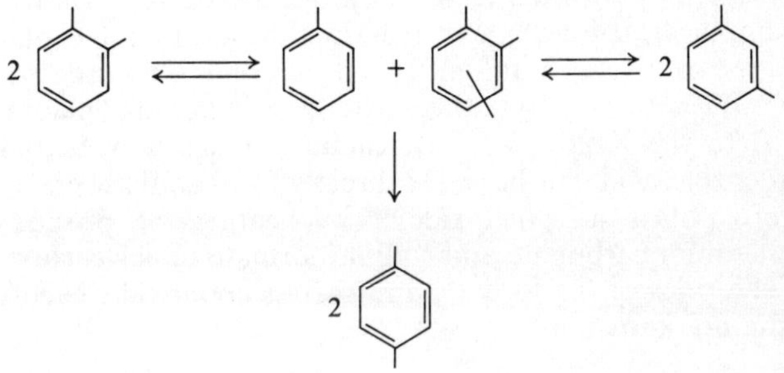

$$2 \quad \rightleftharpoons \quad + \quad \rightleftharpoons 2 \qquad (3.43)$$

$$\downarrow$$

$$2$$

aryl carbanions can be formed. Multiple steps of these kinds are summarised in equation (3.43), which shows that dipotassium phthalate leads to potassium benzoate, benzenetricarboxylates, isophthalate, and terephthalate. The final product is the terephthalate because of its insolubility. The reviewer raised the question some years ago about whether an equilibrium mixture of salts would be achieved if all the salts could be kept in solution[1]. Ratusky has provided the answer by using molten potassium cyanate as the solvent. An example is given in equation (3.44) which shows the mixture of salts obtained if the terephthalate does not crystallise out. It is interesting

*The ring-substituents in all equations are $-CO_2K$

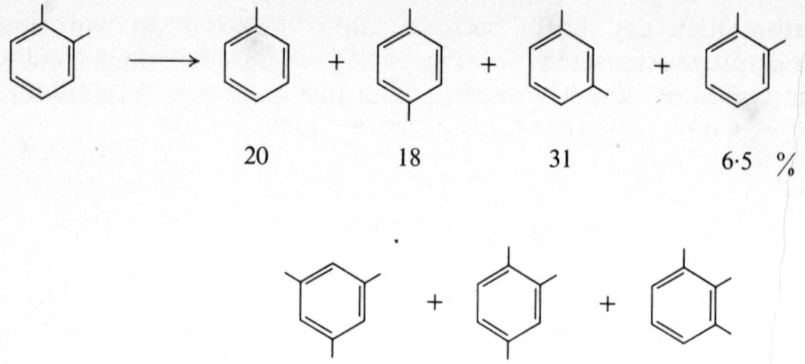

20 18 31 6·5 %

(3.44)

8 12 2 %

to note again that the *m*-isomer predominates among the disubstituted benzenes (see p. 69).

Ratusky has shown further that among the naphthalene dicarboxylates the 2,6-isomer crystallizes out of a melt, but if potassium benzoate and potassium naphthoate are heated together, the products are eventually dipotassium terephthalate and naphthalene.

Similar work has been carried out by Furuyama[102] and by Dozen[103]. Dozen has found with the use of toluic acid salts[103] and naphthoates[104], that the counter ion influences the distribution of products. For example, potassium *o*-toluate gives mainly potassium 3-methylphthalate and 4-methylisophthalate, while sodium *o*-toluate gives mainly sodium *m*-toluate. Further, potassium 2-naphthoate gives the 2,6-dicarboxylate, while the sodium salt gives the 2,3-dicarboxylate. Dozen interprets these results as a restriction to *o*-isomerisation for sodium salts and a freedom of isomerisation of potassium salts[103]. The analogy with the Kolbe–Schmitt reaction is striking[1].

The remaining feature of these reactions which is not explained is the function of the catalyst. Cadmium(II) is most commonly used; zinc metal is also used[101]. Without a catalyst, these reactions do not take place readily[1, 101].

In regard to the Kolbe–Schmitt reaction, it has now been found that *p*-hydroxybenzoic acid can be made directly from sodium phenoxide[105] (as well as from potassium phenoxide)[106] by heating the phenoxide with a mixture of sodium carbonate and sodium formate under carbon monoxide pressure. The roles of the formate and carbon monoxide, each of which is essential, are not known.

Note added in proof

Ring protonation

A new approach to the nitramine (Section 3.4.8) and benzidine rearrangements was proposed recently by Allan (based on an unpublished theory of M. E. Lupes)[107]. Allan proposes that the transition state in the one-proton benzidine rearrangement has the form originally suggested by Hammick and Mason (144a).

A similar configuration is proposed for the nitramine rearrangement (144b). The novelty applied to the benzidene rearrangement is the proposal that the second proton in a two-proton rearrangement attaches at the ring

(144c) rather than the second nitrogen atom. Scission via path *a* would engender disproportionation (Section 3.4.1.1), whereas rearrangement would occur via path *b*, particularly within a boat-like transition state (144d).

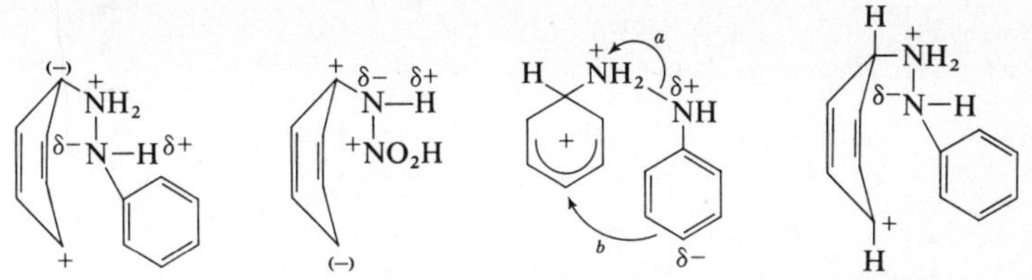

(144a) (144b) (144c) (144d)

The proposal for the nitramine rearrangement does not adjust well to White's recent experimental results, particularly to viscosity effects and intra-/intermolecularities (Section 3.4.8).

As for ring-protonation in the benzidine rearrangement, firm experimental evidence needs to be sought in its support. At this stage proposals for intermediates such as (144c) are an intriguing speculation.

Acknowledgement

The reviewer is indebted to the Robert A Welch Foundation for support of this and other work in aromatic rearrangements.

References

1. Shine, H. J. (1967). *Aromatic Rearrangements.* (Amsterdam: Elsevier Publishing Company)
2. Dewar, M. J. S. and Worley, S. D. (1969). *J. Chem. Phys.*, **50,** 654
3. Turner, D. W., Baker, C., Baker, A. D. and Brundle, C. R. (1970). *Molecular Photoelectron Spectroscopy.* (New York: Wiley-Interscience)
4. Streitwieser, Jr., A. and Reif, L. (1964). *J. Amer. Chem. Soc.*, **86,** 1988
5. Csicsery, S. M. (1969). *J. Org. Chem.*, **34,** 3338
6. Olah, G. A. (1972). *J. Amer. Chem. Soc.*, **94,** 808
7. Olah, G. A., Schlosberg, R. R., Potter, R. D., Mo, Y. K., Kelly, D. P. and Mateescu, G. D. (1972). *J. Amer. Chem. Soc.*, **94,** 2034
8. Jakubetz, W. and Schuster, P. (1971). *Angew. Chem. Int. Edn.*, **10,** 497
9. Dewar, M. J. S. (1963). *Molecular Rearrangements,* Vol. 1. (P. De Mayo, editor) (New York: Interscience)
10. Erykalov, Yu. G., Chirtulov, V. G. and Filatov, V. M. (1969). *J. Org. Chem. USSR,* **5,** 1402
11. Brouwer, D. M. (1968). *Rec. Trav. Chim.*, **87,** 335, 342, 611
12. Gore, P. H., Thorburn, S. and Weyell, D. J. (1971). *J. Chem. Soc. C,* 2372
13. Kooyman, E. C. (1963). *J. Pure and Appl. Chem.*, **7,** 193
14. Augustin, G. J. P. and Kooyman, E. C. (1967). *Rec. Trav. Chim.*, **86,** 122
15. Hart, H. and Janssen, J. F. (1970). *J. Org. Chem.*, **35,** 3637
16. Perrin, C. L. and Skinner, G. A. (1971). *J. Amer. Chem. Soc.*, **93,** 3389
17. Hopff, W. H. and Dreiding, A. S. (1965). *Angew. Chem. Int. Edn.*, **4,** 690
18. Futaki, R. (1967). *Tetrahedron Lett.*, 2455
19. Bell, K. H. (1967). *Tetrahedron Lett.*, 397
20. Shine, H. J. and Schoening, C. E. (1972). *J. Org. Chem.*, in press
21. Carlin, R. B. and Sivaramakrishnan, K. P. (1970). *J. Org. Chem.*, **35,** 3368

22. Vitullo, V. P. (1970). *J. Org. Chem.*, **35,** 3976
23. Miller, B. (1970). *J. Amer. Chem. Soc.*, **92,** 432
24. Brooke, G. M. (1971). *Tetrahedron Lett.*, 2377
25. Jansen, H.-J. and Schmid, H. (1969). *Chemistry in Britain,* 111
26. Jansen, H.-J. and Schmid, H. (1970). *Chimia,* **24,** 89
27. Fräter, Gy. and Schmid, H. (1970). *Helv. Chim. Acta,* **53,** 269
28. Fräter, Gy., Habich, A., Jansen, H.-J. and Schmid, H. (1969). *Helv. Chim. Acta,* **52,** 335
29. Scheinmann, F. Barner, R. and Schmid, H. (1968). *Helv. Chim. Acta,* **51,** 1603
30. Kwart, H. and Schwartz, J. L. (1969). *Chem. Commun.*, 44
31. Makisumi, Y. and Murabayashi, A. (1969). *Tetrahedron Lett.*, 1971, 2449, 2453
32. Mortensen, J. Z., Hedegaard, B. and Lawesson, S.-O. (1971). *Tetrahedron,* **27,** 3831
33. Kwart, H. and Cohen, M. H. (1968). *Chem. Commun.*, 319
34. Kwart, H. and Johnson, N. (1970). *J. Amer. Chem. Soc.*, **92,** 6064
35. Kataev, E. G., Chmutova, G. A., Musina, A. A. and Yarkova, E. G. (1969). *Proc. Acad. Sci. USSR,* **187,** 673
36. Zsindely, J. and Schmid, H. (1968). *Helv. Chim. Acta,* **51,** 1510
37. Kwart, H. and George, T. J. (1970). *Chem. Commun.*, 433
38. Sonnenberg, F. M. (1970). *J. Org. Chem.*, **35,** 3166
39. Dewar, M. J. S. and Hart, L. S. (1970). *Tetrahedron,* **26,** 973, 1001
40. Spanninger, P. A. and von Rosenberg, J. L. (1970). *Chem. Commun.*, 795
41. Shine, H. J. (1969). *Mechanisms of Molecular Migrations,* Vol. 2. 191–247 (B. S. Thyagarajam, editor)
42. Banthorpe, D. V. and co-workers (1968). *J. Chem. Soc. B,* 605, 609, 615, 618, 624, 627
43. Kurihara, M. and Yoda, N. (1967). *Bull. Chem. Soc. Jap.*, **40,** 2429
44. White, W. N. and Moore, E. E. (1968). *J. Amer. Chem. Soc.*, **90,** 526
45. Banthorpe, D. A., Cooper, A. and Ingold, C. K. (1967). *Nature,* **216,** 232. Ingold, C. K. (1969). *Structure and Mechanism in Organic Chemistry,* 2nd edn., 940. (Ithaca: Cornell University Press)
46. Welzel, P. (1970). *Ber.*, **103,** 1318; (1971). **104,** 808
47. Shine, H. J. and Bear, J. L. (1957). *Chem. and Ind. (London)*, 565
48. Miller, B. and Han, C.-H. (1970). *Chem. Commun.*, 623
49. Hogg, D. R., Smith, J. H. and Vipond, P. W. (1968). *J. Chem. Soc. C,* 2713
50. Davis, F. A., Wetzel, R. B., Devon, T. J. and Stackhouse, J. F. (1970). *Chem. Commun.*, 678
51. Cox, Jr., J. R. and Dunn, M. F. (1963). *Tetrahedron Lett.*, 985
52. Sheradsky, T. and Salemnick, G. (1971). *Tetrahedron Lett.*, 645
53. Robinson, B. (1963). *Chem. Rev.*, **63,** 387
54. Elgersma, R. H. C. and Havinga, E. (1969). *Tetrahedron Lett.*, 1735
55. Schiess, P. and Grieder, A. (1969). *Tetrahedron Lett.*, 2097
56. Ahlbrecht, H. (1971). *Tetrahedron Lett.*, 545
57. Owellen, R. J., Fitzgerald, J. A., Fitzgerald, B. M., Welsh, D. A., Walker, D. M. and Southwick, P. L. (1967). *Tetrahedron Lett.*, 1741
58. Bajwa, G. S. and Brown, R. K. (1968). *Can. J. Chem.*, **46,** 3105; (1969); **47,** 785; (1970); **48,** 2293
59. Lyle, R. E. and Skarlos, L. (1966). *Chem. Commun.*, 644
60. Illy, H. and Funderburk, L. (1968). *J. Org. Chem.*, **33,** 4283
61. Palmer, M. H. and McIntyre, P. S. (1969). *J. Chem. Soc. B,* 446
62. Sheradsky, T. (1966). *Tetrahedron Lett.*, 5225
63. Kaminsky, D., Shavel, Jr., J. and Meltzer, R. I. (1967). *Tetrahedron Lett.*, 859
64. Mooradian, A. and DuPont, P. E. (1967). *Tetrahedron Lett.*, 2867
65. Yamada, T. (1969). *Bull. Chem. Soc. Jap.*, **42,** 3565
66. El Nadi, A. H., Hickinbottom, W. J. and Wasif, S. (1970). *J. Chem. Soc. B,* 1131
67. Gassman, P. G., Campbell, G. and Frederick, R. (1968). *J. Amer. Chem. Soc.*, **90,** 7377
68. Gassman, P. G. and Campbell, G. A. (1971). *J. Amer. Chem. Soc.*, 2567
69. Baliga, B. T. (1970). *J. Org. Chem.*, **35,** 2031
70. Morgan, T. D. B. and Williams, D. L. H. (1970). *Chem. Commun.*, 1671
71. Belyaev, E. Yu. and Nikulicheva, T. I. (1970). *Organic Reactivity USSR,* **7,** 165
72. White, W. N. and co-workers. (1970). *J. Org. Chem.*, **35,** 737, 965, 1803, 2759. *J. Amer. Chem. Soc.*, **92,** 4477
73. Welzel, P. (1971). *Ber.*, **104,** 808

74. Biddle, B. N., Davidson, R. S., Harvey, S. H. and Singleton, A. F. (1968). *Chem. Commun.*, 311
75. Spillane, W. J. and Scott, F. L. (1967). *Tetrahedron Lett.*, 1251. (1968). *J. Chem. Soc. B*, 779
76. Scott, F. L. and Spillane, W. J. (1967). *Tetrahedron Lett.*, 2707
77. Spillane, W. J. and Scott, F. L. (1968). *Tetrahedron*, **24**, 5011
78. Vrba, Z. and Allan, Z. J. (1968). *Tetrahedron Lett.*, 4507. (1968). *Coll. Czech. Chem. Commun.*, **33**, 2502
79. Allan, Z. J. and Vrba, Z. (1969). *Coll. Czech. Chem. Commun.*, **34**, 272
80. Scott, F. L., Scoggin, C. B. and Spillane, W. J. (1969). *Tetrahedron Lett.*, 3675
80a. Fischer, P. B. and Zollinger, H. (1970). *Helv. Chim. Acta*, **53**, 1306
81. Buncel, E. and co-workers. (1971). *Tetrahedron Lett.*, 113; (1968). *Mechanisms of Molecular Migrations*. Vol. 1. (B. S. Thyagarajan, editor); (1968). *Tetrahedron*, **24**, 6671; (1970). *Can. J. Chem.*, **48**, 377; (1969). ibid, **47**, 911; (1969). *Chem. Commun.*, 765
82. Hendley, E. C. and Duffey, D. (1970). *J. Org. Chem.*, **35**, 3579
83. Duffey, D. and Hendley, E. C. (1968). *J. Org. Chem.*, **33**, 1918
84. Truce, W. E., Kreider, E. M. and Brand, W. W. (1970). *Organic Reactions*, **18**, 99–215
85. Drozd, V. N. and co-workers. (1967). *J. Org. Chem. USSR*, **3**, 359, 531, 2026; (1968). **4**, 1025, 1894; (1969). **5**, 309, 313, 920, 1220, 1416, 1238, 1409; (1970). **6**, 154, 821, 827, 1071; (1971). **7**, 419
86. Truce, W. E., Robbins, C. R. and Kreider, E. M. (1966). *J. Amer. Chem. Soc.*, **88**, 4027
87. Truce, W. E. and Brand, W. W. (1970). *J. Org. Chem.*, **35**, 1828
88. Pine, S. H. (1967). *Tetrahedron Lett.*, 3393
89. Pine, S. H. and Sanchez, B. L. (1969). *Tetrahedron Lett.*, 1319
90. Lepley, A. R. and co-workers. (1969). *J. Amer. Chem. Soc.*, **91**, 1237; *Chem. Commun.*, 1460; (1971). *J. Amer. Chem. Soc.*, **36**, 1217
91. Schöllkopf, U., Ostermann, G. and Schossig, J. (1969). *Tetrahedron Lett.*, 2619
92. Schöllkopf, U. Ludwig, U., Ostermann, G. and Patsch, M. (1969). *Tetrahedron Lett.*, 3415
93. Jemison, R. W. and Morris, D. G. (1969). *Chem. Commun.*, 1226
94. Baldwin, J. E., Erickson, W. F., Hackler, R. E. and Scott, R. M. (1970). *Chem. Commun.*, 576
95. Iwamura, H., Iwamura, M., Nishida, T., Yoshida, M. and Nakayama, J. (1971). *Tetrahedron Lett.*, 63
96. Zimmerman, H. E. (1963). *Molecular Rearrangements*, Vol. 1. (P. De Mayo, editor) (New York: Interscience)
97. Bunnett, J. F. and Moyer, Jr., C. E. (1971). *J. Amer. Chem. Soc.*, **93**, 1183
98. Bunnett, J. F. and Scorrano, G. (1971). *J. Amer. Chem. Soc.*, **93**, 1190
99. McLennan, D. J. and Bunnett, J. F. (1971). *J. Amer. Chem. Soc.*, **93**, 1198
100. Bunnett, J. F. and Feit, I. N. (1971). *J. Amer. Chem. Soc.*, **93**, 1201
101. Ratusky, J. and co-workers. (1966). *Chem. and Ind. (London)*, 1798; (1967). 1093; (1967). *Coll. Czech. Chem. Commun.*, **32**, 1719, 2504; (1968). **33**, 2346; (1969). *Tetrahedron Lett.*, 325; (1970). *J. Labld. Compds.*, **6**, 124; (1970). *Chem. and Ind. (London)*, 1347
102. Furuyama, S. (1967). *Bull. Chem. Soc. Jap.*, **40**, 1212
103. Dozen, Y. (1967). *Bull. Chem. Soc. Jap.*, **40**, 1218
104. Dozen, Y. (1968). *Bull. Chem. Soc. Jap.*, **41**, 664
105. Yasuhara, Y. and Nogi, T. (1969). *Chem. and Ind. (London)*, 77
106. Yasuhara, Y. and Nogi, T. (1968). *J. Org. Chem.*, **33**, 4513
107. Allan, Z. J. (1971). *Tetrahedron Lett.*, 4225

4
Photochemical Reactions of Aromatic Compounds

D. BRYCE-SMITH and A. GILBERT
University of Reading

4.1 INTRODUCTION

This review is largely concerned with publications which appeared during the period July 1969 to June 1971, but certain prior and subsequent work

Sections 4.6.3–4.6.7 were contributed by Professor H. J. Shine (see Chapter 3).

has also been included when necessary for purposes of comparison. The volume of photochemical publications is so great these days that the selection of key or seminal papers, which is the purpose of this review, becomes very much a personal choice; but we have made some effort to be objective, although not to the extent of omitting reference to our own work. The emphasis is on the chemistry of photoprocesses, and photophysical aspects such as energy transfer from aromatic molecules have not been covered except where they appear specially relevant to an understanding of the chemistry involved. Photoreactions of aromatic compounds have been excluded from consideration when only changes in substituents are involved and there are no grounds to suspect involvement of the aromatic ring as such. Likewise, reactions of aromatic compounds with photochemically generated free radicals, carbenes, etc., have been omitted except in cases where the chemistry is unconventional.

4.2 PHOTO-ISOMERISATION REACTIONS OF AROMATIC RINGS

Since the original report in 1957 of the photo-isomerisation of benzene to fulvene[1], many further reports of this transformation have appeared, and it has become increasingly recognised that the photo-isomerisation of monocyclic benzenoid rings to fulvene and other non-aromatic products is a quite common type of photochemical process. With benzene itself, the products which can be formed include fulvene[1], benzvalene[2], Dewar-benzene[3], and hexa-1,3-dien-5-yne[2]. The recent demonstration[4a] that Dewar-benzene can arise directly from the S_2 state of benzene ($^1B_{1u}$) provides the first example of a non-dissociative reaction from an upper excited singlet state, and is in accord with orbital-symmetry considerations[3d, 4b]. Irradiation of hexafluorobenzene appears to give only the Dewar isomer, but the excited state involved has not yet been identified[3c, d]. The absence of the fulvene and benzvalene isomers is odd, but might be a consequence of their high thermal lability. Pyridine has been shown to be converted into Dewar-pyridine (2-azabicyclo[2,2,0]hexa-2,5-diene) on irradiation in the liquid phase at 254 nm. This isomer has a half-life of 2 min at 25 °C[5]. Photoreduction of pyridine in the presence of aqueous sodium borohydride gives 2-azabicyclo[2,2,0]hex-5-ene (1), and photohydration gives 5-aminopenta-2,4-dienal (2), both via the Dewar isomer[5]. The formation of prismane from benzene has not yet been reported, although tri-t-butyl-[6a], hexamethyl-[6b], and hexakis(perfluoroalkyl)-[6c] derivatives are known to be formed by irradiation of the corresponding Dewar isomers.

Photochemical interconversions of the xylenes and other polyalkylbenzenes are largely intramolecular processes which mainly involve 1,2-transpositions of ring carbon atoms rather than migration of methyl groups, and probably occur via the formation and re-aromatisation of benzvalenes[6a, 7]. Quenching and other studies have indicated that these reactions proceed via the S_1 state of the aromatic compound[8]. As mentioned, 1,2-shifts of substituents normally predominate greatly over those of 1,3-type, and this fact is consistent with the idea of benzvalene rather than prismane intermediates (which on re-aromatisation would tend to give both 1,2- and

1,3-shifts). However, the photo-isomerisation of 2,6-di-t-butylphenol to the 2,5-isomer might well involve a ketone tautomer rather than a benzvalene intermediate since the methyl ether of the phenol does not undergo the rearrangement[9].

A striking *thermal* analogue of the photo-isomerisations to Dewar-benzenes is provided by the report that perfluorohexamethylbenzene (3) yields the *para*-bonded isomer (4) in a flow pyrolysis apparatus giving a contact time of 1 s at 400 °C. Compound (4) was re-aromatised by heating at 140 °C for 6 h[10]. Note that the formation of isomer (4) and its aromatisation are formally symmetry forbidden, so may be presumed to occur via a *para*-di-radical intermediate.

The photo-isomerisation of 1,2-dihydronaphthalene to benzobicyclo-[3,1,0]hexene has been studied with the deuterium-labelled olefins (5) and (6); the products (7) and (8) respectively are formed[11]. The reaction clearly proceeds by rupture of the 1,2-bond, and the *o*-quinodimethane intermediate (9) is probably involved. In the cases of the substituted derivatives (10) and (11), initial formation of the quinodimethane (12) is followed by a 1,7-sigmatropic hydrogen shift to give compounds (13) and (14)[12]. It will be

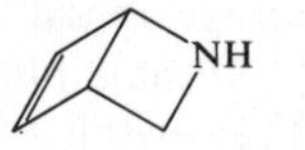

NH

$H_2NCH{=}CH{-}CH{=}CHCHO$

(1) (2)

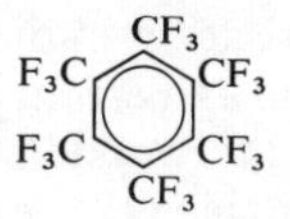

CF_3, F_3C, CF_3, F_3C, CF_3, CF_3

(3)

CF_3, F_3C, CF_3, F_3C, CF_3, CF_3

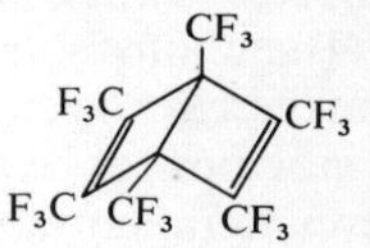

(4)

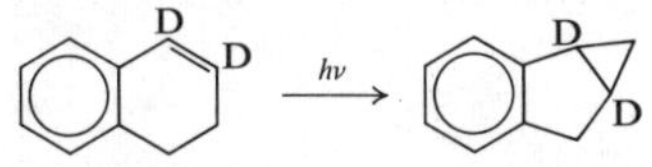

D D $\xrightarrow{h\nu}$ D D

(5) (7)

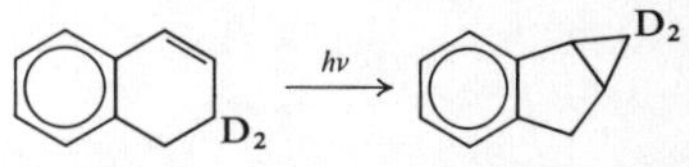

D_2 $\xrightarrow{h\nu}$ D_2

(6) (8)

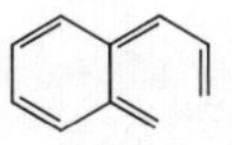

$\xrightarrow{S_0 \rightarrow S_1}$ $\xrightarrow[\text{of H}]{\text{1, 7-shift}}$

R Me · Me R · Me R

(9) 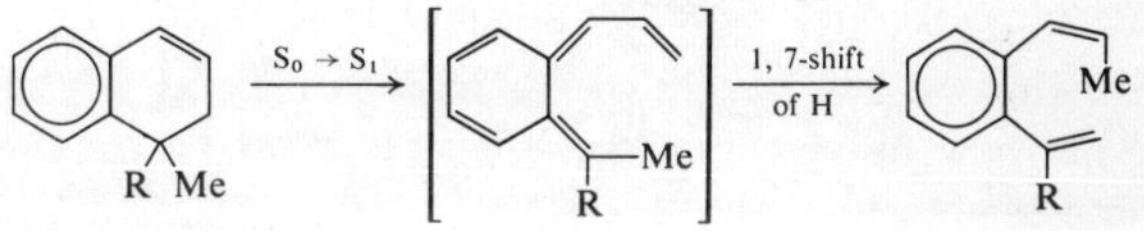 (10) R = Ph (12) (13) R = Ph

(11) R = Me (14) R = Me

recalled that *o*-quinodimethane enols are produced as intermediates in the photo-isomerisation of *o*-methylaryl ketones[13]. Systems of this type often exhibit photochromism, but the quantum efficiency appears to be markedly susceptible to the presence of impurities. The competitive processes of photo-enol formation and photo-pinacol reduction have also been studied[15].

Apart from the above-mentioned formation of Dewar-pyridine[4], there has been considerable recent interest in the photo-isomerisation of picolines, pyrazines, and related heterocyclic systems. Thus vapour phase irradiation of 2-picoline in the $\pi \rightarrow \pi^*$ absorption region has been found to give both the 3- and 4-isomers. The quantum yield for isomerisation increases with decreased wavelength of the radiation and, rather surprisingly, with increase in intensity, but is generally lower than 0.002[16]. A triplet state of the picoline may be involved, for the quantum yields are reduced both in the presence of oxygen and *cis*-but-2-ene. The occurrence of both 1,2- and 1,3-shifts and the probability of a triplet mechanism contrast with the previously-mentioned photo-isomerisations of xylenes and related benzenoid systems[7–9]. It is suspected that the rearrangement occurs via an azaprismane intermediate. On the other hand, comparable studies by French workers of 3-picoline and the lutidines have suggested a more complex situation, with interconversion between carbon atoms at positions 2 or 6 and 4, and between those at positions 3 and 5 or 2 and 6, together with some concurrent true migration of methyl groups; thus the picolines gave some pyridine, and 2-picoline gave some 2,4-lutidine[17]. The mechanism of these processes in the heterocyclic field remains slightly unclear, although it seems evident that azabenzvalene intermediates are not involved, in contrast with the probable involvement of benzvalenes in comparable rearrangements in the benzene series. There seems to be one case at least where an azaprismane is produced; irradiation of solutions of pentakis(pentafluoroethyl)pyridine at 200 nm gives the azaprismane (15)[18]. The highly strained isomer (15) is remarkably stable, and survives heating for several hours at 160 °C.

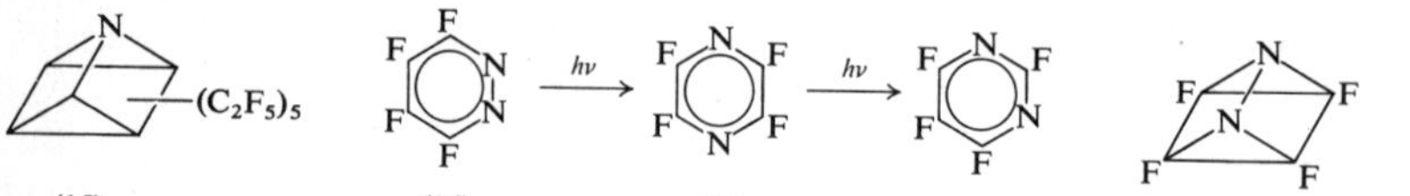

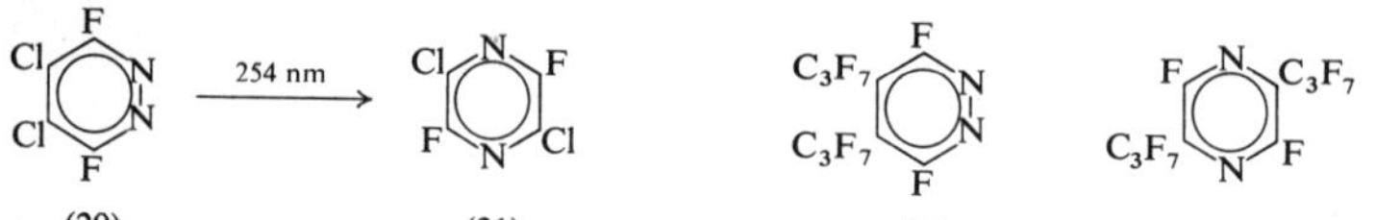

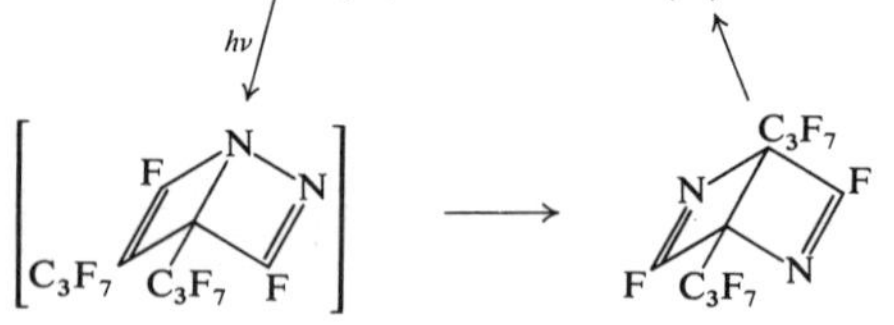

The photo-isomerisation of diaza-heterocycles has proved a fruitful area, but there is no general agreement yet on the mechanisms which are operating. Irradiation of tetrafluoropyridazine (16) in the vapour or solution phase gives tetrafluoropyrazine (17) which then undergoes slow photo-isomerisation to tetrafluoropyrimidine (18)[19]. The second step which is equivalent to a 1,2-shift is analogous to the behaviour of the parent pyrazine[19], but the first step has no known analogy in the parent systems and was originally supposed to involve the diazaprismane (19) as an intermediate[20]. On the other hand, the report that the pyridazine (20) undergoes photo-isomerisation in Freon 114 to the single pyrazine (21) appears inconsistent with a simple diazaprismane mechanism[21]. Other workers[22] have observed that the vapour-phase photolysis of the pyridazine (22) gives the pyrazine (23) via the Dewar isomer (24) which they suppose to be formed by rearrangement of an initially formed Dewar isomer (25). It is odd that the authors of reference 21 had been unable to detect the formation of Dewar intermediates even at 77 K, but in the absence of bulky stabilising groups it is just possible that such intermediates were too thermally or photochemically labile to be detected even at such a low temperature.

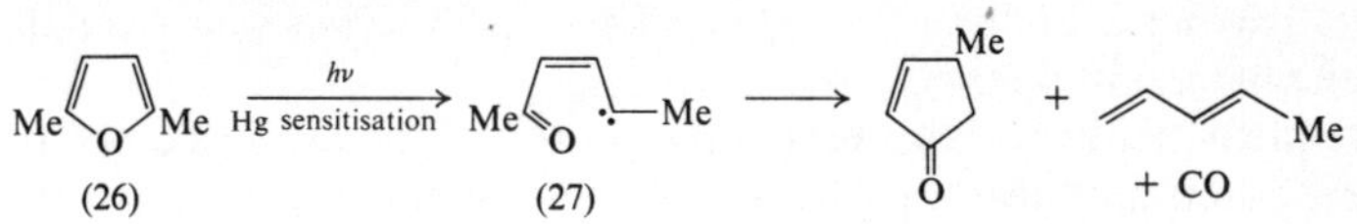

(26) (27)

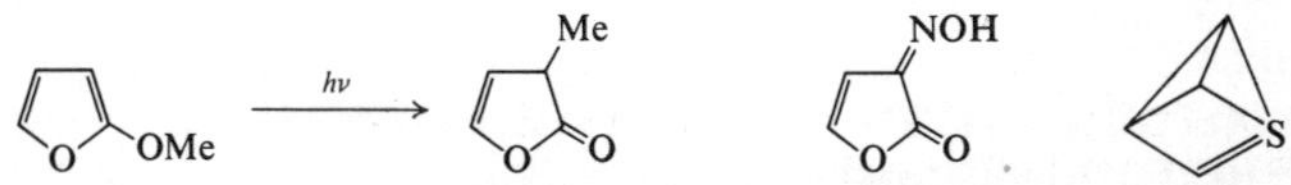

(28) (29) (30)

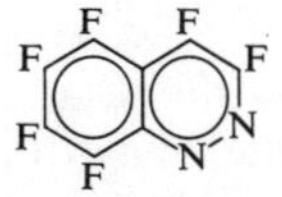

(31)

The photo-isomerisation of five-membered aromatic heterocyclic molecules has been the subject of an excellent review[23]. The intermediacy of cyclopropenecarboxaldehyde in the photolysis of furans is now well established[24]. Mercury-sensitised vapour-phase irradiation of 2-methyl-, 3-methyl-, and 2,4-dimethyl-furans gives 3-methyl-, 1-methyl-, and 1,3-dimethyl-cyclopropenes respectively as the main products, and 3-methylfuran also gives the 2-methyl isomer. On the other hand, direct vapour-phase irradiation of the 2- and 3-methylfurans tends mainly to give ring-opened isomers of the methylcyclopropene, although small quantities of the cyclic furan isomers are also formed[25]. Mercury-sensitised vapour-phase photolysis of 2,5-dimethylfuran (26) gave a complex mixture of products, but a major primary process appeared to be cleavage of a C—O bond to give a carbene (27)[26]. Some of the other products appeared to have a free-radical origin. The tendency for C—O cleavage resembles that of vinyl ethers[27]. Irradiation of 2-methoxyfuran gives the isomer (28) as a major product[28]. The process provides an extraordinary example of a 1,3-shift of a methyl group, and can be rationalised in terms of an initial homolysis of the O—Me bond, followed by radical recombination. Irradiation of 2-nitrofuran gives 3-hydroxyimino-2-oxo-2,3-dihydrofuran (29)[29].

Irradiation of 2-cyanopyrrole in methanol gives the 3-isomer as a major product in a process which is photochemically irreversible and without a thermal counterpart[30]. The photo-isomerisation of 2- to 3-arylthiophens occurs with retention of structural integrity of the migrating substituent and seems to involve an irreversible transposition of ring atoms[31]. A similar rearrangement also occurs less efficiently with alkylthiophens[32]. The irradiation of thiophens in the presence of an excess of primary amines leads to *N*-substituted pyrroles[33]. The reaction probably involves a thioketone or thioaldehyde intermediate, rather than 'Wynberg' intermediates such as (30). The vapour-phase photolysis of thiophen itself has been carefully investigated[34]. Both direct and mercury-sensitised irradiation gave acetylene, allene, methylacetylene, vinylacetylene, carbon disulphide and a polymer. The intermediates $\cdot CH:CHCH:CHS\cdot$ and $C_2H_2S\cdot$ were postulated, together with excited CS species.

Few photo-isomerisations involving fused bicyclic aromatic systems have yet been described, although hexafluorocinnoline (31) appears to form the corresponding quinazoline by means of a 1,2-shift[35]. Azulene photo-isomerises to naphthalene, but very inefficiently ($\Psi = <10^{-5}$)[36].

4.3 PHOTO-ADDITION, PHOTO-OXIDATION AND PHOTODIMERISATION

4.3.1 Photo-addition reactions of olefins and acetylenes to the benzene ring

Following the original observation that maleic anhydride forms a 2:1 photo-adduct (32) with benzene[37], numerous different types of photo-addition to the benzenoid ring have come to light; these include 1,2-, 1,3-, and 1,4-cyclo-addition and also the 1,4-addition of amines to give cyclohexa-1,4-diene derivatives and of carboxylic acids and alcohols to give bicyclo [3,1,0]hexenes. The last-named products appear, however, to arise from a *thermal* acid-catalysed reaction of the addends to benzvalene[38].

Various known and unknown concerted cyclo-additions to the benzene ring have been analysed on an orbital symmetry basis[39]. The reactions considered include 1,2-, 1,3-, and 1,4-cyclo-additions of olefins and dienes to give compounds (33), (34), and (35), and (36), (37), and (38), respectively. The dimerisations of benzene to the tetraenes (39), (40), and (41) have been treated in a similar manner. The results of the orbital symmetry analysis have been tabulated for the following cases: (a) excitation of the aliphatic addends, (b) excitation to the S_1 ($^1B_{2u}$), T_1 ($^3B_{1u}$), and S_2 ($^1B_{1u}$) states of benzene, (c) charge transfer excitation, and (d) reactions occurring through the electronic ground states. This analysis has indicated that although the Woodward–Hoffmann Rules apply to the thermal additions, they do not in general apply to the photochemical reactions. The inapplicability to the photochemical reactions is a consequence of the orbital degeneracy and the fact that the B_{2u} and B_{1u} states of benzene each comprise two orbital components having different symmetry properties, either of which can be used to establish a valid correlation with an excited state of an adduct. The principles underlying the Woodward–Hoffmann Rules are of course as equally valid for aromatic systems as for the aliphatic systems for which they were originally derived. The results of this analysis are generally in accord with existing experimental observations, and have permitted various predictions. The reader is referred to reference 39 for full details of the correlations, but certain points of special interest will be mentioned here.

The predominant tendency for simple olefins to undergo 1,3-cyclo-addition to benzene, giving adducts of type (28), can be rationalised through the orbital symmetry analysis because only additions of this type are symmetry allowed from S_1 benzene plus S_0 olefin; and that of course is the situation which obtains when mixtures of benzene and simple olefins are

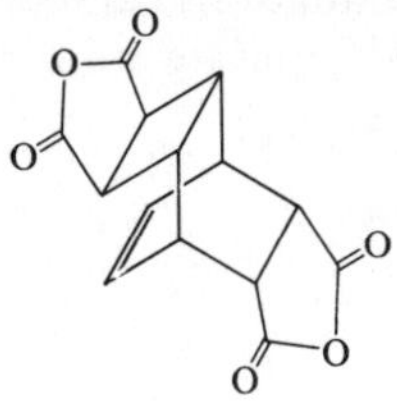

(32)

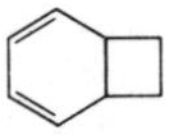

(33) 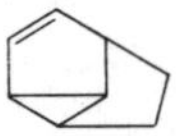 (34) 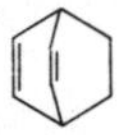 (35) 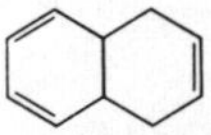 (36)

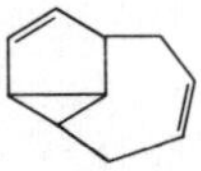

(37) 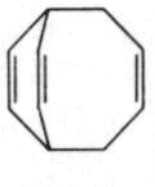 (38) 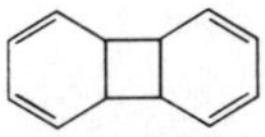 (39)

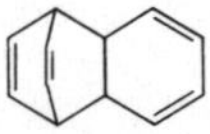

(40) 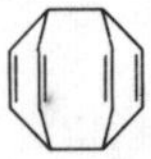 (41)

irradiated at 2537 Å. Sterically feasible concerted cyclo-additions of olefins to give adducts of type (33) and (35) are only allowed from S_0 benzene plus S_1 olefin, and such a system is not in practice attainable since the S_1 states of simple olefins have energies much higher than those of S_1 benzene. On the other hand, when the ethylenic bond bears substituents which reduce the energy of the S_1 state below that of S_1 benzene, as is the case with maleimide for example, photo-addition occurs to give 1,2-adducts of type (33) which may be isolated as such or as their Diels–Alder adducts[40]. The corresponding photo-addition of maleic anhydride to benzene had originally been supposed to involve the intermediate formation of a 1,2-adduct of type (33), but it appears reasonably certain now that no such intermediate is in fact produced in this case. The photo-addition occurs via photo-excitation of a charge-transfer transition[41], and is believed to involve an intermediate zwitter-ion (42) which can be trapped as phenylsuccinic anhydride if the irradiation is carried out in the presence of proton donors[42]. As expected, phenylmaleimide is not formed by irradiation of benzene and maleimide in the presence of proton donors. Solutions containing benzene and *p*-

benzoquinone show charge-transfer absorption at *c.* 280 nm, but no adduct is formed on irradiation. But irradiation in the presence of a proton donor gives the 1:1 adduct 4-phenoxyphenol, presumably via protonation/deprotonation of the zwitterionic intermediate (43)[42]. In the case of dimethyl acetylenedicarboxylate and benzene, a zwitterionic intermediate (44) also seems to be involved since irradiations carried out in the presence of proton-donors give dimethyl phenylfumarate and -maleate; but in this case the

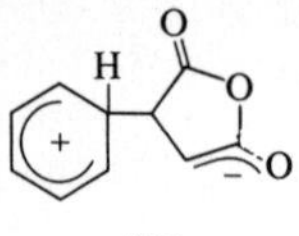

(42) 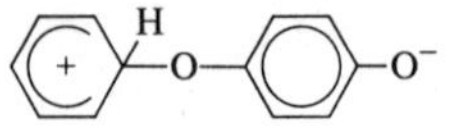 (43) 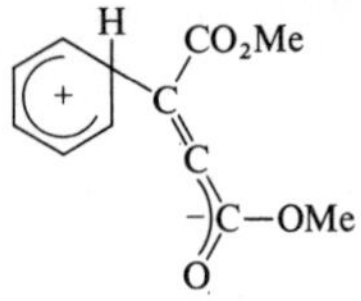 (44)

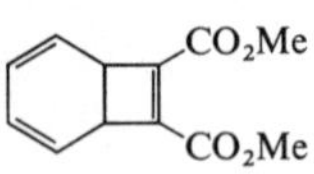

(45) 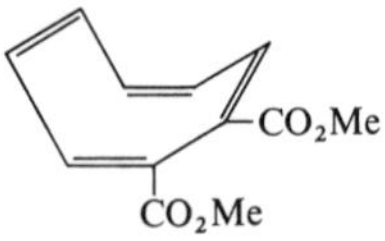 (46) 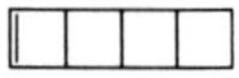 (47)

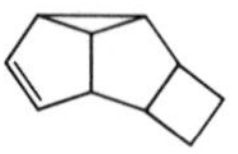

(48) 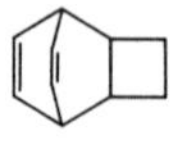 (49) 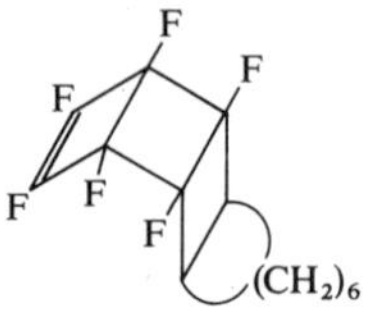 (50)

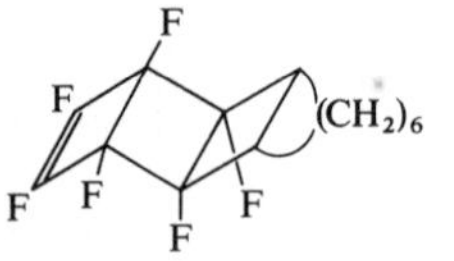

(51) 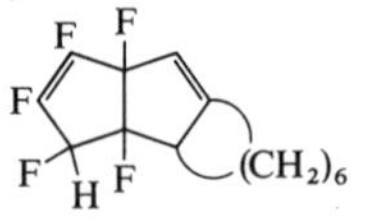 (52) 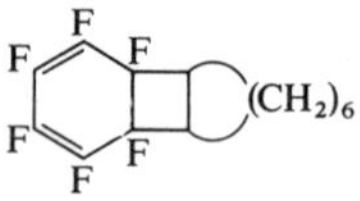 (53)

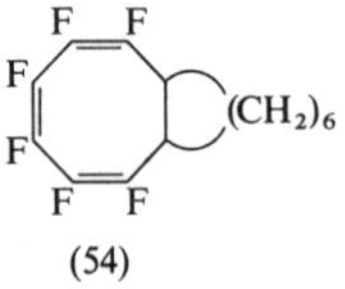

(54)

zwitterion (44) also has a tendency to undergo ring closure to form the bicyclo-octatriene (45) since this could be intercepted as the Diels–Alder adduct with tetracyanoethylene *en route* to the cyclo-octatetraene (46) which is the normal product of this reaction[42].

The foregoing acid-catalysed photoreactions are of interest on two counts; firstly, because they provide evidence for polarised intermediates such as (42) in photoreactions involving donor–acceptor systems, and secondly because they provide examples of photo-electrophilic substitution reactions into the aromatic ring. One may note the similarity in structure between intermediates (42), (43) and (44) and the Wheland intermediate for conventional electrophilic aromatic substitution reactions.

Until recently, the only known examples of 1,2-cyclo-additions to benzene involved olefinic or acetylenic addends bearing electron-withdrawing groups. Simple acetylenes have now been observed to undergo 1,2-cyclo-addition to benzene to give the corresponding cyclo-octatetraenes[43]. 1,2-Photo-adducts of simple olefins with hexafluorobenzene have also been reported[44]. Even with benzene itself and simple olefins, it now appears that photo-adducts of types (33) and (35) can be formed together with the normal 1,3-adducts (34), particularly if the reaction is carried out in the presence of a large excess of the olefin[45]. There are preliminary indications that the 1,2- and 1,4-additions are stereospecific with respect to the olefin, as for the 1,3-cyclo-addition processes. The occurrence of stereospecific 1,2- and 1,4-cyclo-additions of olefins to benzene would be rather surprising because the concerted addition of simple S_0 olefins to S_1 benzene in this manner is formally symmetry forbidden. The abnormal 1,2-photo-addition appears to be favoured by increasing methylation of the olefin, which would of course increase its donor character. It remains to be seen whether these formally symmetry-forbidden processes occur through the involvement of polar factors, or whether some relaxation of orbital-symmetry requirements occurs through mixing of benzene and olefin excited states. Although polar factors might be thought relatively unimportant in reactions between hydrocarbons, at least one case is known where the course of such a reaction is sensitive to the presence of proton donors[46].

The photoproduct from benzene and cyclobutene, previously assigned structure (47)[47], is now known to be largely the 1,3-adduct (48) together with a small proportion of the 1,4-adduct (49)[48]. The cyclobutane ring in adduct (49) is *cis*-fused[49], contrary to expectations based on orbital-symmetry relationships[39]. This again may prove to be a reaction in which polar factors are important. As mentioned above, all the 1,3-cyclo-additions of olefins to benzene which have so far been investigated have proved to be stereospecific with respect to the olefin[50], and these reactions are also insensitive to the presence of proton donors and polar solvents[50b].

The photo-addition of *cis*-cyclo-octene to hexafluorobenzene is a much more complex process than the corresponding addition to benzene[51]. Apart from the expected 1,3-adduct of type (34), the five 1:1 adducts (50)–(54) were also identified. The Dewar isomer of hexafluorobenzene appeared not to be the direct precursor of adducts (50) and (51). Hexafluorobenzene is less prone than benzene to undergo reactions which involve transannular *meta*-bonding, and the formation of a proportion of a 1,3-adduct with cyclo-octene provides the only similarity yet known between the photochemistry of benzene and hexafluorobenzene. The unexpected adduct (52) could have been formed by a thermally allowed suprafacial 1,5-hydrogen shift in the 1,3-cyclo-adduct; but rather surprisingly, the proportion of adduct (52) relative to the other products did not appear to depend upon the time of irradiation.

The irradiation of solutions of ketene in benzene does not appear to give simple adducts, but rather cycloheptatriene and toluene, the products of conventional carbene-insertion reactions[52].

Two groups of workers have reported that dichlorovinylene carbonate undergoes acetone- and acetophenone-sensitised photo-addition to benzene[53, 54]; 1,2- and 1,4-cyclo-adducts are formed, together with various other products.

4.3.2 Photo-oxidation of benzene

The dialdehydes (55) and (56) have been reported to be formed by photo-addition of oxygen to benzene in the liquid phase[55]. Photo-oxidation of benzene in aerated dilute aqueous solution at pH 5–7 gives 2-formyl-4*H*-pyran (57) in a quantum yield of *c.* 0.1 [56]. Kinetic and mechanistic studies

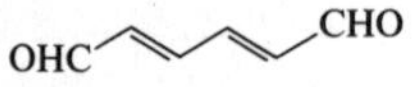

OHC–CH=CH–CH=CH–CHO (55) OHC(CH=CH)$_5$CHO (56) (57) (58)

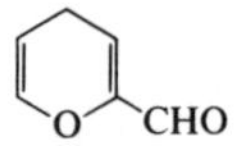

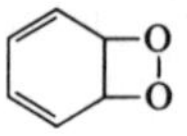

have indicated that the species which reacts directly with oxygen may be T_1 benzene. There is evidence for the intermediate formation of a highly unstable peroxide (58?) which rapidly isomerises to the pyran (57).

4.3.3 Photo-additions to naphthalene, phenanthrene, etc.

Photo-addition of methyl cinnamate to 2-acetylnaphthalene gives adduct (59) in a reaction which although extremely slow is remarkably stereospecific in view of the rapid attainment of a *cis*:*trans* photo-stationary state of the addend[57]. Previous work[58] on the formation of adduct (60) from photo-addition of diphenylacetylene to naphthalene has been extended[59]. An intermediate exciplex derived from S_1 naphthalene and S_0 diphenylacetylene appears to be involved. The reaction is retarded by oxygen, and is completely inhibited by benzophenone. There are two photochemical steps: the first involves photo-excitation of naphthalene, and the second involves intramolecular cyclisation of the intermediate adduct (61). If this mechanism is correct, some sensitivity to proton-donors might be expected; but there is no information on this point.

Whereas the photo-addition of dimethyl acetylenedicarboxylate to benzene gives the corresponding cyclo-octatetraene as virtually the only product, the corresponding photo-addition of this acetylene to naphthalene follows a much more complex course[60]. The expected benzcyclo-octatetraene (62) is apparently not formed as a direct photo-adduct, but can be produced by secondary photo-isomerisation of the thermal 1:1 adduct (63) which tends to be produced in a slow concurrent reaction. The main photoproducts are the adducts (64)–(67). Adduct (67) arose from photochemical cyclisation of adduct (64), and was the major 1:1 adduct formed in methanol as solvent. The formation of adduct (66) may provide the first genuine example of the 1,3-photo-addition of an acetylene to an aromatic ring. It has been suggested[61] that adducts (64) and (65) (and hence adduct (67)) are being formed by proton transfer from methanol in a process analogous to that demonstrated for benzene and dimethyl acetylenedicarboxylate[42].

Photo-addition of acrylonitrile to naphthalene leads to the cyclo-adduct (68) together with the substitution products (69) and (70)[62, 63]. From com-

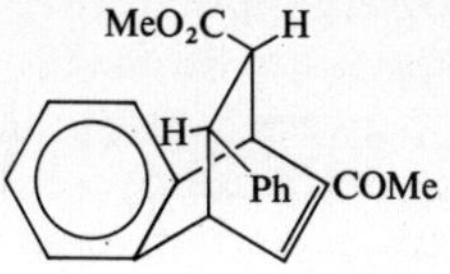

MeO$_2$C H H Ph COMe

(59)

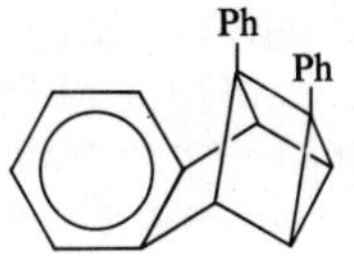

Ph Ph

(60)

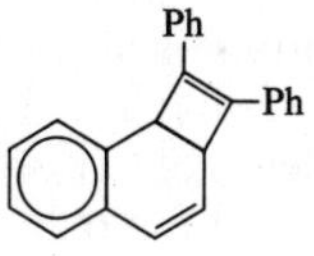

Ph Ph

(61)

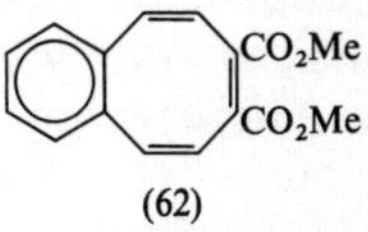

CO$_2$Me CO$_2$Me

(62)

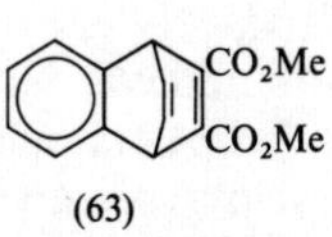

CO$_2$Me CO$_2$Me

(63)

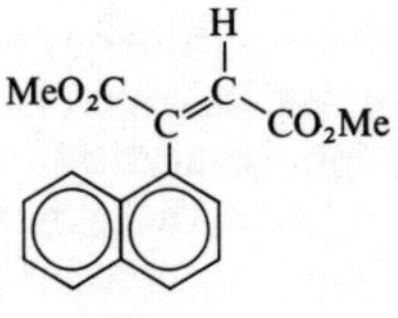

H MeO$_2$C C C CO$_2$Me

(64)

[65 = *cis*-(64)]

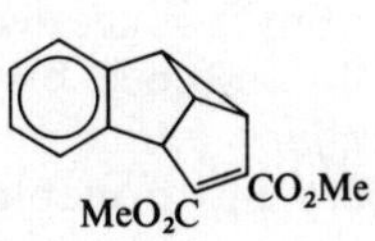

CO$_2$Me MeO$_2$C

(66)

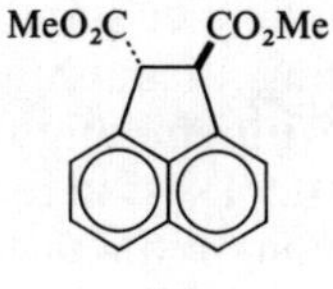

MeO$_2$C CO$_2$Me

(67)

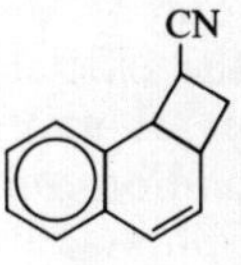

CN

(68)

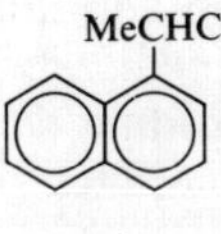

MeCHCN

(69)

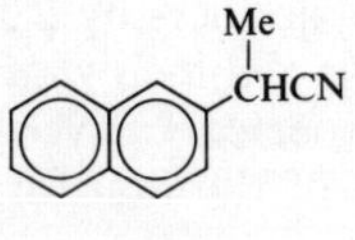

Me CHCN

(70)

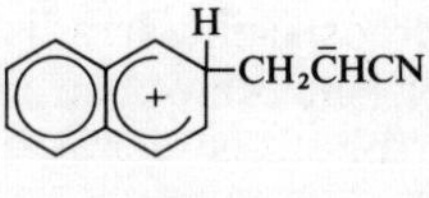

H + CH$_2\bar{C}$HCN

(71)

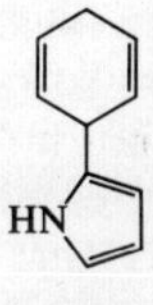

HN

(72)

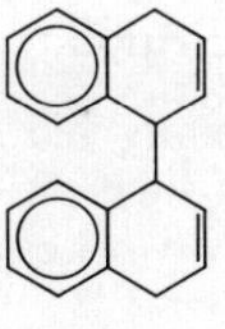

(73)

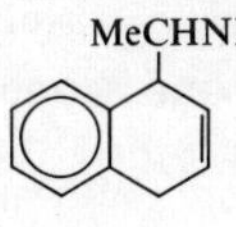

MeCHNEt$_2$

(74)

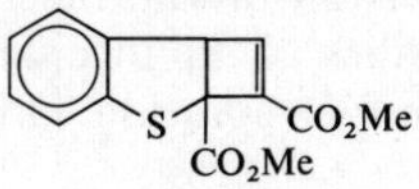

S CO$_2$Me CO$_2$Me

(75)

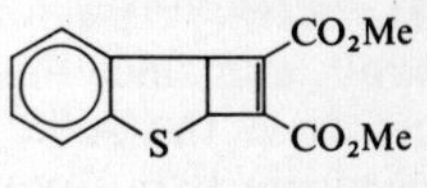

CO$_2$Me S CO$_2$Me

(76)

CO$_2$Me CO$_2$Me S·

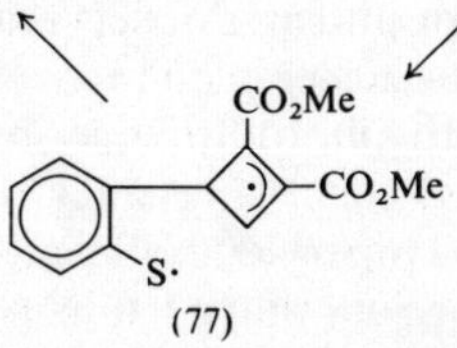

(77)

parison of the rate of quenching of naphthalene fluorescence by acrylonitrile with the quantum yield of cyclo-addition, it is concluded that the S_1 state of naphthalene is the reactive species. Solvent effects have provided evidence for the formation of an exciplex between S_1 naphthalene and S_0 acrylonitrile[63]. The substitution products seem to arise by proton transfer from solvent methanol to the exciplex, and deuterium labelling established that the transferred proton appears in the methyl group of the product. This last observation seems to rule out the involvement of a zwitterionic species (71) analogous to that proposed for the benzene/maleic anhydride system (42). The sensitivity to a proton donor implies the involvement of a polarised intermediate, but it would appear that the charge transfer from naphthalene to acrylonitrile within the exciplex is of a type which tends to localise negative charge on C-1 rather than C-2 of the nitrile. Possibly some form of π-complexing is involved.

Phenanthrene undergoes 9,10-photo-addition of acrylonitrile, dimethyl maleate, and *cis*- and *trans*-dichloroethylenes to give cyclobutane adducts[64]. The corresponding addition of maleic anhydride to phenanthrene has been known for some time[65]. The photo-addition of diphenylacetylene to the 9,10-position of phenanthrenes gives the expected cyclobutenes[66]. None of the products showed any tendency to isomerise to dibenzocyclo-octatetraenes.

Whereas the irradiation of benzene in the presence of pyrrole gives the 1,4-adduct (72)[67], the corresponding reaction with naphthalene leads to both 1,4- and 1,2-adducts[68]. A singlet state mechanism is suggested on the basis of the efficient quenching of naphthalene fluorescence by pyrrole. Irradiation of naphthalene with an aliphatic tertiary amine in methyl cyanide solution containing water leads to reduction to dihydronaphthalenes (a type of photo-Birch reduction), reductive dimerisation to compounds of type (73), and the formation of minor quantities of a 1:1 adduct (74) (from Et_3N)[69].

Olefins such as *cis*- and *trans*-dichloroethylenes undergo photosensitised addition to benzothiophen to give 1:1 adducts by 2,3-cyclo-addition. The reaction is not stereospecific. Irradiation of dimethyl acetylenedicarboxylate in the presence of benzothiophen unexpectedly gave the product (75), presumably through photo-isomerisation of the first formed adduct (76) via the diradical (77)[70].

4.3.4 Photo-dimerisation and related reactions of aromatic rings

The 9,10-, 9,10-photodimerisation of anthracene was reported many years ago[71], and since this time studies of the mechanism and scope of such processes in the solid phase and in solution have occupied many workers, particularly in France.

One of the more unusual observations in this area concerns 9-phenylanthracene which has earlier been reported to be photostable[72]. This anthracene is now reported to photodimerise, but only when piperylene is present[73]. This result is most surprising since anthracene dimerisation is normally a singlet process and piperylene efficiently quenches the fluorescence of anthracene. It begins to appear that a triplet pathway may be available for the photodimerisation of some anthracenes.

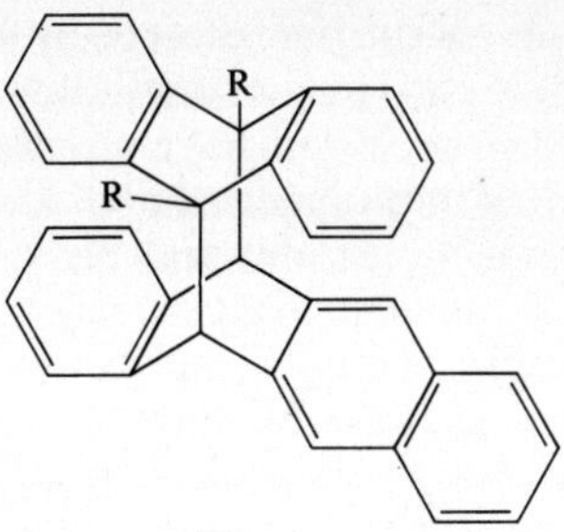

R
R

(78)

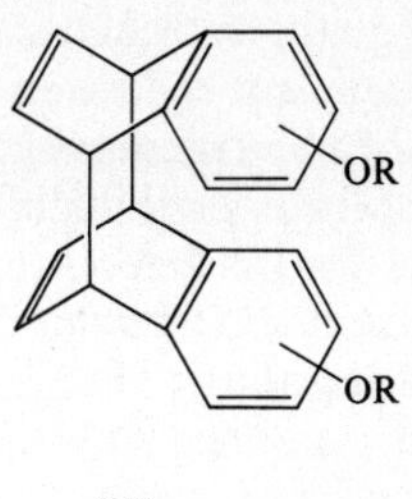

OR
OR

(79)

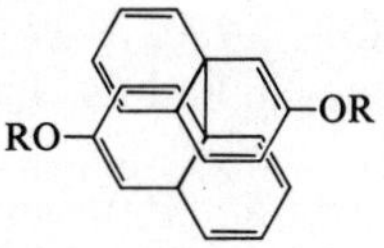

RO OR

(80)

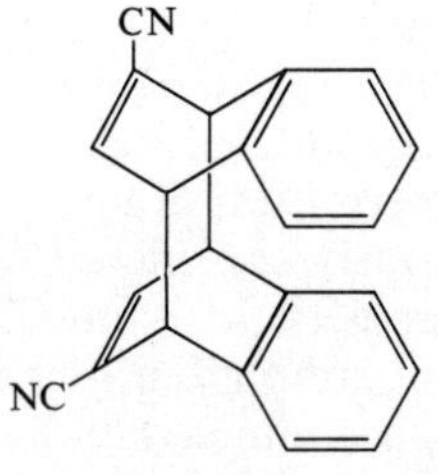

CN
NC

(81)

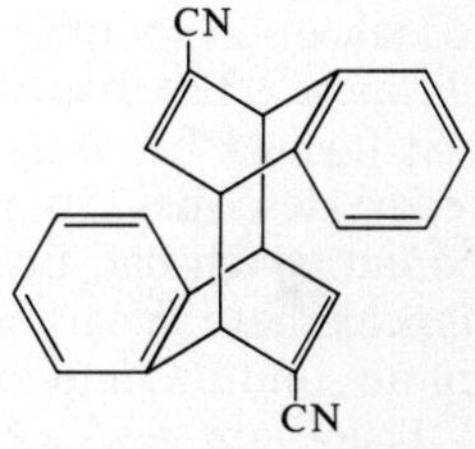

CN
CN

(82)

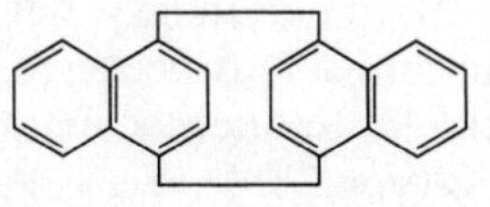

(83)

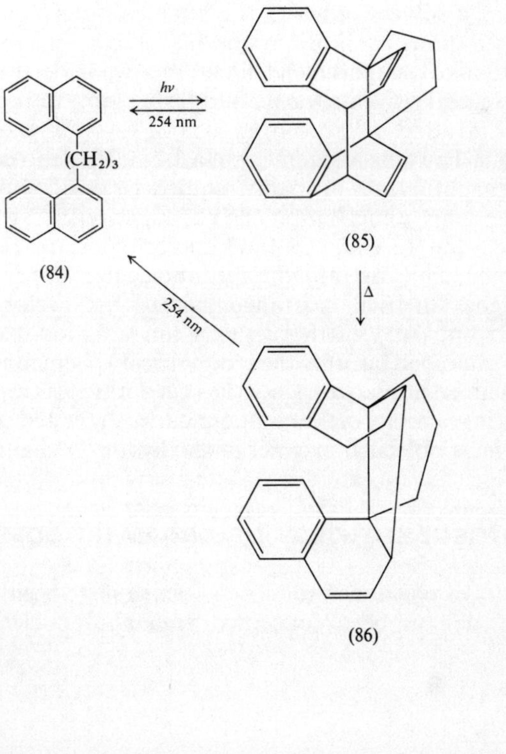

$h\nu$
254 nm

$(CH_2)_3$

(84)

(85)

Δ

254 nm

(86)

The effect of topochemical control in solid phase dimerisations has been explored with anthracenes. 9-Cyanoanthracene is unusual in forming a 'head-to-tail' dimer from a monomer lattice in which the molecules have a 'head-to-head' arrangement[74]. The idea of monomer rotation within ordered regions of the crystal is difficult to accept, so the photoreaction may well be centred at lattice defects or at crystal boundaries. For example, partial dislocations on (211) planes may produce a *trans* arrangement of molecules across contiguous planes and may thus account for the observed product.

'Mixed' dimerisations of anthracenes have been studied by several groups of workers. The electron donor 9,10-dimethylanthracene readily adds to the acceptor 9-cyanoanthracene[75], and although there is no spectroscopic evidence for complexation between the addends, both are known to form excimers[76]. The involvement of exciplexes is also suggested in the cross dimers (78) formed between tetracene and various anthracenes[77], and the same report describes the photodimerisation of 9,10-dimethylanthracene, a compound which had been previously considered to be photostable. It would appear that the earlier workers with 9,10-dimethylanthracene had used too dilute solutions and insufficient lamp intensities[78].

There are few reported photodimerisations of aromatic compounds other than those involving anthracene rings. In particular naphthalene does not appear to photodimerise, and until recently the only known dimers of naphthalenes were those of the 2-alkoxy derivatives[79,80]. In the original report, the structure of the dimer had been reasonably suggested to be (79), i.e. a 1,4-, 1,4-adduct[79], but subsequently evidence was offered which best fitted an adduct of 1,2-, 1,2-type (80)[80]. Very recently, the dimer of 2-cyanonaphthalene has been prepared following the realisation that crystal forces may contribute significantly to dimer stability[81]; the irradiation was performed using a solvent in which the dimer was only sparingly soluble, and the monomer–dimer equilibrium was displaced by precipitation of the dimer. Unlike the 2-alkoxynaphthalene dimer, this cyano derivative was sufficiently soluble in several solvents for its n.m.r. spectrum to be recorded. Only the two dimers (81) and (82), resulting from a 1,4-, 1,4-reaction, are consistent with the data. Thus the assignment of a 1,2-,1,2-type structure for the dimers of 2-alkoxynaphthalenes probably requires reconsideration.

Intramolecular 'dimerisation' is known to occur with the *para*-cyclophane (83)[82], and recently 1,3-bis(1′-naphthyl)propane (84) has been converted photochemically into the intramolecular 'dimer' (85)[83]. Compound (85) undergoes thermal rearrangement to the cyclobutane isomer (86), and both (85) and (86) yield (84) with 254 nm radiation. Various dinaphthylalkanes were examined but only the symmetrical 1,3-dinaphthylalkanes showed a strong 'excimer' interaction since they can form overlapping sandwich pairs. Differences in reactivity of the compounds are suggested to be due to stabilisation of the intramolecular excimer of (84) by the C_3 chain.

4.4 PHOTOSUBSTITUTION OF AROMATIC COMPOUNDS

The majority of reactions in this class which have been reported within the review period have been concerned with photonucleophilic substitution.

Photoelectrophilic substitution processes have been described which are pertinent to the mechanism of certain addition reactions, and are discussed in the section dealing with these. Reactions which simply involve attack of photochemically generated radicals on the aromatic nucleus are not considered here.

Despite published evidence for the involvement of excited singlet states of nitro aromatic compounds in their photonucleophilic substitution processes[84], it has been suggested that the multiplicity of the reactive state is still an open question and further attempts to sensitise such processes have been made[85]. The reaction chosen was that of *p*-nitroanisole with hydroxyl ion, sensitised by benzophenone. That the same products are formed in the same proportions from both the sensitised and unsensitised reaction strongly indicates that the triplet state of the nitroaromatic is the reactive species.

Both solvent and substituent groups may have a pronounced effect on the rate of these bimolecular photoreactions[86]. Such solvent effects are considered to resemble those observed in the photoreduction of aromatic ketones in that the nature of the lowest level ($n\pi^*$, $\pi\pi^*$, charge transfer) varies in different solvents. The substituent effects are demonstrated by the light-induced reactions of *p*-nitroanisole, 1-nitronaphthalene, and 4-methoxy-1-nitronaphthalene with cyanide ion. Since nitrobenzene is inert in this reaction, a clear activation for nucleophilic attack on the benzene ring is uncharacteristically shown by the methoxy group with the formation of 2-cyano-4-nitroanisole from nitroanisole. On the other hand, the methoxy group may also deactivate in photosubstitutions as evidenced by the photostability of 4-methoxy-1-nitronaphthalene towards cyanide ion in contrast with the ready formation of 1-cyanonaphthalene from 1-nitronaphthalene. Displacement of the nitro by the cyano group had not been previously reported. '*meta*-Activation', well known in photosolvolysis processes[87], has again been reported with 3,5-dimethoxybenzyl phosphate[88] and further demonstrated by the reaction of the three nitroanisoles with potassium cyanide[89].

Photosubstitution of the halogen in halogenated phenols, and aryl amines and ethers, by a nitro group has been reported by Russian workers. The reactions were carried out in 50% methanol containing sodium nitrite (4%), and yields of the order of 10–33% were obtained[90].

The effect of distant substituents on the photonucleophilic reaction has been investigated by studying the reactions of (87) and (88) with aqueous pyridine[91]. Electron donor substituents (X) markedly retard the process. In contrast to (88) two types of process were observed with (87). Path *a* is unusual in that the nucleophile attacks a phenyl ring which does not have a nitro substituent. Following on from this work, a preliminary investigation into intermolecular effects on photosubstitution has been made and the reaction of *p*-nitroanisole with hydroxyl ion has been found to be inhibited by 1,4-dimethoxybenzene. This interesting result should receive further attention.

Havinga and his co-workers are well known for their contributions to this area of photochemistry. They have now investigated substitution reactions with uncharged nucleophiles such as ammonia, and ammonia–methanol and ammonia–water[92]. In these reactions, it seems that aromatic compounds divide into two classes. The first class comprises nitroanisoles, and substitution of the methoxy group by amino occurs; the reaction is thus comparable with the photohydrolysis of these compounds. Nitrobenzene, dinitrobenzene, and chloronitrobenzene constitute the second class, and here substitution occurs at the *ortho* and *para* positions but not the *meta* position. These latter reactions are found to be more complicated than those of the first class and

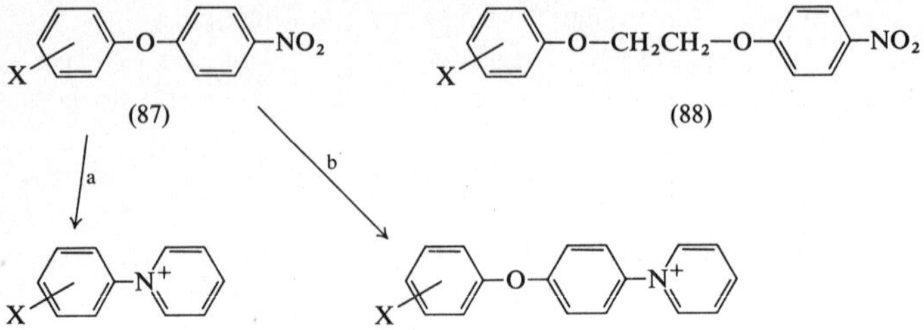

(87)

(88)

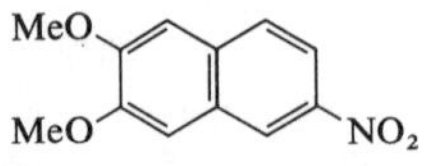

(89)

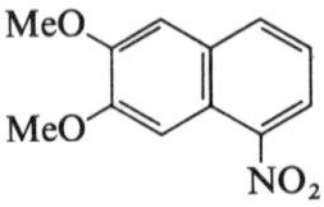

(90)

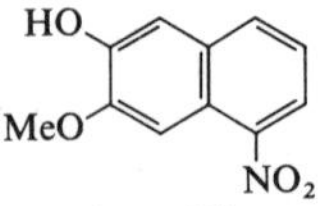

(91)

both singlet and triplet states are postulated as the starting levels for the amination process. That the *para* isomer is the major product from nitrobenzene shows that charge densities are not the sole factor for determining the *o*:*m*:*p* ratios of the product, particularly in the case of uncharged nucleophiles. It should be borne in mind that amines are known to undergo 1,4-photo-addition to the aromatic ring[93]. We suggest that a 'substitution' mechanism involving photo- or thermal oxidation of 1,4-adducts by nitro-compounds should be considered for the photoreactions of nitroaromatic compounds with ammonia and amines.

Photonucleophilic substitution and solvolysis reactions have been investigated with naphthalene derivatives (see reference 86 also), and a type of '*meta*-activation' has been observed with a series of methoxynitronaphthalenes in the replacement of methoxy by hydroxy[94]. In order to confirm this activation, the substitution products of (89) and (90) were examined. In both cases the methoxy group '*meta*' to nitro was replaced, i.e. 3-methoxy-7-nitronaphth-2-ol and 3-methoxy-5-nitro-naphth-2-ol(91) were obtained from (89) and (90) respectively. The nuclear Overhauser effect was used to assign structure (91) to the product from (90)[95].

Photohydrolysis of the three isomeric hydroxybenzotrifluorides is known to yield the corresponding hydroxybenzoic acids[96]. Seiler and Wirz have now examined the process with eight isomeric trifluoromethylnaphthols, thereby increasing the number of isomers such that theoretical prediction of the feasibility of the reaction and experimental observation are less subject to coincidence[97]. As in the previous work on naphthalene photosubstitutions, sensitisation and quenching data implicate the S_1 state of the aromatic. Only two of the isomers failed to give the desired hydroxy-acids, but with the other isomers, the correlation between the reaction rates and the calculated charge densities for the α- and β-naphtholate anion in the first excited singlet state is convincing.

Frato and Havinga have reported a new type of photosubstitution reaction in the naphthalene series[98]. Thus 1-nitronaphthalene in the presence of alkyl chlorides, or in mixtures of HCl with either chloroform, carbon tetrachloride, acetic acid or hexane give the chloro-substituted product. The quantum yields for the acid mixtures are *c.* 0.1–0.5, and in some cases the product yield of 80–90 % suggests that these reactions may be of synthetic utility. Nitrobenzenes did not give the desired product, and 2-nitronaphthalenes reacted only slowly to give unidentified products and low yields of the

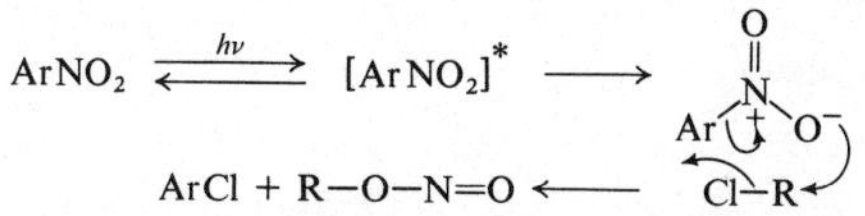

$$ArNO_2 \underset{}{\overset{h\nu}{\rightleftarrows}} [ArNO_2]^* \longrightarrow \text{Ar}\overset{+}{\text{N}}(=O)O^- \cdots Cl{-}R$$

$$ArCl + R{-}O{-}N{=}O \longleftarrow$$

Figure 4.1

chloro-compound. In contrast with the previous studies on photosubstitution of naphthalenes, there is evidence that T_1 naphthalene is involved in this reaction. The authors consider the present process to be of a new concerted type, as shown schematically in Figure 4.1.

Light-induced substitution reactions of 1-nitroazulene have also been reported and while the nitro group is seemingly readily replaced by methoxy, and to a lesser extent by cyano, other nucleophiles (hydroxy, pyridine and cyanate) which undergo reaction with benzenoid compounds, were surprisingly unreactive in the azulene system[99].

4.5 FORMATION OF POLYCYCLIC AROMATIC SYSTEMS BY INTRAMOLECULAR PHOTOCYCLISATION REACTIONS

The formation of phenanthrenes from the irradiation of stilbene derivatives in the presence of an oxidising agent has been known for many years[100]. Each year more and more examples of this general process are reported and together with oxidative cyclisation reactions in which the phenyl groups are separated by moieties other than ethylene, this area is one of the most extensively studied in aromatic photochemistry.

But not all stilbenes undergo this process. Two groups of workers have pointed out that for reaction to occur, the sum of the free-valence indices (ΣF^*) for the first excited state at atoms between which the new bond is formed, must be greater than unity[101, 102]. Other factors are of course important, and stilbenes substituted by groups such as acetyl and nitro which promote intersystem crossing do not form phenanthrenes[101]. Similarly, steric factors are important[103]. If more than one cyclisation route is available, then that with the highest ΣF^* is followed, and when $\Delta\Sigma F^*$ is smaller than 0.1, two products are occasionally formed. Such considerations only hold if no planar or non-planar products can be formed and in cases where both planar and non-planar compounds may be produced, the major product is usually the planar one, even if ΣF^* for its formation is lower than for other cyclisations. The applicability of these considerations in planning synthetic routes to polycyclic aromatic compounds has been illustrated by the preparation of several new compounds, for example the four isomeric benzohexahelicenes. A double hexahelicene (92) has also been prepared by this cyclisation route from 2,6-bis-(3-phenanthrylethenyl)naphthalene (93)[104].

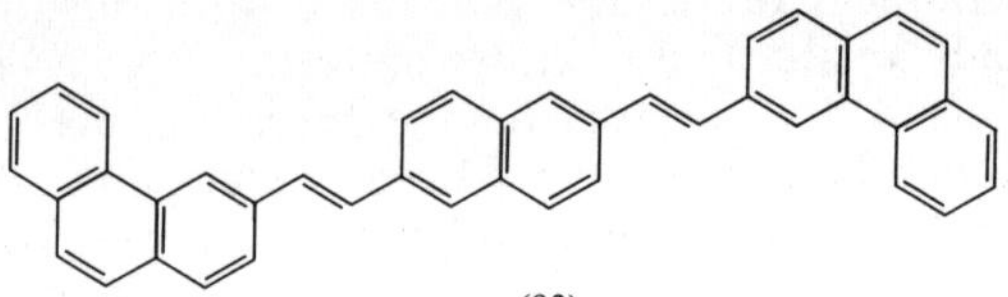

(93)

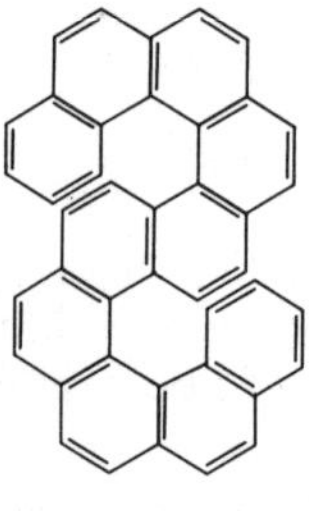

(92)

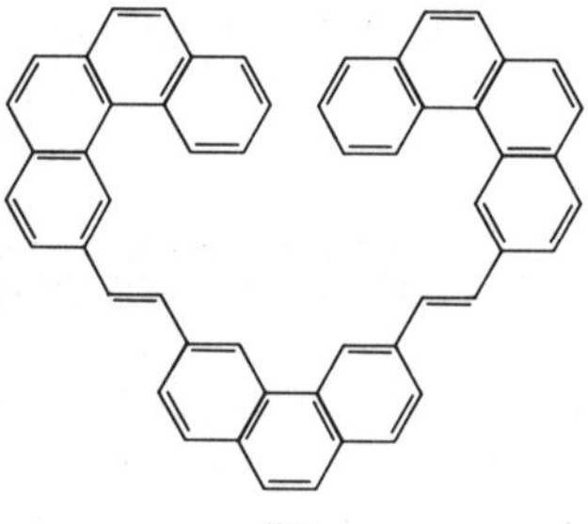

(95)

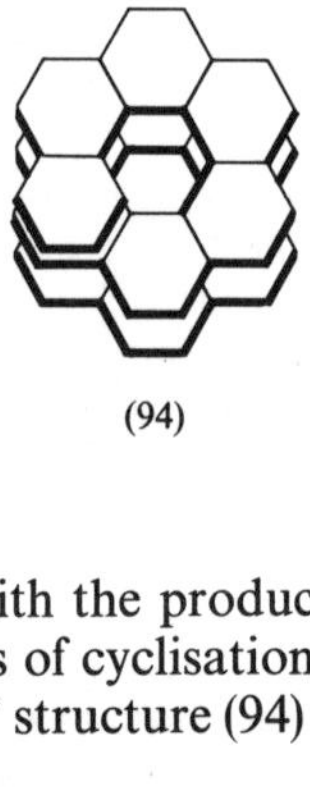

(94)

Other workers have used mass spectrometry with the products of partially deuterated starting materials to assign positions of cyclisation in such complex examples and have thus assigned the $\alpha\alpha'/\alpha\alpha'$ structure (94) to the cyclisation product of (95)[105].

Although the conversion of stilbene to phenanthrene is reasonably considered to proceed via the 4a,4b-dihydrophenanthrene, isolation of the intermediate has never been achieved. Timmons and co-workers have attempted to form such intermediates from sterically hindered naphthalenes[106]. The most stable dihydrophenanthrene observed in their studies was found to be that of (96) from 1,2-di-(2-naphthyl)ethylene. This is readily understood since dibenzo [*c*, *g*] phenanthrene (97), the expected product of the cyclodehydrogenation process, has considerable steric overcrowding and loss in delocalisation energy due to distortion of the ring system from planarity. Conversion of (96) to (97) is slow, and (97) undergoes further oxidative cyclisation of a type common in such systems to yield benzo [g, h, i] perylene (98). The same workers also report cyclisation of compounds (99) in which the positions of attack are blocked. In no case was migration of the substituent observed but elimination occurred preferentially to yield the expected chrysene derivative.

Some success has been achieved in the isolation of a dihydro-intermediate, but the example is rather specialised and owes its stability to formation of the stable diketone (100). A report in 1961 had described the formation of (100)

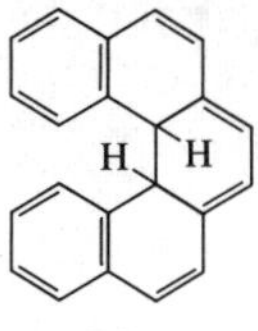

(96) 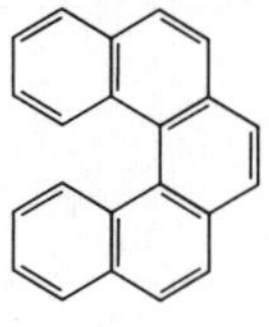 (97) 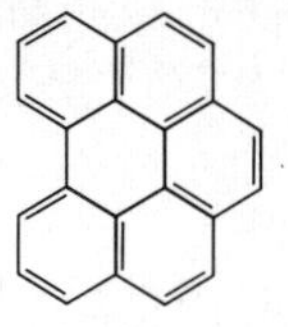 (98)

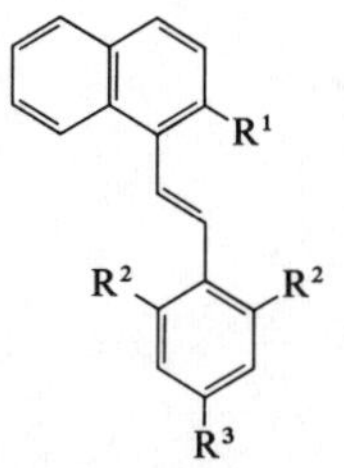

(99) R's = H, Me or Cl

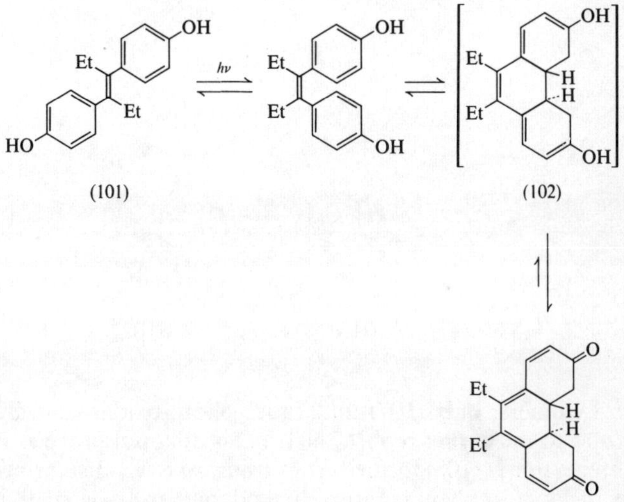

(101) (102)

(100)

from *trans*-αα′-diethyl-4,4′-stilbenediol (101)[107], but now analytical evidence for (100) is offered, and the mechanism of its formation has been reformulated in terms of the intermediate dihydrophenanthrene (102)[108].

The photocyclisation possibilities from the three distyrylbenzenes are many. Laarhoven and co-workers have examined these systems and in a most comprehensive report have related the many observed products to the sum of the free valence indices for each particular reaction[102]. Formation of cyclobutane dimers and regeneration of isomeric monomers are postulated to account for the results. The complexity of such processes is here illustrated by reference to the irradiation of 1,4-distyrylbenzene (Figure 4.2). The reaction with the 1,3-distyryl isomer has been further investigated by other workers who have carried out a detailed analysis of the products and their relation to each other[109]. Compounds (103), (104), (105), and (106) were observed in the reaction, and (107), although not detected, is clearly implicated in the process by the formation of (106). It is suggested that of the

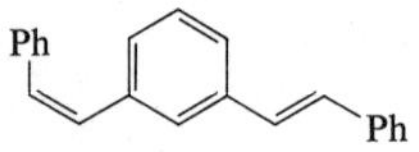

(103) 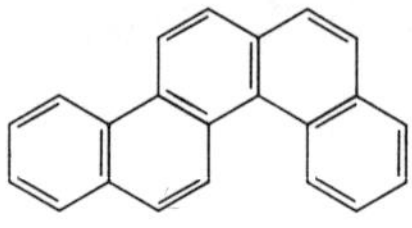 (104)

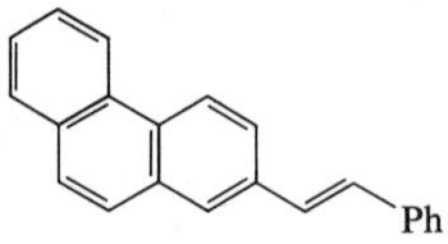

(105) 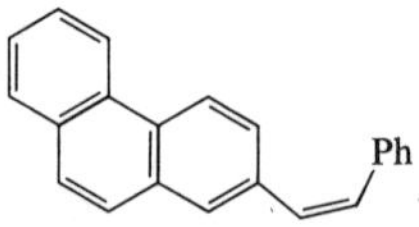 (106)

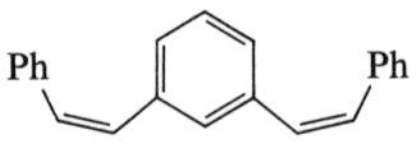

(107) 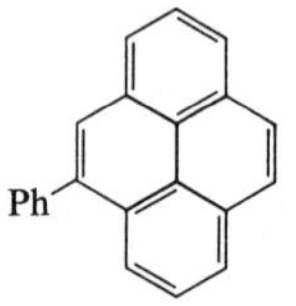 (108)

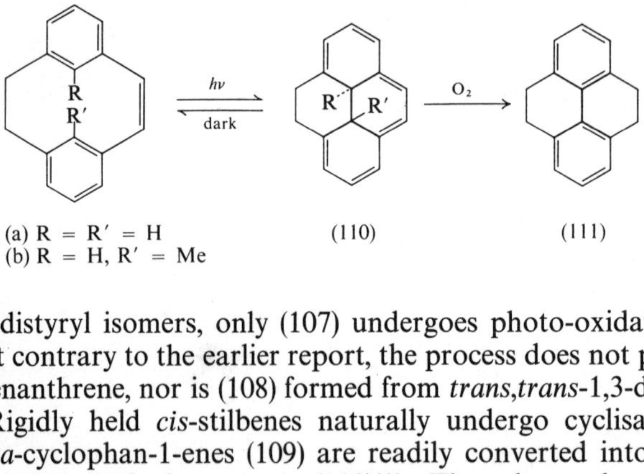

(109) (a) R = R′ = H
(b) R = H, R′ = Me

(110) (111) (112)

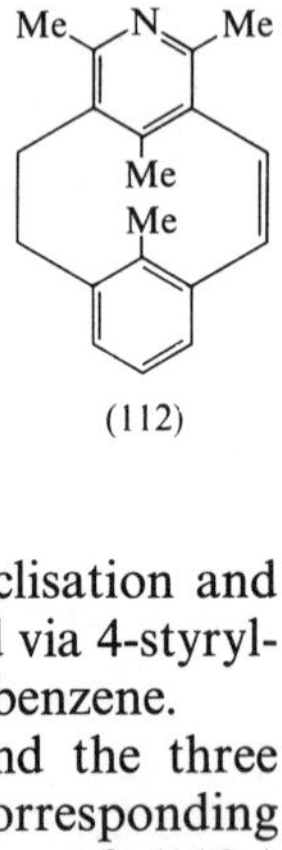

1,3-distyryl isomers, only (107) undergoes photo-oxidative cyclisation and that contrary to the earlier report, the process does not proceed via 4-styrylphenanthrene, nor is (108) formed from *trans,trans*-1,3-distyrylbenzene.

Rigidly held *cis*-stilbenes naturally undergo cyclisation and the three *meta*-cyclophan-1-enes (109) are readily converted into the corresponding 4,5,15,16-tetrahydropyrenes (110)[110]. The observed sensitivity of (110a) to oxygen is not surprising, and 4,5-dihydropyrene results; but surprisingly the monomethyl compound (110b) is almost as sensitive as (110a) to oxygen and loses its methyl group by an unknown mechanism to yield (111). The dimethyl compound is stable and does not undergo further reaction[111].

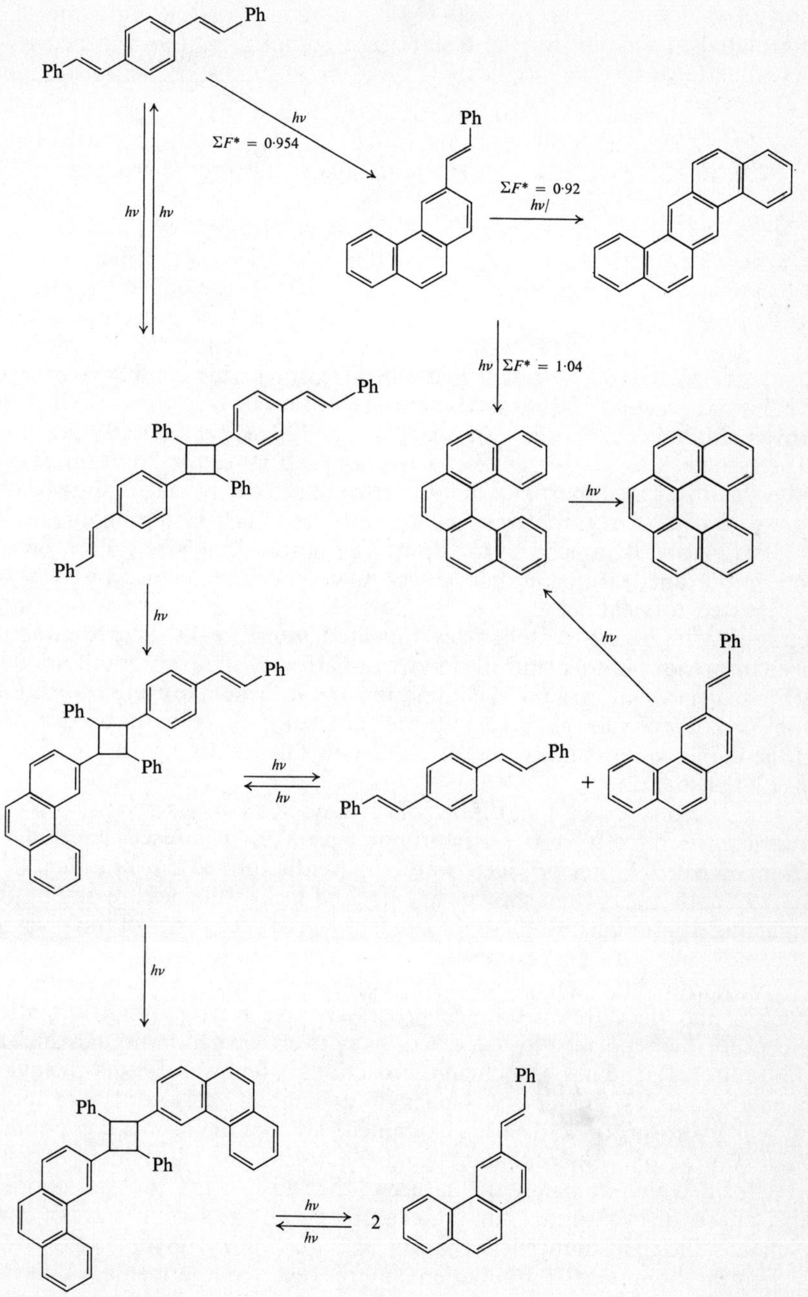

Ph Ph hν ΣF* = 0·954 Ph ΣF* = 0·92 hν/ hν hν hν ΣF* = 1·04 Ph Ph Ph Ph hν hν hν Ph Ph Ph Ph hν hν Ph + hν Ph Ph Ph hν hν 2

Figure 4.2

Such reversible cyclisation has also been observed with the hetero-compound (112)[112].

This general photo-oxidative cyclisation has, contrary to an earlier report[113], been observed with hetero-aromatic compounds, and both 2- and 4-styrylpyridines yield the appropriate benzoquinolines. Various substituted compounds (113) have been examined in this respect, and in most cases the yields of the substituted benzo[*f*]quinolines are useful synthetically[114]. In the case of (114), both phenyl-substituted benzoquinoline and pyridylphenanthrene are formed[115].

Such reactions with heterocyclic compounds has been used as a key step in a new synthetic route to the aporphine alkaloids, nuciferine (115) and glaucine (116)[116]. The products from (117) and (118) are converted into the alkaloids in a two-step reduction. But attempts to use the process as a route to the ergoline ring system (119) in ergot alkaloids failed and Kumler and Dybas were led to investigate the structural features which inhibit or prevent the desired reaction[117]. It appears that cyclisation does not occur when the ethylenic bond is exocyclic to a 5-ring, as in (120), (121) or (122); but other processes such as *cis* → *trans* isomerisation and ethylenic bond migration occur readily. The importance of the constraints of the 5-ring are thus clearly shown and further emphasised by the fact that such 6-ring analogues as (123), (124) and (125) lead to the desired products. Ring size cannot be the sole important feature in such cases, however, since irradiation of (126) is reported to yield the cyclised product (127)[118]. It is known that neither tetraphenylfuran nor tetraphenylcyclopentadienone yield the corresponding phenanthrenes photochemically; but irradiation of the *o*-chloranil adducts of these dienes leads to a novel elimination and stilbene-like cyclisation under non-oxidative conditions, giving the phenanthrenes[119]. Again the unreactivity of the dienes is not simply due to the five-membered ring since irradiation of (128) yields (129)[120].

It has been reported that diphenylacetylene in alcoholic solvents may be transformed directly into phenanthrene via a light-induced internal re-arrangement[121]. Other products were apparently formed but not identified. Subsequently it has been shown that *cis*- and *trans*-stilbenes are among the unidentified products, so these may in fact be the precursors of the phenanthrene[122].

The photocyclisation of 1,4-diarylbutadienes was briefly described in 1962[123], and since this time such processes have received scant attention except for the reported closure reactions of overcrowded molecules such as (130) and (131)[124]. The photochemical reactions of both substituted-phenyl-4-phenylbutadienes and 1-aryl-4-phenylbutadienes (where aryl is a poly-nuclear hydrocarbon) have been examined[125]. From the former compounds, 1-substituted phenylnaphthalenes and 7-substituted-1-phenylnaphthalenes were formed and seemingly the nature of the substituents does not appreciably affect the cyclisation. Among examples with the second class of compound is the light-induced formation of 3,4,8,9-dibenzopyrene (132) from 1-(9′-anthryl)-4-phenyl-1,3-butadiene; in this case, the intermediacy of benzo [10] annulene (133) is suggested.

The light-induced formation of triphenylenes from *o*-terphenyls in the presence of iodine is a very well-known process[126]. The electronic effects of

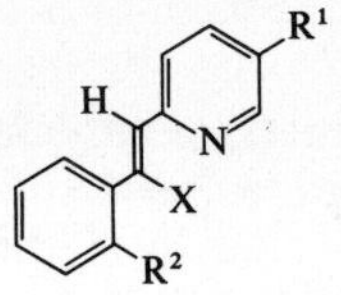

(113) X = H, CN, or CO_2Me
R^1 = H or OMe
R^2 = H, NO_2, or NHCOMe

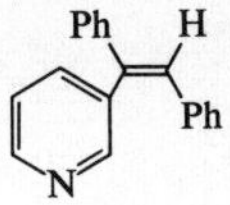

(114)

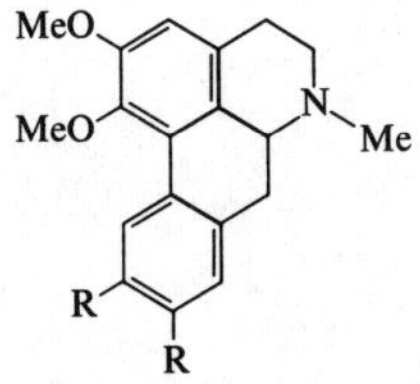

(115) R = H
(116) R = OMe

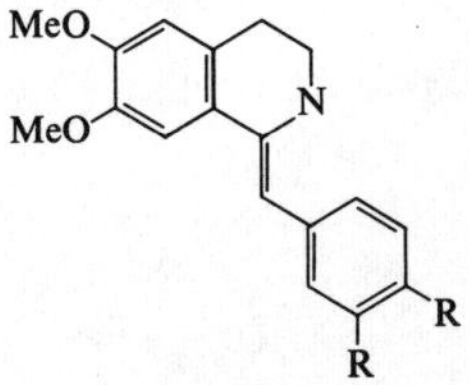

(117) R = H
(118) R = OMe

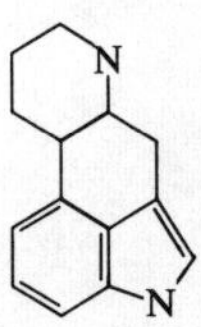

(119)

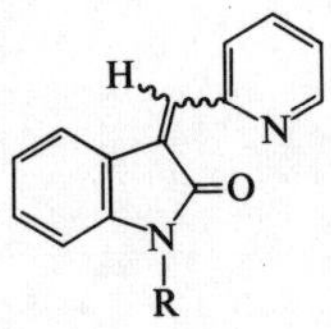

(120)

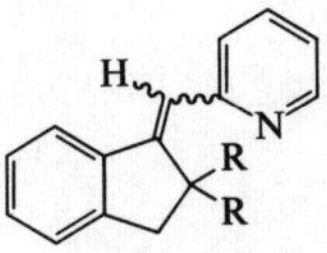

(121) R = H
(122) R = Me

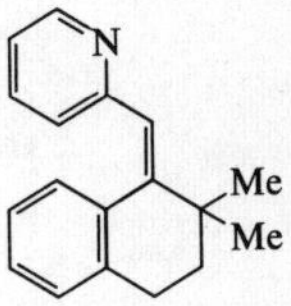

(123)

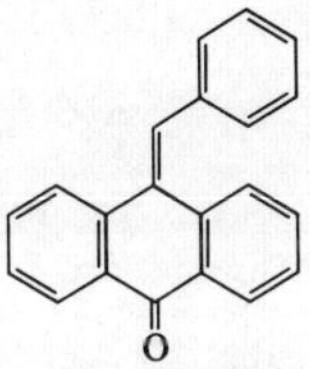

(124)

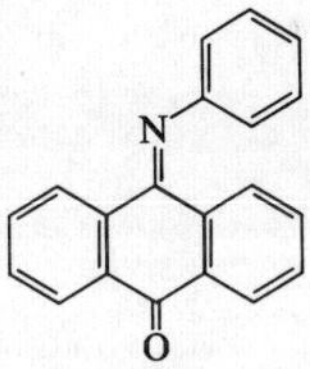

(125)

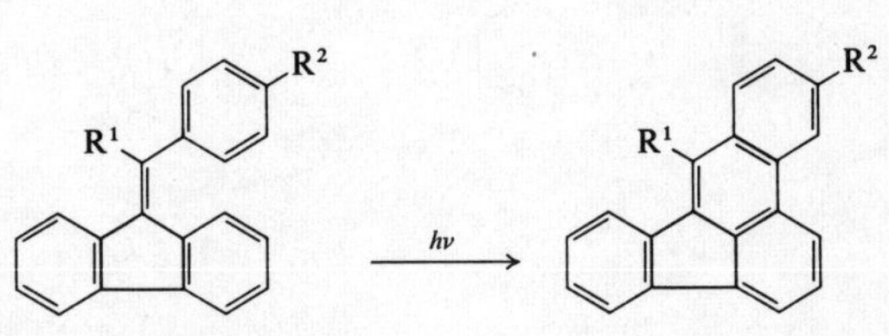

R^1, R^2 = H, Ph, Cl, OMe
(126)

$h\nu$

(127)

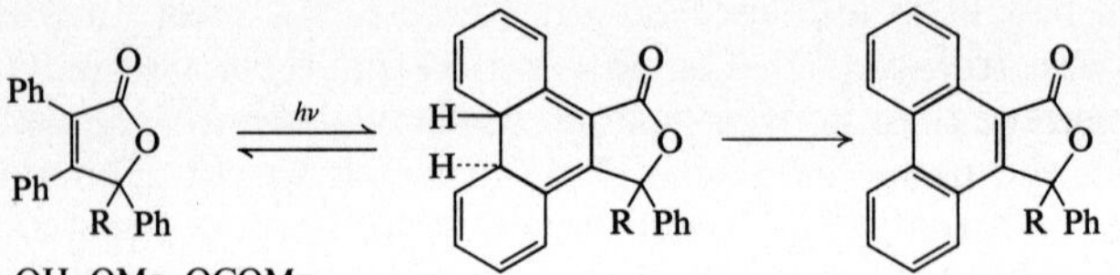

Ph Ph R Ph $\overset{h\nu}{\rightleftharpoons}$ H H R Ph → R Ph

R = H, OH, OMe, OCOMe

(128)

(129)

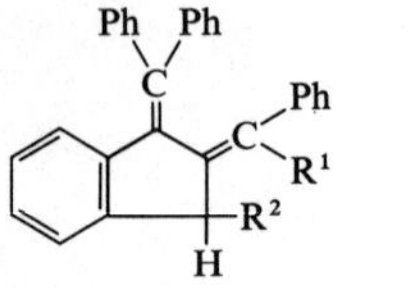

(130) $R^1 = R^2 = H$
(131) $R^1 = Ph$, $R^2 = Me$

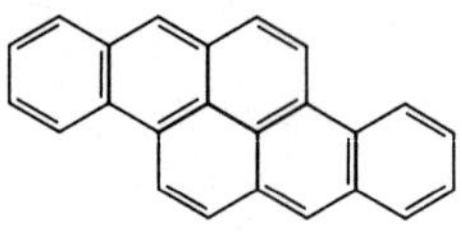

(132)

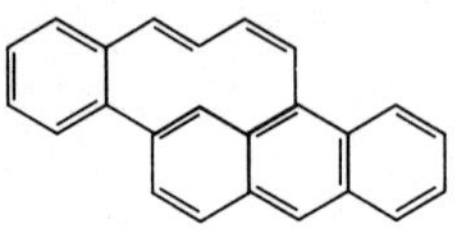

(133)

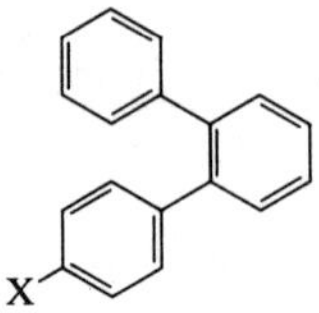

(134)

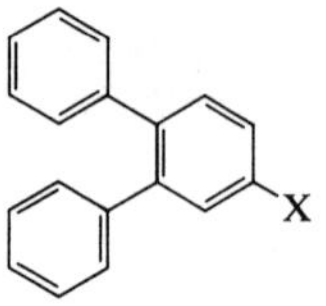

(135)

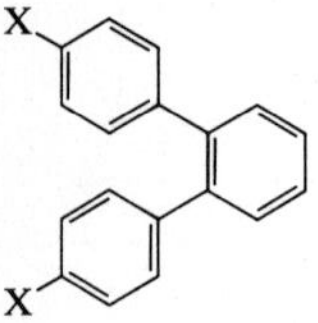

(136)

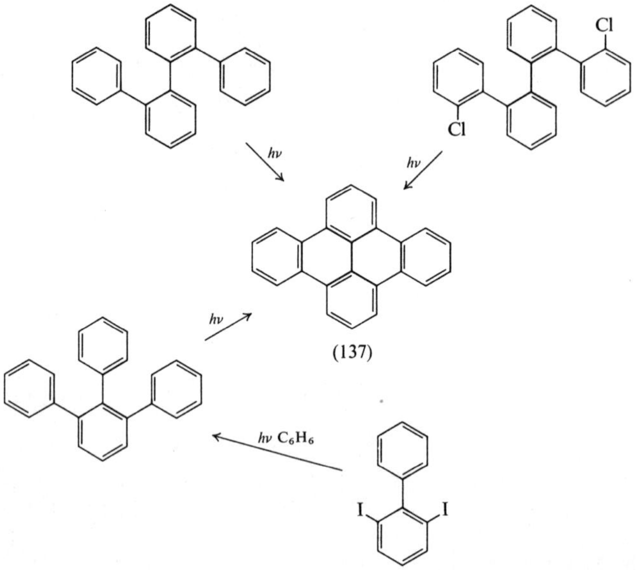

(137)

$h\nu$ C_6H_6

substituents on the process have been studied by reference to the photochemistry of (134), (135) and (136)[127]. Methoxy, fluoro, chloro and bromo substituents had little influence on the process, but electron-withdrawing groups had a retarding effect. The observation that the reaction occurs more readily in benzene than in cyclohexane led the workers to suggest aromatic sensitisation; the necessary study of the wavelength effect on the process was, however, not reported. One of the many examples of this process is the formation of dibenzo[*fg*, *op*]naphthacene (137), and the variety of starting materials used clearly demonstrates the versatility of the process[128].

Other cyclisation reactions which are noteworthy are that observed with the sodium enolate of diethyl *o*-biphenylyl malonate (138)[129], and the formation of *trans*-1,2,3,4,4a,5,6,10β-octahydrophenanthridines (139) by the stereoselective closure of cyclohexanone *N*-benzoyl enamines (140)[130].

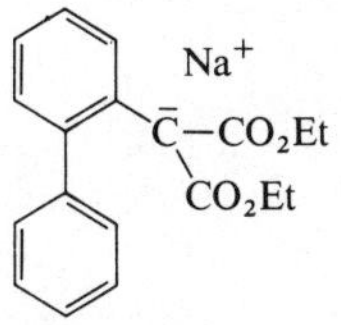

(138) 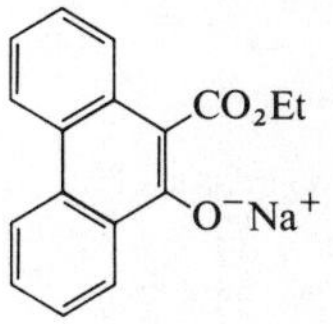 (139)

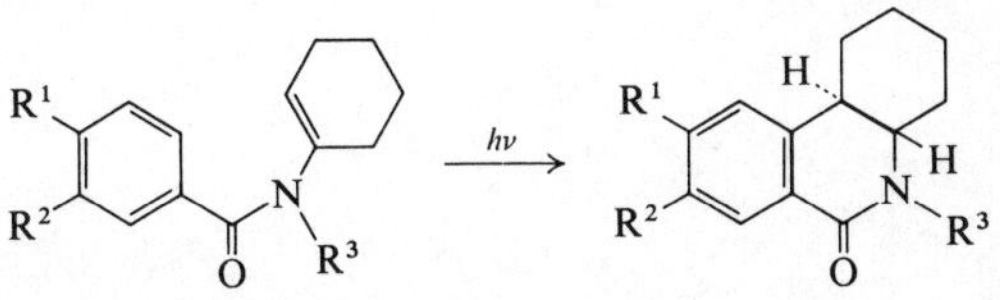

(140)

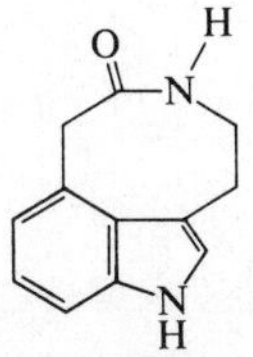

(141) 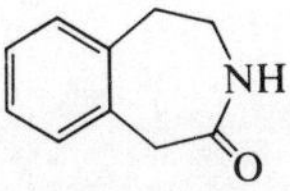 (142) 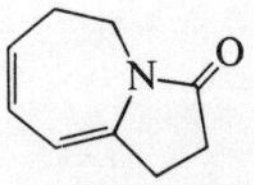 (143)

Compound (139) has the basic structure of a variety of alkaloids and many pharmacologically active compounds, so the reaction may prove interesting to pharmaceutical chemists.

Cyclisation with the loss of the elements of halogen hydracid is a common reaction and has been observed with a variety of systems (see for example reference 131). Witkop and co-workers have worked for several years in this area and have investigated the ring closure reactions which arise from the photolysis of *N*-chloroacetyl derivatives of a variety of pharmacodynamic amines and amino acids. Typical products of such processes are tricyclic indoles (141), benzazepinones (142), and azazulenes (143). A summary of their results with *N*-chloroacetyl-3,4-dimethoxyphenylethylamine (144) is presented schematically in Figure 4.3[132]. Compounds (145), (146) and (147) are formed in a H_2O–EtOH system, whereas in tetrahydrofuran, the major product is (148). A cyclisation process of such compounds in the naphthalene series has also been noted[133]. The mechanism of these reactions has been discussed by Hammond and Witkop and their co-workers[134]. It is suggested that the products arise from a common precursor, probably the aromatic S_1 state, in which there is strong coupling with the C—Cl bond. The process is thus suggested to be initiated by partial electron transfer in which the excited aromatic moiety is the donor and the chlorinated amide is the acceptor. Consistent with this proposal, the room-temperature fluorescence of *N*-acetyltryptamine is *c*. 15 times that of the *N*-chloro-derivative and the fluorescence of (144) is considerably less than the unchlorinated compound. Such

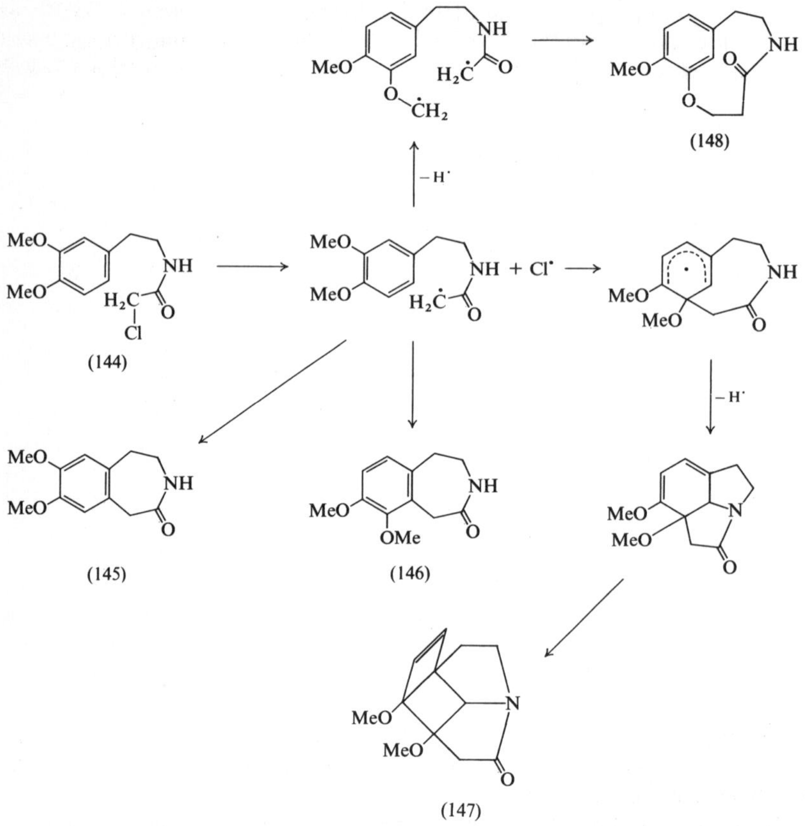

Figure 4.3

fluorescence quenching also occurs intermolecularly and that of electron-rich aromatic compounds is quenched efficiently by methyl chloroacetate and chloroacetamide. The fluorescence of 1,4-dimethoxybenzene is similarly quenched by benzyl chloride and allyl chloride. Thus the S_1 state of the aromatic is being quenched by compounds having no low lying excited singlet states. Charge-transfer seems to be involved: the failure of ethyl acetate and acetamide to quench the fluorescence is consistent with this idea. An intermolecular analogue of the above photoreaction has been observed using anisole and chloroacetamide, and 1,4-dimethoxybenzene and methyl chloroacetate[135].

4.6 LATERAL-NUCLEAR REARRANGEMENTS

4.6.1 Photo-Fries rearrangement

Although the photo-Fries reaction has been known for a number of years, certain aspects of the mechanism are still under discussion. With *p*-tolyl acetates, it is considered that the process occurs from an excited singlet state, or from a short-lived triplet, since common triplet quenchers and sensitisers have no effect[136]. Such results were confirmed with phenyl benzoate but in this case it was felt that the energy of the ketonic sensitiser was probably insufficient for efficient energy transfer to occur to the benzoate ($T > 74$ kcal mol^{-1})[137]. Consistent with this was sensitisation of the rearrangement by triphenylene which has a triplet energy of 81 kcal mol^{-1}. Since both the unsensitised and the sensitised processes yielded the same product distribution, it is tempting to suggest that triplet intermediates are involved in both reactions. The lack of triplet quenching of the process, however, requires comment but may be explained if the rate of intramolecular rearrangement is sufficiently close to the rate of diffusion to preclude quenching at 'normal' quencher concentrations.

There has been much discussion in the past as to whether the photo-Fries is a radical reaction occurring in a solvent cage and phenol production results from radical escape from the cage, or whether the process is a concerted rearrangement and formation of the phenol involves a separate side-reaction. Evidence for or against such mechanisms has largely depended upon the effects of solvent viscosity and polarity on quantum yields[138]. Meyer and Hammond have examined the reaction of phenyl acetate in the vapour phase where solvent cages are not possible[139]. Six products, the major (65%) being phenol, were formed: no *o*-, and *p*-hydroxyacetophenones, which are the main products in solution, were detected. Thus, at least in the vapour phase, a concerted mechanism does not operate and also the absence of toluene and anisole shows that processes of decarboxylation and decarbonylation do not occur. Similar results have also been reported for acetanilide in the vapour phase and from work with the same compound in matrices at -196 °C and in PVC sheet, the importance of a solvent cage in the reaction has been further emphasised[140].

Chalcones and their 2-hydroxy-substituted derivatives occur widely in plants and are precursors of flavanoid and isoflavanoid pigments. Thus the reports of formation of such systems via a photo-Fries reaction from phenyl cinnamates is of great interest[141]. The process was studied with the esters of phenol, resorcinol, guaiacol, or phloroglucinol with cinnamic acid or its 2-hydroxy, 4-hydroxy, 4-methoxy, 3,4-dihydroxy, 3-methoxy-4-hydroxy and 3,4-dimethoxy derivatives, and thereby a variety of chalcones having the complex substitution patterns found in naturally occurring compounds, has been synthesised[142]. The same workers have also shown that the 254 nm radiation of benzene solutions of phenyl epoxycinnamate leads to the hitherto unknown 2′-hydroxyepoxy chalcone (149) as a primary product[143].

The photo-Fries reaction has been studied for the three pyridyl benzoates: the reaction proceeds smoothly except in the case of the 4-isomer when benzoic acid is the product[144]. For the first time the rearrangement of aromatic dicarboxylic esters has been observed[145]. Thus (150) undergoes two consecutive rearrangements, and as expected the phenol (151) is also formed.

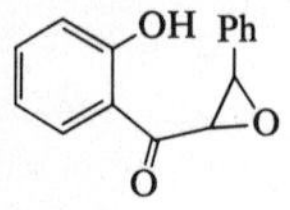

(149)

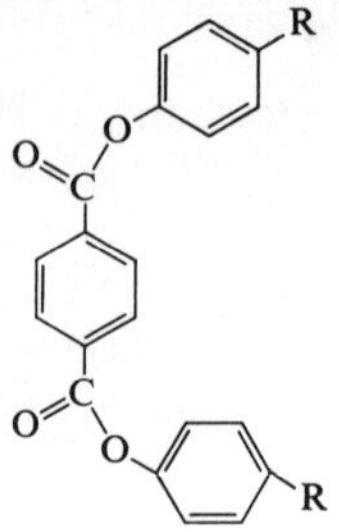

(150) R = $-C(Me)_2-CH_2-C(Me)_2-Me$

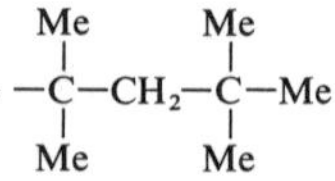

(151) R = $-C(Me)_2-CH_2-C(Me)_2-Me$

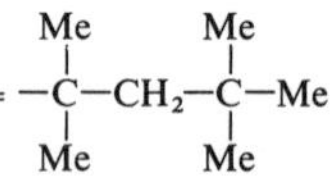

Ph—S—S—Ph

(152)

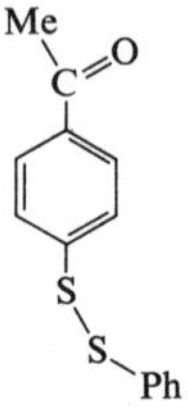

(153)

Although *p*-tolyl thiolacetate is photolabile, it does not undergo a light-induced photo-Fries rearrangement[146]. In contrast, a recent report describes the formation of Fries products from the irradiation of phenyl thiolacetate; thiophenol, methyl phenyl sulphide, (152), and (153) were also formed[147].

4.6.2 Photorearrangements of aryl ethers

The excited states involved in the formation of phenols and rearrangement products from the irradiation of aryl allyl ethers have previously received scant attention. Following investigations into the effects of solvents, sensitisers, and quenchers on the process, the direct irradiation has been concluded to proceed via the S_1 state of the ether[148]. In the sensitised process, however, carbonyl and aromatic hydrocarbon sensitisers were found to behave in different ways. The former are suggested to initiate a triplet reaction of the ether which is excited at the olefinic moiety, and they lead to increased phenol formation, whereas the aromatic hydrocarbon transfers singlet energy and the phenyl moiety of the ether is excited giving rise to the same distribution of products as in the unsensitised reaction. Simple alkyl aryl[149] and diaryl[150] ethers also undergo light-induced lateral-nuclear migration to yield substituted phenols by an intramolecular process involving a solvent cage and the formation of 2-(*p*-tolyl)-4-methylphenol from *p*-ditolyl ether adds another example to the list of light-induced [1,3]-sigmatropic shifts[151].

4.6.3 Photo-benzidine rearrangement

Hydrazobenzenes undergo dehydrogenation and reductive scission when irradiated in solution. They do not appear to rearrange, although trace amounts of *o*- and *p*-semidine have been detected from irradiation of hydrazobenzene[152]. In contrast, *N*,*N*′-dimethylhydrazobenzene gives *N*,*N*′-dimethyl-*p*-semidine[152] while the corresponding *o*-semidines are obtained in good yield from *N*,*N*′-dimethylhydrazo-*p*-toluene, -hydrazoanisole, and -hydrazobiphenyl[154].

4.6.4 Photo-Hofmann–Martius rearrangement (N-Alkylanilines)

The first report of this photo-rearrangement describes rearrangements of *N*-phenylbenzylamine into *o*- (29%) and *p*-aminodiphenylmethane (14%), and of *N*-α-phenylethylaniline (154) into *o*- (26%) and *p*-α-phenylethylaniline (8%). Use of optically-active (154) gave products having some optical activity. A radical-pair pathway was proposed, in which recombination within and outside of the solvent cage occurred[155].

4.6.5 Photo-Wallach rearrangement

This is a convenient but not accurate title. Irradiation of azoxybenzenes gives *o*- rather than *p*-hydroxyazobenzenes. The rearrangement has been reviewed recently[156]. It has also been found that while rearrangement of *trans*-azoxybenzene is almost quantitative in methanol solution, none occurs in cyclohexane solution. It is thought that diphenyloxadiaziridine is formed instead and, itself, undergoes subsequent photodecomposition[157].

4.6.6 Photo-Smiles rearrangement

The first report of such rearrangements (e.g. of (134) to (155)) has been made[158]. Yields were not given.

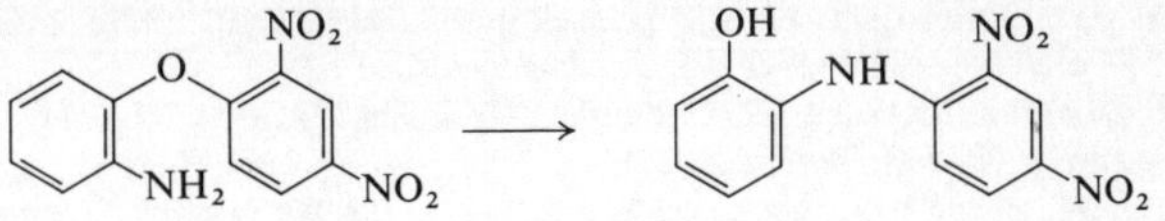

(154) → (155)

4.6.7 Summary

Research into the mechanisms of all of these photochemical rearrangements, except the Smiles, has been carried out. All of the rearrangements appear to involve caged radicals, although concerted rearrangement has also been proposed. The rearrangements are influenced by solvent polarity and by lability of hydrogen atoms in the solvent molecules. Side reactions occur in all of the rearrangements and these are attributable to differences in excited states (e.g. photo-Wallach) or hydrogen abstraction reactions leading to phenols (e.g. aryl ethers).

References

1. Blair, J. M. and Bryce-Smith, D. (1957). *Proc. Chem. Soc.*, 287; Angus, H. J. F., Blair, J. M. and Bryce-Smith, D. (1960). *J. Chem. Soc.*, 2003; Kaplan, L., Walch, S. P. and Wilzbach, K. E. (1968). *J. Amer. Chem. Soc.*, **90,** 5646; Ward, H. R. and Wishnok, J. S. *J. Amer. Chem. Soc.*, **90,** 5347; Ward, H. R. and Barta, C. I.; *J. Amer. Chem. Soc.*, **90,** 5353
2. Wilzbach, K. E., Ritscher, J. S. and Kaplan, L. (1967). *J. Amer. Chem. Soc.*, **89,** 1031; Kaplan, L. and Wilzbach, K. E. (1968). *J. Amer. Chem. Soc.*, **90,** 3291; Wilzbach, K. E., Harkness, A. L. and Kaplan, L. *J. Amer. Chem. Soc.*, **90,** 1116
3. (a) Ward, H. R. and Wishnok, J. S. (1968). *J. Amer. Chem. Soc.*, **90,** 5353. (b) Bryce-Smith, D., Gilbert, A. and Robinson, D. A. (1971). *Angew. Chem.*, **83,** 803. (c) Camaggi, C., Gozzo, F. and Cevidalli, G. (1966). *Chem. Commun.*, 313; Bergomi, A. and Gozzo, F. (1968). *Chimica e Industria,* **50,** 745. (d) Haller, I. (1967). *J. Chem. Phys.*, **47,** 1117
4. (a) Bryce-Smith, D., Gilbert, A. and Robinson, D. A. (1971). *Angew. Chem.*, **83,** 803. (b) Bryce-Smith, D. and Longuet-Higgins, H. C. (1966). *Chem. Commun.*, 593; Hoffmann, R. and Woodward, R. B. (1968). *Accounts Chem. Res.*, **1,** 17
5. Wilzbach, K. E. and Rausch, D. J. (1970). *J. Amer. Chem. Soc.*, **92,** 2178
6. (a) Wilzbach, K. E. and Kaplan, L. (1965). *J. Amer. Chem. Soc.*, **87,** 4004. (b) Schäfer, W., Criegee, R., Askani, R. and Grüner, H. (1967). *Angew. Chem. Int. Edn.*, **6,** 78; Lemal, D. M. and Lokensgard, J. P. (1966). *J. Amer. Chem. Soc.*, **88,** 5934. (c) Barlow, M. G., Haszeldine, R. N. and Hubbard, R. (1969). *Chem. Commun.*, 202
7. Kaplan, L., Wilzbach, K. E., Brown, W. G. and Yang, S. S. (1965). *J. Amer. Chem. Soc.*, **87,** 675
8. Cundall, R. B. and Voss, A. J. R. (1968). *Chem. Commun.*, 902; (1969). *Chem. Commun.*, 116; Matsuura, T. and Kitaura, Y. (1969). *Tetrahedron,* **25,** 4501
9. Matsuura, T. and Kitaura, Y. (1969). *Tetrahedron,* **25,** 4501
10. Clifton, E. D., Flowers, W. T. and Haszeldine, R. N. (1969). *Chem. Commun.*, 1216
11. Cookson, R. C., Costa, S. M. de B. and Hudec, J. (1969). *Chem. Commun.* 1272
12. Kleinhuis, H., Wijting, R. L. C. and Havinga, E. (1971). *Tetrahedron Lett.*, 255
13. Yang, N. C. and Rivas, C. (1961). *J. Amer. Chem. Soc.*, **83,** 2213
14. Gore, P. H., Hoskins, J. A., Lott, K. A. K. and Waters, D. N. (1970). *Photochem. Photobiol.*, **11,** 551
15. Heindel, N. D., Sarver, E. W. and Pfau, M. A. (1968). *Tetrahedron Lett.*, 3579
16. Roebke, W. (1970). *J. Phys. Chem.*, **74,** 4198
17. Caplain, S. and Lablache-Combier, A. (1970). *Chem. Commun.*, 1247
18. Barlow, M. G., Dingwall, J. G. and Haszeldine, R. N. (1970). *Chem. Commun.*, 1580
19. Lahmani, F. and Ivanoff, N. (1967). *Tetrahedron Lett.*, 3913
20. Allison, C. G., Chambers, R. D., Cheburkov, Yu. A., McBride, J. A. H. and Musgrave, W. K. R. (1969). *Chem. Commun.*, 1200
21. Johnson, D. W., Austel, V., Field, R. S. and Lemal, D. M. (1970). *J. Amer. Chem. Soc.*, **92,** 7505
22. Chambers, R. D., Musgrave, W. K. R. and Srivastava, K. C. (1971). *Chem. Commun.*, 264
23. Lablache-Combier, A. and Remy, M. A. (1971). *Bull. Soc. Chim. Fr.*, 679
24. Srinivasan, R. (1968). *Pure Appl. Chem.*, **16,** No. 1, 65. van Tamelen, E. E. and Whitesides, T. H. (1968). *J. Amer. Chem. Soc.*, **90,** 3894
25. Hiraoka, H. (1970). *J. Phys. Chem.*, **74,** 574
26. Boue, S. and Srinivasan, R. (1970). *J. Amer. Chem. Soc.*, **92,** 1824
27. Murad, E. (1961). *J. Amer. Chem. Soc.*, **83,** 1327
28. Srinivasan, R. and Hiroaka, H. (1969). *Tetrahedron Lett.*. 2767

29. Hunt, R. and Reid, S. T. (1970). *Chem. Commun.*, 1576
30. Hiraoka, H. (1970). *Chem. Commun.*, 1306
31. Wynberg, H., Kellogg, R. M., van Driel, H. and Beekhuis, G. E. (1967). *J. Amer. Chem. Soc.*, **89**, 3501, and references therein
32. Kellogg, R. M., Dik, J. K., van Driel, H. and Wynberg, H. (1970). *J. Org. Chem.*, **35**, 2737
33. Couture, A. and Lablache-Combier, A. (1971). *Tetrahedron*, 1059
34. Wiebe, H. A. and Heicklen, J. (1969). *Can. J. Chem.*, **47**, 2965
35. Chambers, R. D., McBride, J. A. H. and Musgrave, W. K. R. (1970). *Chem. Commun.*, 739
36. Olmstead, J. (1969). *Mol. Photochem.*, **1**, 331
37. Angus, H. J. F. and Bryce-Smith, D. (1959). *Proc. Chem. Soc.*, 326; (1960). *J. Chem. Soc.*, 4791
38. Kaplan, L., Ritscher, J. S. and Wilzbach, K. E. (1966). *J. Amer. Chem. Soc.*, **88**, 2881; Farenhorst, E. and Bickel, A. F. (1966). *Tetrahedron Lett.*, 5911: cf. Berson, J. A. and Hasty, N. M. (1971). *J. Amer. Chem. Soc.*, **93**, 1549; Wilzbach, K. E., Ritscher, J. S. and Kaplan, L. (1967). *J. Amer. Chem. Soc.*, **89**, 1031
39. Bryce-Smith, D. (1969). *Chem. Commun.*, 806
40. Bryce-Smith, D. (1968). *Pure Appl. Chem.*, **16**, 47
41. Bryce-Smith, D. and Lodge, J. E. (1962). *J. Chem. Soc.*, 2675; Bryce-Smith, D. and Gilbert, A. (1965). *J. Chem. Soc.*, 918; Bryce-Smith, D. and Hems, M. A. (1968). *J. Chem. Soc.*, 812; Bryce-Smith, D., Connett, B. E. and Gilbert, A. (1968). *J. Chem. Soc.*, p. 816
42. Bryce-Smith, D., Deshpande, R., Gilbert, A. and Grzonka, J. (1970). *Chem. Commun.*, 561; Bryce-Smith, D., Gilbert, A. and Grzonka, J. (1970). *Chem. Commun.*, 498; Grovenstein, E., Campbell, T. C. and Shibata, T. (1969). *J. Org. Chem.*, **34**, 2418; Bryce-Smith, D. and Lodge, J. E. (1963). *J. Chem. Soc.*, 695; Büchi, G., Atkinson, J. G., Ayer, D. E. and Roff, E. W. (1963). *J. Amer. Chem. Soc.*, **85**, 2257; Liu, R. S. H. and Krespan, C. G. (1969). *J. Org. Chem.*, **34**, 1271
43. Miller, R. D. and Abraitys, V. Y. (1971). *Tetrahedron Lett.*, 891
44. Bryce-Smith, D., Gilbert, A. and Orger, B. H. (1969). *Chem. Commun.*, 800
45. Wilzbach, K. E. and Kaplan, L. (1971). *J. Amer. Chem. Soc.*, **93**, 2073
46. Bryce-Smith, D., Foulger, B. E., Gilbert, A. and Twitchett, P. J. (1971). *Chem. Commun.*, 794
47. Srinivasan, R. and Hill, K. A. (1965). *J. Amer. Chem. Soc.*, **87**, 4653
48. Srinivasan, R. (1971). *I.B.M. J. Res. Develop.*, **15**, 34
49. Bryce-Smith, D., Gilbert, A. and Foulger, B. E. Unpublished observation
50. (a) *cis*- and *trans*-but-2-enes; Morikawa, A., Brownstein, S. and Cvetanovic, R. J. (1970). *J. Amer. Chem. Soc.*, **92**, 1471. (b) *cis*- and *trans*-cyclo-octenes; Bryce-Smith, D., Gilbert, A. and Orger, B. H. Unpublished studies. (c) Orger, B. H. (1968). *Ph. D. Thesis*, University of Reading
51. Bryce-Smith, D., Gilbert, A. and Orger, B. H. (1969). *Chem. Commun.*, 800
52. Hwang, D. Y. and Ho, S. Y. (1970). *J. Chin. Chem. Soc. Taipei*, **17**, 26
53. Hesse, G. and Lechtken, P. (1971). *Angew. Chem.*, **83**, 143
54. Scharf, H.-D. and Klar, R. (1971). *Tetrahedron Lett.*, 517
55. Kew Wei, Mani, J.-C. and Pitts, J. N. (1967). *J. Amer. Chem. Soc.*, **89**, 4225
56. Farenhorst, G. (1968). *Tetrahedron Lett.*, 4835; Luria, M. and Stein, G. (1970). *Chem. Commun.*, 1650
57. Arnold, D. R., Gillis, L. B. and Whipple, E. B. (1969). *Chem. Commun.*, 918
58. Sasse, W. H. F., Collin, P. J. and Sagowdze, G. (1965). *Tetrahedron Lett.*, 3373
59. Sasse, W. H. F. (1969). *Austral. J. Chem.*, **22**, 1257
60. Grovenstein, E., Campbell, T. C. and Shibata, T. (1969). *J. Org. Chem.*, **34**, 2418
61. Bryce-Smith, D., Gilbert, A., Harspool, W. M. and Phillips, D. (1970). *Photochemistry*, Vol. 2, 502. (London: The Chemical Society)
62. Bowman, R. M. and McCullough, J. J. (1970). *Chem. Commun.*, 948
63. Bowman, R. M. Chamberlain, T. R., Huang, C. W. and McCullough, J. J. (1970). *J. Amer. Chem. Soc.*, **92**, 4106
64. Miyamoto, T. Mori, T. and Odaira, Y. (1970). *Chem. Commun.*, 1598
65. Bryce-Smith, D. and Vickery, B. (1961). *Chem. and. Ind.*, 429
66. Sugowdz, G., Collins, P. J. and Sasse, W. H. F. (1969). *Tetrahedron Lett.*, 3843: cf. Sargent, M. V. and Timmons, C. J. (1964). *J. Chem. Soc.*, 5544

67. Bellas, M., Bryce-Smith, D. and Gilbert, A. (1967). *Chem. Commun.*, 263
68. McCullough, J. J., Huang, C. W. and Wu, W. S. (1970). *Chem. Commun.*, 1368
69. Barltrop, J. A. and Owers, R. J. (1970). *Chem. Commun.*, 1462
70. Neckers, D. C., Dopper, J. H. and Wynberg, H. (1970). *J. Org. Chem.*, **35**, 1582
71. Luther, R. and Weigert, F. (1905). *Z. Phys. Chem.*, **51**, 397
72. Bouas-Laurent, H., Lapouyade, R. and Faugere, J. (1967). *Compt. Rend.*, **265C**, 506
73. Campbell, R. O. and Liu, R. S. H. (1970). *Chem. Commun.*, 1191
74. Cohen, M. D. and Schmidt, G. M. J. (1964). *J. Chem. Soc.*, 1966; Cohn, M. D., Ludmer, Z., Thomas, J. M. and Williams, J. O. (1969). *Chem. Commun.*, 1172
75. Bouas-Laurent, H. and Lapouyade, R. (1969). *Chem. Commun.*, 817
76. Bouas-Laurent, H. and Lapouyade, R. (1964). *Compt. Rend.*, **264C**, 1061; Barnes, R. L. and Birks, J. B. (1966). *Proc. Roy. Soc. (London)*, **A291**, 570
77. Bouas-Laurent, H. and Castellan, A. (1970). *Chem. Commun.*, 1648
78. Lapouyade, R., Castellan, A. and Bouas-Laurent, H. (1969). *Compt. Rend.*, **268C**, 217
79. Bradshaw, J. S. and Hammond, G. S. (1963). *J. Amer. Chem. Soc.*, **85**, 3953
80. Christie, J. and Selinger, B. K. (1969). *Photochem. and Photobiol.*, **9**, 471
81. Mattingly, T. W., Lancaster, J. E. and Zweig, A. (1971). *Chem. Commun.*, 595
82. Wassermann, H. H. and Keehn, P. M. (1969). *J. Amer. Chem. Soc.*, **91**, 2374; (1967). *ibid*, **89**, 2270
83. Chandross, E. A. and Dempster, C. J. (1970). *J. Amer. Chem. Soc.*, **92**, 703
84. Havinga, E., de Jongh, R. O. and Kronenberg, M. E. (1967). *Helv. Chim. Acta*, **50**, 2550
85. Letsinger, R. L. and Steller, K. E. (1969). *Tetrahedron Lett.*, 1401
86. Letsinger, R. L. and Hautala, R. R. (1969). *Tetrahedron Lett.*, 4205
87. See Havinga, E. and Kronenberg, M. E. (1968). *Pure Appl. Chem.*, **16**, 137
88. Clark, V. M., Hobbs, J. B. and Hutchinson, D. W. (1970). *Chem. Commun.*, 339
89. Letsinger, R. L. and McCain, J. H. (1969). *J. Amer. Chem. Soc.*, **91**, 6425
90. El-Tsov, A. V., Frolov, A. N. and Kul-Bitskaya, O. V. (1970). *Zh. Org. Khim.*, **6**, 1943
91. Steller, K. E. and Letsinger, R. L. (1970). *J. Org. Chem.*, **35**, 308
92. Van-Vliet, A., Kronenberg, M. E., Corvelisse, J. and Havinga, E. (1970). *Tetrahedron*, **26**, 1061; Van-Vliet, A., Cornelisse, J. and Havinga, E. (1969). *Rec. Trav. Chim.*, **88**, 1339
93. Bryce-Smith, D., Clarke, M. T., Gilbert, A., Klunklin, G. and Manning, C. (1971). *Chem. Commun.*, 916; Bellas, M., Bryce-Smith, D. and Gilbert, A. (1967). *Chem. Commun.*, 263 862
94. van Henegouw, G. M. J. B. and Havinga, E. (1970). *Rec. Trav. Chim.*, **89**, 907
95. Lugtenburg, J. and Havinga, E. (1969). *Tetrahedron Lett.*, 1505
96. Grinter, R., Heilbronner, E., Petrzilka, T. and Seiler, P. (1968). *Tetrahedron Lett.*, 3845
97. Seiler, P. and Wirz, J. (1971). *Tetrahedron Lett.*, 1683
98. Frater, G. and Havinga, E. (1969). *Tetrahedron Lett.*, 4603; Frater, G. and Havinga, E. (1970). *Rec. Trav. Chim.*, **89**, 273
99. Lok, C. M., Lugtenburg, J., Cornelisse, J. and Havinga, E. (1970). *Tetrahedron Lett.*, 4701
100. See for example Blackburn, E. V. and Timmons, C. J. (1969). *Quart. Rev. Chem. Soc.*, **23**, 482; Sato, T. (1969). *J. Soc. Org. Synth. Chem. Jap.*, **27**, 715
101. Blackburn, E. V. and Timmons, C. J. (1970). *J. Chem. Soc. C*, 172
102. Laarhoven, W. H., Cuppen, T. J. H. M. and Nivard, R. J. F. (1970). *Tetrahedron*, **26**, 1069
103. Laarhoven, W. H., Cuppen, T. J. H. M. and Nivard, R. J. F. (1970). *Tetrahedron*, **26**, 4865
104. Laarhoven, W. H. and Cuppen, T. J. H. M. (1971). *Tetrahedron Lett.*, 163
105. Martin, R. H. and Schurter, J. J. (1969). *Tetrahedron Lett.*, 3679; Martin, R. H., Morren, G. and Schurter, J. J. (1969). *Tetrahedron Lett.*, 3683
106. Blackburn, E. V., Loader, C. E. and Timmons, C. J. (1970). *J. Chem. Soc. C*, 163
107. Banes, D. (1961). *J. Assoc. Offic. Agric, Chemists*, **44**, 323
108. Doyle, T. D., Filipesch, N., Benson, W. R. and Banes, D. (1970). *J. Amer. Chem. Soc.*, **92**, 6371
109. Morgan, D. D., Horgan, S. W. and Orchin, M. (1970). *Tetrahedron Lett.*, 4347
110. Ramey, C. E. and Boekelheide, V. (1970). *J. Amer. Chem. Soc.*, **92**, 3681
111. Muszkat, K. A., Gegiou, D. and Fischer, E. (1965). *Chem. Commun.*, 447
112. Boekelheide, V. and Pepperdine, W. (1970). *J. Amer. Chem. Soc.*, **92**, 3684
113. Wood, C. S. and Mallory, F. B. (1964). *J. Org. Chem.*, **29**, 3373
114. Kumler, P. L. and Dybas, R. A. (1970). *J. Org. Chem.*, **35**, 125

115. Galiazzo, G., Bortolus, P., Cauzzo, G. and Mazzucato, U. (1969). *J. Heterocyclic Chem.*, **6**, 465; Cauzzo, G., Galiazzo, G., Bortolus, P. and Coletta, F. (1971). *Photochem. and Photobiol.*, **13**, 445
116. Cava, M. P., Mitchell, M. J., Havlicek, S. C., Lindert, A. and Spangler, R. J. (1970). *J. Org. Chem.*, **35**, 175
117. Kumler, P. L. and Dybas, R. A. (1970). *J. Org. Chem.*, **35**, 3825
118. Scholz, M., Dietz, F. and Muehlstaedt, M. (1970). *Tetrahedron Lett.*, 2835
119. Horspool, W. M. (1971). *J. Chem. Soc. C*, 400; Anderson, D. T. and Horspool, W. M. (1971). *Chem. Commun.*, 615
120. Rio, G. and Hardy, J. C. (1970). *Bull. Soc. Chim. Fr.*, 3578
121. Templeton, W. (1970). *Chem. Commun.*, 1412
122. Roberts, T. D. (1971). *Chem. Commun.*, 352
123. Fonken, G. J. (1962). *Chem. and Ind.*, 1327
124. Heller, H. G. and Salisbury, K. (1970). *J. Chem. Soc. C*, 399
125. Leznoff, C. C. and Hayward, R. J. (1970). *Can. J. Chem.*, **48**, 1842; Hayward, R. J. and Leznoff, C. C. (1971). *Tetrahedron*, **27**, 2085
126. Sato, T., Goto, Y. and Hata, K. (1967). *Bull. Chem. Soc. Jap.*, **40**, 1994
127. Sato, T., Shimada, S. and Hata, K. (1969). *Bull. Chem. Soc. Jap.*, **42**, 766
128. Sato, T., Shimada, S. and Hata, K. (1970). *Chem. Commun.*, 766
129. Yang, N. C., Liu, L. C., Shani, A. and Yang, S. S. (1969). *J. Org. Chem.*, **34**, 1845
130. Ninomiya, I., Naito, T. and Kiguchi, T. (1970). *Tetrahedron Lett.*, 4451
131. Mondon, A. and Krohn, K. (1970). *Tetrahedron Lett.*, 2123; Horii, Z., Iwata, C., Wakawa, S. and Nakashita, Y. (1970). *Chem. Commun.*, 1039; Hey, D. H., Leonard, J. A. and Rees, C. W. (1963). *J. Chem. Soc.*, 5266; Hey, D. H., Jones, G. H. and Perkins, M. J. (1971). *J. Chem. Soc. C*, 116; Horii, Z., Nakashita, Y. and Iwata, C. (1971). *Tetrahedron Lett.*, 1167
132. Yonemitsu, O., Okuno, Y., Kanaoka, Y. and Witkop, B. (1970). *J. Amer. Chem. Soc.*, **92**, 5686
133. Ong, H. H. and May, E. L. (1970). *J. Org. Chem.*, **35**, 2544; Foltz, C. M. (1971). *J. Org. Chem.*, **36**, 24
134. McCall, M. T., Hammond, G. S., Yonemitsu, O. and Witkop, B. (1970). *J. Amer. Chem. Soc.*, **92**, 6991
135. Yonemitsu, O. and Naruto, S. (1969). *Tetrahedron Lett.*, 2387
136. Sander, M. R., Hedaya, E. and Trecker, D. J. (1968). *J. Amer. Chem. Soc.*, **90**, 7249
137. Plank, D. A. (1969). *Tetrahedron Lett.*, 4365
138. Sander, M. R. and Trecker, D. J. (1967). *J. Amer. Chem. Soc.*, **89**, 5725; Bradshaw, J. S., Loveridge, E. L. and White, L. (1968). *J. Org. Chem.*, **33**, 4127
139. Meyer, J. W. and Hammond, G. S. (1970). *J. Amer. Chem. Soc.*, **92**, 2187
140. Shizuka, H. and Tanaka, I. (1969). *Bull. Chem. Soc. Jap.*, **42**, 909
141. Obara, H., Takahashi, H. and Hirano, H. (1969). *Bull. Chem. Soc. Jap.*, **42**, 560; Bhatia, V. K. and Kagan, J. (1970). *Chem. and Ind.*, 1203
142. Ramakrishnan, V. T. and Kagan, J. (1970). *J. Org. Chem.*, **35**, 2901
143. Ramakrishnan, V. T. and Kagan, J. (1970). *J. Org. Chem.*, **35**, 2898
144. le-Goff, M. T. and Beugelmans, M. R. (1970). *Tetrahedron Lett.*, 1355
145. Lappin, G. R. and Zannucci, J. S. (1970). *J. Org. Chem.*, **35**, 3679
146. Grunwell, J. R. (1969). *Chem. Commun.*, 1437
147. Loveridge, E. L., Beck, B. R. and Bradshaw, J. S. (1971). *J. Org. Chem.*, **36**, 221
148. Shimamura, N. and Sugimori, A. (1971). *Bull. Chem. Soc. Jap.*, **44**, 281
149. Houser, J. J. and Chen, M. C. (1970). *Chem. Commun.*, 1447
150. Hageman, H. J., Louwerse, H. L. and Mijs, W. J. (1970). *Tetrahedron*, **26**, 2045
151. Ogata, Y., Takagi, K. and Ishino, I. (1970). *Tetrahedron*, **26**, 2703
152. Hashimoto, S., Sunamoto, J. and Nishitani, S. (1968). *Bull. Chem. Soc. Jap.*, **41**, 623
153. Sullivan, J. F., Hailey, K. and Shine, H. J. (1970). *Tetrahedron Lett.*, 2007
154. Shine, H. J. and Cheng, J. D. (1971). *J. Org. Chem.*, **36**, 2787
155. Ogata, Y. and Takagi, K. (1970). *J. Org. Chem.*, **35**, 1642
156. Spence, G. G., Taylor, E. C. and Burchardt, O. (1970). *Chem. Rev.*, **70**, 231
157. Mauser, H., Gauglitz, G. and Stier, F. (1970). *Annalen*, **739**, 84

5
Theoretical Aromatic Chemistry

ISRAEL AGRANAT
The Hebrew University of Jerusalem

5.1 INTRODUCTION

5.1.1 Aromaticity as a theoretical notion

The problem of *Aromaticity* has always been one of the most difficult and yet one of the most fascinating problems in chemistry. Ever since Kekulé's intuitive and brilliant idea on the structure of the benzene molecule in 1865[1–2], aromatic chemistry[3] has been a challenge both to the theoretician and to the synthesist, providing an ideal and fruitful testground for the interaction of theory and experiment.

The present review, which is by no means comprehensive, attempts to survey the numerous theoretical and experimental approaches to the problem of aromaticity[4–19]. It stems from the conviction that the term aromaticity, which has lost none of its vitality, can be defined in a meaningful way and should not be eliminated from the vocabulary of the chemist.

The central theme of the present review is the realisation that '*aromaticity*' is a theoretical notion and as such, its meaning is *theory dependent*. Any attempt to define explicitly the notion of aromaticity should take into consideration the following general remark made by the logician Ernst Nagel[20]:

'In these cases in which a given theoretical notion is made to correspond to two or more experimental ideas, it would be absurd to maintain that the theoretical concept is explicitly defined by each of the two experimental ones in turn.'

5.1.2 Ground state v. transition state definitions of aromaticity

The wealth of definitions, attributes, criteria, prefixes, meanings, connotations and synonyms ascribed to the notion of aromaticity allows one to draw a significant distinction between two independent, 'static' types of definitions (etc.):

(a) The definitions of aromaticity in the *ground state.*

(b) The definitions of aromaticity in the *transition state.*

This basic dichotomy should carefully and strictly be maintained in the future. It should be noted that a theory of aromaticity may be applied to both ground states and transition states, but this is not an essential requisite. A ground state definition of aromaticity need not necessarily apply to the transition state, and vice versa.

The ambiguities encountered in evaluating the various notions of aromaticity were in many instances due to overlaps and discrepancies of the two types of definitions and a failure to notice the distinction and differences between them[21]. Originally the concept of aromaticity developed as a means of characterising a certain type of organic molecules that was inclined to substitution and disinclined to addition reactions, and was thermally stable[22]. The emphasis has been on chemical reactivity. The outcome was a 'chemical' definition of aromaticity: a compound is considered aromatic if it has 'a chemistry like that of benzene[23]'. This definition clearly concerns not only the properties of the ground state of the molecule but also the properties of the transition state. It has been known for many years that aromatic compounds possess typical physical properties in the ground state (e.g. the anisotropy of the diamagnetic susceptibility[24]) and in the excited state v. the ground state (e.g. electronic absorption spectra[25]). With the advent of quantum chemistry, quantitative approximation methods, the valence bond (VB) method and the molecular orbitals (MO) method, were developed and these permitted the calculation of ground state properties of conjugated systems, e.g. resonance energy[7], and relate them to the notion of aromaticity. According to one 'physical' definition, a compound is aromatic if it has 'a low ground-state enthalpy'[23]. There has thus been a continuous process of transforming the meaning of aromaticity from the 'chemical' definition to the 'physical' viewpoint, which underlines the properties of the molecules in the ground state[11]. This trend has been strengthened in recent years by the theoretical and experimental developments related to the thermodynamic and magnetic criteria for aromaticity.

In the classical aromatic compounds, the benzenoid aromatics, there is in fact a good correlation between the two types of definitions. However, this is only an empirical finding; it has no theoretical justification[18]. The ambiguity in the notion of aromaticity is especially pronounced in the non-benzenoid aromatic compounds[12, 26–34]. The cyclopentadienyl anion (prepared by Thiele in 1900[35] but not recognised as such), which is the prototype of the non-benzenoid aromatic compounds, is certainly aromatic according to the ground-state criteria (see below) but is very reactive chemically and is very far from behaving like benzene.

At present, the ground state definitions are undoubtedly more established than the transition state definitions, and certainly more popular. Their theoretical foundations are stronger, and their experimental basis is more comprehensive and of a more direct nature. In principle, however, the ground state definitions are not superior to those of the transition state. It might be that in the future, the situation will be reversed, and the transition state definitions will be more effective both in their explanatory and in their predictive powers.

5.2 THERMODYNAMICAL AND RELATED ASPECTS OF AROMATICITY

5.2.1. Hückel's $4n + 2$ aromaticity rule

The leading role in the theory of aromaticity has been played by *Hückel's* $4n+2$ *aromaticity rule*[36–39] which states that 'those monocyclic co-planar systems of trigonally-hybridised atoms which contain $4n+2$ π-electrons will possess relative electronic stability' ($n = 0, 1, 2, \ldots$)[10]. Hückel's rule originated from theoretical considerations. It was derived from the HMO method[40]., with the additional assumption that a π-electron system is aromatic only if it contains a closed shell of electrons. This assumption probably originated from an intuitive comparison with the especially stable electronic shell of the inert gases. In a closed electronic shell all the π-electrons occupy only bonding orbitals, which are all fully occupied. Application of the HMO method to an $[N]$annulene with equal bond lengths gives the following analytical expression for the energy of the π-molecular orbitals:

$$E_k = -2\cos(2k\pi/N) \qquad (k = 0, 1, 2, \ldots N-1).$$

This solution, coupled with the idea of a closed shell of electrons leads to the magic number of $4n+2\pi$ electrons.

Considering the success of Hückel's rule both in explaining a vast variety of observations and in predicting new ones[10–12], it is curious that its theoretical foundations were so weak[41,42]. The discrepancy between theory and experiment is most pronounced by comparing the $[4N]$annulenes with the respective $[4N+2]$annulenes. Neither the VB method nor the resonance theory distinguishes between the two types of annulenes[15,42]. The HMO method indicated that in the lower members of the series, the former type should have lower resonance energies per atom than the latter, but it did not imply that the difference between the two types would be as large as the experimental facts suggest. This is even more striking in the higher members of the $[4N]$annulenes, which are predicted to have much greater delocalisation energy than the lower members of the $[4N+2]$annulenes, and thus should be extraordinarily stable[15].

5.2.2 Hückel systems v. Möbius systems

From a topological point of view, Hückel's $4n+2$ rule can be shown to apply only to normal *Hückel systems*, in which the p atomic orbitals (AOs) can be chosen so that they all overlap in phase. By contrast, *Möbius systems*[19, 43–49], termed also the *anti-Hückel systems*[49], where there will always be at least one phase dislocation, no matter how the phases of the AOs are chosen, do not obey Hückel's rule. A ring is of a Hückel type if the number of phase dislocations in it is even, of a Möbius type if the number is odd. In Möbius systems, aromaticity is associated with $4n$ π-electrons.

5.2.3 Local symmetry. Homomorphic v. heteromorphic systems

Apart from the topological classification into Hückel and Möbius systems, a second classification, according to local symmetry types of atomic orbitals, has been considered[19, 48, 50–51]. There are two systems of delocalisation: in one the participating orbitals are antisymmetric to reflection in the molecular plane (π_a system) and in the other they are symmetric (π_s system)[48]. In $(AB)_n$ rings of alternating atoms, e.g. the phosphonitrilic halides, the distinction is between *homomorphic* systems in which all the AOs have the same local symmetry type (e.g., p_π–p_π or p_π–d_{yz}), and *heteromorphic* systems, in which the AOs are alternately of one type and the other (p_π–d_{xz})[6, 19, 48, 50]. This same classification may be made within the π_a and π_s delocalisation types. In heteromorphic systems, there is no analogue of Hückel's rule. The importance of the local symmetry properties of the interacting AOs as conditions for aromaticity has been demonstrated experimentally.

5.2.4 Jahn–Teller distortions

The theoretical foundations of Hückel's rule were further weakened by considering the *Jahn–Teller distortions effects*[15, 52, 53]. These could be separated into two distinct types:

(a) Distortions of non-linear molecules whose ground electronic state is orbitally degenerate (e.g., odd cyclic polyene radicals). These molecules distort in such a way that the degeneracy is removed. A characteristic feature of these distortions, which are called (proper) *Jahn–Teller* effects, is the small energy gain which they generally produce, and hence the experimental evidence of their existence is scarce. One can distinguish between a (rare) *static* Jahn–Teller effect in which the molecules are permanently distorted and a *dynamic* Jahn–Teller effect in which there is a rapid interconversion between two equivalent forms.

(b) Distortions which occur in molecules without an orbitally degenerate ground state, but with a very low energy excited electronic state. A distortion of proper symmetry can mix ground and excited states, and thereby lower the ground-state energy. In considering these distortions, which are called *pseudo-Jahn–Teller effects*, two distinct classes of molecules should be noticed: (i) Molecules whose ground state, although not truly degenerate, is degenerate in the Hückel scheme. The false degeneracy is lifted by electron repulsion, but the lowest states remain nearly degenerate. Typical examples are the open-shell $(4N]$annulenes, in which the energy stabilisation due to the pseudo Jahn–Teller effect is of the same order of magnitude as in the proper Jahn–Teller effect. (ii) Molecules which even in Hückel's scheme have non-degenerate ground states. The closed shell $(4N+2]$annulenes belong to this class. Longuet-Higgins and Salem have shown that the HMO method in which σ-bond compression has been included leads unambiguously to *bond alternation* in $[4N+2]$ annulenes, for N sufficiently large[54]. The only difference from the pseudo-Jahn–Teller distortion of $[4N]$annulenes is one of order of magnitude. It appears that bond alternation in $[4N]$annulenes decreases with N while that in $[4N+2]$annulenes increases[15,55].

5.2.5 Thermodynamic definitions of aromaticity, anti-aromaticity, and non-aromaticity

Inherent in Hückel's rule, although not explicitly expressed, is the thermodynamic point of view of aromaticity, according to which aromatic compounds are thermodynamically more stable, i.e. contain less energy, than their related non-aromatic reference compounds. Thus, the following thermodynamic definition of aromaticity has been put forward: 'Cyclic conjugated systems are considered aromatic if cyclic delocalisation of electrons makes a *negative* contribution to their heat of formation[56,57].

This definition allows the introduction, by a logical extension, of the concepts of anti-aromaticity[11,56–61] and non-aromaticity[58]. Cyclic conjugated systems are considered *anti-aromatic* if cyclic delocalisation of electrons is destabilising, i.e. makes a positive contribution to their heat of formation. *Non-aromaticity* means lack of stabilisation due to cyclic electron delocalisation[15]. These definitions throw light upon the fundamental question in organic chemistry which relates to the net energy change associated with electron delocalisation. Thus, either energy stabilisation, or no net energy change, or energy destabilisation may be associated with electron delocalisation[62]. Two representative cases which were predicted to possess anti-aromatic character are the cyclopropenyl anion[57,61] and cyclobutadiene[63,64].

5.2.6 Polycyclic systems

Hückel's rule strictly applies only to monocyclic systems. Several attempts were made to broaden the rule to polycyclic systems. These include the periphery modification of Platt[65], the polycyclic modification of Vol'pin[9], Breslow's fusion approach[11,61,66,67] and Badger's broad definition of aromaticity[3]. Platt proposed that $4n+2$ polycyclic systems can be treated in terms

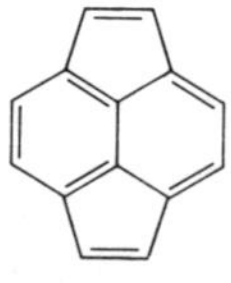

(1)  (2)

of their $4n+2$ peripheries by considering the cross-links as small perturbations[65]. This model has been extended to include anti-aromatic systems, e.g. cyclopent[*fg*]acenaphthylene(pyracylene) (1)[62,68] and applied also to non-alternant hydrocarbons containing no benzene moiety at all, e.g. dicyclopenta[*ef,kl*]heptalene (2)[69,70]. According to Vol'pin's polycyclic modification, any plane fused system containing no atoms common to more than two rings will be aromatic if the number of π-electrons in it is equal to $4n+2$[9].

The magic number of $4n+2$ π-electrons may be attained in polycyclic molecules by fusion of two anti-aromatic rings with $4n$ π-electrons: $(4n')+(4n'')\rightarrow 4n+2$. On the other hand, fusion of an anti-aromatic ring to an aromatic ring may produce an anti-aromatic system: $(4n')+(4n''+2)\rightarrow 4n$[11,61]. Examples of the two types of fusion are octalene (3)[66] and cyclobutadienocyclopentadienyl anion (4)[67] respectively. The experimental evidence for the

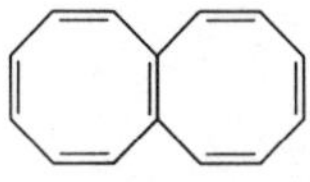

(3) 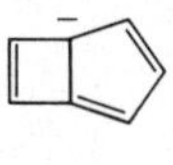 (4)

fusion approach so far, is still ambiguous[61]. Badger's broad definition of aromaticity, which includes also polycyclic systems is formulated as follows[3]:

'An unsaturated cyclic or polycyclic molecule or ion (or part of a molecule or ion) may be classified as aromatic if all the annular atoms participate in a conjugated system such that, in the ground state, all the π-electrons (which are derived from atomic orbitals having axial orientation to the ring) are accommodated in bonding molecular orbitals in a closed (annular) shell.'

5.2.7 Non-alternant systems

In contrast to the annulenes[15], the application of Hückel's rule to the non-alternant hydrocarbons[18,71] was only intuitive. For example, the problem of aromaticity in pentaheptafulvalene[72–74] and azulene[8,12], could be analysed in terms of possible contributions of the 'aromatic,' Hückel-type dipolar structures (5) and (6) in the ground state. Such a dipolar structure contains two oppositely charged conjugated rings, each of which alone satisfies the requirement of Hückel's rule. These considerations may apply not only when the two oppositely charged rings are linked directly (5) or fused (6), but also when they are concentric, as in corannulene (dibenzo[*ghi,mno*]fluoranthene (7)[75–77]. In this system, which illustrates also a unique confrontation between the factors of strain and aromaticity[76], the dipolar structure (7a) contains the inner cyclopentadienyl anion and the outer cyclopentadecaheptenyl cation, each alone satisfying the requirements of Hückel's rule.

To improve the theoretical foundation of the rule, it was deemed necessary to apply a MO theory that would take into account explicitly the effects of the interelectronic and internuclear repulsions[18]. One attempt, based on Pople's method[78,79] was made by Fukui and his co-workers[80], who suggested guidelines for the synthesis of new aromatic compounds: they proposed to choose suitable substituents to be attached to the ring in such a way that the ring will acquire $4n+2$ electrons. The 'push–pull' approach towards the synthesis of stable cyclobutadienes[81–83], e.g. diethyl 2,4-bis(diethylamino)cyclobuta-1,3-diene-1,3-dicarboxylate (8)[83,84], of 10-dimethylaminononafulvene (9)[85,86] and of 1,3-bis(dimethylamino)pentalene (10)[87,88] is based on this suggestion. Additional examples are the heterofulvalenes[89,90], e.g. 2,6-dimethyl-4-(1-cyanocyclopentadienylidene)thi-in(11) and 10,9-borazaronaphthalene (12)[91,92].

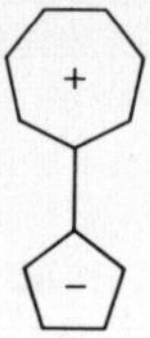

(5) 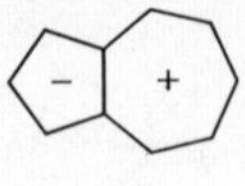 (6) 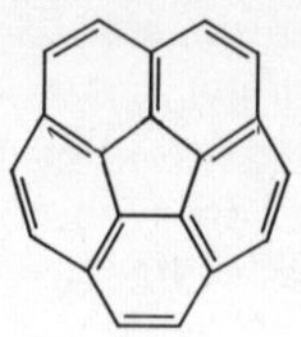 (7) 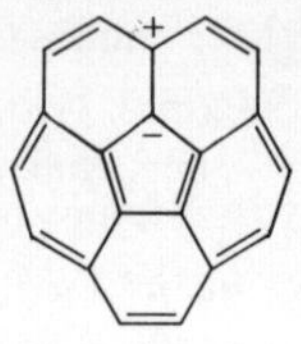 (7a)

(8) 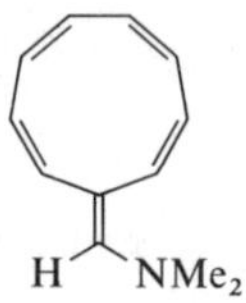 (9) 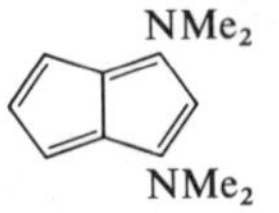 (10) 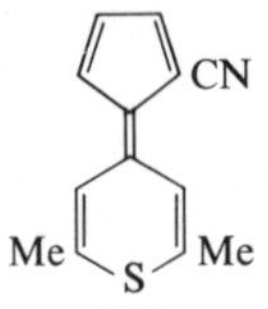 (11) 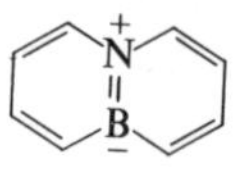 (12)

The aromaticity problem in ferrocene, and in metallocenes in general, must take into consideration the significant role played by the central metal atoms[93].

5.2.8 The PMO theory of aromaticity

The first treatment that could justify Hückel's rule was the PMO method, introduced by Coulson and Longuet-Higgins[94–98] and extended into a general theory by Dewar[99] who applied it to the problem of aromaticity[42, 49, 56, 100].

In the PMO method, the energies of two very similar systems are compared by a perturbational approach, based on the HMO method. In the past, 'resonance energy' was defined by reference to idealised structures with pure single and double bonds. It is better, however, to compare the heat of formation of a cyclic conjugated molecule with that of the corresponding classical structure. The idea is to compare the total π-bond energy of the cyclic conjugated hydrocarbon with that of an open chain analogue by relating both systems to open chain radicals or ions; thus benzene can be related to hexatriene by constructing both molecules by 'union' of a pair of allyl radicals.

Application of the PMO method to both Hückel and anti-Hückel systems gives the following set of general rules for aromaticity[49].

(a) Replace all heteroatoms by carbon atoms with an equivalent number of p(π) electrons.

(b) Delete all essential single and double bonds, i.e. bonds that are respectively single or double in all classical (unexcited resonance) structures for the molecule.

(c) The aromaticity of each remaining ring is indicated in Table 5.1:

The rules have been extended to radical cations, and applied to the analysis of mass spectrometric reactions[101]. Furthermore, the PMO derivation of the rules has also been extended to polycyclic systems[49]. As indicated by Dewar and Longuet-Higgins[102], the PMO method also gave a valid interpretation to the resonance theory. Their analysis of resonance formalism showed that the simple counting rules of resonance may be applied to even alternant systems composed of rings of $4n+2$ atoms and chains. Modified rules have also been constructed for simple or fused $4n$-rings[103, 104] and for non-alternant molecules[105].

The PMO theory of aromaticity is not limited to *ground states*, but may also be applied to *transition states*, e.g. the transition state of pericyclic reactions[49] and to excited electronic states[106].

The aromaticity of a given transition state is defined, according to Dewar[42], by reference to the isoconjugate hydrocarbon, i.e. the cyclic or acyclic conjugated hydrocarbon whose π-system shows the same topology as the system of delocalised orbitals in the transition state.

Table 5.1 Rules for aromaticity

Type of ring	*Type of conjugated system* *Hückel*	*Anti-Hückel*
Neutral even-numbered rings		
$4n$ atoms	anti-aromatic	aromatic
$4n+2$ atoms	aromatic	anti-aromatic
Odd-numbered rings		
Cations with $4n+1$ atoms	anti-aromatic	aromatic
Cations with $4n+3$ atoms	aromatic	anti-aromatic
Anions with $4n+1$ atoms	aromatic	anti-aromatic
Anions with $4n+3$ atoms	anti-aromatic	aromatic
Radicals	non-aromatic	non-aromatic

It has been argued that any theory of aromaticity is *ipso facto* a theory of pericyclic reactions[49]. Pericyclic reactions formally resemble the conversion of one Kekulé structure into another; the transition state may be 'aromatic', 'non-aromatic' or 'anti-aromatic'. Thermally-induced pericyclic reactions proceed preferentially via aromatic transition states (Evans principle)[42, 56, 107] whereas their photochemical counterparts lead to products that are formed via anti-aromatic transition states[49]. It should be noted that the energy factors favouring reaction with aromatic transition states are by no means overwhelming. Resonance stabilisation or destabilisation of the transition state is only one of the factors reflecting the rate of a pericyclic reaction. Thus anti-aromatic reactions are not necessarily 'forbidden'[49, 108].

In addition to the PMO method, Dewar developed a more refined approach, based on semi-empirical self-consistent field (SCF) MO treatment which enabled the heats of formation of conjugated systems to be calculated accurately[41, 109–111]. The resonance energy was calculated following the definition of Mulliken and Parr[112], the reference compound for benzene was a cyclohexatriene with alternating carbon–carbon bonds (1.48 and 1.34 Å respectively) and not a Kekulé benzene with alternating bonds of equal length (1.4 Å). In this treatment both the magnitude and the sign of the resonance energy might serve as a criterion for aromaticity. The results confirmed Hückel's rule in the annulene series for $n<5$, taking into account realistic geometries of the molecules, and pointed to the existence of a critical value of n for which a $4N+2$ annulene would not be aromatic.

Hess and Schaad have recently argued that the failure of the HMO method to predict aromatic or anti-aromatic character in cyclic conjugated hydrocarbons is due not to a fault of the Hückel wave functions but instead to an erroneous identification of various 'delocalisation energies' with aromaticity[113]. They defined resonance energy per π electron (REPE) as the difference between the Hückel π energy and the π energy of a 'localised structure' obtained in an additive manner using empirical π bond energies, divided by the number of π electrons. Molecules with a positive, zero and negative REPE should be aromatic, non-aromatic, and anti-aromatic respectively. It appears that the REPE correlates well with experimental aromaticity both in the alternant[114] and in the non-alternant series[115].

5.2.9 The theory of bond fixation

The importance of bond alternation in the study of the problem of aromaticity prompted Heilbronner and Binsch to introduce the theory of bond fixation[116–119]. This theory distinguishes between two effects: first-order bond fixation and second-order bond fixation. Consider a system with a second-order geometry consisting of a plane σ-bonded skeleton of sp^2-hybridised carbon atoms with all C—C bonds equal to R_0 and an empty set of orbitals perpendicular to the plane. A distortion of this geometry is effected by filling in the π-electrons. The theory is based on developing the total energy of a system into a Taylor Series about R_0 in terms of bond distortions ΔR, assuming a complete σ–π separation and independent σ bonds but allowing interactions between the π-electrons.

$$E = E_\sigma(R_0)+E_\pi(R_0)+\sum\left[\left(\frac{\partial E_\sigma}{\partial R_{\mu r}}\right)_{R_0}+\left(\frac{\partial E_\pi}{\partial R_{\mu r}}\right)_{R_0}\right]\Delta R_{\mu r}+$$
$$+\tfrac{1}{2}\sum_{\mu<r}\left(\frac{\partial^2 E_\sigma}{\partial R_{\mu r}^2}\right)\Delta R_\mu^2+\tfrac{1}{2}\sum_{\mu<r}\sum_{k<\lambda}\left(\frac{\partial^2 E_\pi}{\partial R_{\mu r}\partial R_{k\lambda}}\right)\Delta R_{\mu r}\,\Delta R_{k\lambda}+\cdots$$
$$(+\text{ higher terms})$$

The favoured distortion is the one that minimises the energy. Considering only first-order terms, a linear relationship between 'first-order bond lengths' $R_{\mu r}$ and the bond orders $P_{\mu r}$ is obtained:

$$R_{\mu r} = A - P_{\mu r}$$

Large bond orders correspond to short bonds and small bond orders to long bonds. The difference in such lengths between adjacent bonds are then a measure of first-order double-bond fixation. First-order distortions are such that the symmetry of the system is not lowered. Information about second-order bond fixation may be obtained by investigating the eigenvalues and eigenvectors of a matrix whose elements are the second partial derivatives of π-electron energy with respect to the bond lengths ($\partial^2 E_\pi/\partial R_{\mu r}\partial R_{k\lambda}$). This matrix has been shown to be the bond-bond polarisability matrix.

The most negative eigenvalue (λ_{max}) of this matrix corresponds to the energetically most favourable second-order bond distortions. Its magnitude is a measure of the π-electron energy given. If λ_{max} exceeds a critical value, λ_{crit}, equal to the negative curvature of the σ energy, the molecule will suffer second-order bond distortions which will in general result in the destruction of the symmetry of the σ framework.

A very strong tendency for second-order double-bond fixation was calculated for pentalene (13), indicating that it consists of an equilibrating mixture of two iso-energetic forms of lower symmetry, rather than by a 'resonance arrow'. First-order double-bond fixation turns out to be insignificant in pentalene (as in azulene).

The theory of double bond fixation, prompted Binsch to propose the following theoretical criterion of aromaticity[119, 121].

'A conjugated π-electron system is called aromatic if it shows neither strong first-order nor second-order double bond fixation.'

It should be noted that this criterion does not follow either from the theory of double-bond fixation, or from chemical experience or from numerical results. It should rather be viewed upon as a postulate. By definition, it is a structural criterion, related to the ground state and not to reactivity. The double-bond fixation criterion has the advantage that it is closely related to a genuine physical phenomenon: the tendency of π-electrons to cluster in certain bonds. In contrast to the criteria based on resonance energies, it is capable of pinpointing the lack of π-electron delocalisation in a particular segment, and of detecting the dynamic distortions in certain π systems which hamper electron delocalisation.

Comparison of the double-bond fixation criterion with the ring current criterion shows (*vide infra*) that both lead to the same qualitative predictions for monocyclic conjugated hydrocarbons. However, for some polycyclic

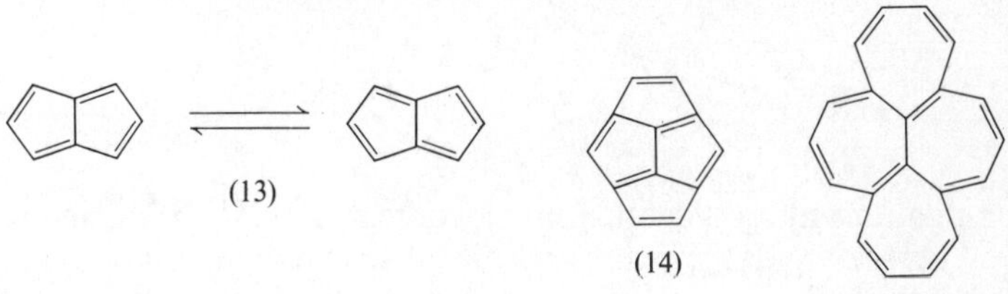

(13)

(14)

(15)

systems, the prediction of the two criteria are at variance[119]. While the ring-current criterion predicts dicyclopenta[*cd,gh*]pentalene (14) and dicyclo hepta[*ef,kl*]heptalene (15) to be aromatic[122], they are predicted to be anti-aromatic due to their second-order double-bond fixation[119]. In these molecules, the very low-lying Ψ_1 configurations are inaccessible by magnetic dipole transitions, with a component perpendicular to the plane of the π system; the ring-current criterion is therefore incapable of detecting the important effects these states are expected to exert on their chemical structure (14) and (15)[119].

Nakajima considered further the aromaticity of conjugated systems from the point of view of bond alternation and double-bond fixation[123–125]. Using the semi-empirical SCF MO theory in conjunction with the variable bond-length technique, a general method was developed for predicting the energetically most favourable molecular symmetries and geometrical structures of conjugated hydrocarbons without presuming the type of bond distortion. Application of the method to non-alternant hydrocarbons revealed that bond alternation is a common phenomenon in these systems. The double-bond fixation criterion allowed in some cases to distinguish between aromatic and non-aromatic regions in a molecule. In such molecules, a clear-cut distinction between aromatic and non-aromatic character is impossible.

5.2.10 Experimental methods

5.2.10.1 Thermochemical measurements

The thermodynamic definitions of aromaticity and anti-aromaticity are meaningful if it becomes possible to measure the energies of suitable systems and to correct for any differences between the cyclic and the reference systems which are not due to conjugation effects[61]. The limitations of the thermodynamic criteria arise in the thermochemical data, from which it is experimentally derived. As pointed out by Turner[126], 'the total stabilisation represented by thermochemical data is certainly not only related to the delocalisation of π-electrons'. Neutral aromatic compounds were in fact classically characterised by heat of combustion or heat of hydrogenation measurements[127]. A recent illustrative example is cyclohept[*fg*]acenaphthylene (16)[128]. From the heats of hydrogenation of (16) and its 1,2,5,6,7,8-hexahydroderivative, the 'conjugation energy' of (16) was deduced to be 80.7 kcal mol^{-1}.

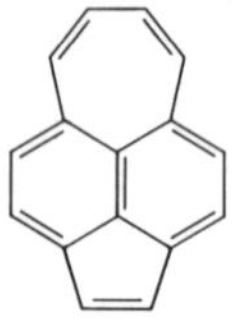

(16)

The conjugation energy is defined, in this context, as the difference between the observed heats of hydrogenation and the values predicted from hypothetical models having isolated non-resonating double bonds. It is of course the sum of various terms, including π-electron delocalisation energy, σ-bond energy, bond compression energy, steric strain energy, etc. A more significant value is the conjugation energy of the three non-napthalene double bonds in (16), estimated at 18.9 kcal mol^{-1} [128].

The methods of heat of combustion and heat of hydrogenation involve drastic alternation of the compound. Under such circumstances, it is difficult to remove factors such as strain energy for the interpretation of the energy measurements. It is thus preferable to apply thermodynamic measurements that would convert an aromatic into a non-aromatic system with a minimum total change in structure. A method of choice for carbonium ions, which has been applied successfully in the cyclopropenyl cation series[61], is the determination of the pK_{R^+} values of the reaction $R^+ + H_2O \rightarrow ROH + H^+$. Likewise, the method of tautomeric equilibria has been applied as a quantitative energy estimate of the aromatic stabilisation energy, giving a value of 25 kcal mol^{-1} for 2-pyridone[128a].

5.2.10.2 Energy barriers of restricted rotation and inversion

Dynamic nuclear magnetic resonance (d.n.m.r.) spectroscopy is a powerful tool for determining the energy barriers of restricted rotations and inversions.

In conjugated systems, these barriers may lead to conclusions which are pertinent to the stabilisation and hence to the problem of aromaticity. One should only be aware whether and to what extent these conclusions refer to ground states, and/or to transition states of the systems under study.

A quantitative d.n.m.r. study (by complete n.m.r. line-shape analysis) of the conformational mobility of [18]annulene (17), in which the inner and outer protons exchange their chemical (magnetic) environment, gives an enthalpy of activation of 16.1 kcal mol^{-1} [130]. This fairly high value is at least 7 kcal mol^{-1} greater than the respective value observed in [16]annulene (18)[131]. The difference is probably due to the fact that the π-bonds in [18]annulene (17) are delocalised so that any bonds around which rotation occurs, have a π-character. In [18]annulene (17) the π-bonds are localised and rotation takes place around pure single bonds. Thermochemical considerations show that the activation enthalpy of [18]annulene (17) can

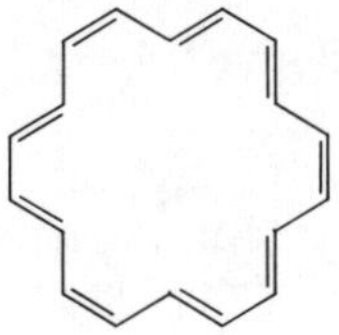

(17)

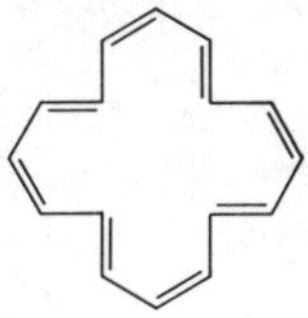

(18)

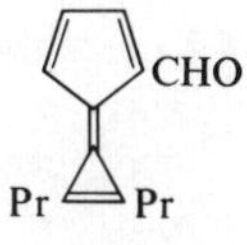

Bu^t CHO Bu^t Bu^t Bu^t NMe$_2$

Pr Pr Me Me Me Me H

(19) 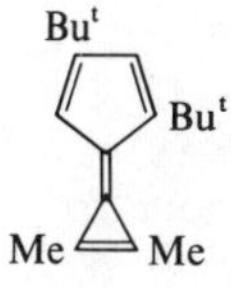 (20) 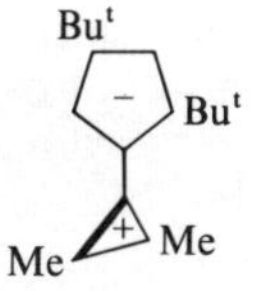 (20a) 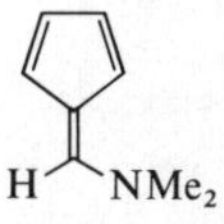 (21)

Me Ph P Pr^i CMe$_2$OH D D D CMe$_2$OH D H D D

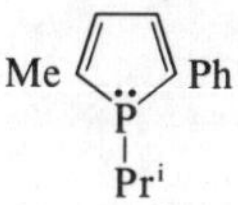

(22) 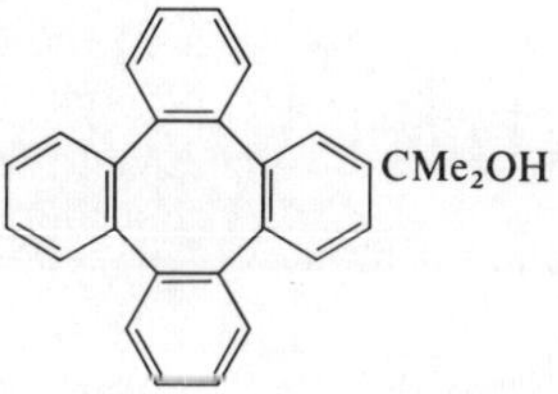 (23) 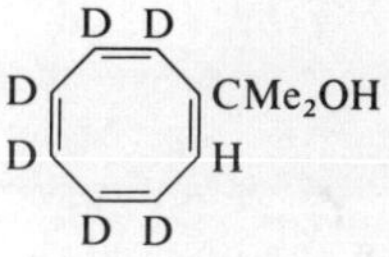 (24)

formally be represented as the sum of the resonance energy and the strain of the transition state with respect to the planar Kekulé structure. The latter was estimated to be greater than -3 kcal mol^{-1}. This led to the conclusion that the resonance energy of the aromatic [18]annulene cannot be greater than 19 kcal mol^{-1}. This value is in complete disagreement with the value of 100 kcal mol^{-1} deduced from heat of combustion measurements of (17)[132]. The latter proved to be incorrect[130].

The treatment of the problem of 'aromaticity' in fulvenes should take into account the extent of the contribution of dipolar structures which imbue these systems with an aromatic character. The central double bond ('pinch') of the fulvalenes and the exocyclic double bond of fulvenes may be attributed with a certain amount of single bond character. This may lower the barrier so much that rotation becomes amenable to a d.n.m.r. study. An abnormally low rotational barrier ($\Delta G^{\ddagger} = 49\,\text{kcal mol}^{-1}$) has been observed in 3-formyl-1,2-dipropyltriapentafulvalene (19)[133]. Presumably this indicated a strikingly low π-bond order of the pinch and a significant 'cyclopropenylium-cyclopentadienide aromaticity'. From the modest but consistent decrease in the rotational barriers of (19) with increase in solvent dielectric constant, a transition state more polar than the ground state has been implied. D.n.m.r. studies of 3,5-di-t-butyl-1,2-dimethyltriapentafulvalene (20) showed an analogous behaviour[74]. It has been suggested, as in the case of (19), that the most reasonable transition state for the rotation is the 'aromatic cyclopropenium-cyclopentadienide' structure (20a) in which the two halves of the molecule are perpendicular to each other. It appears that (20) is the hydrocarbon with a cross conjugated π-electron system exhibiting the lowest barrier to rotation around a carbon–carbon double bond.

It is not yet clear whether or not and to what extent the dynamic behaviour of (20) depends on a deviation from co-planarity and a consequent higher contribution of the dipolar form (20a). Another example is 6-dimethylaminopentafulvene (21), in which, from the study of the energy barriers for restricted rotation about the C-6—N and C-1—C-6 bonds, the delocalisation energy was estimated as 25–32 kcal mol^{-1}[134].

For rings containing $4n+2$ electrons, extended π-delocalisation in the transition state enhances considerably the rate of inversion. The barrier $\Delta G^{\ddagger}$ for pyramidal inversion at phosphorus in 1-isopropyl-2-methyl-5-phenylphosphole (22) has the extraordinarily low value of 16 kcal mol^{-1}[135,136]; i.e. 23 kcal mol^{-1} less than that of the corresponding saturated ring compounds. This dramatic lowering of the inversion barrier may be ascribed to increased (3p–2p)π delocalisation (aromaticity) of (22) in the planar transition state of inversion, relative to the pyramidal ground state.

An experimental proof of the anti-aromaticity of the planar 8-π electron ring system was derived from the low value of the barrier for ring inversion found in 2-(1-hydroxy-1-methylethyl)tetraphenylene (23) (5.7 kcal mol^{-1})[137]. The respective value for 2-(1-hydroxy-1-methylethyl)cyclo-octatetraene-2,3,4,5,6,7-d_6 (24) is 14.7 kcal mol^{-1}[138]. This difference was ascribed to the decreased anti-aromaticity (as compared to cyclo-octatetraene itself) of the eight-membered ring in the postulated planar transition state of the inversion.

5.2.10.3 The dipole-moment method

Dipole moments may serve as a means for determining the 'aromaticity' of non-alternant conjugated systems and hetero-molecules. In non-alternant hydrocarbons, in contrast to alternant hydrocarbons, the charge distribution is not even, leading to non-negligible π dipole moments in the ground state. The dipole moment method is particularly attractive due to its role as an interplay of theory and experiment: dipole moments can be determined experimentally as well as calculated to various approximations by the existing quantum mechanical theories[139]. A typical example is azulene. Its experimental dipole moment is 1.0 D. The direction of the moment as calculated from the moment of 2-chloroazulene (2.69 D) is from the seven- to the five-membered ring. The HMO method predicts for azulene a π-moment of 6.4 D. Yet the simple molecular orbital theory considerably overestimates the π-moments of non-alternant hydrocarbons. An SCF calculation gave a value of 0.85 D for the moment of azulene[140].

Another illustrative case is the dipole moment of the triapentafulvalene system (25). The hexaphenyl- and the 1,2-dimethyl-3,4,5,6-tetraphenyl-derivatives (26)[141] and (27)[142] have the highest experimental dipole moments ever recorded for a hydrocarbon, 6.3 and 6.0 D respectively. These values agree well with the calculated value of 6.04 D reported by Nakajima for the hexaphenyl derivative (26)[143]. Experimental estimations using these and other experimental moments gave a value of 5.6 D for the parent system (25)[73] in agreement with the theoretical moments (5.70 D[143], 5.56 D[144]).

(25)  (25a)  (25b) 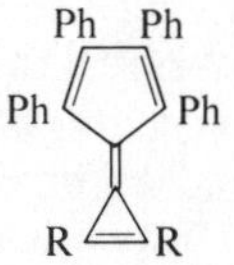 (26) R = Ph (27) R = Me

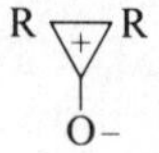

(28) 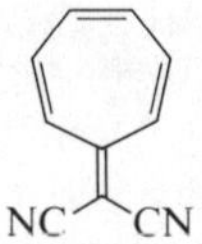 (29) 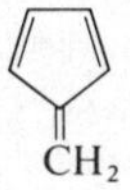 (30) 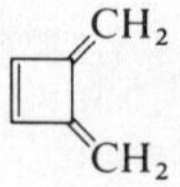 (31)

This value certainly does not imply a large contribution of the dipolar 'aromatic' structure (25a) but is consistent with a marked bond alternation in the ground state. The question is whether the high value of 5.6 D can be explained without assuming any contribution of an 'aromatic' dipolar structure. Dewar has argued that 'the existence of large dipole moments cannot in itself be taken as evidence that a molecule has a highly delocalised aromatic character[56, 145]. However, in the example cited, a certain deviation from a complete bond-alternation picture towards 'aromaticity' is called for. It might be that this deviation in the direction of 'aromaticity' will be limited only to the three-membered ring, as reflected in the dipolar structure (25b).

The evaluations of observed dipole moments and their comparison with theoretical values may often lead to qualitative results. The measured moment is a global, overall property of the molecule; its breakdown to contributions of the different parts of the molecule is not necessarily unequivocal. Sometimes it does not permit to pinpoint the π-electron delocalisation in a particular region of the molecule. It is always essential to consider proper reference compounds. A re-evaluation of the dipole moment (4.30 D) of tropone by comparison with those of 2,6-cycloheptadien-1-one (4.04 D) and related model compounds indicated that tropone is devoid of aromatic character[146].

The distance over which charge separation takes place should be considered. It has been shown that this factor is responsible for the high dipole moments (4.7–5.1 D) of cyclopropenones[147]. Their fractional degree of charge separation is in fact comparable to that in ordinary ketones; it is not necessary to assume a high contribution of the 'aromatic' cyclopropenium oxide structure (28) in the ground state. The special feature of cyclopropenone apparently lies in the ability of the cyclopropenium nucleus to delocalise the (modest) positive charge over all the carbon atoms of the ring.

It should be noted that the dipole moment of a polar substituted parent system is not necessarily the sum of the moments of the parent system and of the polar substituents[148]. The attempt[149] to derive the moment of heptafulvene from the moment of 8,8-dicyanoheptafulvene (29) (7.5 D) led to the erroneous value of 3 D. This is not theoretically valid; a contribution of a dipolar structure in the dicyano derivative does not necessarily mean a similar contribution in the case of the parent system.

The dipole moment of the fundamental members of the fulvene series deserves a special comment. Recently, Brown has reported that the dipole moment of pentafulvene (30) determined by microwave spectroscopy, is only 0.49 D[150], whilst the value obtained from solution measurements is 1.1 D[151]. The new value shows the absence of any significant contribution of a dipolar aromatic character in the ground state of pentafulvene. It is interesting and somewhat surprising that the formally alternant 3,4-dimethylenecyclobutene (31) which is an isomer of the non-alternant pentafulvene, has a high moment, namely, 0.62 D[152]. These results called for a re-evaluation of the alternant: non-alternant classification[14,153,154]. This classification implies that non-neighbour interactions are not significant. In (31) the non-neighbour (C-1—C-3) distance in the ring is unusually short (2.0 Å), so that it is not justified to ignore the corresponding core-resonance integrals. The molecule may thus be expected to possess non-alternant properties. This is the basis for the designation of (31) as *pseudo-alternant*[14,153,154].

5.2.10.4 Diffraction methods

(a) *The x-ray method*—x-Ray crystallographic diffraction is formally one of the best experimental methods for determining bond alternations as well as deviations from planarity of a conjugated system.

The following cases illustrate how powerful the x-ray method has been in evaluating the notion of aromaticity. (i) [18]annulene (17)[155,156] *v.* [16] annulene (18)[157]. A centrosymmetric molecule was indicated in (17) with only a small deviation from planarity in the solid state, but two types of carbon–carbon bonds of different length: six long outer bonds (1.419 Å) and twelve short inner bonds (1.382 Å). By contrast, [16]annulene (18) proved to be non-planar with almost complete π-electron localisation. Two sets of C(sp^2)—C(sp^2) bonds were observed: eight single bonds (1.454 Å) with alternating *trans* and *gauche* configuration and eight double bonds (1.333 Å) alternating between *cis* and *trans*. (ii) Bridged [14]annulenes[158]. Comparison between the molecular structures of *syn*- and *anti*-double-bridges[14]annulene derivatives showed the former (e.g. 1,6:8,13-propanediylidene[14]annulene

(32)[158] to be aromatic and the latter (e.g. 7-methoxycarbonyl-*anti*-1,6:8,13-bismethano[14]annulene (33)[159] to possess a polyolefinic structure. (iii) 3,5,8,10-Tetramethylcyclopenta[*e,f*]heptalene(34)[88,160]. A symmetrical planar structure and bond alternation (between 1.451 and 1.364 Å) in the periphery of the 7-membered rings were indicated. (iv) Fulvenes. In the fulvene series, the x-ray method may throw light upon the contribution of dipolar 'aromatic' structures in the ground state. The evidence of bond lengths and angles from the molecular structure of 2-formyl-6-dimethylaminopentafulvene is in

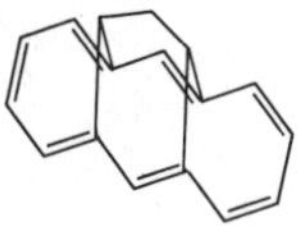

(32) 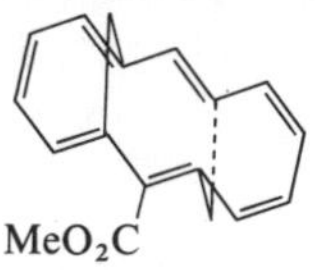 (33) 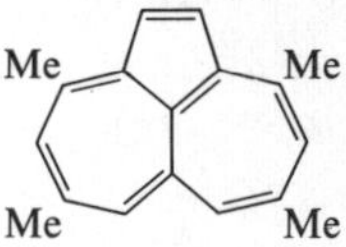 (34)

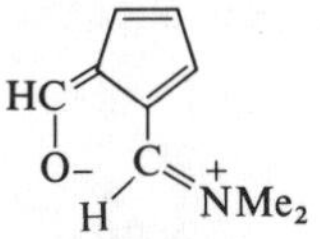

(35) 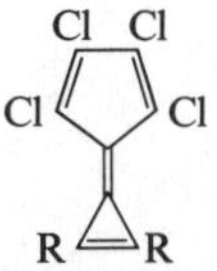 (36) R = Ph (37) R = Pr (38) 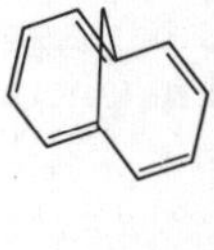 (39)

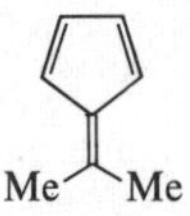

favour of the dipolar structure (35) 8,8-Dicyanoheptafulvene (29) proved to be planar with a long exocyclic C-7—C-8 double bond (1.42 Å)[162], while in its 1,6-dimethyl derivative the ring adopts a boat form, and the length of the exocyclic double bond is considerably shorter (1.37 A)[163]. Finally, in 1,2,3,4-tetrachloro-5,6-diphenyltriapentafulvalene (36), and in the 5,6-dipropyl analogue (37) the x-ray investigation indicated a marked bond-length alternation. It is noteworthy, however, that in the three-membered ring of (36) the bond lengths are considerably shorter than the theoretical values[164,165].

(b) *Electron diffraction* – The electron-diffraction method provides, similarly to the x-ray method, direct evidence on molecular geometry, including bond alternation and deviation from planarity in conjugated systems. The method is applied to molecules with sufficient vapour pressure. It is thus 'complementary' to the x-ray method and is suitable in particular to 'parent' systems which are in the gas or liquid state.

Following are a few examples which illustrate the results obtained by the electron diffraction method.

The most reliable information on the 'aromaticity' of pentafulvenes came from the electron-diffraction study of 6,6-dimethylpentafulvene (38)[166]. This study indicated lack of 'aromatic' character: a typical bond alternation in the ring with interatomic distances similar to those present in butadiene. Likewise, the analysis of 3,4-dimethylenecyclobutene (31)[167] revealed a pronounced single bond—double bond alternation. An electron diffraction investigation of 1,6-methano[10]annulene (39) showed that the hydrocarbon in the gas phase possesses a near planar C_{10}-perimeter, and qualifies as an aromatic molecule[168].

(c) *Neutron diffraction*—Due to the fact that neutrons are scattered by atomic nuclei, neutron diffractions are not subject to the errors of x-rays, which are scattered by electrons. The method which is frequently applied in combination with the x-ray method often leads to a better location of the hydrogen atoms and a better charge density picture of the molecule under study[169]. Neutron-diffraction studies of phenanthrene[170,171] showed that the distance between the angular overcrowded hydrogen atoms is 2.01–2.04 Å, considerably less than the usual van der Waals distance.

It is hoped that the diffraction methods will be more accessible to the organic chemist for the purpose of evaluating the notion of aromaticity in aromatic compounds.

5.3 MAGNETIC ASPECTS OF AROMATICITY

5.3.1 Introduction

When a diamagnetic molecule is placed in a magnetic field $\boldsymbol{H}$, it acquires an induced magnetic moment $\boldsymbol{M}$:

$$\boldsymbol{M} = \chi \boldsymbol{H}$$

where χ is the molecular susceptibility. The molar susceptibility is given by

$$\chi_{\mathrm{M}} = N_{\mathrm{A}}$$

where N_{A} is Avogadro's number.

For molecules devoid of cyclid π-electron delocalisation, the molar susceptibility is, according to Pascal's rule, the sum of atomic susceptibilities and certain structural correctionsλ[172].

$$\chi_{\mathrm{M}} = \Sigma \chi_{\mathrm{at}} + \Sigma \lambda$$

χ_{M} of a compound in a fluid system can be expressed as the average of three orthogonal components:

$$\chi_{\mathrm{M}} = \tfrac{1}{3}(\chi_x + \chi_y + \chi_z)$$

Aromatic molecules exhibit two striking features with respect to their diamagnetic susceptibilities:

(a) Enhanced diamagnetic susceptibility. The molar susceptibilities are larger than one would expect on the basis of the additivity rules. For example, for Kekulé benzene, $\chi_{\mathrm{M}}^{\mathrm{calc}} = -39.66 \times 10^{-6}$ c.g.s. while $\chi_{\mathrm{M}}^{\mathrm{ob}} = -55.1 \times 10^{-6}$ c.g.s.

(b) Enhanced diamagnetic anisotropy. The measured susceptibility of χ_z normal to the molecular plane is always much greater than those measured in the plane of the molecule, i.e.

$$\chi_x \approx \chi_y \ll \chi_z$$

The diamagnetic anisotropy of a planar molecule is defined by

$$\Delta\chi_{\mathrm{M}} = \chi_z - \tfrac{1}{2}(\chi_x + \chi_y)$$

In general,

$$\Delta\chi_{\mathrm{M}} = \chi_{zz} - \tfrac{1}{2}(\chi_{xx} + \chi_{yy})$$

Thus, aromatic molecules exhibit an enhanced diamagnetic anisotropy, i.e. $\Delta\chi_M \gg 0$. In benzene, $\Delta\chi_M = -59.7 \times 10^{-6}$ (Ref. 173). These effects in aromatic molecules have been attributed to the presence of appreciable cyclic delocalisations of π-electrons ('*ring currents*') in these systems.

5.3.2 The ring-current effect

5.3.2.1 *The ring-current model and the chemical shift*

The simplest ring-current model is due to Pauling[24]. According to Pauling's free electron model, the π-electrons of an aromatic molecule, under the influence of the applied external magnetic field $\boldsymbol{H}$, flow freely from one carbon atom to another and thus form interatomic currents around the ring. If $\boldsymbol{H}$ is directed upward through the molecular plane, the ring current circulates clockwise around $\boldsymbol{H}$, and gives rise to a large induced moment, also perpendicular to the ring, but which opposes $\boldsymbol{H}$. This induced moment causes an additional negative susceptibility perpendicular to the plane, thus explaining the anomalous exaltations and anisotropies. The induced current in Pauling's model corresponds to one electron. In a multiple-electron system, it is multiplied by the number of π-electrons, and is independent of the ring size. Thus it should be equal to benzene, cyclopentadienyl anion and tropylium cation. The quantum mechanical treatment of Pauling's model, due to London[174] assumed planarity, conjugation, molecular orbitals as linear combination of gauge-invariant atomic orbitals of the molecule and a uniform $\boldsymbol{H}$.

Another important magnetic property of aromatic molecules which may be explained by the existence of ring currents, is the large *chemical shift* of the high-resolution magnetic resonance signal of aromatic compounds[14, 172, 175–178]. While ethylenic proton in a ring (cyclohexa-1,3-diene) appears at δ 5.77 p.p.m., the protons of benzene appear at δ 7.27 p.p.m. The shift may be of either sign: in [18]annulene[17], the external protons appear at $\delta =$ 8.9 p.p.m. while the internal protons appear at $\delta = 1.8$ p.p.m.[179].

According to Pople[180] the effect of the induced current in the benzene ring is assumed to be equivalent to that of a point dipole placed at the centre of the ring. Applying Pauling's model under these conditions leads to the following equation for the chemical shift in polycyclic hydrocarbons:

$$\delta_{\mathrm{H}} = \frac{e^2 a^2}{2mc^2} \sum_i R_i^{-3}$$

where R_i is the distance to the proton from the centre of the ith ring and a is the radius of the ring current.

More generally, the chemical shift of a proton in the vicinity of the rings is given by

$$\delta(\text{p.p.m.}) = 2\beta(2\pi e/hc)^2(S^2/a^3) \sum_\mu J_\mu K_\mu$$

where S is the area of the benzene ring, β is the benzene exchange integral, a is the length of the benzene C—C bond, J_μ is the ring current intensity of the ring μ, and K_μ is a distribution function of the effect of the anisotropy in space[15, 172, 175]. Accordingly, the benzenoid hydrocarbons are strongly deshielded compared with the respective olefins.

5.3.2.2 *Induced ring current as a criterion of aromaticity*

Elvidge and Jackman[181, 182] considered the chemical shift as a quantitative measure of the ring current and consequently defined an *aromatic compound as a compound which will sustain an induced ring current*. The magnitude of the ring current will be a function of π-electron delocalisation around the ring, and therefore a quantitative measure of aromaticity. On the basis of this criterion, pyrid-4-one possesses 35% of the aromatic character of benzene[183]. One of the problems encountered by this method is the choice of proper model compounds for comparison with the 'aromatic' molecule under study. The controversy in estimating the degree of aromaticity of furan and thiophen is partially due to this difficulty[184, 185]. It was emphasised that the ring current criterion is most helpful in the detection rather than in the quantitative measurement of aromaticity.

The ring-current criterion of aromaticity is not free of limitations[14]. Experimental[186] and theoretical[187, 188] studies have shown that the low-field shift of the ring protons of aromatic molecules, as compared to that of the respective ethylenes, is due not only to the ring current, but also to local induced atomic currents[172]. Electron densities also play a role, as exemplified in the chemical shifts of cyclo-octatetraene and its aromatic dianion which are almost identical[189]. The theoretical basis of the ring current criterion has been criticised by Musher[190, 191], who argued that the chemical shift of aromatic hydrocarbons, generally attributed to π-electron ring currents, can be correctly represented as a sum of contributions from localised electrons of both π and σ characters. According to Musher, any connection between the large chemical shifts and magnetic anisotropies of aromatic hydrocarbons and the 'aromatic character' of these molecules is only indirect, i.e. through the *geometry*. Thus the definition of aromaticity in terms of unusually large chemical shifts seems to be without justification. Instead, 'aromaticity' should be defined as that class of physical and chemical properties possessed by roughly planar cyclic molecules in which the bond angles at the carbon atoms are roughly 120 degrees (240 degrees); i.e. the properties of benzene, condensed hydrocarbons, etc.[190]. This 'theoretical' controversy is still unsettled[14, 192, 193].

5.3.2.3 *Experimental evidence for the ring-current effect*

(a) *Diamagnetic and paramagnetic ring current effects*—The most convincing experimental evidence in favour of the ring-current criterion comes from Sondheimer's n.m.r. studies in the annulene series[14, 15, 194–198]. A clear distinction was revealed between [$4N+2$]annulenes and [$4N$]annulenes. This significant phenomenon was rationalised theoretically[199–201]. It was pointed out that in cyclic conjugated molecules, the π-electronic ring current is the difference between a *diamagnetic part* which depends only upon the electron density in the ground state and a *paramagnetic part* which is a sum of contributions from the electronically excited states. The paramagnetic ring current will be particularly large if the molecule possesses suitable very low lying excited states, e.g. the [4*N*]annulenes, under the Jahn–Teller effect. In the [4*N*+2]annulenes, the predominantly diamagnetic ring-current shifts the outer protons downfield and the inner protons upfield. In the [4*N*] annulenes, the reverse trend operates: the predominantly paramagnetic ring-current shifts the outer protons to a higher field and the inner protons to a lower field. These conclusions led to a reformulation of the ring current criterion of aromaticity and to a new definition of *anti-aromaticity*, in terms of magnetic criteria. The presence of an induced *diamagnetic ring current* is an indication of aromatic character, while anti-aromaticity is shown by the presence of an induced *paramagnetic ring current*. On the basis of these magnetic criteria, the [4*N*]annulenes and the [4*N*+2]annulenes should be considered as anti-aromatic and aromatic molecules respectively. The n.m.r. spectrum of 5-bromo-1,9-bisdehydro[12]annulenes (40) provides a striking confirmation of these ideas[202]. The spectrum contains a signal at 16.4 p.p.m. attributed to the intra-annular proton (H-6). Due to the size of the bromine

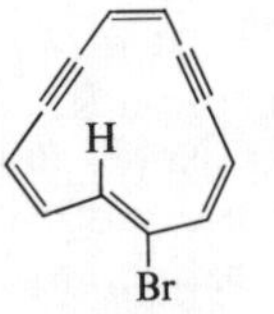

H
Br
(40)

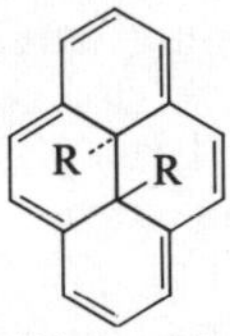

R R
(41) R = Me
(42) R = H

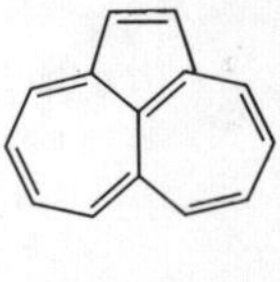

(43)

atom, no conformational interconversion of this system can occur, so that the spectrum represents a single conformer. It should be noted that in addition to the paramagnetic ring current effect, the anisotropy of the sp-hybridised carbon atoms and the neighbouring π-electrons of the triple bonds in the same plane contribute somewhat to the chemical shift of H-6.

The concept of the ring current has been successfully applied by Boekelheide to aromatic molecules having substituents within the cavity of the π-electron cloud[203–205], e.g. *trans*-15,16-dimethyl-15,16-dihydropyrene (41). The n.m.r. spectrum of (41) consists of bands at 7.98–8.67 p.p.m. due to the annular ring protons and a high-field singlet at δ −4.25 p.p.m. due to the intra-annular methyl protons. In the unsubstituted 15,16-dihydropyrene (42), the observed chemical shift (−5.49 p.p.m.) of the internal protons is remarkably higher, and must be very close to the region of the highest ring current effect[205].

Bond alternation causes a quenching of the ring current in the annulenes[199, 201]. The larger the ring, the more effective is the quenching for a given amount of bond alternation. The ring current intensities per unit benzene area of both the [4*N*+2]annulenes and the [4*N*]annulenes decrease with the ring size. In [*N*]annulenes with *N* sufficiently large, bond alternation predominates and the presence of an induced ring current will be negligible. Under such circumstances, both the [4*N*]annulenes and the [4*N*+2] annulenes should be considered non-aromatic[15]. Diamagnetic and paramagnetic ring currents in potential aromatic and anti-aromatic molecules have recently been studied by application of solvent effects[206].

Calculations of the π-electron ring currents in a series of tri- and tetracyclic non-alternant conjugated hydrocarbons, coupled with new evidence led in some cases to ambiguous results[122]. For example, in cyclopent[*ef*]heptalene (43), diamagnetic and a paramagnetic ring currents are induced by the five- and seven-membered rings respectively[88, 122]. Consequently, there exist π-electron systems which sustain diamagnetic ring currents in some rings and concomitantly paramagnetic ring currents in others. For such systems, aromaticity cannot be defined in terms of the ring current concept.

(b) *Carbon*-13 *magnetic resonance studies* – The ring-current criterion may be reflected also in ^{13}C chemical shifts[14, 172]. Diamagnetic ring current contributions to carbon-13 shifts have been postulated to account for the high-field shift relative to benzene (+3.6 p.p.m.) of the internal carbon atoms in pyrene[207]. The observed ^{13}C shifts in biphenylene were explained in terms of a paramagnetic ring current in the four-membered ring[14, 208].

The value of ^{13}C chemical shifts as an insight into the electronic structure of aromatic compounds, and in particular its dependence upon the charge density, has been established[209]. A plot of the ^{13}C chemical shift against charge density in cyclic conjugated ions ($(CH)_n^{m\pm}$) gives a straight line, indicating that the diamagnetic ring current has little effect on the position of the ^{13}C resonances[210]. The large $J(^{13}C—H)$ (265 Hz) observed in the aromatic cyclopropenyl cation corresponds to sp hybridisation at the carbon[211]. ^{13}C Chemical-shift ranges provide a marked distinction between alternant and non-alternant systems (ranges of <10 p.p.m. and 14–22 p.p.m. respectively). Both π- and σ-electronic variations must be considered in analysing the relationship between ^{13}C shifts and π-electron density[209]. The aromatic stability of cyclohept[*fg*]acenaphthylene (16) is reflected in the narrow ^{13}C chemical shift range[212]. The high-field shifts of the intra-annular quaternary carbon atoms imply a peripheral $4n+2$ electron ring-current flow.

(c) *N.M.R. coupling constants* – Theoretical studies by Karplus[213] indicate that n.m.r. *vicinal* coupling constants should decrease linearly with increasing length of the intervening carbon–carbon bond, provided constant angles and hybridisation are maintained. *Vicinal* coupling constants may thus serve as a means of evaluating the extent of bond alternation and consequently of aromatic character. Linear relations between π-bond order and *vicinal* coupling constants have in fact been noted in a series of five-, six-, and seven-membered rings[214–216] and polycyclic benzenoid hydrocarbons[217, 218]. The following examples illustrate the application of the coupling constants method. The correlation of the coupling constants of tropone and 8,8-dicyanoheptafulvene (29) and those of related systems indicate that they exhibit bond alternation characteristic of polyenes and polyenones and thus do not exhibit appreciable aromatic character[219]. Comparison of the vicinal coupling constants in 3-formyl-1,2-dipropyltriapentafulvalene (19) ($J_{2,3}$ = 2.51, $J_{3,4}$ = 5.05 Hz) with those of 1-formyl-6,6-diphenylpentafulvene (44) ($J_{2,3}$ = 3.25, $J_{3,4}$ = 4.21 Hz) indicate that the bond alternation is much less pronounced in the former compound[220]. In the pentafulvene series in general, the results imply a strong bond alternation[14, 151]. The availability of accurate bond-length data from x-ray measurements of some pentafulvenes allowed a consideration of various n.m.r. coupling constant (J)–bond length (d) relationships in the pentafulvene series[221]. It appears that the linearity of the $\Delta d(d_b - d_a)$ v. $\Delta J(J_a - J_b)$ plot is better than that of the d v. J plot. The former line indicates that a ΔJ value of *c.* 1 Hz corresponds to essentially equal C—C distances in a and b.

It should be noted that the coupling constants method may give faulty predictions, e.g. 4,8-dihydrodibenzo[*cd*, *h*]pentalene (45)[222]. Empirical cor-

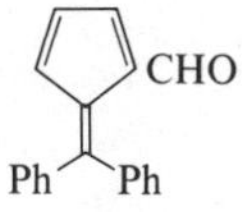

CHO

Ph Ph

(44)

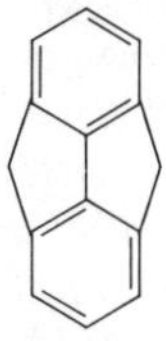

(45)

relations between n.m.r. coupling constants and various dihedral angles suggest that the molecule is cup-shaped. According to x-ray studies, however, the molecule is planar, with considerable bond length and angle deformations. Thus, it appears to be a dangerous practice to apply n.m.r. correlations when the molecule is considerably strained.

5.3.3 Enhanced diamagnetic susceptibility anisotropy as a criterion of aromaticity

Previous attempts to apply directly the diamagnetic anisotropy of cyclic conjugated hydrocarbons as a criterion of aromaticity were limited by the difficult experimental problems encountered in the direct determination of the anisotropies. Particularly limiting was the requirement of single crystals in which the orientation of the molecule was known in advance[223]. Several attempts to estimate anisotropies from more readily available molar-magnetic susceptibility data did not succeed[14, 224]. Recently, however, theoretical[225] and experimental[226, 227] methods have been developed of using gas-phase microwave spectra to measure the magnetic-susceptibility anisotropies in any molecule which has a microwave spectrum. These methods, which are based on the observations of the rotational Zeeman effect, are capable of measuring directly and quite accurately the magnetic susceptibility anisotropy tensor elements. By comparing the $\Delta\chi_M$ values of cyclic conjugated molecules obtained by this direct method with the respective values of their open-ring analogues, a criterion of aromatic character may be established[228]. A dramatic effect was observed by comparing benzene and cyclohexa-1,3-diene; the anisotropy for the latter (-7.4×10^{-6} erg G^{-2} mol^{-1}) is only *c.* 12% that of benzene. This was interpreted as due to both non-local and local contributions. The local contribution in benzene was estimated to be one half or less of the observed anisotropy[229]. The large anisotropy in benzene relative to cyclohexa-1,3-diene was interpreted as strong evidence for the existence of a substantial ring current (non-local contribution) in aromatic compounds[229].

5.3.4 Diamagnetic susceptibility exaltation as a criterion of aromaticity

It has been pointed out that it is unnecessary to determine the anisotropy in order to discover if the electronic characteristics causing it are present[230–232]. Cyclic delocalisation manifests itself not only by the presence of a large perpendicular anisotropy but also by an increase in the average susceptibility over that expected for a non-delocalised structure. This is the basis of the introduction of the concept of diamagnetic susceptibility exaltation Λ which is defined as follows:

$$\Lambda = \chi_M - \chi'_M$$

where χ_M is the experimental molar susceptibility of a compound and χ'_M is the susceptibility estimated for a cyclopolyene of that structure, i.e. omitting any 'corrections' for ring current. The values of χ'_M may be obtained by applying the magnetic-susceptibility-increment system of Haberditzl[233]. All compounds which possess cyclically delocalisable electrons will exhibit $\Lambda \gg 0$, while for all other compounds, $\Lambda \approx 0$. Thus, *exhibition of diamagnetic susceptibility exaltation by a compound implies that it is aromatic*[230–232].

According to this criterion, benzene ($\Lambda = 13.7$) is aromatic, while pentafulvene (30) ($\Lambda = 1.1$) is non-aromatic. Observation of appreciable negative exaltations for [16]annulene (18) and heptalene (46) ($\Lambda = -5$ and -6

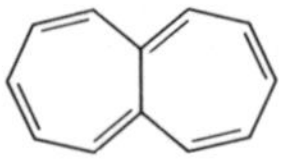

(46)

respectively) is consistent with the theory of induced paramagnetic ring current. It is interesting to compare the values obtained for naphthalene, azulene and 1,6-methano[10]annulene (39) ($\Lambda = 30.5$, 29.6 and 36.8 respectively). The exaltation of the 'simple' non-cross-conjugated derivative of [10]annulene (39) is considerably larger than that of the two 'bridged' [10]annulenes, naphthalene and azulene.

The examples of cycloheptatriene and cyclopentadiene ($\Lambda = 8.1$ and 6.5 respectively) are particularly intriguing: both are classified as aromatics[230–232]. Crystallographic[234], electron diffraction[235] and thermochemical[126] studies indicate a certain degree of π-electrons delocalisation in cycloheptatriene. This conclusion was rationalised in terms of overlap of the indented 1–6 π-orbital lobes of the buckled, boat-shaped molecule. Cyclic delocalisation via hyperconjugative π–σ interactions has been suggested as the origin for the high exaltation value of cyclopentadiene[230–232]. It is somewhat supported by the susceptibility anisotropies of cyclopentadiene and isoprene obtained recently by the rotational Zeeman effect[229]. In any event, the case of cyclopentadiene should be considered 'anomalous'.

The importance of the diamagnetic susceptibility exaltation as a criterion of 'aromaticity' stems from the realisation that it is capable of combining theory with experiment. On the one hand, it is an empirical, experimental criterion which can relatively easily be measured. On the other hand, it is directly related to an easily calculated theoretical quantity, the London diamagnetism (K_L) which gives the contribution of ring current to susceptibility. It can be shown that the two quantities are proportional: $\Lambda \propto K_L$[231, 232].

5.3.5 The Faraday effect

The Faraday effect reflects the rotation of the plane of polarisation of light by molecules in a magnetic field. For molecules with cyclic π delocalisation, the experimental value is always higher than the theoretical value, i.e. the figure calculated by neglecting the delocalisation of electrons. The difference E between the experimental and the theoretical values is the magnetic conjugation excess. E(aromatics) − E(aliphatics) expresses the existence of a ring current in aromatic molecules. On the basis of a linear relationship between E and the gradient of the $(\sigma+\pi)$ electronic charges in conjugated molecules belonging to the same group, Labarre has proposed the following definition of aromaticity[236–238].

'A cyclic molecule is aromatic when it contains a distribution of delocalised π electrons susceptible to be set in ordered movement when submitted to a magnetic field, following the absence of an important gradient of localised $(\sigma+\pi)$ charges or the atoms of the ring, to give rise to a Pauling–Pople current.' This aromaticity will either be measured experimentally by means of the Faraday effect (measure of E), or calculated theoretically from the gradient of the $(\sigma+\pi)$ electronic charges.

In spite of this definition, Labarre considered the term 'aromaticity' to be ambiguous and outdated. Instead, he proposed to use several new words so that each one will refer to one of the multiple chemical or physical properties that were intended to be expressed by 'aromaticity'. For the phenomenon measured by the Faraday effect and by n.m.r. in cyclic molecules that are the seat of π electron delocalisation, Labarre suggested the term '*potential strobilism*' (*Gk.* 'children's merry-go-round').

5.4 CRAIG RULES OF AROMATICITY AND THE CONCEPT OF PSEUDO-AROMATICITY

The concept of pseudo-aromaticity refers, in the broad sense, to all non-benzenoid aromatic compounds. The term 'pseudo-aromatic' has been attributed to any non-benzenoid system for which the problem of the presence or the absence of an aromatic character arises.

In the more specific connotation, the concept of pseudo-aromaticity is associated with Craig's rules for aromaticity[6, 7, 10, 14, 16, 18–19, 239, 240]. Craig's rules are empirical guidelines for the classification of cyclic conjugated hydrocarbons. Basically, Craig's rules relate aromatic character to high resonance energy: they try to predict on the basis of certain symmetry features whether the resonance energy between the cannonical structures, representing the ground state of a conjugated hydrocarbon is high or low.

Craig's rules distinguish between molecules with a valence bond wave function totally symmetrical in the ground state and molecules with a valence bond wave function not totally symmetrical in the ground state.

The former were designated aromatic while the latter were designated pseudoaromatic. A non-totally symmetric ground state is indicative of an abnormal π-electron system. (a) In an *aromatic molecule*, the valence-bond wave function in the ground state is *totally symmetrical*, in the sense that it is unchanged under all the symmetry operations that transform the nuclear framework of the molecule into itself. (b) In a *pseudo-aromatic molecule*, the valence-bond wave function in the ground state is *non-totally symmetrical*, in the sense that it changes sign under some of the operations of the nuclear framework.

The necessary element of symmetry for the application of Craig's rules is a symmetry axis of order two passing through two or more π centres and transforming one Kekulé structure into itself or into another Kekulé structure which belongs to the same cannonical set. The application of Craig's rules to a molecule which fulfils this condition is done as follows: The molecule is labelled with an equal number of spin symbols α and β (analogously to the 'starring procedure') in such a way as to give opposite spins to the ends of the double bonds on one Kekulé structure, but with a minimum number of identical neighbouring spins. A symmetry operation which transforms this structure (A) into itself or into another of the same cannonical set (B) is characterised by two numbers, f and g; f is the number of interchanges of π-electron centres affected by the symmetry operation and g is the number of interchanges of spin symbols required to restore the original labelling scheme from the transformed one. According to the valence-bond method, the symmetry of the ground-state of a molecule can be deduced from the character χ of its symmetry operations, where

$$\chi = (-1)^{f+g}$$

The transformation is $A \rightarrow (-1)^{f+g}B$.

A molecule for which $f+g$ is even, has a totally symmetrical valence-bond wave function in the ground state and therefore is aromatic, while a molecule for which $f+g$ is odd, has a non-totally symmetric valence-bond wave function in the ground state and is pseudo-aromatic. It should be noted that the characters are invariant to the original labelling of the spins, while g is not invariant to this labelling. Thus, for any given structure, a correct application of the rules should show unequivocally whether $f+g$ is even or odd.

According to Craig's rules, benzene is symmetric ($\chi = +1$) while cyclobutadiene and heptalene (46) are antisymmetric ($\chi = -1$).

Craig's rules have been criticised[18, 116, 241–244] for several reasons: they are limited by the fact that in many cases the symmetry conditions necessary for their application are not fulfilled; furthermore, they have occasionally led to ambiguous or faulty predictions. The ambiguities encountered in the application of the rules may often be removed by recalling, in labelling a molecule, that a structure and its symmetry transform must belong to the same cannonical set. The case of cyclopenta[*ef*]heptalene (43)[88, 243] illustrates this point. Structures (47a), (47b) and (47c) (as labelled) give $f+g$ values of 8, 7 and 12 respectively. It is obvious that structures (47a) and (47b) belong to different group representation and both cannot contribute to the same linear contribution valence-bond resonance hybrid. This indicates little stabilisation due to interactions of (47a) and (47b). As to the different characters of (47b) and (47c), it should be noted that the labelling of (47c) is not acceptable since its symmetry transform (47d) has similar spins at the ends of two of the double bonds. In any event, if a molecule may be represented by any Kekulé structures that lead to a negative character of its symmetry operations, it is abnormal from the valence-bond point of view[10,19]. An alternative explanation is to consider cyclopenta[*ef*]heptalene as cyclohepta[*cd*]azulene (47e),

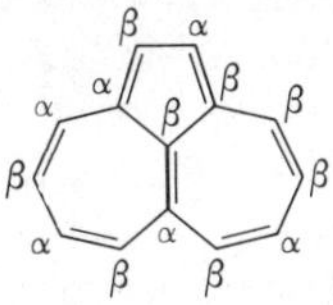

(47a) 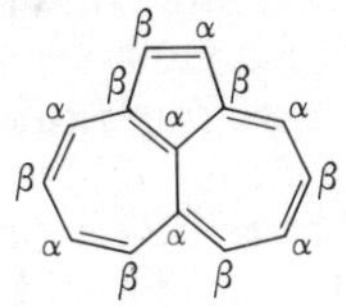 (47b) 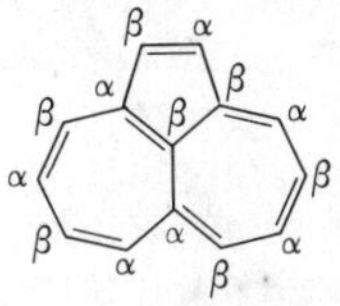 (47c)

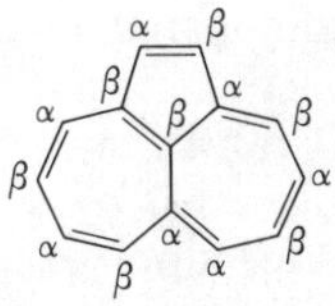

(47d)  (47e)

in which the two seven-membered rings may not be equivalent. A pseudo-aromatic compound could thus be viewed upon as a derivative of a normal aromatic compound[10]. Several cyclopenta[*ef*]heptalene derivatives have been synthesised and proved to possess aromatic character[88].

For dicyclopenta[*ef,kl*]heptalene (2)[69] an even $f+g$ value has been claimed for any of its Kekulé structures, with either of the two axes of symmetry; however, a careful application of the rules leads to two cannonical sets of structures, one with an even $f+g$ value, the other with an odd $f+g$ value. In such borderline cases, where the rules give ambiguous predictions (mostly in molecules with carbon atoms that are not part of the periphery), energy calculations might reveal the symmetry of the valence-bond wave function in the ground state[19].

A noteworthy example for which Craig's rules led to faulty predictions is pentafulvene (30). In this case, a totally symmetric valence-bond wave function in the ground state is predicted; the rules formally designate pentafulvene as 'aromatic'[7,116,242,244]. This prediction contrasts sharply with all the available experimental evidence, which demonstrates beyond doubt that the molecule is non-aromatic[151,166].

In order to try and clear this discrepancy, it is important to realise that for Craig's rules to have any significant value from the point of view of aromaticity, they should be applied only to those molecules for which there exists a problem of resonance between Kekulé structures in the ground state, that is, only to non-classical conjugated hydrocarbons. Accordingly, the rules could be reformulated in the following manner[16]. An *aromatic* molecule is a non-classical conjugated molecule that is represented in the ground state by a totally symmetric valence bond wave function. A *pseudo-aromatic* molecule is a non-classical conjugated molecule that is represented in the ground state by a non-totally symmetrical valence-bond wave function. Craig's rules may serve as a criterion to distinguish between aromatic and pseudo-aromatic molecules only when applied to non-classical conjugated molecules. In a classical conjugated molecule, the problem of resonance between distinct Kekulé structures does not exist, so that one cannot predict, on the basis of Craig's rules, whether such a molecule is aromatic, whatever may be the degree of symmetry of its valence-bond wave function[245]. The general case of the fulvenes both of [*n*]fulvenes and [*n,m*]fulvalenes, may illustrate this point: they are all classical conjugated molecules which satisfy the necessary symmetry conditions for applying Craig's rules. It can be proved (applying the detailed analytical procedure of the rules) that they have totally-symmetric valence-bond wave functions. Nevertheless, one cannot predict on the basis of Craig's rules that they should be aromatic. The interesting point is that the two properties, the classical character of a conjugated molecule and its representation by a totally-symmetric valence bond wave function, are mutually dependent. A non-classical conjugated molecule may be represented by a valence-bond wave function that is neither necessarily totally symmetric nor non-totally symmetric. It can be shown, however, that each Kekulé structure that is transformed into itself by a symmetry operation, does so with a positive sign[245]. In Craig's terminology,

$$A \rightarrow (-1)^{f+g}A, \quad (f+g \text{ are even})$$

Consequently, for a conjugated molecule, the classical character is a sufficient, but not a necessary condition for being represented in the ground state by a totally symmetric valence bond wave function*.

The concept of pseudoaromaticity has recently acquired the following new definition[246]: 'A neutral pseudo-aromatic molecule is considered to be one in which $(4n+2)$ p orbitals, consisting of olefinic and radical components, are coupled in relatively strain-free fashion with the resultant stabilisation in

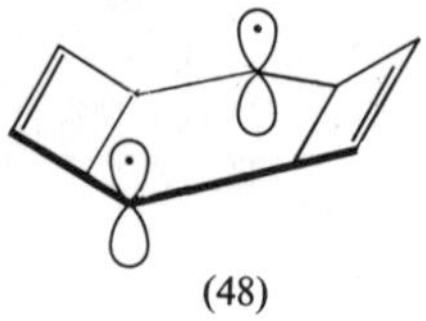

(48) 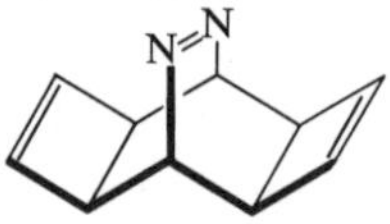 (49)

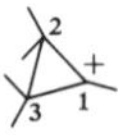

(50) 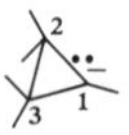 (51)

excess of that available in the absence of interaction'. In an uncharged six-electron system, the partition may take one of the following forms: (5+1), (4+2), (4+1+1), (3+3), (3+2+1), (3+1+1+1), (2+2+2), (2+2+1+1), (2+1+1+1+1) and (1+1+1+1+1+1). Partition (4+2) is representative of the transition state of the Diels–Alder reaction, (2+2+2) is benzene, and (3+3) is representative of the transition state of the Cope rearrangements. Partition (2+1+1+2) is representative of the diradical (48) which can be

*The author does not intend to advocate here the use of Craig's rules as a criterion of aromaticity, to claim their validity or to deny them. It is merely an attempt to clarify the original rules and to point out some of their limitations.

considered as a 'stretched benzene'. The analogous diradicals obtained from the thermal decomposition of derivatives of *exo, exo*-11,12-diazatetracyclo [4.4.2.0.^{2,5}0^{7,10}]dodeca-3,8,11-triene (49) did not show any pseudo-aromatic stabilisation.

Finally, pseudo-aromaticity has been defined in the case of cyclopropyl cation (50)[247] as the decrease in energy on allowing the delocalisation of the pseudo-π-system (constructed from $2p_z$ orbitals at C-2 and C-3 and the corresponding anti-symmetric combination of hydrogen 1s-orbitals) into the vacant $2p_z$ orbital of π-symmetry at C-1. Analogously, *pseudo-anti-aromaticity*[247] has been defined in the case of cyclopropyl anion (51) as the increase in energy on allowing the delocalisation of the electrons in the doubly occupied orbital symmetry at C-1 into the pseudo-π-system associated with $2p_z$ orbitals at C-2 and C-3, and the corresponding antisymmetric combinations of hydrogen 1s-orbitals.

5.5 TOPOLOGY AND AROMATICITY

Very recently, the concept of topology has been introduced as a basic condition for stabilising a system of interacting orbitals[248]. The fundamental building block of topology is a ribbon, defined as an intact conjugated polyene segment (subject to some constraints on its structure and its mode of interaction excluding also Möbius ribbons[43]). The symmetry of any ribbon is defined as pseudo-p or pseudo-d according to whether the phase relations at the termini resemble those of a p or a d orbital.

Four topologies were considered: *pericyclic, spirocyclic*[249–252], *longicyclic* and *laticyclic.*

Perturbation theory points to the crucial roles of the highest occupied molecular orbital (HOMO) and the lowest unoccupied molecular orbital (LUMO). There are only four patterns possible for the HOMO–LUMO symmetry. These may be represented by one of the four integers 0, 1, 2, 3 computed as $(n-z)$ *modulo* 4, where n is the number of centres and z is the electron occupancy.

The following topological survey is limited to ribbons of the even modes, 0 and 2. Both in the pericyclic topology, and in the spirocyclic topology, stabilisation is accessible to a mode 2 ribbon $(0+2)$, i.e., to one that contains $4q+2$ π electrons. Conversely, neither $(0+0)$ nor $(2+2)$ are stabilised. In the pericyclic topology, this is simply another version of Hückel's rule. It should be noted that only orbitals of the same pseudo-symmetry can interact. In the spirocyclic topology, orbital interaction is further limited to that between two pseudo d levels. In the longicyclic topology, the partitions (0, 2, 2) and (0, 0, 2) are stabilising while the two other partitions (0,0,0) and (2,2,2), are not. In contrast to both the pericyclics and the spirocyclics, the stabilising total number of π electrons can be either $4q$ or $4q+2$. In principle, more stabilisation is available in a longicyclic topology than in a pericyclic one. In the laticyclic topology, the partitions $(0+0+0)$ and $(2+2+2)$ lack stabilisation, $(0+0+2)$ and $(2+2+0)$ are stabilised by virtue of a single interaction, while $(0+2+0)$ and $(2+0+2)$ are stabilised by two interactions. In general, stabilisation can come only for $(0+2)$ interactions.

On the basis of this analysis, Goldstein and Hoffmann put forward the following *topological definition of aromaticity*[248] as 'the possession of that enhanced stabilisation which is achieved whenever an appropriate acyclic ribbon closes to an isoconjugate pericyclic one.'

A set of ribbons may be stabilised in one topology and destabilised in another. Among the sets of three ribbons, (2, 2, 2) contains $4q+2$ π electrons and is stabilised in a pericyclic topology but destabilised in a longicyclic and a laticyclic one, while (2, 2, 0) contains $4q$ π electrons and is stabilised as a longicyclic or a laticyclic but is destabilised in a pericyclic topology. Thus it

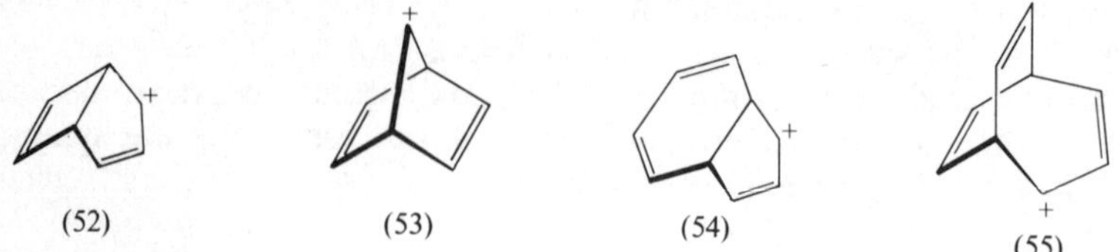

(52) (53) (54) (55)

becomes possible to predict the preferred direction of topological interconversions. For example, the destabilised pericyclic bicyclo[3.2.0]heptadienyl cation (52) is transformed into the stabilised longicyclic 7-norbornadienyl cation (53)[253] and the destabilised longicyclic bicyclo[3.2.2]nonatrienyl cation (54) is transformed to the stabilised pericyclic dihydroindenyl cation (55)[254].

The topological definition of aromaticity covers only the pericyclic topology. For the laticyclic and longicyclic topologies, the terms '*bicyclo-aromatic*'[255] and '*anti-bicyclo-aromatic*' have been introduced and defined as follows: bicyclo-aromatic hydrocarbons are the stabilised laticyclic and longicyclic molecules which contain $4q$ π electrons; anti-bicyclo-aromatic laticyclic and longicyclic molecules are those that are both destabilised and contain $4q+2$ π electrons. These terms do not cover the stabilised longicyclic and laticyclic molecules which contain $4q+2$ π electrons (2,0,0) and the destabilised ones which contain $4q$ π electrons (0,0,0). The distinction between (2,2,0) and (2,0,0) (as well as between (2,2,2) and (0,0,0)) is not based upon any presumed difference in electronic stabilisation (or destabilisation).

5.6 INDICES AND VARIATIONS ON THE THEME OF AROMATICITY

The overlaps and discrepancies in the various meanings of the term 'aromaticity' prompted a demand for a clear-cut, 'black and white' definition[256]. Within this context, new terms, including prefixes as well as quantitative indices were proposed to describe the phenomenon of aromaticity and its variations. The terms anti-aromaticity, non-aromaticity, and pseudo-aromaticity, and some others, have already been considered in the previous sections. The present section adds a few to the list and surveys the mushrooming of various indices associated with aromaticity.

5.6.1 Quasi-aromaticity

The term 'quasi-aromatic' and later 'regenerative' or meneidic[12, 257, 258] has been used by Lloyd to describe the chemical behaviour of some non-benzenoid compounds, e.g. 2,3-dihydro-1,4-diazepinium salts (56). These compounds and terms are discussed in another chapter of this volume[259].

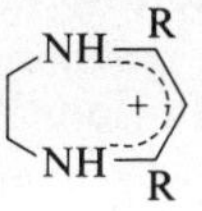

(56)

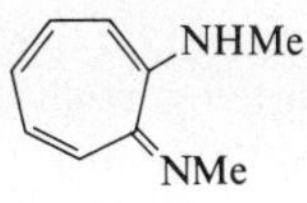

(57)

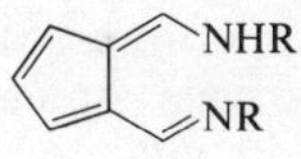

(58)

5.6.2 Non-classical aromaticity

The term 'non-classical aromaticity' in hydrogen-bonded systems has been chosen as a synonym for the assumed interaction between the hydrogen donor and acceptor atoms, allowing a cyclic delocalisation of the π-electron system[260, 261]. A typical illustration is *N,N′*-dimethyl-1-amino-7-imino-cyclohepta-1,3-5-triene (57) for which the x-ray determination[262] ruled out a fast tautomeric equilibrium, and indicated instead a symmetrical structure with a delocalised ten π-electron system. By contrast, the physical properties of 6-aminopentafulvene-2-aldimines (58) show that these molecules represent non-aromatic rather than non-classical aromatic systems[261].

5.6.3 Homoaromaticity

The basic idea of homoaromaticity[263, 264] is that a certain amount of delocalisation in a π-electron system may be achieved even 'where the cycle of overlapping orbitals has to be completed across an intervening group, such as a methylene, by the overlapping of p_π-orbitals not on adjacent carbon atoms'[19]. A typical illustration is the homotropylium ion (59)[265]. Under such circumstances, when the continuous polyene conjugation has been interrupted,

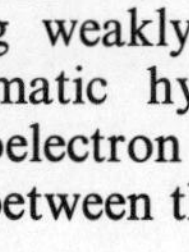

(59)

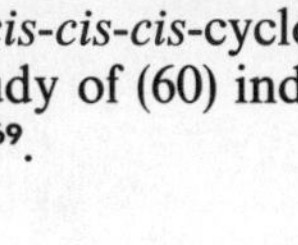

(60)

Hückel's rule still applies. Homoaromaticity concerns the delocalisation of π-electrons among weakly interacting double bonds. An example of a neutral homoaromatic hydrocarbon is *cis-cis-cis*-cyclononatri-1,4,7-ene (60)[266–268]. A photoelectron spectroscopy study of (60) indicated a through-space interaction between the π-orbitals[251, 269].

5.6.4 Indices

(a) Balaban[270, 271] proposed to define as aromatic all cyclic compounds with continuous conjugation satisfying the structural requirements for possessing a closed π-electron shell. The quantitative implications were included in an overall aromatic constant *K* expressing the electrophilicity or nucleophilicity relatively to benzene ($K = 0$). The types of first-row atoms which may take part in an aromatic system were ascribed each a numerical constant k. K could be obtained by summation of k values over the whole aromatic ring.

(b) In order to estimate the degree of 'aromaticity' of a system, Zahradník and co-workers proposed the following '*stability*' *index* (P), constructed by a proper combination of indices that reflect the necessary conditions of aromaticity[18, 272]:

$$P = [273(E_{D,sp} - 0.2796)(0.552 - \Delta q)(0.586 - F)]/S_r,$$

where $E_{D,sp}$ is the specific delocalisation energy, Δq is the difference between the maximum and minimum π-electron densities (related to the reactivity toward nucleophilic and electrophilic agents), F is the maximum free valence (related to the radical reactivity) and S_r is the maximum radical superdelocalisability (related to radical reactivity, oxidisability and reducibility). The P index has been used successfully in the non-alternant hydrocarbon series.

(c) A description of delocalisation, and thus implicity of aromaticity, in terms of bond index has been put forward[273]. The delocalisation index[274],

$$D = 1 - \frac{2}{m} \sum_{r=1}^{m} P - P_r$$

where P is the average bond order, P_r the bond order of the rth bond and m the number of peripheral bonds, gives a value of $D = 1$ for benzene, $D = 0$ for the Kekulé structure of benzene and $D = 0.57$ for pentafulvene (30).

(d) On the basis of a quantum chemical model of a chemical definition of aromaticity, a numerical index KK defining the susceptibility of a given π-electron compound to addition, has been proposed[275]:

$$KK = \min|L_{r,s}|,$$

where

$$L_{r,s} = E_A^{\pi}(rs) - E_0^{\pi} - 2\alpha.$$

E_0^{π} and E_A^{π} are the π-electron energy of the system before and after the reaction, respectively.

In aromatic compounds $KK \gg 3$. For benzene, $KK = 3.53$, compared with the values of 1.53, 2.52 and 2.99 obtained for Kekulé's benzene, hexa-1,3,5-triene, and pentafulvene (30), respectively. A classification of π-electron compounds on the basis of KK values has been attempted[276].

(e) *Ab initio* calculations on the benzene and napthalene isomers using the molecular-fragment approach, prompted Christoffersen[277] to define the aromaticity (A) of a given molecule (M) as

$$A = \left[\frac{E_{\pi}(\text{M}) - E_{\pi}(\text{ref})}{E_{\pi}(\text{ref})}\right] \times 100$$

where $E_{\pi}(\text{M})$ and $E_{\pi}(\text{ref})$ are the sum of the occupied π-orbital energies for the molecule in question and the open-chain reference molecule respectively. For benzene, $A = 11.073$; for hexatriene, $A = 0.000$; for pentafulvene (30), $A = 1.032$, while for 3,4-dimethylenecyclobutene (31), $A = 3.506$.

(f) Häfelinger defined aromatic compounds as planar cyclically conjugated π systems with completely filled π-electron shells, that are further characterised by extremely high thermodynamic and kinetic ground-state stability because of cyclic delocalisation which can be characterised in terms of the HMO-π-electron model by various energy criteria[278].

(g) The aromaticity degree A proposed by Julg[279, 280] takes into account simultaneously the uniformity of the lengths of the peripheral bonds and the resistance to the circulation of the electrons along the rings:

$$A = [1-225/n \sum_{(rs)} (1-d_{rs}/d^2] \prod_{(ij)} [1-(\Delta q_{ij}/d_{ij})^2]$$

where n is the number of peripheral bonds (rs); d_{rs} is their lengths, d is their mean length; d_{ij} is the distance between atoms i and j; and Δq_{ij} is the difference in the π charge on these atoms. For benzene $A = 1.00$, for pentafulvene (30) $A = 0.62$. The formula can express the 'gradation' in aromatic charter in quantitative terms as exemplified in the difference between the A values of azulene (0.85) and naphthalene (0.90). From the A values of pentalene (13) and heptalene (46) it appears that thc C_{2h} structure should be preferred over the D_{2h} one[281].

5.7 CONCLUSION

An attempt was made to survey the various approaches to the problem of 'aromaticity'. In many instances, when two well-defined criteria are in harmony, the agreement between them is at the most, qualitative in nature. Hopefully, a universal definition of 'aromaticity' will emerge that will allow simple, qualitative, as well as significant comparisons between molecules. Perhaps it is appropriate to end this review with the following optimistic accord made by the eminent scientist Sherlock Holmes[282]:

'When you follow two separate chains of thought, Watson, you will find some point of intersection which should approximate to the truth.'

Acknowledgement

I am deeply grateful to my teacher, Professor Ernst David Bergmann, who introduced me to the problems of 'aromaticity', for his constant interest, advice, and encouragement. I am also very much indebted to Mrs. Ayala Barak for her invaluable assistance and suggestions in the preparation of this manuscript.

References

1. Kekulé, A. (1865). *Bull. Soc. Chim. France,* **3,** 98
2. Snyder, J. P. (1969). *Nonbenzenoid Aromatics,* Vol. I (J. P. Snyder, editor) 1 (New York and London: Academic Press)
3. Badger, G. M. (1969). *Aromatic Character and Aromaticity* (Cambridge: Cambridge University Press)
4. Baker, W. and McOmie, J. F. W. (1955). *Progress in Organic Chemistry,* Vol. 3 (J. W. Cook, editor) 44 (London: Butterworths Scientific Publications)

5. Baker, W. (1956). *Perspectives in Organic Chemistry* (A. Todd, editor), 28 (New York: Interscience Publishers)
6. Craig, D. P. (1959). *Theoretical Organic Chemistry (Papers presented to the Kekulé Symposium)*, 20 (London: Butterworths Scientific Publications)
7. Craig, D. P. (1959). *Non-Benzenoid Aromatic Compounds,* (D. Ginsburg, editor), 1, (New York: Interscience Publishers)
8. Heilbronner, E. (1959). *Non-Benzenoid Aromatic Compounds,* (D. Ginsburg, editor), 171 (New York: Interscience Publishers)
9. Vol'pin, M. E. (1960). *Russ. Chem. Rev.*, **29,** 129
10. Streitweiser, A. Jr. (1961). *Molecular Orbital Theory for Organic Chemists,* (New York: Wiley)
11. Breslow, R. (1965). *Chem. Eng. News,* **43,** (26), 90
12. Lloyd, D. (1966). *Carbocyclic Non-Benzenoid Aromatic Compounds,* (London: Elsevier)
13. Aromaticity (Special Publication, No. 21). (1967). (London: The Chemical Society)
14. Jones, A. J. (1968). *Rev. Pure Appl. Chem.*, **18,** 253
15. Figeys, H. P. (1969). *Topics in Carbocyclic Chemistry,* Vol. 1, (D. Lloyd, editor), 269, (London: Logos Press)
16. Bergmann, E. D. and Agranat, I. (1971). *'Aromaticity, Pseudo-Aromaticity, Anti-Aromaticity', The Jerusalem Symposium on Quantum Chemistry and Biochemistry,* III (E. D. Bergmann and B. Pullman, editors), 9. (Jerusalem: The Israel Academy of Sciences and Humanities)
17. Baker, W. (1965). *Chem. Brit.*, **1,** 191, 250
18. Zahradník, R. (1971). *Nonbenzenoid Aromatics,* Vol. 2 (J. P. Snyder, editor), 1, (New York and London: Academic Press)
19. Chandler, G. S. and Craig, D. P. (1971). *Rodd's Chemistry of Carbon Compounds,* 2nd edition, (S. Coffey, editor), Vol. 3A, 5, (London: Elsevier)
20. Nagel, E. (1961). *The Structure of Sciences—Problems in the Logic of Scientific Explanation,* 99 (London: Routledge and Kegan Paul)
21. Robinson, R. (1959). *Non-benzenoid Aromatic Compounds,* (D. Ginsburg, editor), v, (New York: Interscience Publishers)
22. Robinson, R. (1958). *Tetrahedron,* **3,** 323
23. Peters, D. (1960). *J. Chem. Soc.*, 1274
24. Pauling, L. (1936). *J. Chem. Phys.*, **4,** 673
25. Dörr, F. (1971). DMS UV Atlas of Organic Compounts, Vol. V, E_1, (London: Butterworths Scientific Publications)
26. Baker, W. (1945). *J. Chem. Soc.*, 258
27. Ginsburg, D. (1959). *Non-Benzenoid Aromatic Compounds,* (New York: Interscience Publishers)
28. Hafner, K. (1964). *Angew. Chem. Int. Edn.*, **3,** 165
29. Garratt, P. J. and Sargent, M. V. (1969). Advances in Organic Chemistry – Methods and Results, Vol. 6 (E. C. Taylor and H. Wynberg, editors), 1, (New York: Interscience Publishers)
30. Thyagarajan, B. S. (1970). *Intra-Science Chem. Reports,* **4,** 1
31. Snyder, J. P. (1969). *Nonbenzenoid Aromatics,* Vol. 1, (New York and London: Academic Press)
32. Snyder, J. P. (1971). *Nonbenzenoid Aromatics,* Vol. 2, (New York and London: Academic Press)
33. International Symposium on the Chemistry of Nonbenzenoid Aromatic Compounds (held at Sendai, Japan, on 24th–28th August 1970). (1971). *Pure Appl. Chem.*, **28,** 111–398
34. *International Symposium on the Chemistry of Nonbenzenoid Aromatic Compounds* (held at Sendai, Japan, on 24th-28th August, 1970). Abstracts of Papers.
35. Thiele, J. (1900). *Ber. Deutsch. Chem. Ges.*, **33,** 666
36. Hückel, E. (1931). *Z. Physik,* **70,** 204; **72,** 310
37. Hückel, E. (1932). *Z. Physik.*, **76,** 628
38. Hückel, E. (1937). *Z. Elektrochem.*, **43,** 752
39. Hückel, E. (1938). *Grundzuge der Theorie ungesattigter und aromatischer Verbindungen* (Berlin: Verlag Chemie)
40. Heilbronner, E. and Bock, H. (1968). *Das HMO-Modell und seine Anwendung,* (Weinheim/Bergstr., Germany: Verlag Chemie)
41. Chung, A. L. H. and Dewar, M. J. S. (1965). *J. Chem. Phys.*, **42,** 756

42. Dewar, M. J. S. (1966). *Tetrahedron*, Suppl. **8**, Part 1, 75
43. Heilbronner, E. (1964). *Tetrahedron Letters*, 1923
44. Mason, S. F. (1965). *Nature (London)*, **205**, 495
45. Zimmerman, H. E. (1966). *J. Amer. Chem. Soc.*, **88**, 1564, 1566
46. Zimmerman, H. E. and Iwamura, H. (1970). *J. Amer. Chem. Soc.*, **92**, 2015
47. Zimmerman, H. E. (1971). *Accounts Chem. Res.*, **4**, 272
48. Craig, D. P. and Paddock, N. L. (1971). *Nonbenzenoid Aromatics*, Vol. 2, (J. P. Snyder, editor), 273, (New York and London: Academic Press)
49. Dewar, M. J. S. (1971). *Angew. Chem. Int. Edn.*, **10**, 761
50. Craig, D. P. (1959). *J. Chem. Soc.*, 997
51. Ashe III, A. J. (1968). *Tetrahedron Letters*, 359
52. Jahn, H. A. and Teller, E. (1937). *Proc. Roy. Soc.*, **A161**, 220
53. Salem, L. (1966). *The Molecular Orbital Theory of Conjugated Systems*, Chapter 8 (New York: Benjamin)
54. Longuet-Higgins, H. C. and Salem, L. (1959). *Proc. Roy. Soc.*, **A251**, 172
55. Figeys, H. P. (1971). *Tetrahedron*, **26**, 5225
56. Dewar, M. J. S. (1967). *Aromaticity (Special Publication, No. 21)*, 177, (London: The Chemical Society)
57. Breslow, R., Brown, J. and Gajewski, J. J. (1967). *J. Amer. Chem. Soc.*, **89**, 4383
58. Dewar, M. J. S. (1965). *Advances in Chemical Physics*, Vol. 8, (R. Daudel, editor), 65, (New York: Interscience Publishers)
59. Breslow, R. (1968). *Angew. Chem. Int. Edn.*, **7**, 565
60. Rabinovitz, M. (1971). *'Aromaticity, Pseudo-Aromaticity, Anti-Aromaticity' The Jerusalem Symposia on Quantum Chemistry and Biochemistry III*, (E. D. Bergmann and B. Pullman, editors), 208 (Jerusalem: The Israel Academy of Sciences and Humanities)
61. Breslow, R. (1971). *Pure Appl. Chem.*, **28**, 111
62. Trost, B. M., Bright, G. M., Frihart, C. and Brittelli, D. (1971). *J. Amer. Chem. Soc.*, **93**, 737
63. Cava, M. P. and Mitchell, M. J. (1967). *Cyclobutadiene and Related Compounds* (New York: Academic Press)
64. Breslow, R., Grubbs, R. and Murahashi, S.-I. (1970). *J. Amer. Chem. Soc.*, **92**, 4139
65. Platt, J. R. (1954). *J. Chem. Phys.*, **22**, 1448
66. Breslow, R., Horspool, W., Sugiyama, H. and Vitale, W. (1966). *J. Amer. Chem. Soc.*, **88**, 3677
67. Breslow, R. and Washburn, W. (1970). *J. Amer. Chem. Soc.*, **92**, 427
68. Trost, B. M. and Bright, G. M. (1967). *J. Amer. Chem. Soc.*, **89**, 4244
69. Anderson, Jr., A. G., MacDonald, A. A. and Montana, A. F. (1968). *J. Amer. Chem. Soc.*, **90**, 2993
70. Jutz, C. and Schweiger, E. (1971). *Angew. Chem. Int. Edn.*, **10**, 808
71. Zahradník, R. (1965). *Angew. Chem. Int. Edn.*, **4**, 1039
72. Prinzbach, H. and Knöfel, H. (1969). *Angew. Chem. Int. Edn.*, **8**, 881
73. Prinzbach, H., Knöfel, H. and Woischnik, E. (1971). *'Aromaticity, Pseudo-Aromaticity, Anti-Aromaticity', The Jerusalem Symposia on Quantum Chemistry and Biochemistry* III, (E. D. Bergmann and B. Pullman, editors), 269 (Jerusalem: The Israel Academy of Sciences and Humanities)
74. Prinzbach, H. (1971). *Pure Appl. Chem.*, **28**, 281
75. Barth, W. E. and Lawton, R. G. (1966). *J. Amer. Chem. Soc.*, **88**, 380
76. Barth, W. E. and Lawton, R. G. (1971). *J. Amer. Chem. Soc.*, **93**, 1730
77. Gleicher, G. L. (1967). *Tetrahedron*, **23**, 4257
78. Pople, J. A. (1953). *Trans. Faraday Soc.*, **49**, 1375
79. Brickstock, A. and Pople, J. A. (1954). *Trans. Faraday Soc.*, **50**, 901
80. Fukui, K., Imamura, A., Yonezawa, T. and Nagata, C. (1960). *Bull. Chem. Soc. Japan*, **33**, 1591
81. Roberts, J. D. (1958). *Developments in Aromatic Chemistry (Special Publication, No. 12)*, 111 (London: The Chemical Society)
82. Breslow, R., Kivelevich, D., Mitchell, M. J., Fabian, W. and Wendel, K. (1965). *J. Amer. Chem. Soc.*, **87**, 5132
83. Gompper, R. and Seybold, G. (1971). *'Aromaticity, Pseudo-Aromaticity, Anti-Aromacity', The Jerusalem Symposia on Quantum Chemistry and Biochemistry* III (E. D. Bergmann and B. Pullman, editors), 215 (Jerusalem: The Israel Academy of Sciences and Humanities)

84. Gompper, R. and Seybold, G. (1968). *Angew. Chem. Int. Edn.*, **7,** 824
85. Hafner, K. and Tappe, H. (1969). *Angew. Chem. Int. Edn.*, **8,** 593
86. Hafner, K. (1971). *'Aromaticity, Pseudo-Aromaticity, Anti-Aromaticity', The Jerusalem Symposia on Quantum Chemistry and Biochemistry* III, (E. D. Bergmann and B. Pullman, editors), 256 (Jerusalem: The Israel Academy of Sciences and Humanities)
87. Hafner, K., Bangert, K. F. and Orfanos, V. (1967). *Angew. Chem. Int. Edn.*, **6,** 451
88. Hafner, K. (1971). *Pure Appl. Chem.*, **28,** 153
89. Bergmann, E. D. (1968). *Chem. Rev.*, **68,** 41
90. Seitz, G. (1969). *Angew. Chem. Int. Edn.*, **8,** 478
91. Dewar, M. J. S. and Jones, R. (1968). *J. Amer. Chem. Soc.*, **90,** 2137
92. Davis, F. A., Dewar, M. J. S., Jones, R. and Worley, S. D. (1969). *J. Amer. Chem. Soc.*, **91,** 2094
93. Cais, M. (1971). *'Aromaticity, Pseudo-Aromaticity, Anti-Aromaticity', The Jerusalem Symposia on Quantum Chemistry and Biochemistry* III, (E. D. Bergmann and B. Pullman, editors), 96, (Jerusalem: The Israel Academy of Sciences and Humanities)
94. Coulson, C. A. and Longuet-Higgins, H. C. (1947). *Proc. Roy. Soc.*, **A191,** 39
95. Coulson, C. A. and Longuet-Higgins, H. C. (1947–1948). *Proc. Roy. Soc.*, **A192,** 16
96. Coulson, C. A. and Longuet-Higgins, H. C. (1948). *Proc. Roy. Soc.*, **A193,** 447, 456
97. Coulson, C. A. and Longuet-Higgins, H. C. (1948–1949). *Proc. Roy. Soc.*, **A195,** 188
98. Longuet-Higgins, H. C. (1950). *J. Chem. Phys.*, **18,** 265, 275, 283
99. Dewar, M. J. S. (1952). *J. Amer. Chem. Soc.*, **74,** 3341, 3345, 3350, 3353, 3355, 3357
100. Dewar, M. J. S. (1969). *The Molecular Orbital Theory of Organic Chemistry*, 191, (New York: McGraw-Hill)
101. Dougherty, R. C. (1968). *J. Amer. Chem. Soc.*, **90,** 5780, 5788
102. Dewar, M. J. S. and Longuet-Higgins, H. C. (1952). *Proc. Roy. Soc.*, **A214,** 482
103. Wilcox, Jr. C. F. (1968). *Tetrahedron Lett.*, 795
104. Dewar, M. J. S. and Trinajstic, N. (1969). *Tetrahedron Lett.*, 2129
105. Wilcox, Jr. C. F. (1969). *J. Amer. Chem. Soc.*, **91,** 2732
106. Dougherty, R. C. (1971). *J. Amer. Chem. Soc.*, **93,** 7187
107. Evans, M. G. (1939). *Trans. Faraday Soc.*, **35,** 824
108. Dewar, M. J. S. and Kirschner, S. (1971). *J. Amer. Chem. Soc.*, **93,** 4292
109. Dewar, M. J. S. and Gleicher, G. J. (1965). *J. Amer. Chem. Soc.*, **87,** 685
110. Dewar, M. J. S. and Gleicher, G. J. (1965). *J. Amer. Chem. Soc.*, **87,** 692
111. Dewar, M. J. S. and de Llano, C. (1969). *J. Amer. Chem. Soc.*, **91,** 789
112. Mullican, R. S. and Parr, R. G. (1951). *J. Chem. Phys.*, **19,** 1271
113. Hess, Jr., B. A. and Schaad, L. J. (1971). *J. Amer. Chem. Soc.*, **93,** 305
114. Hess, Jr., B. A. and Schaad, L. J. (1971). *J. Amer. Chem. Soc.*, **93,** 2413
115. Hess, Jr., B. A. and Schaad, L. J. (1971). *J. Org. Chem.*, **36,** 3418
116. Binsch, G., Heilbronner, E. and Murrell, J. N. (1966). *Molec. Phys.*, **11,** 305
117. Binsch, G. and Heilbronner, E. (1968). *Structural Chemistry and Molecular Biology* (A. Rich and N. Davidson, editors), 815, (San Francisco: Freeman)
118. Binsch, G. and Heilbronner, E. (1968). *Tetrahedron*, **24,** 1215
119. Binsch, G. (1971). *'Aromaticity, Pseudo-Aromaticity, Anti-Aromaticity', The Jerusalem Symposia on Quantum Chemistry and Biochemistry* III, (E. D. Bergmann and B. Pullman, editors), 25 (Jerusalem: The Israel Academy of Sciences and Humanities)
120. Binsch, G., Tamir, I. and Hill, R. D. (1969). *J. Amer. Chem. Soc.*, **91,** 2446
121. Binsch, G. and Tamir, I. (1969). *J. Amer. Chem. Soc.*, **91,** 2450
122. Jung, D. E. (1969). *Tetrahedron*, **25,** 129
123. Nakajima, T. (1964). *Molecular Orbitals in Chemistry, Physics and Biology*, (P. O. Löwdin and B. Pullman, editors), 451 (New York: Academic Press)
124. Nakajima, T., Toyota, A. and Yamaguchi, H. (1971). *'Aromaticity, Pseudo-Aromaticity, Anti-Aromaticity', The Jerusalem Symposia on Quantum Chemistry and Biochemistry,* III, (E. D. Bergmann and B. Pullman, editors), 227 (Jerusalem: The Israel Academy of Sciences and Humanities)
125. Nakajima, T. (1971). *Pure Appl. Chem.*, **28,** 219
126. Turner, R. B. (1969). *Theoretical Organic Chemistry (Papers presented to the Kekulé Symposium)*, 67. (London: Butterworths Scientific Publications)
127. Wheland, G. W. (1955). *Resonance in Organic Chemistry* (New York: John Wiley)
128. Turner, R. B., Lindsay, W. S. and Boekelheide, V. (1971). *Tetrahedron*, **27,** 3341
128a. Cook, M. J., Katritsky, A. R., Linda, P. and Tack, R. D. (1971). *Chem. Commun.*, 510

129. Binsch, G. (1968). *Topics in Stereochemistry,* Vol. 3, (E. L. Eliel and N. L. Allinger, editors), 97, (New York: Interscience Publishers)
130. Gilles, J.-M., Oth, J. F. M., Sondheimer, F. and Woo, E. P. (1971). *J. Chem. Soc.(B),* 2177
131. Oth, J. F. M. and Gilles, J.-M. (1968). *Tetrahedron Lett.,* 6259
132. Beezer, A. E., Mortimer, C. T., Springall, H. D., Sondheimer, F. and Wolovsky, R. (1965). *J. Chem. Soc.,* 216
133. Kende, A. S., Izzo, P. T. and Fulmor, W. (1966). *Tetrahedron Lett.,* 3697
134. Downing, A. P., Ollis, W. D., Nógrádi, M. and Sutherland, I. O. (1971). *'Aromaticity, Pseudo-Aromaticity, Anti-Aromaticity', The Jerusalem Symposia on Quantum Chemistry and Biochemistry* III, (E. D. Bergmann and B. Pullman, editors), 296 (Jerusalem: The Israel Academy of Sciences and Humanities)
135. Egan, W., Tang, R., Zon, G. and Mislow, K. (1970). *J. Amer. Chem. Soc.,* **92,** 1442
136. Egan, W., Tang, R., Zon, G. and Mislow, K. (1971). *J. Amer. Chem. Soc.,* **93,** 6205
137. Figeys, H. P. and Dralants, A. (1971). *Tetrahedron Lett.,* 3901
138. Anet, F. A. L., Bourn, A. J. R. and Lin, Y. S. (1964). *J. Amer. Chem. Soc.,* **86,** 3576
139. Salem, L. (1966). *The Molecular Orbital Theory of Conjugated Systems,* Chapter 3 (New York: Benjamin)
140. Julg, A. and François, P. (1962). *J. Chim. Phys.,* **59,** 339
141. Bergmann, E. D. and Agranat, I. (1965). *Chem. Commun.,* 512
142. Prinzbach, H. and Weischnik, E. (1969). *Helv. Chim. Acta,* **52,** 2472
143. Nakajima, T., Kohda, S., Tajiri, A. and Karasawa, S. (1967). *Tetrahedron,* **23,** 2189
144. Yamaguchi, H., Nakajima, T. and Kunii, T. L. (1968). *Theor. Chim. Acta (Berl.),* **12,** 349
145. Dewar, M. J. S. (1969). *The Molecular Orbital Theory of Organic Chemistry,* 181–182 (New York: McGraw-Hill)
146. Bertelli, D. J. and Andrews, Jr., T. G. (1969). *J. Amer. Chem. Soc.,* **91,** 5280
147. Tobey, S. W. (1971). *'Aromaticity, Pseudo-Aromaticity, Anti-Aromaticity', The Jerusalem Symposia on Quantum Chemistry and Biochemistry* III, (E. D. Bergmann and B. Pullman, editors), 351 (Jerusalem: The Israel Academy of Sciences and Humanities)
148. Weiler-Feilchenfeld, H., Agranat, I. and Bergmann, E. D. (1966). *Trans. Faraday Soc.,* **62,** 2084
149. Yamakawa, M., Watanabe, H., Mukai, T., Nozoe, T. and Kubo, M. (1960). *J. Amer. Chem. Soc.,* **82,** 5665
150. Brown, R. D., Burden, F. R. and Kent, J. E. (1968). *J. Chem. Phys.,* **49,** 5542
151. Yates, P. (1968). *Advances in Alicyclic Chemistry,* Vol. 2, (H. Hart and G. L. Karabatsos, editors), 59 (New York: Academic Press)
152. Brown, R. D., Burden, F. R., Jones, A. J. and Kent, J. E. (1967). *Chem. Comm.,* 808
153. Coller, B. A. W., Heffernan, M. L. and Jones, A. J. (1968). *Aust. J. Chem.,* **21,** 1807
154. Brown, R. D. and Burden, F. R. (1966). *Chem. Comm.,* 448
155. Bregman, J., Hirshfeld, F. L., Rabinovich, D. and Schmidt, G. M. J. (1965). *Acta Crystallogr.,* **19,** 227
156. Hirschfeld, F. L. and Rabinovich, D. (1965). *Acta Crystallogr.,* **19,** 235
157. Johnson, S. M., Paul, I. C. and King, G. S. D. (1970). *J. Chem. Soc.(B),* 643
158. Vogel, E. (1971). *Pure Appl. Chem.,* **28,** 355
159. Gramaccioli, C. M., Mimun, A., Mugnoli, A. and Simonetta, M. (1971). *Chem. Commun.,* 796
160. Carstensen-Oeser, E. and Habermehl, G. (1968). *Angew. Chem. Int. Edn.,* **7,** 543
161. Ammon, H. L. and Plastas, L. A. (1971). *Chem. Comm.,* 356
162. Shimanouchi, H., Ashida, T., Sasada, Y., Kakudo, M., Murata, I. and Kitahara, Y. (1966). *Bull. Chem. Soc. Japan,* **39,** 2322
163. Kabuto, C., Kitahara, Y., Shimanouchi, H. and Sasada, Y. (1970). *Abstracts of Papers, International Symposium on the Chemistry of Nonbenzenoid Aromatic Compounds,* (Sendai, Japan: 24-28 August 1970), 89
164. Kennard, O., Kerr, K. A., Watson, D. G. and Fawcett, J. K. (1970). *Proc. Roy. Soc.* **A316,** 551
165. Shimanouchi, H., Sasada, Y., Ashida, T., Kakudo, M., Murata, I. and Kitahara, Y. (1969). *Acta Crystallogr.,* **B25,** 1890
166. Chiang, J. F. and Bauer, S. H. (1970). *J. Amer. Chem. Soc.,* **92,** 261
167. Skancke, A. (1968). *Acta Chem. Scand.,* **22,** 3239

168. Montgomery, K. L. and Coetzer, J. (1971). (cited by Vogel, E. (1971) *XXIII International Congress of Pure and Applied Chemistry,* Boston, U.S.A., Vol. 1, 275 (London: Butterworth Scientific Publications)
169. Coppens, P. (1969). *Mol. Cryst. Liq. Crys.,* **9,** 25
170. Jones, D. W. and Yerkess, Y. (1971). *J. Cryst. Molec. Struct.,* **1,** 17
171. Kay, M. I., Okaya, Y. and Cox, D. E. (1971). *Acta Crystallogr.,* **B27,** 27
172. Salem, L. (1966). *The Molecular Orbital Theory of Conjugated Systems,* Chapter 4, (New York: Benjamin)
173. Pacault, A. (1971). *'Aromaticity, Pseudo-Aromaticity, Anti-Aromaticity', The Jerusalem Symposia on Quantum Chemistry and Biochemistry* III, (E. D. Bergmann and B. Pullman, editors), 39, (Jerusalem: The Israel Academy of Sciences and Humanities)
174. London, F. (1937). *J. Phys. Radium,* **8,** 397
175. McWeeny, R. (1958). *Molec. Phys.,* **1,** 311
176. Pople, J. A. (1958). *Molec. Phys.,* **1,** 175
177. Pople, J. A. (1958). *Developments in Aromatic Chemistry (Special Publication, No. 12)* 211, (London: The Chemical Society)
178. Pople, J. A., Schneider, W. G. and Bernstein, H. J. (1959). *High Resolution Nuclear Magnetic Resonance,* 180 (New York: McGraw-Hill)
179. Jackman, L. M., Sondheimer, F., Amiel, Y., Ben-Efraim, D. A., Gaoni, Y., Wolovsky, R. and Bothner-By, A. A. (1962). *J. Amer. Chem. Soc.,* **84,** 4307
180. Pople, J. A. (1956). *J. Chem. Phys.,* **24,** 1111
181. Elvidge, J. A. and Jackman, L. M. (1961). *J. Chem. Soc.,* 859
182. Elvidge, J. A. (1965). *Chem. Commun.,* 160
183. Hall, G. G., Hardisson, A. and Jackman, L. M. (1963). *Tetrahedron,* 19, Suppl. **2**, 101
184. Abraham, R. J., Sheppard, R. C., Thomas, W. A. and Turner, S. (1965). *Chem. Commun.,* 43
185. Abraham, R. J. and Thomas, W. A. (1966). *J. Chem. Soc.(B),* 127
186. Dailey, B. P. (1964). *J. Chem. Phys.,* **41,** 2304
187. Pople, J. A. (1964). *J. Chem. Phys.,* **41,** 2559
188. Ferguson, A. F. and Pople, J. A. (1965). *J. Chem. Phys.,* **42,** 1560
189. Katz, T. J. (1960). *J. Amer. Chem. Soc.,* **82,** 3784, 3785
190. Musher, J. I. (1966). *Advances in Magnetic Resonance,* Vol. 2 (J. S. Waugh, editor), 177 (New York: Academic Press)
191. Musher, J. I. (1965). *J. Chem. Phys.,* **43,** 4081
192. Gaidis, J. M. and West, R. (1967). *J. Chem. Phys.,* **46,** 1218
193. Musher, J. I. (1967). *J. Chem. Phys.,* **46,** 1219
194. Sondheimer, F., Calder, I. C., Elix, J. A., Gaoni, Y., Garratt, P. J., Grohmann, K., DiMaio, G., Mayer, J., Sargent, M. V. and Wolovsky, R. (1967). *Aromaticity (Special Publication, No. 21),* 75, (London: The Chemical Society)
195. Sondheimer, F. (1967). *Proc. Roy. Soc.* **A, 297,** 173
196. Sondheimer, F. (1968). *Proceedings of the Robert A. Welch Foundation Conferences on Chemical Research,* XII, Organic Synthesis, 125, Houston, Texas
197. Sondheimer, F. (1971). *Pure Appl. Chem.,* **28,** 331
198. Metcalf, B. W. and Sondheimer, F. (1971). *J. Amer. Chem. Soc.,* **93,** 6675
199. Pople, J. A. and Untch, K. G. (1966). *J. Amer. Chem. Soc.,* **88,** 4811
200. Longuet-Higgins, H. C. (1967). *Aromaticity (Special Publication, No. 21),* 109, (London: The Chemical Society)
201. Nakajima, T. and Kohda, S. (1966). *Bull. Chem. Soc. Japan,* **39,** 804
202. Untch, K. G. and Wysocki, D. C. (1967). *J. Amer. Chem. Soc.,* **89,** 6386
203. Boekelheide, V. and Phillips, J. B. (1967). *J. Amer. Chem. Soc.,* **89,** 1695
204. Phillips, J. B., Molyneux, R. J., Sturm, E. and Boekelheide, V. (1967). *J. Amer. Chem. Soc.,* **89,** 1704
205. Boekelheide, V. and Mitchell, R. H. (1971). *'Aromaticity, Pseudo-Aromaticity, Anti-Aromaticity', The Jerusalem Symposia on Quantum Chemistry and Biochemistry* III, (E. D. Bergmann and B. Pullman, editors), 150, (Jerusalem: The Israel Academy of Sciences and Humanities)
206. Anet, F. A. L. and Schenk, G. E. (1971). *J. Amer. Chem. Soc.,* **93,** 556
207. Alger, T. D., Grant, D. M. and Paul, E. G. (1966). *J. Amer. Chem. Soc.,* **88,** 5397
208. Jones, A. J. and Grant, D. M. (1968). *Chem. Commun.,* 1670
209. Jones, A. J., Alger, T. D., Grant, D. M. and Litchman, W. M. (1970). *J. Amer. Chem. Soc.,* **92,** 2386

210. Garratt, P. J. and Sargent, M. V. (1971). *Nonbenzenoid Aromatics,* Vol. 2 (J. P. Snyder, editor), 207, (New York: Academic Press)
211. Breslow, R. and Groves, J. T. (1970). *J. Amer. Chem. Soc.,* **92,** 984
212. Jones, A. J., Gardner, P. D., Grant, D. M., Litchman, W. M. and Boekelheide, V. (1970). *J. Amer. Chem. Soc.,* **92,** 2395
213. Karplus, M. (1963). *J. Amer. Chem. Soc.,* **85,** 2870
214. Smith, W. B., Watson, W. H. and Chiranjeevi, S. (1967). *J. Amer. Chem. Soc.,* **89,** 1438
215. Heffernan, M. L. and Jones, A. J. (1966). *Austr. J. Chem.,* **19,** 1813
215. Heffernan, M. L. and Jones, A. J. (1966). *Austr. J. Chem.,* **19,** 1813
216. Bertelli, D. J. (1971). *'Aromaticity, Pseudo-Aromaticity, Anti-Aromaticity', The Jerusalem Symposia on Quantum Chemistry and Biochemistry* III, (E. D. Bergmann and B. Pullman, editors), 326, (Jerusalem: The Israel Academy of Sciences and Humanities)
217. Jonathan, N., Gordon, S. and Dailey, B. P. (1962). *J. Chem. Phys.,* **36,** 2443
218. Cooper, M. A. and Manatt, S. L. (1969). *J. Amer. Chem. Soc.,* **91,** 6325
219. Bertelli, D. J., Andrews, T. G. and Crews, P. O. (1969). *J. Amer. Chem. Soc.,* **91,** 5286
220. Kende, A. S. (1966). *Trans. N. Y. Acad. Sci.,* **28,** 981
221. Ammon, H. L. and Wheeler, G. L. (1971). *Chem. Commun.,* 1032
222. Trost, B. M., Kinson, P. L., Maier, C. A. and Paul, I. C. (1971). *J. Amer. Chem. Soc.,* **93,** 7275
223. Bothner-By, A. A. and Pople, J. A. (1965). *Ann. Rev. Phys. Chem.,* **16,** 43
224. Pullman, B. and Pullman, A. (1952). *Les Theories Electroniques de la Chimie Organique,* 529–533, 541–550 (Paris: Masson and Cie)
225. Hüttner, W. and Flygare, W. H. (1967). *J. Chem. Phys.,* **47,** 4137
226. Hüttner, W., Lo., M.-K. and Flygare, W. H. (1968). *J. Chem. Phys.,* **48,** 1206
227. Flygare, W. H., Hüttner, W. Shoemaker, R. L. and Foster, P. D. (1969). *J. Chem. Phys.,* **50,** 1714
228. Pochan, J. M. and Flygare, W. H. (1969). *J. Amer. Chem. Soc.,* **91,** 5928
229. Benson, R. C. and Flygare, W. H. (1970). *J. Amer. Chem. Soc.,* **92,** 7523
230. Dauben, Jr., H. J., Wilson, J. D. and Laity, J. L. (1968). *J. Amer. Chem. Soc.,* **90,** 811
231. Dauben, Jr., H. J., Wilson, J. D. and Laity, J. L. (1969). *J. Amer. Chem. Soc.,* **91,** 1991
232. Dauben, Jr., H. J., Wilson, J. D. and Laity, J. L. (1971). *Nonbenzenoid Aromatics,* Vol. 2 (J. P. Snyder, editor), 167, (New York: Academic Press)
233. Haberditzl, W. (1966). *Angew. Chem. Internat. Edn.,* **5,** 288
234. Davis, R. E. and Tulinsky, A. (1962). *Tetrahedron Lett.,* 839
335. Traetteberg, M. (1964). *J. Amer. Chem. Soc.,* **86,** 4265
236. Labarre, J. F. and Crasnier, F. (1967). *J. Chim. Phys.,* **64,** 1664
237. Labarre, J. F. (1970). *Bull. Soc. Chim. Fr.,* 2463
238. Labarre, J. F. and Gallais, F. (1971). *'Aromaticity, Pseudo-Aromaticity, Anti-Aromaticity', The Jerusalem Symposia on Quantum Chemistry and Biochemistry* III, (E. D. Bergmann and B. Pullman, editors), 48, (Jerusalem: The Israel Academy of Sciences and Humanities)
239. Craig, D. P. (1950). *Proc. Roy. Soc.* **A, 200,** 390
240. Craig, D. P. (1951). *J. Chem. Soc.,* 3175
241. Berson, J. A., Evleth, Jr., E. M. and Hamlet, Z. (1965). *J. Amer. Chem. Soc.,* **87,** 2887
242. Nakajima, T. and Katagiri, S. (1964). *Molec. Phys.,* **7,** 149
243. Rosowsky, A., Fleischer, H., Young, S. T., Partch, R., Saunders, Jr., W. H. and Boekelheide, V. (1960). *Tetrahedron,* **11,** 121
244. Abramovitch, R. A. and McEwen, K. L. (1965). *Can. J. Chem.,* **43,** 2616
245. Craig, D. P. Personal communication.
246. Paquette, L. A., Short, M. R. and Kelly, J. F. (1971). *J. Amer. Chem. Soc.,* **93,** 7179
247. Clark, D. T. and Armstrong, D. R (1969). *Chem. Comm.,* 850
248. Goldstein, M. J. and Hoffmann, R. (1971). *J. Amer. Chem. Soc.,* **93,** 6193
249. Simmons, H. E. and Fukunaga, T. (1967). *J. Amer. Chem. Soc.,* **89,** 5208
250. Hoffmann, R., Imamura, A. and Zeiss, G. D. *J. Amer. Chem. Soc.,* **89,** 5215
251. Heilbronner, E. (1971). *'Aromaticity, Pseudo-Aromaticity, Anti-Aromaticity', The Jerusalem Symposia on Quantum Chemistry and Biochemistry* III, (E. D. Bergmann and B. Pullman, editors), 58, (Jerusalem: The Israel Academy of Sciences and Humanities)
252. Hoffmann, R. (1971). *Pure Appl. Chem.,* **28,** 181
253. Lustgarten, R. K., Brookhart, M. and Winstein, S. (1967). *J. Amer. Chem. Soc.,* **89,** 6350, 6352
254. Goldstein, M. J. and Odell, B. G. (1967). *J. Amer. Chem. Soc.,* **89,** 6356

255. Goldstein, M. J. (1967). *J. Amer. Chem. Soc.*, **89,** 6357
256. Marchand, A. P. (1965). *Chem. and Ind.*, 161
257. Lloyd, D. M. G. and Marshall, D. R. (1964). *Chem. and Ind.*, 1760
258. Lloyd, D. M. G. and Marshall, D. R. (1971). *'Aromaticity, Pseudo-Aromaticity, Anti-Aromaticity', The Jerusalem Symposia on Quantum Chemistry and Biochemistry* III. (E. D. Bergmann and B. Pullman, editors), 85, (Jerusalem: The Israel Academy of Sciences and Humanities)
259. Lloyd, D., this volume, p. 179
260. Brasen, W. R., Holmquist, H. E. and Benson, R. E. (1961). *J. Amer. Chem. Soc.*, **83,** 3125
261. Müller-Westerhoff, U. (1970). *J. Amer. Chem. Soc.*, **92,** 4849
262. Goldstein, P. and Trueblood, K. N. (1967). *Acta Crystallogr.*, **23,** 148
263. Winstein, S. (1967). *Aromaticity (Special Publication, No. 21)*, 5, (London: The Chemical Society)
264. Winstein, S. (1969). *Quart. Rev.*, **23,** 141
265. Rosenberg, J. L., Mahler, J. E. and Pettit, R. (1962). *J. Amer. Chem. Soc.*, **84,** 2842
266. Radlick, P. and Winstein, S. (1963). *J. Amer. Chem. Soc.*, **85,** 344
267. Untch, K. G. (1963). *J. Amer. Chem. Soc.*, **85,** 345
268. Untch, K. G. and Kurland, R. J. (1964). *J. Molec. Spectrosc.*, **14,** 156
269. Bischof, P., Gleiter, R. and Heilbronner, E. (1970). *Helv. Chim. Acta.* **53,** 1423
270. Balaban, A. and Simon, Z. (1962). *Tetrahedron,* **18,** 315
271. Balaban, A. and Simon, Z. (1965). *Rev. Roumaine Chim.*, **10,** 1059
272. Zahradnik, R., Michl, J. and Pancíř, J. (1966). *Tetrahedron,* **22,** 1355
273. Trindle, C. (1969). *J. Amer. Chem. Soc.*, **91,** 219
274. Kemula, W. and Krygowski, T. M. (1968). *Tetrahedron Lett,* 5135
275. Kruszewski, J. and Krygowski, T. M. (1970). *Tetrahedron Lett.*, 319
276. Krygowski, T. M. (1970). *Tetrahedron Lett.*, 1311
277. Christoffersen, R. E. (1971). *J. Amer. Chem. Soc.*, **93,** 4104
278. Häfelinger, G. (1971). *Tetrahedron,* **27,** 4609
279. Julg, A. and François, P. (1967). *Theor. Chim. Acta,* **8,** 249
280. Julg, A. (1971). *'Aromaticity, Pseudo-Aromaticity, Anti-Aromaticity', The Jerusalem Symposia on Quantum Chemistry and Biochemistry* III, (E. D. Bergmann and B. Pullman, editors), 383, (Jerusalem: The Israel Academy of Sciences and Humanities)
281. François, P. and Julg, A. (1968). *Theor. Chim. Acta.* **11,** 128
282. Doyle, A. C. (1930). *The Disappearance of Lady Frances Carfax, The Complete Sherlock Holmes,* 950, (Garden City, N. Y.:: Doubleday and Co.)

6
Microcyclic Compounds

DOUGLAS LLOYD
University of St. Andrews, Scotland

6.1 INTRODUCTION

This chapter deals with cyclic compounds having rings smaller than benzene rings, i.e. with derivatives of cyclopropene, cyclobutadiene and cyclopentadiene. The name 'microcyclic' was suggested for these compounds by P. Skrabal[1].

Earlier reviews, with references, on these compounds include general accounts of non-benzenoid aromatic compounds[2], and reviews of cyclopropene derivatives[3], cyclopropenium salts[4], chloro-derivatives of cyclopropene, cyclobutadiene and cyclopentadiene[5], methylenecyclopropenes[6], anti-aromatic cyclopropene derivatives[7], cyclobutadiene derivatives[8, 9], biphenylenes[9, 10], cyclobutadiene–metal complexes[11, 12], cyclopentadienide derivatives[13], cyclopentadienylidene derivatives[14], and of fulvenes and fulvalenes[6, 15]. The present chapter in general limits itself to material published in most recent years and does not concern itself with earlier work. Organometallic compounds are not included within the scope of this review.

6.2 DERIVATIVES OF CYCLOPROPENE

6.2.1 Cyclopropenium salts

Salts of the unsubstituted cyclopropenium cation have been prepared by reaction of 3-chlorocyclopropene with Lewis acids such as antimony pentafluoride[16, 17], aluminium chloride[17], or silver tetrafluoroborate[17]. Phosphorus pentachloride is apparently not a strong enough Lewis acid to form a corresponding salt[17]. The hexachloroantimonate was obtained as an analytically-pure white solid, which away from air was stable for some days at room temperature and apparently indefinitely at $-20\,°C$. Exposure to atmospheric moisture caused decomposition. The i.r. and n.m.r. spectra are consistent with the structure of the cation and are the same for each of the salts obtained. The i.r. spectrum is very simple, consisting only of four bands, as expected for a molecule with this symmetry. Two of the bands are C—H stretching bands and two correspond to skeletal bands previously reported for the trichlorocyclopropenium cation[18, 19]. The n.m.r. spectrum shows a sharp singlet in the region τ -1.1 to -1.2. The ^{13}C—H coupling constant is very large, indicating a high degree of s-character in the C—H bonds. From this result it was suggested that each carbon atom uses an sp-orbital to bond to hydrogen, two sp^3-orbitals in the plane of the ring for the bent carbon–carbon bonds and a p-orbital for the π-system. The pK_R+ value, determined by n.m.r. spectroscopy, was -7.4. When this value is compared with the estimated pK_R+ for allyl cations, it suggests that the cyclopropenium cation is the more stabilised by *c.* 18 kcal mol^{-1}. This value agrees well with that of 17–18 kcal mol^{-1} determined in the gas phase from appearance potentials. The formation of the cyclopropenium cation has also been observed when the pyrolysis product of trimethyl tricyclo[3,2,2,$0^{2,4}$]nona-6,8-diene-3,6,7-tricarboxylate, which is said to be methyl cyclopropene-3-carboxylate[20], is passed into chloro- or fluoro-sulphonic acids[21].

By treatment of dichloro-, trichloro- and tetrachloro-cyclopropenes with very strong Lewis acids such as antimony pentachloride, mono-chloro-[22], dichloro-[22] and trichloro-cyclopropenium[18, 19] salts, respectively, have also been prepared. In all these cases it is necessary to use a strong Lewis acid since the resultant cyclopropenium cations are themselves strong acceptors of halide ions. The trichlorocyclopropenium cation reacts readily with phenols or aryl ethers to give, at 0 °C, di-(*p*-substituted aryl)chlorocyclopropenium chlorides, which are readily hydrolysed to the corresponding cyclopropenones[23, 24]. At higher temperatures triarylcyclopropenium chlorides are formed[24, 25]; hydrolysis of the product so obtained from 2,6-di-t-butylphenol provided a quinocyclopropene[25].

Among other cyclopropenium salts which have been prepared in recent years tri-t-butylcyclopropenium perchlorate[26, 27] is of interest. Because of the greater inductive effects of t-butyl groups *vis-à-vis* n-propyl groups it might be expected that this compound would be more stable than the previously described[28] tri-n-propylcyclopropenium salt; both for this reason and because of steric destabilisation of the alcohol derived from the tri-t-butyl compound, one would expect this compound to have the greater pK_{R^+} value. In fact it has not. It has been suggested that stabilising solvation of the t-butyl-substituted cation may be more difficult than for the n-propyl-compound and that this factor may be playing an important role[26, 27].

Cyclopropenium salts are frequently reduced by complex hydrides to cyclopropenes, for example the triphenylcyclopropenium cation is reduced in high yield to triphenylcyclopropene by lithium aluminium hydride[29]. Under usual conditions the diphenylcyclopropenium cation gives, however, only 1,2,4,5-tetraphenylbenzene[30]. It was suggested[30] that diphenylcyclopropene was indeed formed first and then reacted with more cation as follows:

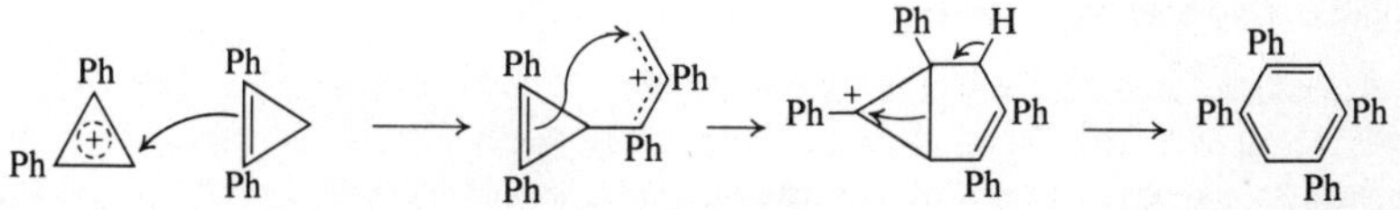

Confirmation of this hypothesis was obtained[31] by carrying out the reduction in an inverse order, by slow addition of the cyclopropenium salt to sodium borohydride. This gave the cyclopropene in high yield and furthermore this product reacted with further cyclopropenium salt to give 1,2,3,5-tetraphenylbenzene. Electrochemical reduction of diphenylcyclopropenium salts also results in the formation of 1,2,4,5-tetraphenylbenzene[32].

Reaction of diphenyl- or triphenyl-cyclopropenium salts with triphenylphosphine in polar[33] or non-polar[34] solvents leads to the formation of cyclopropenyltriphenylphosphonium salts. The diphenylcyclopropenylphosphonium salts are converted by potassium t-butoxide into an ylide which reacts readily with aldehydes or ketones[35]. This has provided a route to trisubstituted cyclopropenium salts by means of the following reaction sequence:

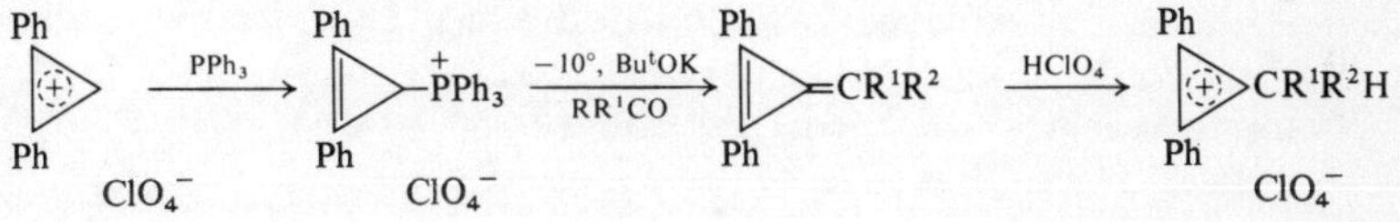

$$\text{Ph}_2\text{C}_3^+\ \text{ClO}_4^- \xrightarrow{\text{PPh}_3} \text{Ph}_2\text{C}_3\text{-}\overset{+}{\text{P}}\text{Ph}_3\ \text{ClO}_4^- \xrightarrow[\text{RR}^1\text{CO}]{-10^\circ,\ \text{Bu}^t\text{OK}} \text{Ph}_2\text{C}_3\text{=CR}^1\text{R}^2 \xrightarrow{\text{HClO}_4} \text{Ph}_2\text{C}_3^+\text{CR}^1\text{R}^2\text{H}\ \text{ClO}_4^-$$

Since the ylide is unstable at normal temperature it is generated in the presence of the carbonyl compound.

There has been an amount of interest in cyclopropenium salts having electron-donating substituents which might be expected to stabilise the cation. Thus tris(dimethylamino)cyclopropenium and other similar tris-amino salts have been prepared[36]. All these salts are stable in air and in particular are non-hygroscopic and very resistant to attack by water. Mono-alkoxycyclopropenium salts are readily attacked by nucleophiles with displacement of the alkoxy-group[37–39]. Thus, for example, enamines give stabilised γ-aminocyclopropenium salts[38, 39];

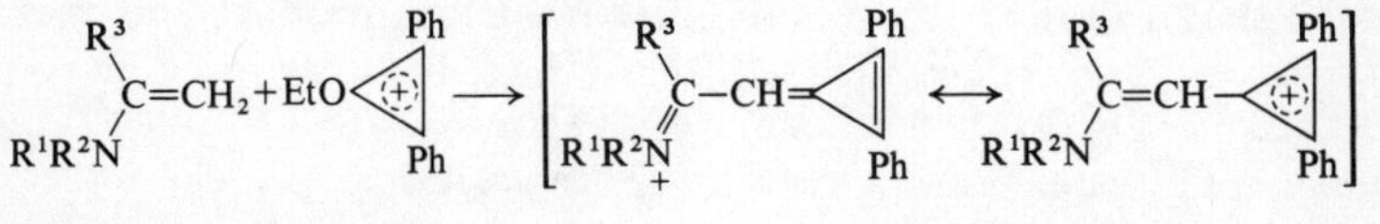

Carbanions, such as those derived from dicyanomethane, cyanacetic esters, etc., attack either of the ring carbon atoms, the site being affected by the identity of the solvent and substituent effects; the products include cyclopropene ethers, methylenecyclopropene derivatives and ring-opened adducts[40].

The symmetry of the triphenylcyclopropenium ring has been nicely shown by labelling experiments[41, 42]. Salts were prepared by reaction of phenylchloro- or phenylbromo-carbene, labelled in either case at the exocyclic carbon atom, with diphenylacetylene. Ring opening of these salts with aqueous potassium hydroxide gave benzoylstilbene equally labelled at each of the non-phenyl carbon atoms.

6.2.2 Cyclopropenones

A standard method for the preparation of cyclopropenones involves the initial addition of a dihalogenocarbene to an acetylene. This method has been shown to give only low ylides of adduct (because of secondary reactions) when dichlorocarbene is generated by the action of hydroxide on chloroform in the presence of catalytic amounts of benzyltriethylammonium chloride[43].

Unsubstituted cyclopropenone has been obtained by hydrolysis of 3,3-dichlorocyclopropene[22, 44] or chlorocyclopropenium salts[44]. It is interesting to note that its infra-red spectrum shows that even in aqueous systems it appears to exist as the carbonyl species and not as a *gem.*-diol; this is in marked contrast to cyclopropanone. It is slowly hydrolysed to acrylic acid. Dichlorocyclopropenone has been prepared similarly from trichlorocyclopropenium tetrachloroaluminate; this product decomposes readily and sometimes even explosively at room temperature[45].

In contrast di-t-butylcyclopropenone is remarkably stable[26, 46]. For example it was recovered in 86% yield after being heated for 2 h in refluxing 25% aqueous sodium hydroxide; under similar conditions di-n-propylcyclopropenone, for example, is totally destroyed. The carbonyl group is not seriously hindered and the stability may be associated with the fact that ring opening involves steric compression of the t-butyl groups, since a *cis*-di-t-butylethylene derivative is the expected hydrolysis product[46]. The t-butyl groups hinder conjugate addition to the carbon–carbon double-bond; thus phenyl lithium reacts with di-t-butylcyclopropenone at the carbonyl group whereas it reacts with diphenylcyclopropenone by conjugate addition[46].

The basicities of different substituted cyclopropenones have been compared by studying their partition between carbon tetrachloride and hydrochloric acid. These results show that alkyl substituents lead to higher basicities than phenyl substituents[47, 48].

Diphenylcyclopropenone undergoes bimolecular reduction by the action of aluminium amalgam or magnesium–magnesium iodide; the product is, however, not the expected pinacol but tetraphenylresorcinol[49]. It was suggested that a radical anion is the first reduction product and that this then dimerises, as in normal pinacol reductions, but that the dimerisation product rearranges, first to a prismane derivative, then to a bicyclohexadiene derivative, and finally into the resorcinol derivative.

A number of workers have investigated the reactions of cyclopropenones, in particular diphenylcyclopropenone, with various amines. Ammonia converts it into an azetidine at room temperature but gives a β-aminovinylaldehyde at $-78\,°C$[50–52]. Trimethylamine gives an azetidine at either temperature[51]; diethylamine[52,53], aziridine[54], and various *N*-heteroaromatic compounds[55] cause ring opening. Enamines undergo cyclo-addition to the carbon–carbon double-bond while dienamines lead to the formation of a

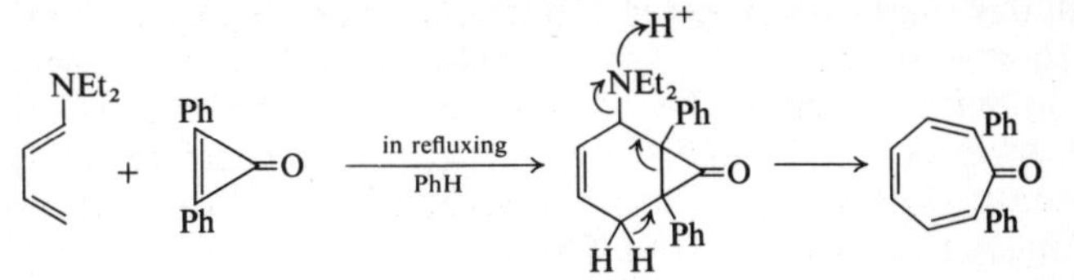

tropone[56]. Several workers[57–61] have shown that cyclopropenones react with acylylides to form α-pyrones, presumably by the following reaction sequence:

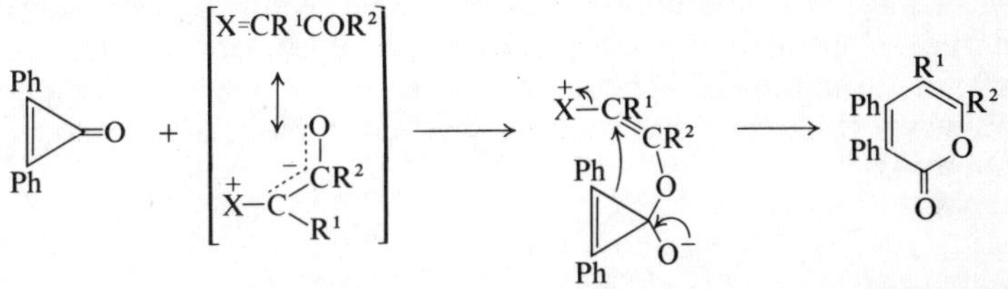

It has been demonstrated that benzocyclopropenones are formed as intermediates in the decomposition of benzotriazin-4-ones, by photolysis of the 3-*p*-tolylsulphaminobenzotriazinone[62] or by oxidation of the 3-aminobenzotriazinone by lead tetra-acetate[63,64]. A nitrene is formed initially which collapses, with loss of nitrogen, to form the benzocyclopropenone:

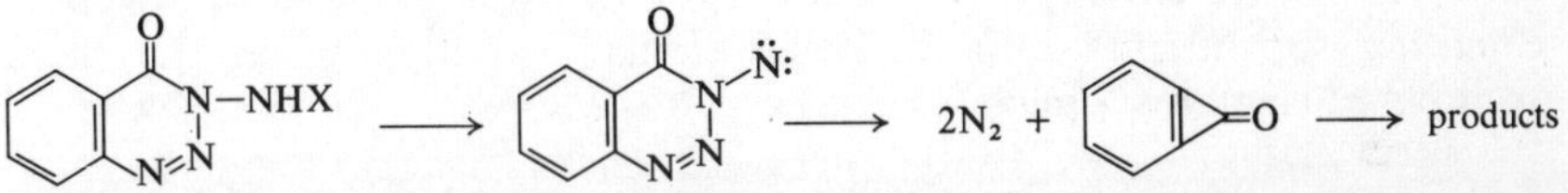

The uncertainty about the assignment of the bands in the infra-red spectra of cyclopropenones has been resolved[65,66]. The band at 1850 cm^{-1} is assigned to an out-of-phase stretching vibration of the two double bonds with a predominance of the carbonyl coordinate, and the band at 1640 cm^{-1} to the corresponding in-phase stretching vibration with a predominance of the olefin coordinate. The band at 880 cm^{-1} is mainly a symmetrical stretching vibration of the carbon–carbon bonds in the ring. Model calculations in connection with the i.r. and Raman spectra suggest a contribution of about 30% from the dipolar form (b) to the ground state[65]:

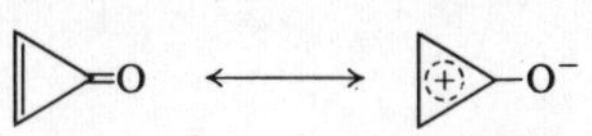

(a) (b)

The importance of the contribution of this dipolar form has been questioned. It has been suggested[66] that the polarity of the carbonyl group is no greater than in other carbonyl groups and that the dipole moment and spectroscopic properties need only the assumption that the small positive charge normally resident on a carbonyl carbon atom is delocalised over the other two atoms in the ring. The high basicity is ascribed to the high polarisability rather than the high polarity of the molecule.

CNDO/2 calculations predict[68], however, that there will be more π-polarisation in cyclopropenone than in tropone. The same authors point out that this might be expected intuitively since the electrostatic work required to achieve the dipolar structure (b) would be considerably less than that required to achieve the corresponding polar structure in tropone. Hence cyclopropenone might be expected to approximate rather more to an 'aromatic' system than does tropone. SCF/MO calculations[69] are in general agreement with the CNDO/2 results.

Other non-empirical LCAO–MO–SCF calculations support the view that cyclopropenone has considerable conjugative stabilisation[70]. These calculations also suggest that decarbonylation of cyclopropenone is thermodynamically very favourable and that the relatively high temperatures required to bring about decarbonylation of cyclopropenones with concomitant formation of acetylenes is due to the concerted reaction being symmetry forbidden. It has been shown that in the related cyclopropenone-imonium salts:

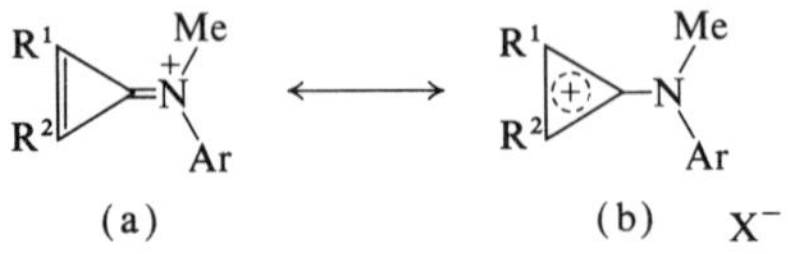

R1 Me +N R2 Ar ⟷ R1 (+) Me N R2 Ar

(a) (b) X^-

there must be some contribution from form (b) since n.m.r. studies indicate that rotation about the carbon–nitrogen double bond is easier than in acyclic imonium salts[71]. Electron-withdrawing substituents on the *N*-aryl group lower the enthalpy of activation for this rotation whereas electron-donating substituents raise it, while similar substituents on the three-membered ring have opposite effects.

6.2.3 Cyclopropenylidene derivatives

Condensations of 3,3-dichlorocyclopropenes[72] or ethoxycyclopropenium salts[73] with metal complexes of β-diketones have produced cyclopropenylidene derivatives which are vinylogues of cyclopropenone, and which, like cyclopropenones, must have some small contribution from a dipolar form:

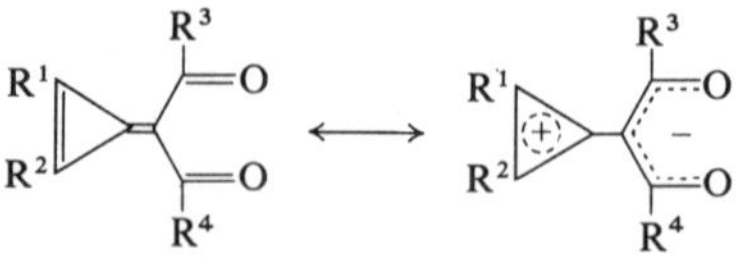

R3 R1 =O R2 =O R4 ⟷ R3 R1 (+) O − R2 O R4

Quinocyclopropenes have been known for some time; interesting recent examples include the dicyano-compound (1)[74,75] and a trisquinonoid derivative (2)[23,25,76]:

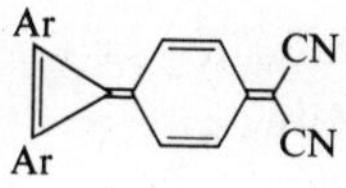

(1)

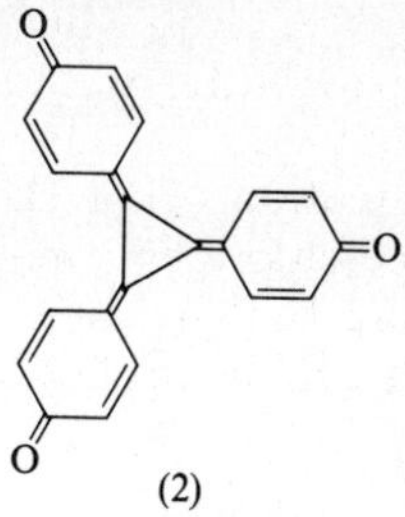

(2)

N.M.R. studies on the triquinocyclopropanes indicate that there is no rotation about the interannular bonds at room temperature[76].

Condensation of diphenylcyclopropenone with heterocyclic molecules such as pyrroles[77], indoles[77,78] indolizines[77], cyclopenta-pyridazines and -oxazines[79,80] gave products whose n.m.r. spectra showed that the positive charge was shared between the heterocyclic and the cyclopropene rings. Examples are:

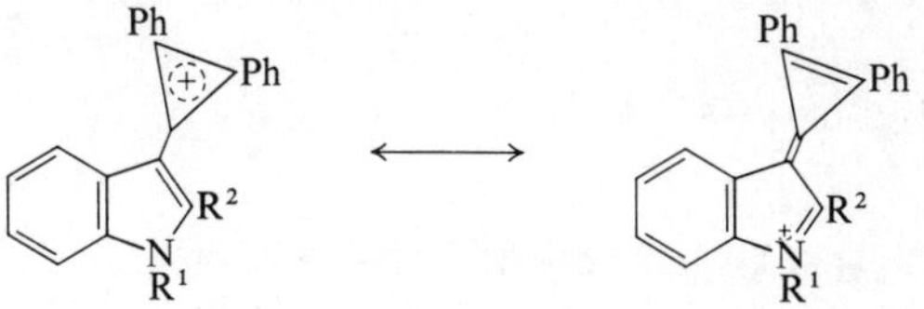

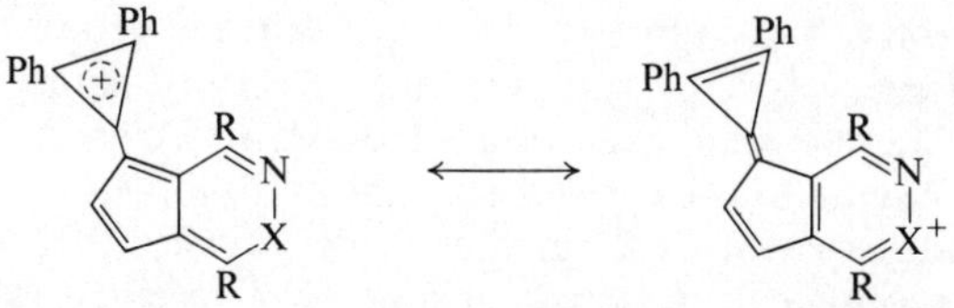

X = O, NH, NPh

Attempts to convert these salts into azacalicenes led only to the formation of polymeric tars. Cyclopropenylidenepyrrolium salts have also been prepared[81] by heating di- and tri-phenylpyrroles with methylthiodiphenylcyclopropenium perchlorate in acetic acid, and in this case unstable azacalicenes were obtained on deprotonation of the salts with potassium t-butoxide.

6.2.4 Cyclopropenide derivatives

A topic of major interest has been the predicted 'anti-aromaticity'[82,83] of cyclopropenide anions; a useful review is available[7]. An indirect assessment of the pK_a for 1,2,3-triphenylcyclopropene makes it *c*. 20 units greater than for triphenylmethane and greater even than some of the estimates for methane[84]. This assessment did not allow for ring-strain, but the factor presumably reflects a substantial contribution attributable to the anti-aromaticity of cyclopropenide anions.

Reaction of cyclopropene with sodium or potassium amides in liquid ammonia has been shown to result in loss of a vinylic proton rather than a methylene proton[85], a result which is in accord with the earlier finding[86] that methyl cyclopropene-l-carboxylate exchanged its vinylic but not its methylene hydrogen atom in base.

The ready formation of an ylide from diphenylcyclopropenyltriphenylphosphonium perchlorate[35], mentioned previously in Section 6.2.1, is perhaps surprising, but any carbanionic nature in the cyclopropene portion is presumably largely neutralised by double-bonding with the phosphorus atom:

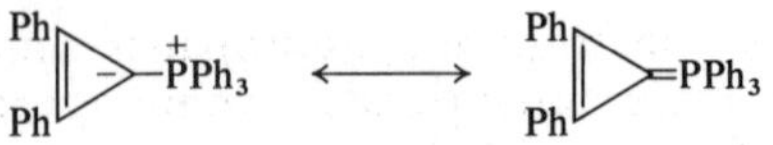

Ph, Ph, $\overset{+}{P}Ph_3$ ⟷ Ph, Ph, $=PPh_3$

This ylide decomposes readily at room temperature so there are obviously, at the very least, no special stabilising influences. Although presumably existing in a largely non-polar form, it reacts readily with aldehydes and ketones. Any small contribution from the polar form will result in high nucleophilic reactivity since in forming the resultant adduct the potentially anti-aromatic structure is lost.

6.3 CYCLOBUTADIENE AND DERIVATIVES

6.3.1 Cyclobutadienes

Considerable interest continues to be taken into the ground-state structure of cyclobutadiene. Most of the evidence, described later in this section, favours a rectangular singlet ground state, but it seems likely that there is also a very low-lying excited triplet state, which should be square. Both the theoretical and pertinent experimental evidence were reviewed in detail in 1969[9]. Very recent calculations[86], using two recently developed semi-empirical SCF–MO procedures, one a π-approximation and the other (MINDO/2) using all valence electrons, again predict cyclobutadiene to have a rectangular singlet ground state with a very low-lying triplet state. They also predict that it will be stable with respect to dissociation into acetylene. Calculations of the heats of atomisation lead to an estimated negative resonance energy of -18 kcal mol^{-1} and suggest that its destabilisation by conjugation should be almost equal to the stabilisation by conjugation in benzene[86]. These results compare well with other estimated values[87, 88].

Early work on cyclobutadiene generated by ceric ion oxidation of cyclobutadienetricarbonyliron had shown that it added stereospecifically to olefins in cycloaddition reactions[89]. In a more recent investigation[90], cyclobutadiene was generated in the presence of a mixture of 1,1-diphenylethylene and either dimethyl fumarate or dimethyl maleate. The adduct isolated in each case was exclusively that formed from the ester. Esters are known to be more reactive dienophiles than diphenylethylene whereas the latter is the more reactive radical trapping reagent. These results thus provide further evidence that cyclobutadiene produced from its iron tricarbonyl complex is singlet in nature. 1,3-Di-t-butylcyclobutadiene was similarly liberated from its iron tricarbonyl complex in the presence of tetracyanoethylene and the resultant adduct was isolated[90]:

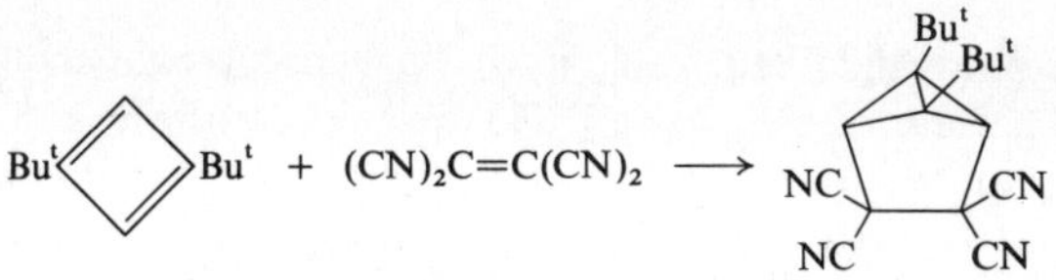

$$\text{Bu}^t\text{-cyclobutadiene-Bu}^t + (CN)_2C{=}C(CN)_2 \longrightarrow$$

Formation of this adduct is readily explicable in terms of orbital-symmetry considerations, provided that the cyclobutadiene possesses a singlet ground state[90].

Further evidence for singlet ground states of cyclobutadienes has been obtained by studies of the reaction of 1,2- and 1,4-diphenylcyclobutadienes[91]:

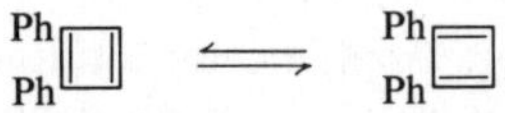

Presumably generation of *vic*-diphenylcyclobutadiene leads to an equilibrating mixture of these two isomers. If this mixture is generated in the presence of an only moderately reactive dienophile such as benzoquinone or *N*-phenylmaleimide, the only adduct formed is that derived from the 1,4-isomer:

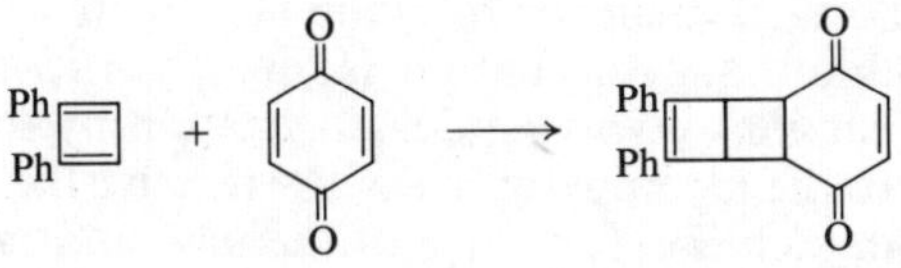

The 1,4-isomer should be more reactive than the 1,2-isomer on steric grounds. If, however, the diphenylcyclobutadiene mixture is generated in the presence of a more reactive dienophile such as tetracyanoethylene or dicyanomaleimide, then two isomeric adducts are formed:

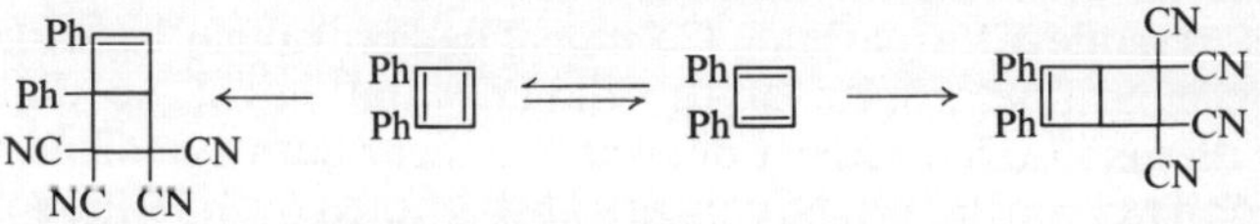

With either of these dienophiles the same ratio of 1,2-diphenyl- and 1,4-diphenyl-cyclobutadiene adducts is obtained, namely 7 : 1. It is suggested that with less reactive dienophiles, equilibration between the isomeric cyclobutadienes proceeds much more rapidly than addition to the dienophile and that the latter reaction only takes place with the more reactive isomer. By contrast, with highly reactive dienophiles the rate of equilibration of the isomers no longer greatly exceeds the rate of the addition reactions and both isomers participate in addition reactions. These results thus provide excellent evidence for the presence of two distinct isomers and thereby for a singlet rectangular structure for these cyclobutadienes.

Cyclobutadiene generated by photolysis of its iron tricarbonyl complex reacts with acetylenes to form benzene derivatives, presumably via an intermediate bicyclohexadiene adduct[92]:

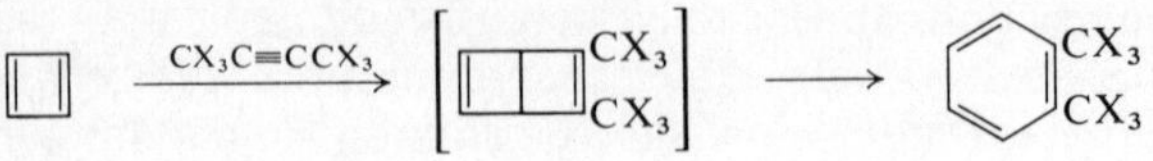

$CX_3C{\equiv}CCX_3$

CX_3 CX_3 CX_3 CX_3

(X = H or F)

It has also been suggested[93] that cyclobutadiene is formed as a transient species in the pyrolysis of its iron tricarbonyl complex, but flash vacuum pyrolysis of this same complex gave vinylacetylene as the only C_4H_4 product[94], and the latter authors believe that vinylacetylene rather than cyclobutadiene was the pyrolytic product also.

Attempts to prepare cyclobutadiene or its derivatives by dehalogenation of halogenocyclobutenes date back to the last century. Evidence has now been produced that when 3,4-dibromo- or 3,4-di-iodo-cyclobutene react with either zinc or lithium amalgam cyclobutadiene is indeed formed as a transient intermediate[95]. It is not formed in any appreciable amount from these reagents and 3,4-dichlorocyclobutene but is formed from the latter compound by the action of sodium- or lithium-amalgam in benzene containing 1,2-bis(dimethylamino)ethane, a chelating agent which is known to increase the anionic nature of lithium–organic systems[95]. Formation of cyclobutadiene was again shown either by the isolation of its dimer, tricyclo[4,2,0,$0^{2,5}$]octa-3,7-diene, or by trapping it with a suitable diene or dienophile. These authors suggest that the failure of earlier attempts to trap cyclobutadiene in reactions involving dehalogenation by sodium amalgam may have been due to the incompatibility of the reactive diene or dienophile trapping agents with sodium. The ratio of the *syn*- and *anti*-forms of the dimer of cyclobutadiene which are produced varies greatly with the identity of the solvent. Thus in dimethylformamide the ratio of *syn*- to *anti*-dimer is 100:1, whereas in acetic acid they are formed in approximately equal amounts. Similar ratios of products are obtained when cyclobutadiene is generated by the oxidation of its iron tricarbonyl complex, so that it appears to be connected with interactions involving the liberated cyclobutadiene rather than its mode of formation[95]. Comment had previously been made that the stereochemical selectivity of the dimerisation might be diminished because of the very high reactivity of cyclobutadiene both as a diene and as a dienophile[89]. This may in turn be considerably affected by differing solvation.

Other methods for the preparation of cyclobutadiene as a reactive intermediate appear to be the flash vacuum pyrolysis of a α-pyrone[96], and the photolysis of the adduct of cyclo-octatetraene and dimethyl acetylenedicarboxylate, dimethyl tricyclo[4,2,2,$0^{2,5}$]deca-3,7,9-triene-7,8-dicarboxylate[97].

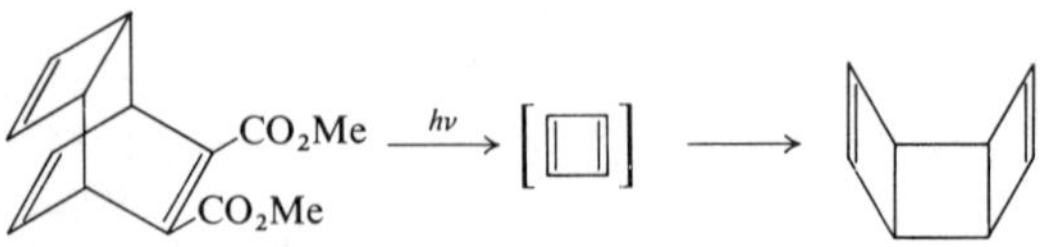

CO_2Me CO_2Me $h\nu$

Tetra-iodocyclobutadiene has been prepared by oxidation of its iron tricarbonyl complex[98]. The action of potassium hydroxide on 1,2,3,3,4-pentachlorocyclobutene appears to give tetrachlorocyclobutadiene; its *syn*-dimer has been isolated and also its adducts with styrene and with cyclohexadiene[99]. Its addition reactions appear to be stereospecific, indicating that it also has a singlet structure. It was suggested[99] that it was more readily available because of stabilisation arising from some contribution from a dipolar form:

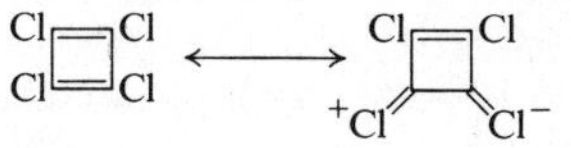

Treatment of hexachlorocyclobutene with lithium amalgam, and ozonolysis of 1,2,3,4-tetrachloro-5,6-dimethylcyclohexa-2,5-diene followed by photolysis of the resultant ozonide, appear to give tetrachlorocyclobutadiene, and tetrabromocyclobutadiene is apparently formed from hexabromocyclobutene by the action of lithium amalgam[100]. Irradiation of the anhydrides of 1,2-dichloro- and 3,4-dichlorocyclobutene-3,4-dicarboxylic acids appear to lead to the formation of dichlorocyclobutadienes and dichlorocyclopentadienones, which then react with one another to give Diels–Alder adducts[101]. Similar irradiation of the anhydride of tetrachlorocyclobutene-3,4-dicarboxylic acid gave octachlorocyclo-octatetraene[101].

When the anhydrides of dimethyl-, trimethyl- or tetramethyl-cyclobutene-3,4-dicarboxylic acids were irradiated in matrices at −196 °C deep yellow products were obtained which were apparently mixtures of cyclobutadienes and cyclopentadienones[102–104]. It seems that carbon dioxide is first lost from the anhydrides; this is then followed either by loss of carbon monoxide to give a cyclobutadiene or by isomerisation to give a cyclopentadienone. When the matrices were warmed, the colours disappeared and diene–ketone adducts and ketone dimers were obtained. It is interesting to note that in the diene–ketone reaction the cyclobutadienes act as the dienophiles and the cyclopentadienones as the dienes. The adducts are *endo*, suggesting that both the cyclobutadienes and cyclopentadienones exist as singlet monomers at −196 °C. By photolysis of a pyridazinobicyclohexene it was possible to produce tetramethylcyclobutadiene without simultaneous formation of a cyclopentadienone[105]:

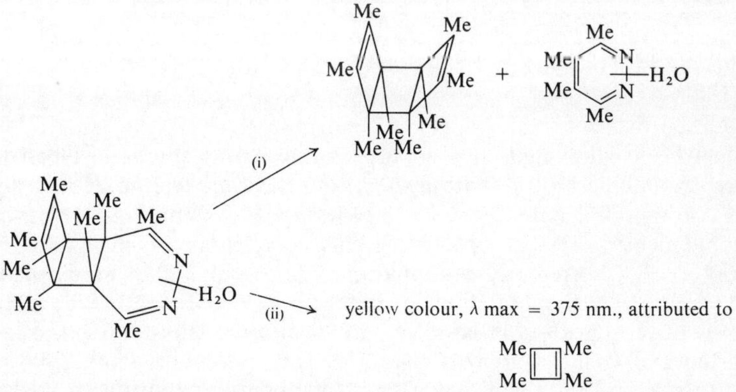

yellow colour, λ max = 375 nm., attributed to

(i) *hν* (ii) *hν*, in matrix, − 196°C

The isolation of the *syn*-dimer once again indicates a singlet monomer; confirmation of this is provided by the low ε-value for the assumed monomer obtained in the matrix at −196 °C. A cyclobutadiene also appears to be formed by γ-radiolysis of dimethylacetylene[106]; among the products isolated were the *syn*-dimer of tetramethylcyclobutadiene and hexamethylbenzene. When the same acetylene is treated with aluminium chloride, a cyclobutadiene complex is formed which reacts with dimethyl acetylenedicarboxylate to give a bicyclohexadiene derivative, probably formed by addition of the ester to liberated tetramethylcyclobutadiene[107]:

MeC≡CMe →(i) [Me, Me, Me, Me cyclobutenyl cation–$AlCl_3$] →(ii) [bicyclohexadiene: Me, Me, Me, Me, CO_2Me, CO_2Me]

(i) $AlCl_3$ (ii) $MeO_2C—C≡C—CO_2Me$ *either* in DMSO—CH_2Cl_2 *or* without solvent followed by addition of DMSO or H_2O

6.3.2 'Push–Pull' cyclobutadienes

The first non-annellated cyclobutadiene derivative to be isolated was a diaminocyclobutadienediester which was obtained as a yellow crystalline product in 1968 as follows[108, 109]:

$Et_3N—C≡C—CO_2Et$ →(i) [H, CO_2Et, NEt_2, Et_2N, CO_2Et cyclobutenyl cation] →(ii) [EtO_2C, NEt_2, Et_2N, CO_2Et cyclobutadiene]

(i) $BF_3—Et_2O$, PhOH, CH_2Cl_2 (ii) NaH

This compound is stabilised by a 'push-pull' effect from the electron-donating amine groups and electron-withdrawing ester groups, e.g.

[EtO_2C, NEt_2, Et_2N, CO_2Et cyclobutadiene] ⟷ [$^-O—C(OEt)$, $\overset{+}{N}Et_2$, $Et_2\overset{+}{N}$, $C(OEt)—O^-$]

It has been calculated that an alternating arrangement of electron-withdrawing and electron-donating substituents should lead to greater stabilisation of the molecule than in an isomeric 1,2-donating-3,4-withdrawing arrangement[109, 110]. The spectra of this compound are in accord with its structure[108, 109]. An x-ray examination[111] shows that the ring is planar and that within experimental error its sides are of equal length. The ring-amine bonds have high double-bond character and the ring-ester bonds are also shortened. Two striking properties of this compound are its thermal stability and its sensitivity towards water and other protonic solvents; in its handling moisture has to be rigorously excluded[109]. Other 'push–pull' cyclobutadienes have also been prepared, including 1,3-diacyl-2,4-diamino-derivatives[112]. The diacyl-compounds are less stable than the diesters, and dimerise thermally to 1,3,5,7-tetra-acyl-2,4,6,8-tetra-aminocyclo-octatetraenes[112]. Di-(alkoxycarbonyl)diaminocyclobutadienes react readily with both electrophiles and nucleophiles. With electrophiles reaction takes place at the ring carbon atom next to an ester group; thus protonation leads to the cyclobutenium salt from which it was prepared[109, 112, 113]. They react similarly with methyl iodide, bromine, acetyl bromide and tropylium perchlorate[109, 113]. With nucleophiles such as water[112], alcohols[109, 112, 113], hydrogen chloride[112] or dimethylamine[112], the initial nucleophilic attack is followed by ring opening, e.g.

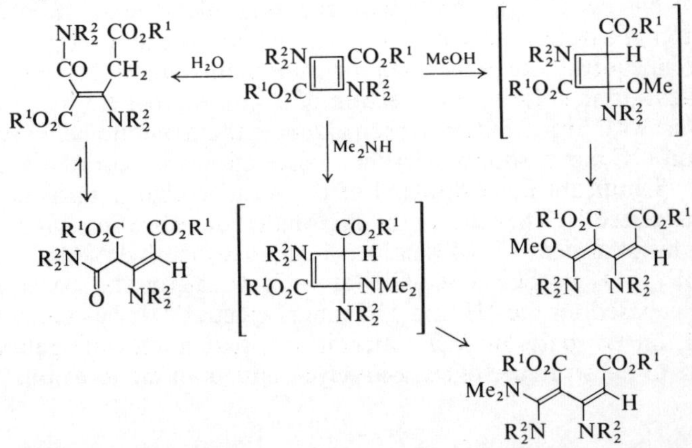

Cyclo-addition reactions with acetylenes lead to benzene derivatives; no bicyclohexadiene derivatives could be detected, even at −80 °C, although they are believed to be intermediates in the reaction[109, 114]. HMO calculations indicate that 'push-pull' bicyclohexadienes should have considerably lower activation energies for their ring opening to form benzene derivatives than bicyclohexadienes without such 'push–pull' substituents. The 'push–pull' cyclobutadienes will not undergo cyclo-addition reactions with non-polar olefins[114], but react readily with electrophilic addends such as tetracyanoethylene or *p*-nitrobenzylidenemalononitrile[109, 114]. Initially dipolar adducts are formed which in the latter case readily undergo conversion to an open-chain product:

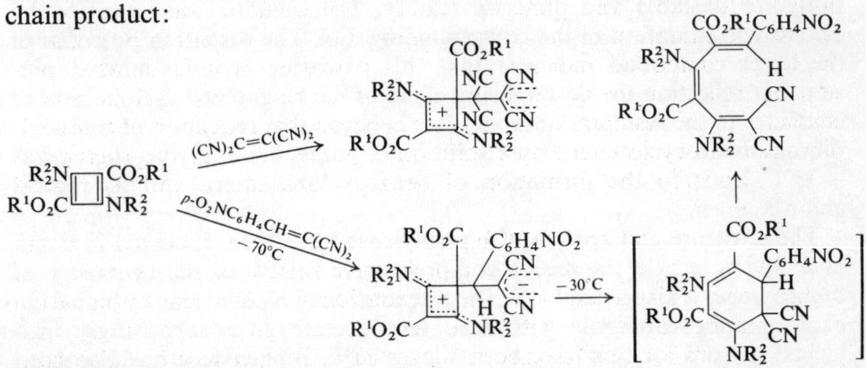

When *p*-nitrobenzaldehyde or nitrosobenzene were used as addends, no intermediate dipolar adduct could be isolated and only the final ring-opened products were obtained[109]. Reactions of the 'push–pull' cyclobutadienes with cyclopropenylidene derivatives (e.g. cyclopropenylidenemalononitrile) resulted in the formation of cycloheptatrienylidene derivatives[109].

6.3.3 Cyclobutenium salts

The formation of a tetraphenylcyclobutenium di-cation was described some years ago, its structure being assigned on the basis of n.m.r. spectra[115]. More recently it has been suggested[116] that exchange phenomena, involving the monocation in a donor–acceptor complex, might instead by responsible for the apparent equivalence of the phenyl groups in this cation. The preparation of this di-cation in solution by reaction of 3,4-dibromo-1,2,3,4-tetraphenylcyclobutene with SbF_5–SO_2 has been claimed, the structure being based on the ^1H and ^{13}C n.m.r. spectra and their correlation with calculated (HMO) spectra[117]. Significant delocalisation of the positive charge into the phenyl rings is suggested[117]. The tetramethylcyclobutenium di-cation has also been obtained by treatment of 3,4-dichloro-1,2,3,4-tetramethylcyclobutene with either SbF_5–SO_2 or SbF_5–SO_2ClF at $-78\,°C$; again the assignment of structure is based on the ^1H and ^{13}C n.m.r. spectra[116]. It has recently been suggested, on the basis of SCF calculations, that homoconjugation must contribute to the structure of tetra-arylcyclobutenium monocations[118].

6.3.4 Benzocyclobutadienes

Further examples of naphthocyclobutadienes have been prepared[119]:

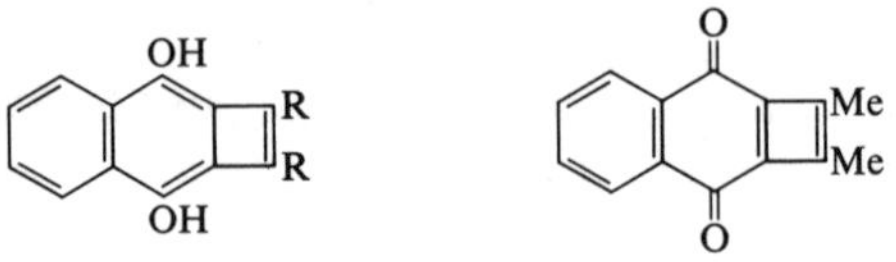

R = Me *or* Ph

Both are unstable and dimerise readily. The quinone was obtained by electrolytic oxidation of the corresponding diol. The oxidation potential of the latter compound indicates that this oxidation is unfavourable, presumably reflecting the destabilising effect of the heightened cyclobutadiene character in the resultant quinone. It is believed that reduction of *trans*-3,4-dibromobenzocyclobutene by a solution of potassium in glyme–diglyme at $-78\,°C$ leads to the formation of benzocyclobutadiene and its radical anion[120].

The structure and spectra of biphenylenes have been discussed in detail in a review article[9]; a recent comprehensive review of the chemistry of biphenylenes is also available[10]. The preparation of biphenylene by oxidation of 1-aminobenzotriazole with lead tetra-acetate gives surprisingly high yields; reasons for this have been suggested[121]. Biphenylene has also been obtained on a preparative scale in up to 20% yield from fluorenone, by feeding its vapour through a high-frequency glow discharge at reduced pressure[122]. Octafluorobiphenylene has been prepared by pyrolysis of tetrafluorophthalic acid derivatives[123, 124].

The benzo-homobiphenylene cation (3) has been prepared in solution[125], and the norbiphenylene anion (4) is also thought to have been prepared[126]:

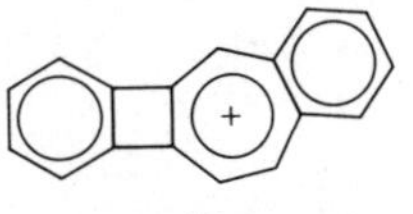

$\overset{+}{N}Me_3\ I^-$ base $\longrightarrow$ $\overset{+}{N}Me_3\ I^-$

(3) 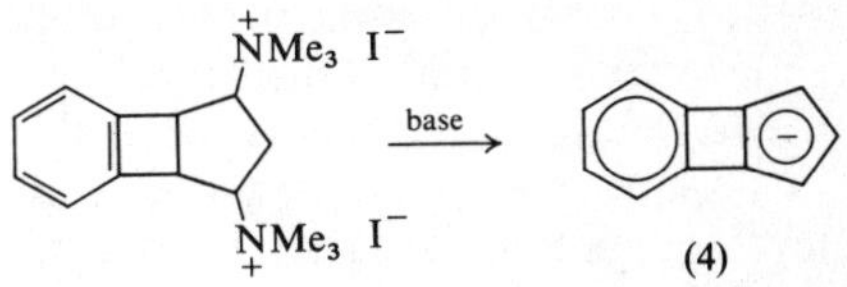 (4)

Aqueous work-up of the anion preparation provides dimers of the protonated form. The u.v. spectrum of the cation indicates that it is more extensively conjugated than the benzotropylium ion.

The ^7Li n.m.r. spectrum of the di-lithium salt of biphenylene has been recorded, and indicates the presence of a diamagnetic ring current in the biphenylene dianion, and also that the negative charge is delocalised over the whole anion[127]. Protodesilylation and protodetritiation studies on appropriate 1- and 2-substituted biphenylenes indicate that the 1-position is unique in polycyclic aromatic compounds in being activated or deactivated according to the demands for resonance in the transition states of reactions[128]. Thus it is activated in hydrogen exchange but deactivated in desilylation, the difference being attributed to the greater demand of resonance stabilisation of the transition state in the latter reaction.

Another example of the trapping of the biradical formed from biphenylene on pyrolysis has been recorded. When biphenylene is heated with selenium at *c.* 275 °C dibenzoselenole is formed[129].

6.4 DERIVATIVES OF CYCLOPENTADIENE

A review of the electrophilic substitution reactions of cyclopentadienide derivatives has been published recently[130].

The relationship between vicinal coupling constants and bond order has been shown to be of great utility in the investigation of delocalisation in potentially 'aromatic' derivatives of cyclopentadiene. Thus it has been used to compare the delocalisation in a series of different cyclopentadiene derivatives such as cyclopentadienylides and fulvenes[131]. It has also been suggested that the difference between the coupling constants $J_{2,3}$ and $J_{3,4}$ in cyclopentadiene derivatives may be used as an indication of the anionic character of such derivatives[132, 133], and this also has proved of great utility.

6.4.1 Cyclopentadienide salts

The ^7Li n.m.r. signal for lithium cyclopentadienide is 8.5 p.p.m. upfield from LiCl, suggesting that the lithium ion is located over the π-cloud of the cyclopentadienide anion, forming a tight or contact ion pair[134]. This salt undergoes an interesting reaction with methyl-lithium in methylene chloride–ether to give a moderate yield of benzvalene, together with benzene[135]. Cyclopentenynide anion, the cyclopentadienide analogue of benzyne, has apparently been prepared by decomposition of the anion of diazocyclopentadiene-2-carboxylic acid; it was trapped by appropriate reagents and, not surprisingly, appeared to be a highly reactive species[136]. Thallium and quaternary ammonium pentachlorocyclopentadienides have been obtained but are unstable at room temperature[137]. N.Q.R. spectroscopy suggests that the negative charge on the ring is repelled by the chlorine lone pairs, causing the C – Cl bonds to be more ionic than usual[137].

Acylcyclopentadienide salts, in which the negative charge is shared between the ring and the oxygen atoms, may undergo electrophilic attack in the ring or at an oxygen atom. Thus the formylcyclopentadienide ion is acylated on oxygen[138, 139], while 1,2-diacylcyclopentadienides react at an oxygen atom with methyl chloroformate and at the ring with reagents such as bromine and diazonium salts[140]. It is interesting that whereas bromination and deuteriation of these 1,2-diacylcyclopentadienes take place at the 4-position, the coupling with diazonium salts gives 3-arylazo-products; it has been suggested that in the latter case the initial attack occurs at an oxygen atom and this is followed by an intramolecular rearrangement to give the final product[140].

6.4.2 Cyclopentadienylides

An elegant new preparation of dimethylsulphonium cyclopentadienylide involves the reaction of trimethylsilylcyclopentadiene with dimethyl sulphoxide[141]. Sulphoxides have also been used in the preparation of sulphonium cyclopentadienylides by condensing them with cyclopentadienes having electron-withdrawing substituents[142]. Arsonium cyclopentadienylides have been prepared similarly using triphenylarsine oxide[140, 143, 144]. A wide variety of heteronium tetraphenylcyclopentadienylides including pyridinium[145, 146], phosphonium[147, 148], arsonium[149], stibonium[150], bismuthonium[151], sulphonium[152], selenonium[153] and telluronium[154] ylides, has been prepared by carbenic decomposition of diazotetraphenylcyclopentadiene in the presence of suitable substrates. Two new preparative routes to cyclopentadienylides start from ferrocene derivatives. Chlorination or bromination of ferrocene in pyridine gave pyridinium tetrachloro- or tetrabromo-cyclopentadienylides[155], while photolysis of a diethylferrocenylmethylammonium derivative produced diethylmethylammonium cyclopentadienylide[156].

A variety of electrophilic substitution reactions have been effected on triphenylphosphonium cyclopentadienylide. Substitution takes place preferentially at the 2(5)-position, as might be expected since this involves a linearly conjugated transition state, whereas substitution at the 3(4)-position must involve a cross-conjugated transition state[157]. Thus it is mononitrated[158], -nitrosated[158] and -acetylated[157], monoformylated (Vilsmeier) at $-15\,^{\circ}\mathrm{C}$ [158] but diformylated at higher temperatures[157, 158]. Bromination gives a tetrabromo-derivative[157, 158]. It also undergoes Reimer–Tiemann formylation[159]. Electron-deficient olefins react with this ylide by electrophilic substitution and not by a cyclo-addition reaction[160–164]. A recent review is available of condensation reactions of triphenylphosphonium cyclopentadienylide which produce cyanine-like dyestuffs[165]. Triphenylarsonium cyclopentadienylide likewise undergoes electrophilic substitution at the 2-position[166].

It has been suggested, on the basis of semi-empirical MO calculations, that the ground state of triphenylphosphonium cyclopentadienylide is more highly polar than it had been considered to be, and that previous estimates based on the dipole moment were low because no account was taken of the σ-moment, which opposes the π-moment[167]. Detailed analyses of the n.m.r. spectra of the dimethylsulphonium[168, 169], triphenylphosphonium and triphenylarsonium[169] cyclopentadienylides show that in each case the bond orders of the 2,3-bonds are greater than those for the 3,4-bonds. Since p_π–d_π overlap is expected to be less effective in a C–As bond than in a C—P or C—S bond, the arsonium ylide would be expected to be the more polar and to have more cyclopentadienide character; this is found to be so[169]. The other two ylides appear to have very similar cyclopentadienide character; it is assumed that the generally greater p_π–d_π overlap expected in a C—P bond compared to a C—S bond is nullified by the replacement of phenyl groups by methyl groups in the sulphonium ylide[169].

Comparative studies on the tetraphenylcyclopentadienylides[170, 171] and on triphenylcyclopentadienylides[172] confirm the supposition that p_π–d_π overlap in a carbon–heteronium atom bond decreases down Groups V and VI of the periodic table, with concomitant increase of cyclopentadienide character in the ylides; this is consonant with the stabilities, spectra, basicities and reactivity towards Wittig-type reagents of these ylides.

An x-ray crystallographic examination of triphenylarsonium 2-acetylcyclopentadienylide indicates that the negative charge is located to a greater extent on the oxygen-atom than on the cyclopentadiene ring and that there is strong interaction between the oxygen and arsenic atoms[173].

6.4.3 Fulvenes

Recent comprehensive reviews on fulvenes are available[6, 15]. An excellent general method for the preparation of fulvenes from sodium cyclopentadienide and *gem.*-acetoxy-chloro-compounds has been described[174, 175]:

$$CH_3COCl + R^1R^2CO \xrightarrow{(i)} R^1\text{–}C(R^2)(Cl)\text{–}OAc \xrightarrow{(ii)} C_5H_5\text{–}C(R^1)(R^2)\text{–}OAc \xrightarrow{(iii)} C_5H_4{=}CR^1R^2$$

(i) Lewis acid

(ii) Na^+ (cyclopentadienide anion)

(iii) Et_3N

Variable temperature n.m.r. studies on 6-dimethylaminofulvenes indicate an appreciable dipolar character[176–180]. 6-Dimethylaminofulvene is deuteriated readily by deuteriomethanol at the 2-, 5-positions and more slowly at the 3-,4-positions[180]. The dipolar contribution is increased by the presence of electron-withdrawing groups in the five-membered ring[177, 178, 181]. An x-ray crystallographic examination of the 2-formyl derivative suggests a major contribution from a dipolar form with the negative charge sited on the oxygen atom rather than on the ring[182]. In bis(dialkylamino)fulvenes the C—N bonds have less double bond character despite the somewhat higher dipole moments, since the positive charge has now to be shared over two nitrogen atoms[183]. The crystal structure of *S,S'*-ethylene-6,6-dimercaptofulvene indicates that there is little contribution from a dipolar form[184]. Physical properties show that although intramolecular hydrogen bonding exists and fast tautomeric proton exchange occurs in the 6-aminofulvene-2-aldimines (5), non-classical aromaticity due to delocalisation through the hydrogen-bond does not exist[185].

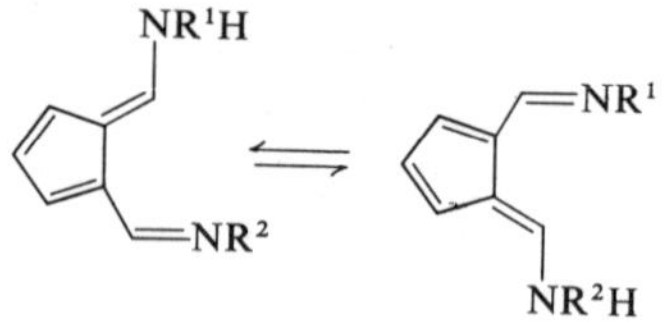

NR^1H / $=NR^2$ ⇌ $=NR^1$ / NR^2H

The chemistry of cyclopentadienylidene-pyrans and -thiopyrans has been reviewed[186]; these compounds are largely non-polar in their ground state. This has been confirmed by x-ray analysis[187].

References

1. Skrabal, P., private communication
2. Lloyd, D. (1966). *Carbocyclic Non-benzenoid Aromatic Compounds* (Amsterdam: Elsevier)
3. Krebs, A. W. (1965). *Angew Chem.*, **77,** 10; *Angew Chem. Int. Edn.*, **4,** 10
4. D'yakonov, I. A. and Kostikov, R. R. (1967). *Russ. Chem. Rev.*, **36,** 557
5. West, R. (1970). *Accounts Chem. Res.*, **3,** 130
6. Bergmann, E. D. (1968). *Chem. Rev.*, **68,** 41
7. Breslow, R. (1968). *Angew Chem.*, **80,** 573; *Angew. Chem. Int. Edn.*, **8,** 565
8. Cava, M. P. and Mitchell, M. J. (1967). *Cyclobutadiene and Related Compounds* (New York: Academic Press)
9. Figeys, H. P. (1969). *Topics in Carbocylic Chemistry,* Vol. 1, 269. (D. Lloyd, editor) (London: Logos Press)
10. Barton, J. W. (1969). *Nonbenzenoid Aromatics,* Vol. 1, 32. (J. P. Snyder, editor) (New York: Academic Press)
11. Maitlis, P. M. (1966). *Advan. Organometal. Chem.*, **4,** 95
12. Maitlis, P. M. and Eberius, K. W. (1971). *Nonbenzenoid Aromatics,* Vol. 2, 360. (J. P. Snyder, editor) (New York: Academic Press)
13. Rybinskaya, M. I. and Korneva, L. M. (1971). *Russ. Chem. Rev.*, **40,** 247
14. Seitz, G. (1969). *Angew Chem.*, **81,** 518; *Angew. Chem. Int. Edn.*, **8,** 478
15. Yates, P. (1968). *Advances in Alicyclic Chemistry,* Vol. 2, 60. (H. Hart and G. J. Karabatsos, editors) (New York: Academic Press)
16. Breslow, R., Groves, J. T. and Ryan, G. (1967). *J. Amer. Chem. Soc.*, **89,** 5048
17. Breslow, R. and Groves, J. T. (1970). *J. Amer. Chem. Soc.*, **92,** 984
18. Tobey, S. W. and West, R. (1964). *J. Amer. Chem. Soc.*, **86,** 1459
19. West, R., Sadô, A. and Tobey, S. W. (1966). *J. Amer. Chem. Soc.*, **88,** 2488
20. Doering, W. von E., Laber, G., Bonderwahl, R., Chamberlain, N. F. and Williams, R. B. (1956). *J. Amer. Chem. Soc.*, **78,** 5448
21. Farnum, D. G., Mehta, G. and Silberman, R. G. (1967). *J. Amer. Chem. Soc.*, **89,** 5048
22. Breslow, R., Ryan, G. and Groves, J. T. (1970). *J. Amer. Chem. Soc.*, **92,** 988

23. Zecher, D. C. and West, R. (1967). *J. Amer. Chem. Soc.*, **89**, 153
24. West, R., Zecher, D. C. and Goyert, W. (1970). *J. Amer. Chem. Soc.*, **92**, 149
25. West, R. and Zecher, D. C. (1967). *J. Amer. Chem. Soc.*, **89**, 152
26. Ciabottoni, J. and Nathan, E. C. (1968). *J. Amer. Chem. Soc.*, **90**, 4495
27. Ciabottoni, J. and Nathan, E. C. (1969). *Tetrahedron Lett.*, 4997
28. Breslow, R., Höver, H. and Chang, H. W. (1962). *J. Amer. Chem. Soc.*, **84**, 3168
29. Breslow, R. and Dowd, P. (1963). *J. Amer. Chem. Soc.*, **85**, 2729
30. Stehouwer, D. M. and Longone, D. T. (1969). *Tetrahedron Lett.*, 5311
31. Longone, D. T. and Stehouwer, D. M. (1970). *Tetrahedron Lett.*, 1017
32. Shono, T., Toda, T. and Oda, R. (1970). *Tetrahedron Lett.*, 369
33. Dulenko, V. I., Semenov, N. C., Baranov, S. N. and Krivun, S. V. (1970). *Zh. Obshch. Khim.*, **40**, 701; *J. Gen. Chem. USSR*, **40**, 672
34. Battiste, M. A. and Sprouse, C. T. (1969). *Tetrahedron Lett.*, 3165
35. Krivan, S. V., Semenov, N. S., Baranov, S. N. and Dulenko, V. I. (1970). *Zh. Obshch. Khim.*, **40**, 1904; *J. Gen. Chem. USSR*, **40**, 1885
36. Yoshida, Z. and Tawara, Y. (1971). *J. Amer. Chem. Soc.*, **93**, 2573
37. Breslow, R., Eicher, T., Krebs, A., Peterson, R. A. and Posner, J. (1965). *J. Amer. Chem. Soc.*, **87**, 1320
38. Eicher, T. and Löschner, A. (1966). *Z. Naturforsch.*, **21b**, 295
39. Eicher, T. and Hansen, A. (1967). *Tetrahedron Lett.*, 4321
40. Eicher, T. and Hansen, A. (1969). *Chem. Ber.*, **102**, 319
41. D'yakonov, I. A., Kostikov, R. R. and Molchanov, A. P. (1969). *Zhur. Org. Khim.*, **5**, 175; *J. Org. Chem. USSR*, **5**, 171
42. D'yakonov, I. A., Kostikov, R. R. and Molchanov, A. P. (1970). *Zhur. Org. Khim.*, **6**, 316; *J. Org. Chem. USSR*, **6**, 314
43. Dehmlow, E. V. and Schönefeld, J. (1971). *Liebigs Ann.*, **744**, 42
44. Breslow, R. and Ryan, G. (1967). *J. Amer. Chem. Soc.*, **89**, 3073
45. West, R., Chickos, J. and Osawa, E. (1968). *J. Amer. Chem. Soc.*, **90**, 3885
46. Ciabottoni, J. and Nathan, E. C. (1969). *J. Amer. Chem. Soc.*, **91**, 4766
47. Dehmlow, E. V. (1966). *Chem. and Ind.*, 1379
48. Dehmlow, E. V. (1968). *Chem. Ber.*, **101**, 410
49. Harrison, E. A. (1970). *Chem. Commun.*, 982
50. Toda, F., Mitote, T. and Akaji, K. (1969). *Chem. Commun.*, 228
51. Toda, F., Mitote, T. and Akaji, K. (1969). *Bull. Chem. Soc. Jap.*, **42**, 1777
52. Dehmlow, E. V. (1969). *Liebigs Ann.*, **729**, 64
53. Dehmlow, E. V. (1967). *Tetrahedron Lett.*, 5177
54. Dehmlow, E. V. (1969). *Chem. Ber.*, **102**, 3863
55. Lown, J. W. and Katsumoto, K. (1971). *Can. J. Chem.*, **49**, 1165
56. Ciabottoni, J. and Berchtold, G. A. (1966). *J. Org. Chem.*, **31**, 1336
57. Eicher, T. and Hansen, A. (1967). *Tetrahedron Lett.*, 1169
58. Eicher, T., von Angerer, E. and Hansen, A. (1971). *Liebigs Ann.*, **746**, 102
59. Eicher, T. and von Angerer, E. (1971). *Liebigs Ann.*, **746**, 120
60. Hayasi, Y. and Nozaki, H. (1971). *Tetrahedron*, **27**, 3085
61. Sasaki, T., Kanematsu, K. and Kakehi, A. (1971). *J. Org. Chem.*, **36**, 2451
62. Ao, M. S., Burgess, E. M., Schauer, A. and Taylor, E. A. (1969). *Chem. Commun.*, 220
63. Adamson, J., Forster, D. L., Gilchrist, T. L. and Rees, C. W. (1969). *Chem. Commun.*, 221
64. Adamson, J., Forster, D. L., Gilchrist, T. L. and Rees, C. W. (1971). *J. Chem. Soc. C*, 981
65. Krebs, A. and Schrader, B. (1967). *Liebigs Ann.*, **709**, 46
66. Krebs, A., Schrader, B. and Höfler, F. (1968). *Tetrahedron Lett.*, 5935
67. Tobey, S. W. (1971). '*Aromaticity, Pseudo-aromaticity, Anti-aromaticity*,' 351. (Jerusalem: The Israel Academy of Sciences and Humanities)
68. Bertelli, D. J. and Andrews, T. G. (1969). *J. Amer. Chem. Soc.*, **91**, 5280
69. Watanabe, A., Yamaguchi, H., Amako, Y. and Azumi, H. (1968). *Bull. Chem. Soc. Jap.*, **41**, 2196
70. Clark, D. T. and Lilley, D. M. J. (1970). *Chem. Commun.*, 147
71. Krebs, A. and Breckwoldt, J. (1969). *Tetrahedron Lett.*, 3797
72. Hecht, S. S. (1970). *Tetrahedron Lett.*, 4385
73. Eicher, T. and Pelz, N. (1970). *Chem. Ber.*, **103**, 2647
74. Eiglmeier, K. and Eicher, T. (1968). *Angew. Chem.*, **80**, 798; *Angew. Chem. Internat. Edn.*, **7**, 809

75. Eicher, T. and Eiglmeier, K. (1971). *Chem. Ber.*, **104,** 605
76. West, R. and Zecher, D. C. (1970). *J. Amer. Chem. Soc.*, **92,** 155
77. Battiste, M. A. and Hill, J. H. M. (1968). *Tetrahedron Lett.*, 5537
78. Battiste, M. A. and Hill, J. H. M. (1968). *Tetrahedron Lett.*, 5541
79. Lloyd, D. and Preston, N. W. (1969). *Chem. and Ind.*, 1055
80. Lloyd, D. and Preston, N. W. (1970). *J. Chem. Soc. C,* 610
81. Gompper, R. and Weiss, R. (1968). *Angew. Chem.*, **80,** 277; *Angew. Chem. Int. Edn.*, **7,** 296
82. Breslow, R. (1965). *Chem. Eng. News,* **43,** No. 26, 90
83. Breslow, R., Brown, J. and Gajewski, J. (1967). *J. Amer. Chem. Soc.*, **89,** 4383
84. Breslow, R. and Balasubramanian, K. (1969). *J. Amer. Chem. Soc.*, **91,** 5182
85. Schipperijn, A. J. (1971). *Rec. Trav. Chim.*, **90,** 1110
86. Dewar, M. J. S., Kohn, M. C. and Trinajstić, N. (1971). *J. Amer. Chem. Soc.*, **93,** 3436
87. Breslow, R. and Washburn, W. (1970). *J. Amer. Chem. Soc.*, **92,** 427
88. Breslow, R., Grubbs, R. and Murahashi, S.-I. (1970). *J. Amer. Chem. Soc.*, **92,** 4139
89. Watts, L., Fitzpatrick, J. D. and Pettit, R. (1966). *J. Amer. Chem. Soc.*, **88,** 623
90. Reeves, P., Henery, J. and Pettit, R. (1969). *J. Amer. Chem. Soc.*, **91,** 5888
91. Reeves, P., Devon, T. and Pettit, R. (1969). *J. Amer. Chem. Soc.*, **91,** 5890
92. Font, J., Barton, S. C. and Strausz, O. P. (1970). *Chem. Commun.*, 499
93. Li, P. H. and McGee, H. A. (1969). *Chem. Commun.*, 592
94. Hedaya, E., Krull, I. S., Miller, R. D., Kent, M. E., D'Angelo, P. F. and Schissel, P. (1969). *J. Amer. Chem. Soc.*, **91,** 6880
95. Schmidt, E. K. G., Brener, L. and Pettit, R. (1970). *J. Amer. Chem. Soc.*, **92,** 3240
96. Hedaya, E., Miller, R. D., McNiel, D. W., D'Angelo, P. F. and Schissel, P. (1969). *J. Amer. Chem. Soc.*, **91,** 1875
97. Miller, R. D. and Hedaya, E. (1969). *J. Amer. Chem. Soc.*, **91,** 5401
98. Amiet, G., Nicholas, K. and Pettit, R. (1970). *Chem. Commun.*, 161
99. Scherer, K. V. and Meyers, T. J. (1968). *J. Amer. Chem. Soc.*, **90,** 6253
100. Criegee, R. and Huber, R. (1970). *Chem. Ber.*, **103,** 1862
101. Maier, G., Fritschi, G. and Hoppe, B. (1971). *Tetrahedron Lett.*, 1463
102. Maier, G. and Mende, U. (1969). *Angew Chem.*, **81,** 114; *Angew. Chem. Int. Edn.*, **8,** 132
103. Maier, G. and Mende, U. (1969). *Tetrahedron Lett.*, 3155
104. Maier, G., Fritschi, G. and Hoppe, B. (1970). *Angew Chem.*, **82,** 551; *Angew Chem. Int. Edn.*, **9,** 529
105. Maier, G. and Schneider, M. (1971). *Angew Chem.*, **83,** 885; *Angew Chem. Int. Edn.*, **10,** 809
106. Whitten, D. G. and Berngruber, W. (1971). *J. Amer. Chem. Soc.*, **93,** 3204
107. Koster, J. B., Timmermans, G. J. and van Bekkum, H. (1971). *Synthesis,* 139
108. Gompper, R. and Seybold, G. (1968). *Angew. Chem.*, **80,** 804; *Angew. Chem. Int. Edn.*, **7,** 824
109. Gompper, R. and Seybold, G. (1971). '*Aromaticity, Pseudo-aromaticity, Anti-aromaticity,*' 215. (Jerusalem: The Israel Academy of Sciences and Humanities)
110. Hoffmann, R. (1969). *Chem. Commun.*, 240
111. Lindner, H. J. and von Gross, B. (1971). *Angew. Chem.*, **83,** 489; *Angew. Chem. Int. Edn.*, **10,** 490
112. Neuenschwander, M. and Niederhauser, A. (1970). *Helv. Chim. Acta,* **53,** 519
113. Gompper, R. and Seybold, G. (1971). *Angew. Chem.*, **83,** 44; *Angew. Chem. Int. Edn.*, **10,** 67
114. Gompper, R. and Seybold, G. (1971). *Angew. Chem.*, **83,** 45; *Angew. Chem. Int. Edn.*, **10,** 68
115. Freedman, H. H. and Young, A. E. (1964). *J. Amer. Chem. Soc.*, **86,** 734
116. Olah, G. A., Bollinger, J. M. and White, A. M. (1969). *J. Amer. Chem. Soc.*, **91,** 3667
117. Olah, G. A. and Mateescu, G. D. (1970). *J. Amer. Chem. Soc.*, **92,** 1430
118. Lodder, A. E., Buck, H. M. and Oosterhoff, L. J. (1970). *Rec. Trav. Chim.*, **89,** 1229
119. Breslow, R., Grubbs, R. and Murahashi, S.-I. (1970). *J. Amer. Chem. Soc.*, **92,** 4139
120. Rieke, R. D. and Hudnall, P. M. (1969). *J. Amer. Chem. Soc.*, **91,** 3678
121. Campbell, C. D. and Rees, C. W. (1969). *J. Chem. Soc. C,* 742
122. Suhr, H. and Weiss, R. I. (1970). *Angew. Chem.*, **82,** 295; *Angew. Chem. Int. Edn.*, **9,** 312
123. Gardner, D. V., McOmie, J. F. W., Albriktsen, P. and Harris, R. K. (1969). *J. Chem. Soc. C,* 1994
124. Sartori, P. and Golloch, A. (1969). *Chem. Ber.*, **102,** 1765

125. Garratt, P. J. and Vollhardt, K. P. C. (1971). *Chem. Commun.*, 1143
126. Cava, M. P., Narasimhan, K., Ziegler, W., Radonovich, L. J. and Glick, M. D. (1969). *J. Amer. Chem. Soc.*, **91,** 2378
127. Cox, R. H., Terry, H. W. and Harrison, L. W. (1971). *Tetrahedron Lett.*, 4815
128. Taylor, R. (1971). *J. Chem. Soc. B,* 536
129. Gaidis, J. M. (1970). *J. Org. Chem.*, **35,** 2811
130. Rybinskaya, M. I. and Korneva, L. M. (1971). *Russ. Chem. Rev.*, **40,** 247
131. Smith, W. B., Watson, W. H. and Chiranjeevi, S. (1967). *J. Amer. Chem. Soc.*, **89,** 1438
132. Kende, A. S., Izzo, P. J. and MacGregor, P. J. (1966). *J. Amer. Chem. Soc.*, **88,** 3359
133. Sofer, H. and Polansky, O. E. (1971). *Monatsh.*, **102,** 256
134. Cox, R. H., Terry, H. W. and Harrison, L. W. (1971). *J. Amer. Chem. Soc.*, **93,** 3297
135. Katz, T. J., Wang, E. J. and Acton, N. (1971). *J. Amer. Chem. Soc.*, **93,** 3782
136. Martin, J. C. and Bloch, D. R. (1971). *J. Amer. Chem. Soc.*, **93,** 451
137. Wulfsberg, G. and West, R. (1971). *J. Amer. Chem. Soc.*, **93,** 4085
138. Hafner, K., Bauer, W. and Schulz, G. (1968). *Angew. Chem.*, **80,** 800; *Angew. Chem. Int. Edn.*, **7,** 806
139. Hafner, K. and Schmidt, F. (1971). *Tetrahedron Lett.*, 2237
140. Lloyd, D. and Preston, N. W. (1968). *J. Chem. Soc. C,* 2464
141. McLean, S. and Reed, G. W. B. (1970). *Can. J. Chem.*, **48,** 3110
142. Seitz, G. (1968). *Chem. Ber.*, **101,** 585
143. Harris, G. S., Lloyd, D., Preston, N. W. and Singer, M. I. C. (1968). *Chem. and Ind.*, 510
144. Lloyd, D. and Singer, M. I. C. (1971). *J. Chem. Soc. C,* 2941
145. Band, I. B. M., Lloyd, D., Singer, M. I. C. and Wasson, F. I. (1966). *Chem. Commun.*, 544
146. Lloyd, D. and Singer, M. I. C. (1971). *J. Chem. Soc. C,* 2939
147. Lloyd, D., Singer, M. I. C., Regitz, M. and Liedhegener, A. (1967). *Chem. and Ind.*, 324
148. Regitz, M. and Liedhegener, A. (1967). *Tetrahedron,* **23,** 2701
149. Lloyd, D. and Singer, M. I. C. (1967). *Chem. and Ind.*, 510
150. Lloyd, D. and Singer, M. I. C. (1967). *Chem. and Ind.*, 787
151. Lloyd, D. and Singer, M. I. C. (1967). *Chem. Commun.*, 1042
152. Lloyd, D. and Singer, M. I. C. (1967). *Chem. and Ind.*, 118
153. Lloyd, D. and Singer, M. I. C. (1967). *Chem. Commun.*, 390
154. Freeman, B. H. and Lloyd, D. (1970). *Chem. Commun.*, 924
155. Nefedov, V. A. (1966). *Zh. Obshch. Khim.*, **36,** 1508; *J. Gen. Chem. USSR,* **36,** 1513
156. Nesmeyanov, A. N., Sazonova, V. A., Romanenko, V. I., Postnov, V. N., Zol'nikova, G. P., Blinova, V. A. and Kalyanova, R. M. (1967). *Dokl. Akad. Nauk SSR.*, **173,** 589
157. Lloyd, D. and Singer, M. I. C. (1971). *Chem. and Ind.*, 786
158. Yoshida, Z., Yoneda, S., Murata, Y. and Hashimoto, H. (1971). *Tetrahedron Lett.*, 1523
159. Yoshida, Z., Yoneda, S. and Yato, T. (1971). *Tetrahedron Lett.*, 2973
160. Rigby, C. W., Lord, E. and Hall, C. D. (1967). *Chem. Commun.*, 714
161. Lord, E., Naan, M. P. and Hall, C. D. (1970). *J. Chem. Soc. B,* 1401
162. Lord, E., Naan, M. P. and Hall, C. D. (1971). *J. Chem. Soc. B,* 213
163. Rigby, C. W., Lord, E., Naan, M. P. and Hall, C. D. (1971). *J. Chem. Soc. B,* 1192
164. Yoshida, Z., Yoneda, S., Hashimoto, H. and Murata, Y. (1971). *Tetrahedron Lett.*, 1527
165. Van Dormael, A. (1969). *Bull. Soc. Chim. Fr.*, 2701
166. Freeman, B. H. and Lloyd, D. (1971). *J. Chem. Soc. C,* 3164
167. Yoshida, Z., Iwata, K. and Yoneda, S. (1971). *Tetrahedron Lett.*, 1519
168. Yoshida, Z., Yoneda, S. and Hazama, M. (1971). *Chem. Commun.*, 716
169. Ernstbrunner, E. E. and Lloyd, D. (1971). *Chem. and Ind.*, 1332
170. Lloyd, D. and Singer, M. I. C. (1968). *Chem. and Ind.*, 1277
171. Freeman, B. H., Lloyd, D. and Singer, M. I. C. (1972). *Tetrahedron,* **28,** 343
172. Lloyd, D. and Singer, M. I. C. (1972). *Tetrahedron,* **28,** 353
173. Ferguson, G., Rendle, D. F., Lloyd, D. and Singer, M. I. C. (1971). *Chem. Commun.*, 1647
174. Neuenschwander, M., Kyburz, R. and Iseli, R. (1970). *Chimia,* **24,** 342
175. Kyburz, R., Schaltegger, H. and Neuenschwander, M. (1971). *Helv. Chim. Acta,* **54,** 1037
176. Downing, A. P., Ollis, W. D. and Sutherland, I. O. (1968). *Chem. Commun.*, 1053
177. Downing, A. P., Ollis, W. D. and Sutherland, I. O. (1969). *J. Chem. Soc. B,* 111
178. Downing, A. P., Ollis, W. D., Nógrádi, M. and Sutherland, I. O. (1971). '*Aromaticity, Pseudo-aromaticity, Anti-aromaticity,*' 296. (Jerusalem: The Israel Academy of Sciences and Humanities)

179. Mannschreck, A. and Kölle, U. (1967). *Tetrahedron Lett.*, 863
180. Mannschreck, A. and Kölle, U. (1969). *Chem. Ber.*, **102,** 243
181. Downing, A. P., Ollis, W. D. and Sutherland, I. O. (1967). *Chem. Commun.*, 143
182. Ammon, H. L. and Plastas, L. A. (1971). *Chem. Commun.*, 356
183. Hartke, K. and Salamon, G. (1970). *Chem. Ber.*, **103,** 133
184. Burzlaff, K., Hartke, K. and Salamon, R. (1970). *Chem. Ber.*, **103,** 156
185. Müller-Westerhoff, U. (1970). *J. Amer. Chem. Soc.*, **92,** 4849
186. Seitz, G. (1969). *Angew. Chem.*, **81,** 518; *Angew. Chem. Int. Edn.*, **8,** 478
187. Burzlaff, K. and Kawada, I. (1969). *Chem. Ber.*, **102,** 3632

7
Troponoids, Heptafulvenes and Azulenes

TETSUO NOZOE*
Tohoku University, Sendai, Japan
and
ICHIRO MURATA
Osaka University, Osaka, Japan

---

*Present Address: No. 811, 2-5-1, Kamiyoga, Setagaya-ku Tokyo 158, Japan

---

7.1 INTRODUCTION

The 'International Symposium on the Chemistry of Non-benzenoid Aromatic Compounds' (ISNA)[1] and the Jerusalem Symposium on 'Quantum Chemistry and Biochemistry'[2] which were held in 1970 reflect the interest in non-benzenoid aromatic compounds.

This chapter presents the principal information available on tropylium compounds—compounds having trigonally-hybridised carbon atoms in a seven-membered ring system—from publications from 1969 to 1971, including some unpublished results. Many Japanese papers which have already been referred to in one of the main lectures of ISNA have been excluded here, and interested readers are recommended to refer to that article[3].

7.2 TROPONOIDS

7.2.1 Tropylium ions

7.2.1.1 Non-condensed tropylium ions

(a) *Tropylium ion*—Tropylium (tropenylium) ion (1) and alkyl tropyl ether react readily with enamines (2)[4a] or dienamines (3)[4b] to form tropyl ketones e.g. (4; R = H or tropyl), (5) and (6). Although the tropylium ion can easily react with *S*-ylides to give tropylated derivatives, rearrangement of the seven-membered ring also takes place producing benzylidene derivatives[5]. Tropylium perchlorate, when allowed to react at room temperature with cyclopentadiene in aqueous dioxane, yielded a complex mixture of alcohols in a

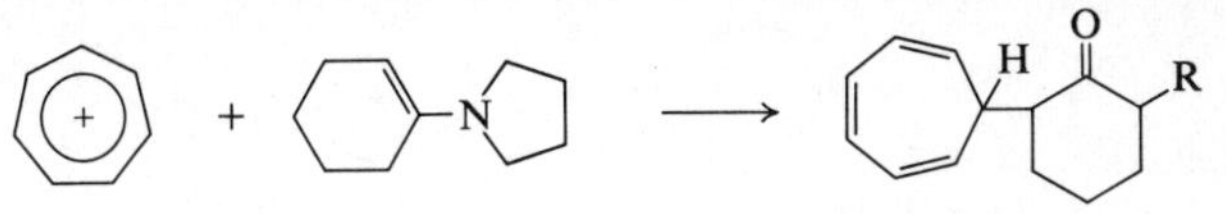

(1) (2) (4)

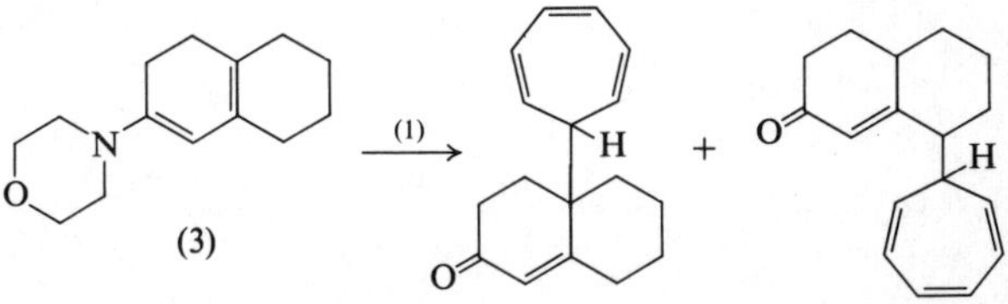

(3) (5) (6)

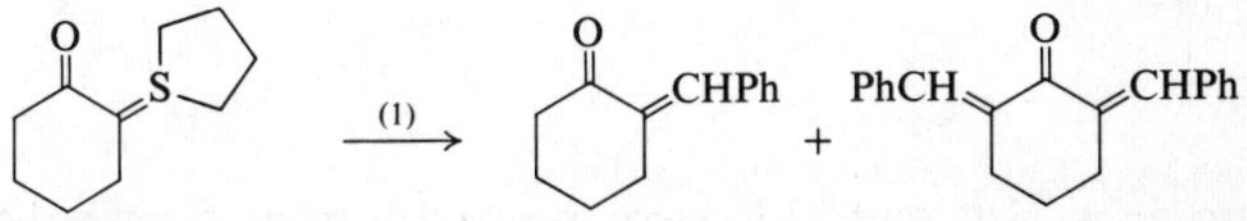

highly stereospecific $[\pi^4 s + \pi^2 s]$ manner. All of the eight possible isomeric alcohols were characterised from the cyclo-addition reaction. It has been interpreted that all of the products of two types (8 and 9; R^1,R^2 = H and OH) are formed from an intermediate such as (7)[6].

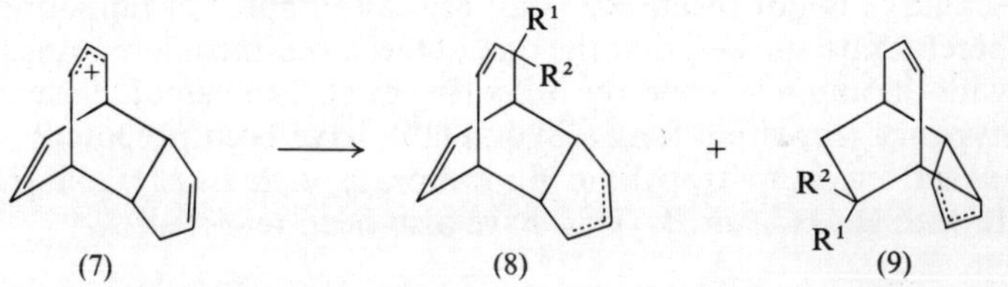

(7) (8) (9)

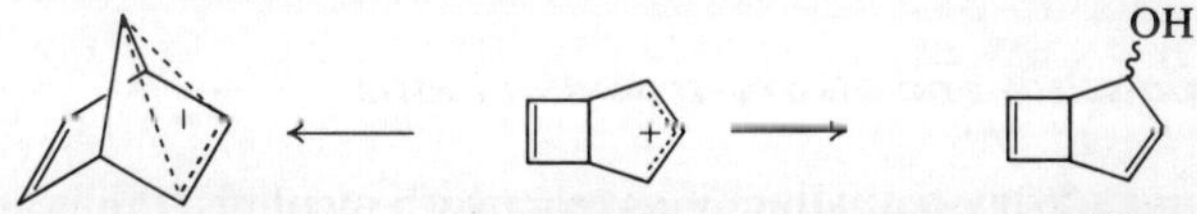

(11) (10) (12)

Irradiation of tropylium fluoroborate in fluorosulphuric acid induced a clean isomerisation of the tropylium ion to the norbornadien-7-yl cation (11)[7]. In aqueous ethanol, the photochemical reaction of (1) afforded ditropyl, photoditropyl and ethoxyditropyl while phenyltropylium ion is converted to diphenyl and its formyl derivatives along with *cis*- and *trans*-2, 2′-diphenylstilbenes[8]. The most reasonable mechanism for this rearrangement is an initial photo-isomerisation of the tropylium ion to give the bicyclic cation (10) followed by a thermal conversion into (11) or (12)[7,8].

It has been recently suggested that the type of mechanism for the oxidation of (1) leading to benzaldehyde can be explained by the intermediate formation of a radical cation derived from the co-existing conjugate base (tropyl alcohol)[9].

(b) *Special derivatives*—Detailed studies on the various spectral data including ^{13}C n.m.r. have been made on the di-cation (14) formed from heptafulvalene (13) with antimony pentachloride[10]. New types of stable

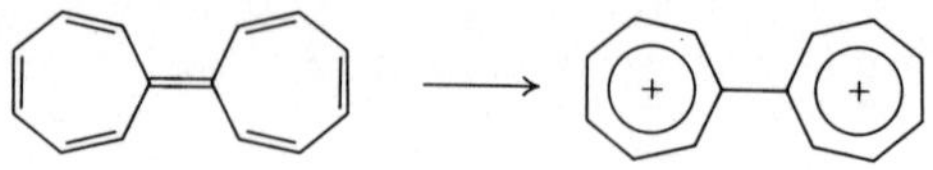

(13) (14)

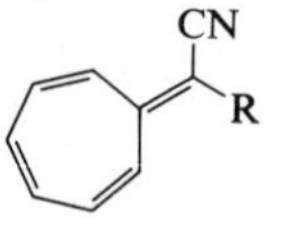

(15) 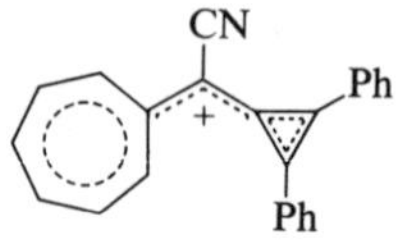 (16) 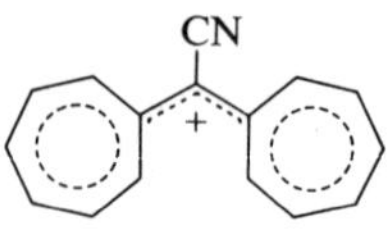 (17)

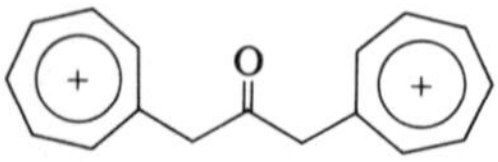

(18) 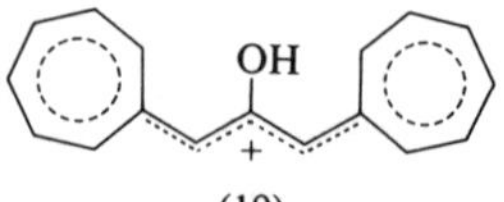 (19)

carbonium ions (16) and (17) were prepared from 8-cyanoheptafulvene (15; R = H)[11]. These cations (16) and (17), possess a substantial thermodynamic stability (pK_{R^+} = 3.41 and 3.44, respectively) which are intermediate between those of tropylium (pK_{R^+} = 4.7) and triphenylcyclopropenyl cation (pK_{R^+} = 2.8). Although the spectral data of (17) indicate that the positive charge is not localised, x-ray crystallographic examination of (17) as the perchlorate showed that the one of the seven-membered rings is associated with the anion whereas the other ring exists as a heptafulvene moiety[12]. Other types of tropylium ions, (18) and (19), have been prepared[13], and the reactions of ferrocenyltropylium fluoroborate with base[14] and tropylium bromide with $B_{10}H_{10}^{2-}$ or $B_{12}H_{12}^{2-}$ have also been reported[15].

7.2.1.2 *Halogeno- and alkoxy-tropylium ions*

Halogeno- and alkoxy-tropylium ions react with alcohols, thiols and amines to give *O*-, *S*- and *N*-substituted cations respectively[16a, 17, 18] and with malononitrile to give 8,8-dicyanoheptafulvene (15: R = CN)[16b]. 1,2-Diethoxytropylium cation has been shown to be a potential intermediate in the synthesis of dithiotropolone (20; X = S, Y = SH), aminotroponeimines (20; X = NR, Y = NHR) and aminothiotropones (20: X = S, Y = NHR)[19].

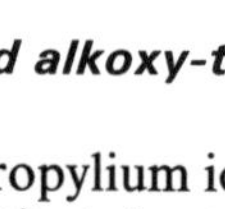

(20) 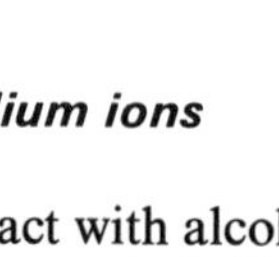 (21)

Octachlorobicyclo[3.2.0]hepta-2,6-diene reacts with aluminium chloride on heating to give heptachlorotropylium salt, which converts to octachlorocycloheptatriene with ice water and is transformed to perchlorotropone (21) with concentrated sulphuric acid[20].

7.2.1.3 Benzologues of tropylium ions

In 5-aryldibenzo[*a,d*]tropylium ions (22)[21] and especially in tribenzo[*a,c,e*] tropylium ion (23)[22] the seven-membered ring must be non-planar due to non-bonded interactions of the neighbouring groups and electronic delocalisation of the positive charge is seriously inhibited. Extremely low stabilities and

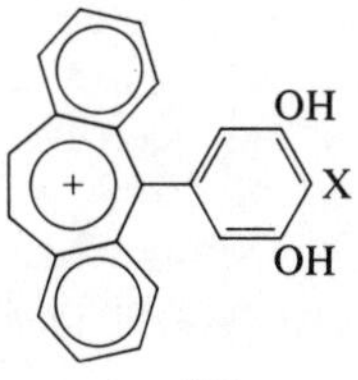

(22) 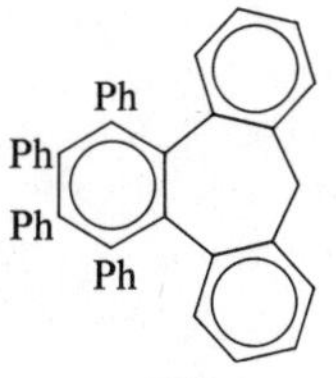 (23) 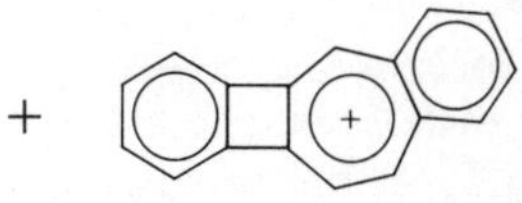 (24)

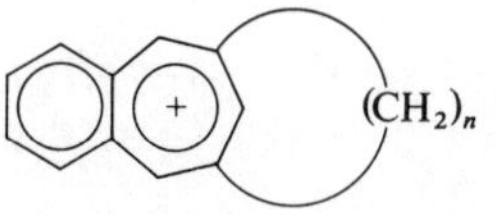

(25) 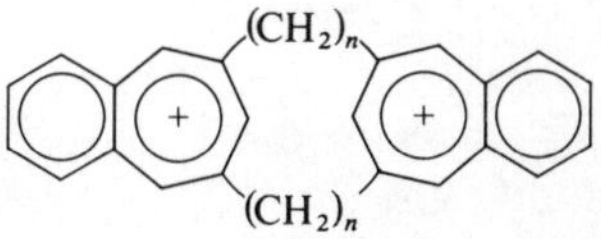 (26)

the reaction of these cations reflect these steric situations[22]. The syntheses of (24)[23], (25; $n = 10, 12$)[24a] and (26; $n = 4, 5, 8$)[24b] have also been accomplished.

7.2.2 Tropones and tropolones

7.2.2.1 Synthesis

Several new or modified synthetic methods for tropones and tropolones have been developed in recent years.

(a) *Bromination–dehydrobromination*—Dibromo derivatives (28: R = H, NO_2, NHAc) of benzosuberone (27) were smoothly dehydrobrominated by lithium chloride (or carbonate) in boiling dimethyl formamide to give the corresponding 2,3-benzotropone (29) in high yield[25]. This method can be applied to tropone and nezukone[26].

Dehalogenation of α,α'-dibromo ketones (30) with iron carbonyls in the presence of various 1,3-dienes provides a new route to cycloheptenones (31), which can be converted to various polysubstituted tropones and disubstituted 4-hydroxytropones (32) by the bromination–dehydrobromination reactions[27].

(b) *From bicyclo*[3.2.0]*heptanes*—The most valuable method[28] for the preparation of tropolone involving cyclo-addition of cyclopentadiene with

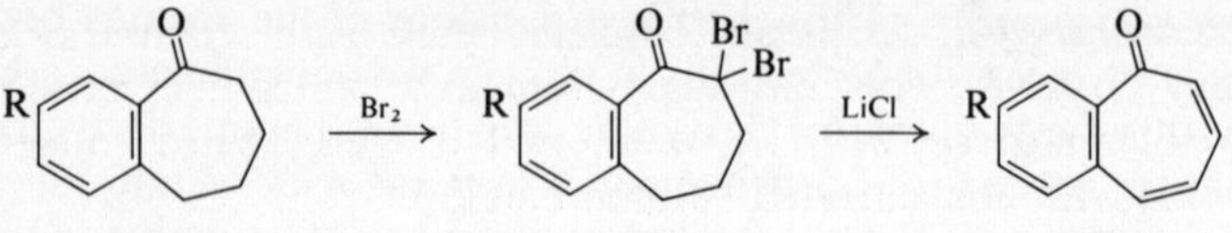

(27) $\xrightarrow{Br_2}$ (28) $\xrightarrow{LiCl}$ (29)

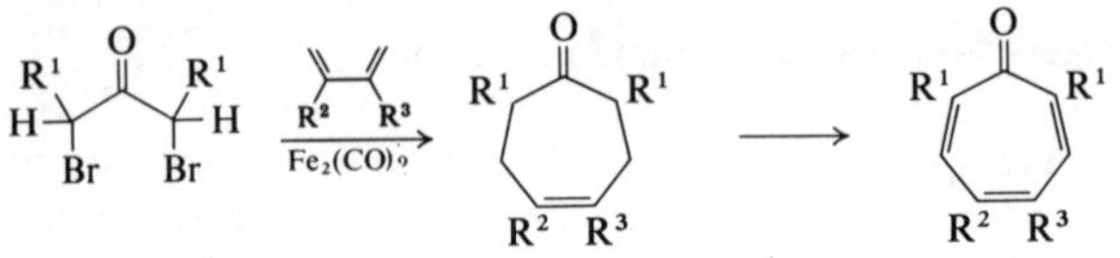

(30) $\xrightarrow[Fe_2(CO)_9]{}$ (31) $\longrightarrow$ (32)

dichloroketene followed by acetolysis is applicable to a wide range of tropolones. The proposed mechanism[29] of this synthesis involves the intermediacy of isomeric chloro-acetates (33: X = Ac) in the acetolysis; these were successfully isolated and hydrolysed to the alcohol (33: X = H) with subsequent collapse to tropolone via the diketone (34). Many experimental results available to date[30a, 31–33] clearly support this mechanistic pathway.

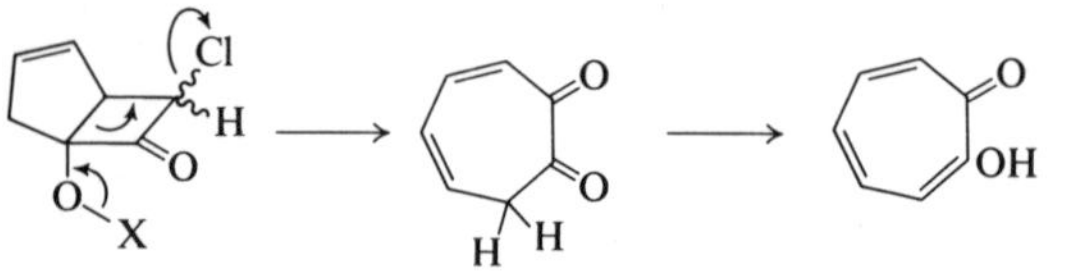

(33) $\longrightarrow$ (34) $\longrightarrow$

One of the isomeric cyclo-adducts (36), which are obtained from (35) and dichloroethylene by irradiation, is converted in one step into 4-hydroxytropone[34] by the action of aqueous methanolic potassium hydroxide at room temperature. Stipitatonic acid (37) was also synthesised using this concept[34].

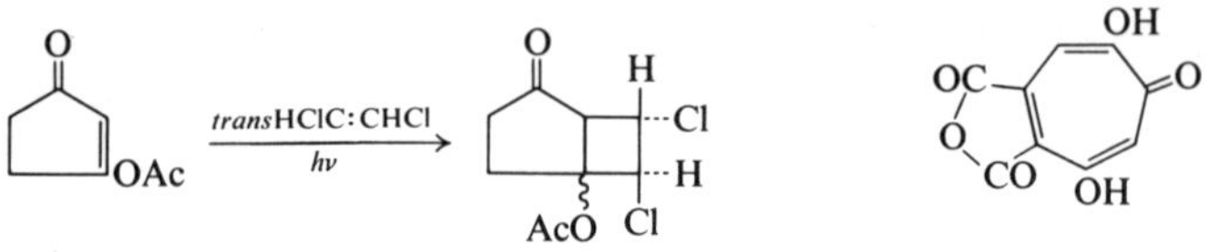

(35) $\xrightarrow[h\nu]{trans\,HClC:CHCl}$ (36) (37)

(c) *Hofmann degradation*—1-Methyl-3-oxidopyridinium (38), reacts with acrylonitrile and methyl acrylate to give cyclo-adducts (39: R = CN; CO_2Me). The methiodides of (39) undergo Hofmann elimination to form the corresponding 4-substituted troponoids (40: R = CN, CO_2Me: X = OH, NMe_2)[35].

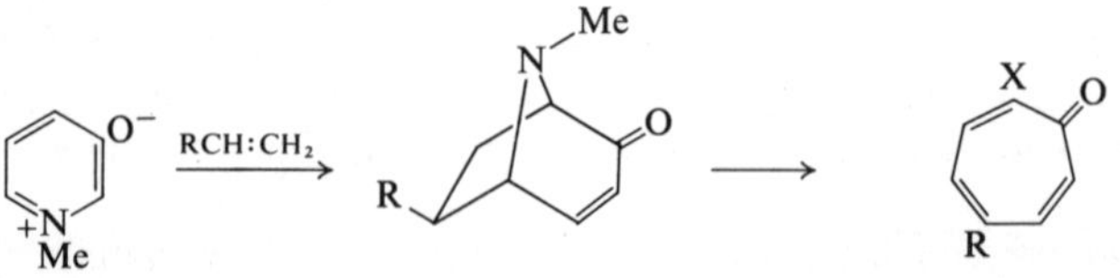

(38) $\xrightarrow{RCH:CH_2}$ (39) $\longrightarrow$ (40)

(d) *Ring enlargement*—Though ring expansions of the various benzenoid compounds by means of carbenoids have been well documented, some new examples are now available for the synthesis of benzotropone derivatives[36–39].

Alternative routes to 2,3-benzotropolones involves action of zinc in acetic acid either with the triketo-bromohydrins (41) or with the hydroxy-diketo-bromohydrin (42) to afford smoothly the new benzotropolones (42: R^1 = OH, R^2 = H) and (43: R^1 = H, R^2 = Me), respectively[40].

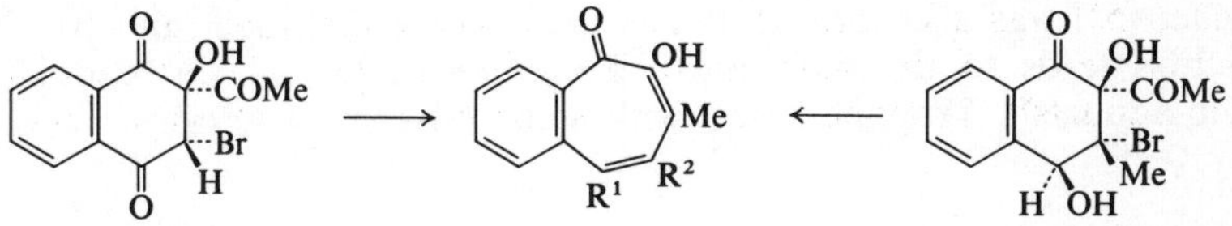

(41) (42) (43)

(e) *Miscellaneous*— α-Oximinobenzosuberone is smoothly converted to 2-acetylamino-4,5-benzotropone and its 7,2,3-isomer[41].

A variety of troponoids, dinaphthotropone (44)[42], benzoditropones (45: R = H or OH)[43,44], fused benzotropone with heptalene skeleton (46)[45], ferrocenotropones (47: R^1,R^2 = H, Me or Ph)[46], and a cyclobutadieno-hydroxytropone complex (48)[47], have been prepared during the last few years.

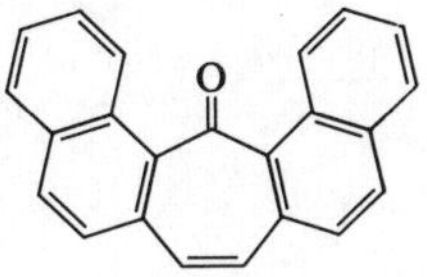

(44) 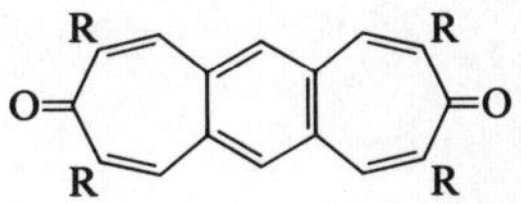 (45)

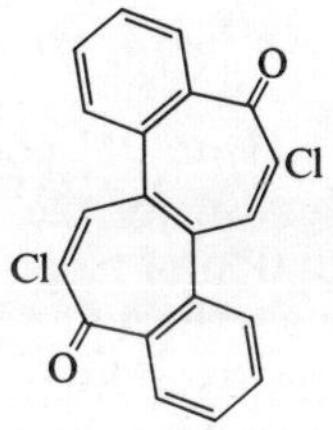

(46) 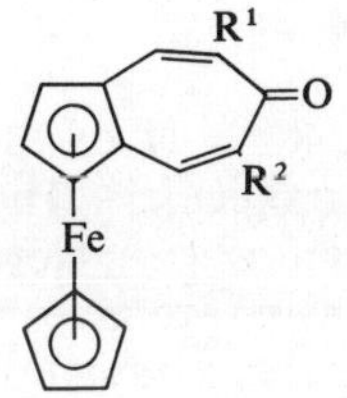 (47) 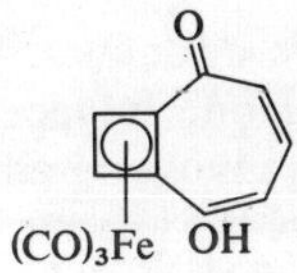 (48)

7.2.2.2 Reactions

(a) *Cyclo-additions*—The study of cyclo-addition reactions of tropone has received increased interest in the past few years because of the Woodward–Hoffmann rules. Many workers have examined the cyclo-addition reactions of tropone in addition to Japanese research efforts which have recently been reviewed by one of the present authors[30b].

Tropone reacts with dimethyl acetylenedicarboxylate[30b,48] and 4-phenyl-1,2,4-triazolin-3,5-dione[30b,49] as a π4-component to give a [4+2]-adduct, whereas the reaction with dehydrobenzene not only reacts as a π4- but also as a π6-component[50].

It has been observed that the course of the cyclo-addition reactions of tropone with some dienes is generally markedly affected by reaction conditions. For example, when tropone and 2,5-dimethyl-3,4-diphenyl-cyclopentadienone (as the dimer) were heated in acetone at 60 °C, the adduct (49) was obtained in quantitative yield, while in benzene at 70 °C a new adduct (50) was also obtained as well as (49); and at above 100 °C, an endo[4+2] adduct (51) was also formed. It has thus been established that the above reaction leads to thermally allowed exo[6+4], [8+2] and endo[4+2] cyclo-adducts[51]. Tropone reacts with some fulvenes in tetrahydrofuran at

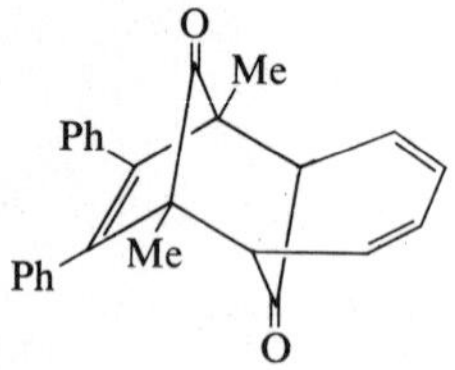

(49) 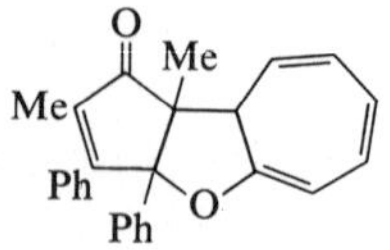 (50) 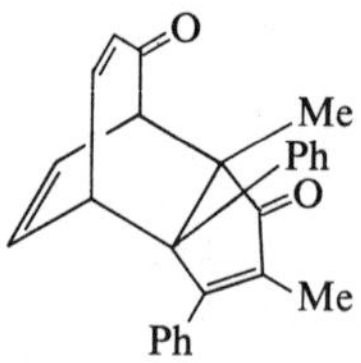 (51)

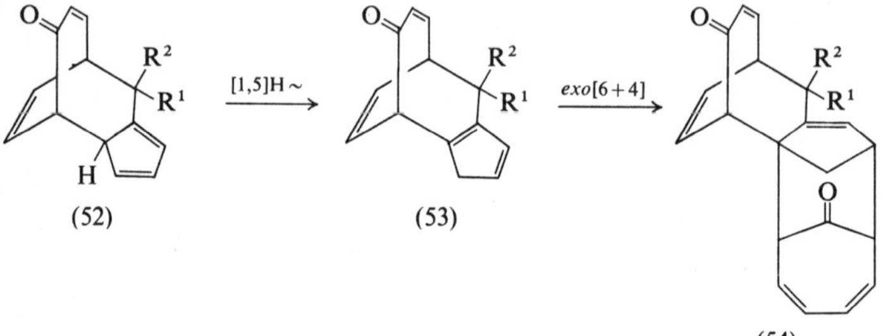

(52) [1,5]H~ (53) *exo*[6+4] (54)

50 °C to give adducts (54. R^1 = H or Ph, R^2 = H or Ph) containing two tropone and one fulvene molecule[52]. The mechanism of the reaction was suggested as involving an initial [6+4]cyclo-addition of tropone [as π4] to the termini of the fulvene [as π6] to give (52) followed by a [1,5]-sigmatropic hydrogen shift to afford the thermodynamically more stable cyclopentadiene derivative (53), which underwent a second *exo*[6+4] cyclo-addition with tropone to form the final product (54)[52]. Thermal reaction of tropone with butadiene in xylene resulted in the formation of [6+4]adduct (55) in addition to an isomeric mixture of [4+2]adducts[53]. Upon dehydrogenation with DDQ, the adduct (55) yielded a hitherto unknown 1,6-methano[10]annulen-11-one (56)[53]. The bridged annulene (57) which had been obtained by an alternative route[54] was also derived from (55)[53].

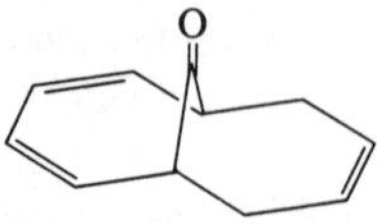

(55) 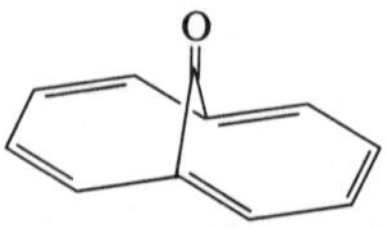 (56) 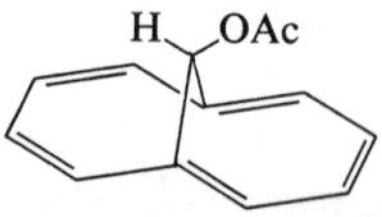 (57)

Reactions of 2-chlorotropone with cyclic enamines (58: n = 1,2 and 3) readily proceed at room temperature to give the corresponding [8+2] cyclo-adducts (59: n = 1,2 and 3) in good yield while the reaction with an acylic enamine gave rise to a [4+2]cyclo-adduct. Treatment of the cyclo-adducts (59) with 2N hydrochloric acid followed by triethylamine leads to the formation of (60: n = 1,2 and 3) in high yield[55].

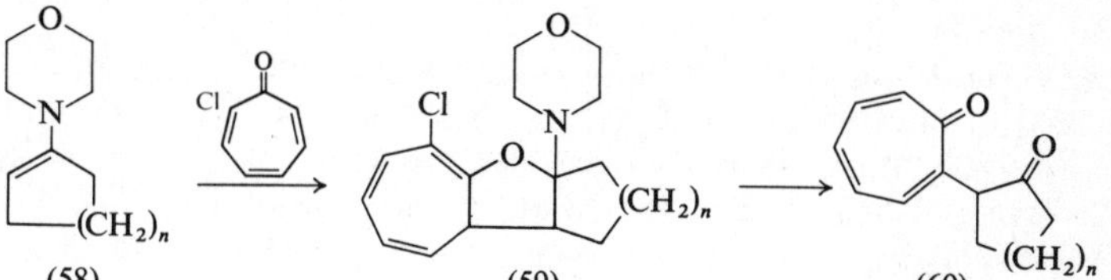

(58) (59) (60)

When diphenylnitrileimine (61), was generated in the presence of tropone, two kinds of adducts (62) and (63 or its isomer) are obtained[56]. Adduct (63) is the result of the usual [4+2] 1,3-dipolar cyclo-addition to the 2,3-double bond of tropone followed by a hydrogen shift, but the spectral evidences of (62) confirmed the [6+4] nature of this adduct[56].

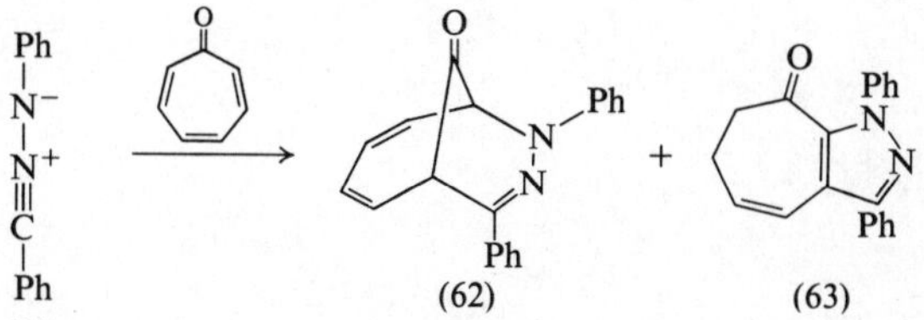

(61) (62) (63)

It has been known that halogeno- and 4-hydroxytropone add diazomethane to give pyrazolotropones[30b]. Tropone itself reacts with diazoalkanes to give either cyclo-octatrienone derivatives (64: R = H, Me) or the bicyclic ketone (65)[57]. On the other hand, if tropone is kept in a solution of diazomethane 2,3-homotropone (66) and N-methylpyrazolotropones (67 and 68) are obtained besides (64: R = H)[58]. Ring enlargement of tropone to a cyclo-octatrienone derivative is also observed in the case of reaction with p-nitrobenzylsulphonium ylide[59].

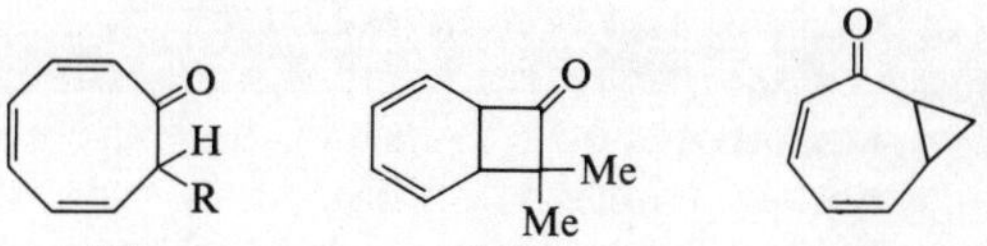

(64) (65) (66)

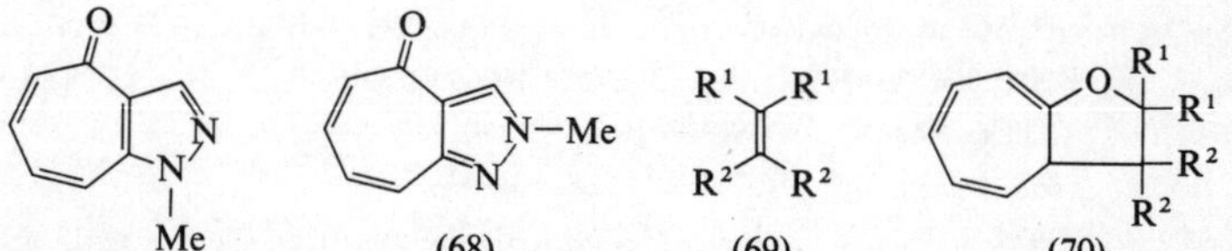

(67) (68) (69) (70)

Although many of the thermal cyclo-additions of tropone to various unsaturated molecules are now well documented and all of the products are predicted to be symmetry allowed by the orbital-symmetry selection rules, little is known about the photochemical reaction of tropone with simple olefins. Irradiation of tropone in the presence of various olefins (69: R = H, R^2 = Me; R^1 = R^2 = Me) gave [8+2] adducts (70)[60]. Detailed experimental results favour the intermediacy of an excited triplet state of tropone in this reaction.

(b) *Photochemical and thermal reactions*—Examples have already been mentioned in another review[30c] of photochemical isomerisations of troponoids. Photochemistry of tropolone itself has now been observed in recent years[61]. Tropolone photo-isomerised in hexane solution via (71) to a mixture of (72) and (73); (71) is also formed when tropolone is irradiated in water and in methanol. The 4,7-dione (73) may be produced from (71) either by a symmetry allowed photochemical supra,supra-[1,3]sigmatropic shift or by

the biradical process through (74), followed by ketonisation. Analogously, the irradiation of tropolone in methanol gave (72) and subsequently the methyl ester (75: R = Me)[61].

2-Bromotropone, when refluxed in benzene containing ethylene glycol or trimethylene glycol and toluene-*p*-sulphonic acid gave 2-bromoethyl or 3-bromopropyl benzoate, respectively[62]. Thermal isomerisation of other derivatives of tropone have also been investigated[63–66].

(c) *Miscellaneous*—Examples of reactions of some troponoids such as the rearrangement of azidotropones[67], 2-diazocycloheptadien-1,3-dione[68], the Claisen rearrangement of trimethyltropolone allyl ethers[69], the hydrogen peroxide oxidation of 1,2-dihydroxy-3,4-benzotropolone and purpurogalloquinone[70] have recently been reported.

The cyclo-addition product (76: R = H) and 8,8-dicyanoheptafulvene (77: R = H) were formed when 2,3-benzotropone (29: R = H) was treated with malononitrile in the presence of acetic anhydride or a base. On the other hand, two isomeric heptafulvene derivatives (77: R = Me and 78: R = Me) were isolated as well as the cyclo-addition product (76: R = Me) when (29: R = Me) was used in the reaction[71]. It appears quite likely that not only the C-1-position but also the C-4-position of 2,3-benzotropone were attacked by the carbanion. Deuterium labelling experiments have revealed that the attack of the malononitrile carbanion occurs equally at both the 1- and 4-positions to produce heptafulvene derivatives (77: R = H) and (78: R = H)[72].

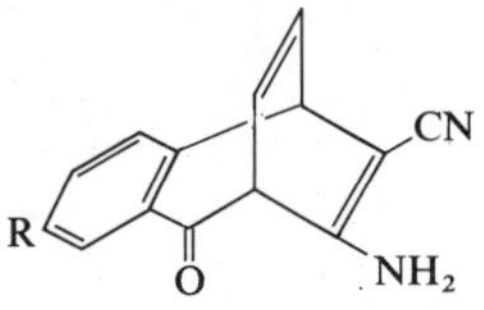

(76) 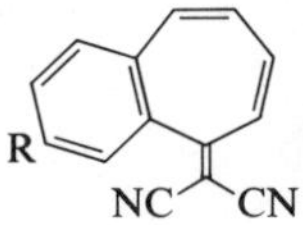 (77) 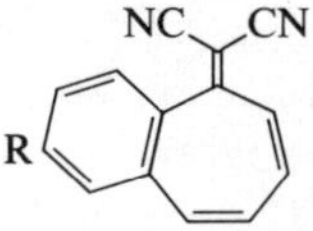 (78)

4,5-Benzotropone and its dimethyl derivative react also with malononitrile in the presence of a base or acetic anhydride to give heptafulvenes by the normal condensation[71]. With dimethyl malonate more complicated reactions occur[73].

Similarly, treatment of 4,5-benzotropone with 1,2,3,4-tetrachlorocyclopentadiene in methanol leads directly to a cage molecule (79) together with (80) as a minor product[74]. The initial reaction is probably another example of Michael type addition followed by an intramolecular Diels–Alder reaction.

Reaction of arylsulphonyloxytropone (81: X = OSO_2Ar) with carbanions afforded (82: R = H, OH) and trisubstituted azulenes (83)[75]. In this reaction the relative yields of the products are markedly influenced by reaction

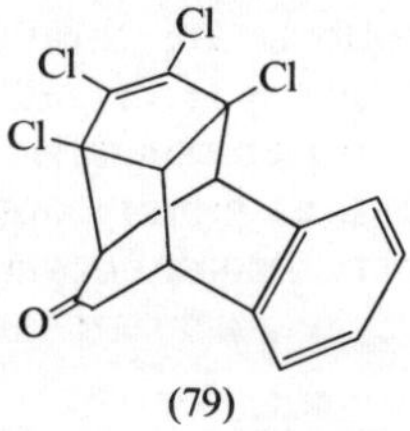

(79) 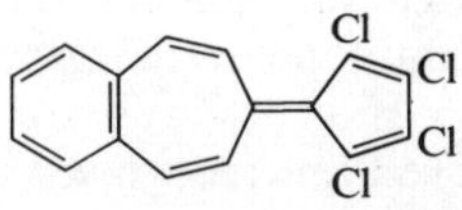 (80)

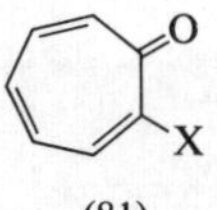

(81) 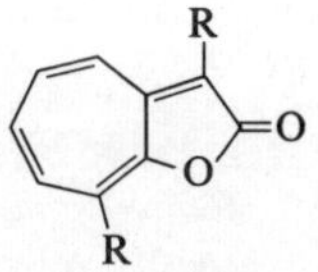 (82) 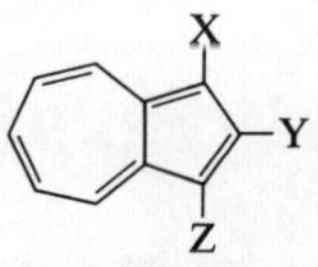 (83)

The structure (84) of the main reaction product of tropone with decacar-conditions and reasonable interpretations are presented[75]. 4- and 5-Styryl tropolones are synthesised by the Wittig reaction utilising either benzylphosphonium ylides or tropolonylphosphonium ylides with formyltropolones[76] or arylaldehydes[77], respectively.

The structure (84) of main reaction product of tropone with decacarbonyldimanganese has been determined by x-ray analysis and shows that its formation involves the cleavage of one molecule of tropone, condensation with a second tropone unit, and formation of a complex with the open chain moiety[78].

With tropone ethylene ketal as the reactant, the irradiation of cyclobutadienetricarbonyliron proceeded smoothly to afford the complex (85)[79]. Starting with (85), the novel cage compound, homopentaprismanone, was prepared[79].

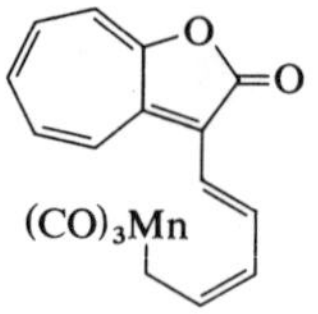

(84) 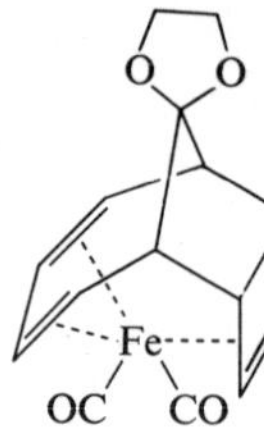 (85)

7.2.2.3 *Natural tropolones*

The pigments characteristic of black tea infusions are of two general types associated with the theaflavin fraction and the thearubigin fraction. From the former fraction at least eight compounds which are benzotropolone derivatives have so far been isolated. Absolute configurations of the pigments have been proved by spectroscopic data (n.m.r., m.s., o.r.d. and c.d.) and also by the comparison with authentic specimens which have been synthesised by the co-oxidation of pairs of catechins and gallocatechins occurring in green tea leaves[80a, 81]. The most abundant components are theaflavin and the 3-gallate, 3′-gallate and 3,3′-digallate (86: R^1,R^2 = H or galloyl). Epimers of the above components (86), theaflavic acid, epitheaflavic acid and its gallate (87: R = H or galloyl) are also found as minor constituents. The thearubigins have been recognised as polymeric proanthocyanidins[80b].

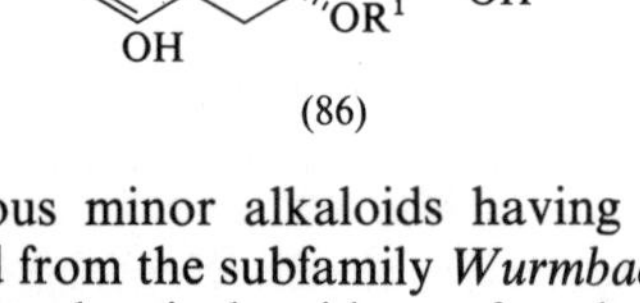

(86) 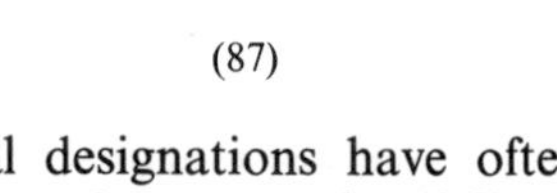 (87)

Various minor alkaloids having special designations have often been isolated from the subfamily *Wurmbaesideae* (*Liliaceae*) to date[82, 83].

Direct chemical evidence for the correlation between the two *N*-formyldesacetyllumicolchicines which were previously related to β- and γ-lumicolchicine respectively has now clearly been demonstrated[84a]. From the tubers of *Gloriosa superba L*, a nitrogen free compound, β-lumicolchicone

(88), was isolated in addition to some other alkaloids of the colchicine and lumicolchicine group[84b]. From a biosynthetic point of view, it is interesting that the structure of speciosine (89) has been proved[85].

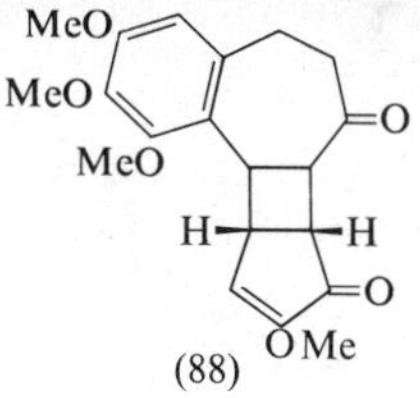

(88)

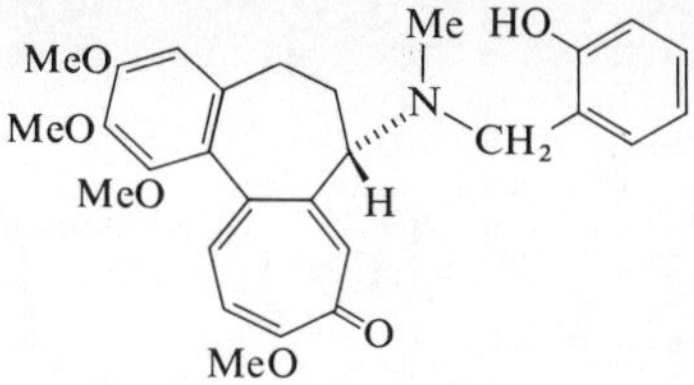

(89)

The isolation of androcymbin (90) together with the alkaloids colchicine, lumicolchicine and related compounds is of interest in connection with the biosynthetic pathway of colchicine[86,87]. Recently, the biosynthesis of the mould metabolites stipitatic acid and stipitatonic acid (37)[88] and sepedonin (91)[89,90] have been established.

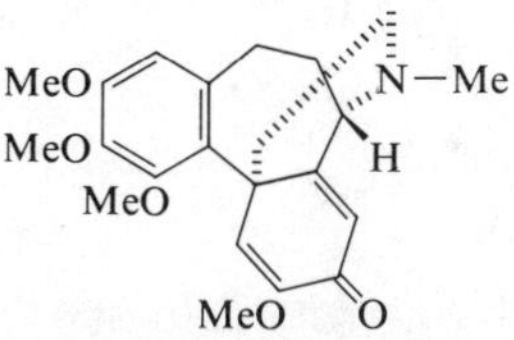

(90)

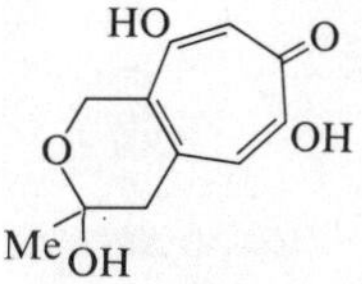

(91)

7.3 HEPTAFULVENES AND HEPTAFULVALENES

7.3.1 Heptafulvene-type compounds

Heptafulvene, a highly-reactive non-benzenoid conjugated hydrocarbon of theoretical interest, has not been isolated to date. Recently, several research groups reported the preparation and structural elucidation of the stable carbonylmetal complex of heptafulvene[91] and an alkyl and aryl derivative[92]. The thermal stability of the complexes produced in this way demonstrates the effectiveness of the tricarbonyliron group as a stabilising unit; however, attempts to liberate the free heptafulvenes from these complexes were unsuccessful.

A comparison of the energy barriers for the conformational inversion of the seven-membered ring in the dibenzoheptafulvenes of the type (92) have been examined in some detail[93]. In general, benzene-annelation is expected to increase the classical character of the seven-membered ring and the heptafulvenes (92) have properties consistent with limited delocalisation of the π-electrons in the seven-membered ring. Thus, the temperature dependence of the n.m.r. spectrum of the prochiral group (92: $R^2 = C(Me)_2OH$) provides a free energy of activation (21.1 kcal mol^{-1} for the conformational inversion of the seven-membered ring, and some of these compounds were obtained in an optically active form[93].

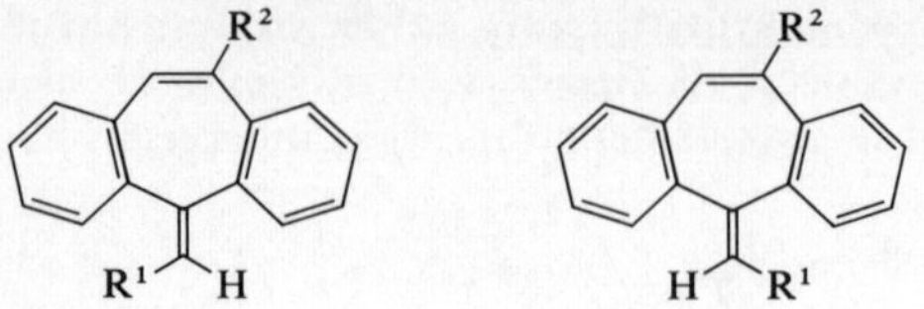

(92)

Some interesting types of heptafulvenes involving extended conjugation, (93: R = CN or CO_2Me)[94–96], (94) and (95)[96] have been synthesised from tropylacetaldehyde.

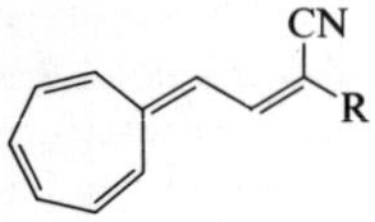

(93) 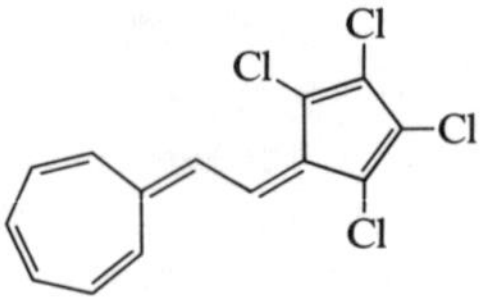 (94) 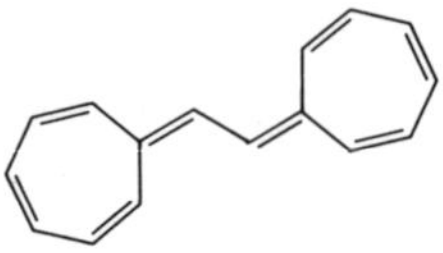 (95)

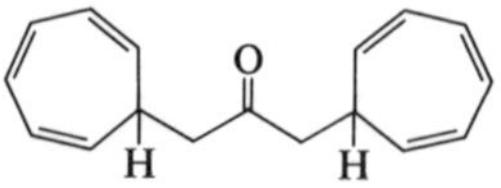

(96) 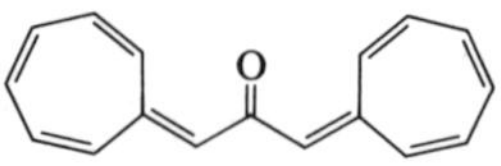 (97)

The reaction of ditropylacetone (96) with trityl tetrafluoroborate followed by treatment with triethylamine furnished a novel heptafulvene derivative, bisheptafulvenylketone (97)[13]. The large bathochromic effect of dilute acid on the long-wavelength adsorption band in the electronic spectrum supports the idea of a highly polarised structure for the ketone (97).

7.3.2 Heptafulvalene-type compounds

Di-t-butyl sesquifulvalene (98) was stable in solution in the absence of impurities and afforded interesting cyclo-adducts. Tetracyanoethylene rapidly added in a [12+2] fashion to give (99), wherein stereochemistry is postulated on the basis of the symmetry allowed ground state $[\pi^{12}s+\pi^{2}s]$ process. On the other hand, however, dimethylacetylenedicarboxylate does not react similarly; the spiro adduct (100) was obtained in a [6+2] fashion[97,98].

Attempts have been made to generate cycloheptatrienylidene (101). Either photolysis or pyrolysis of the sodium salt of tropone tosylhydrazone

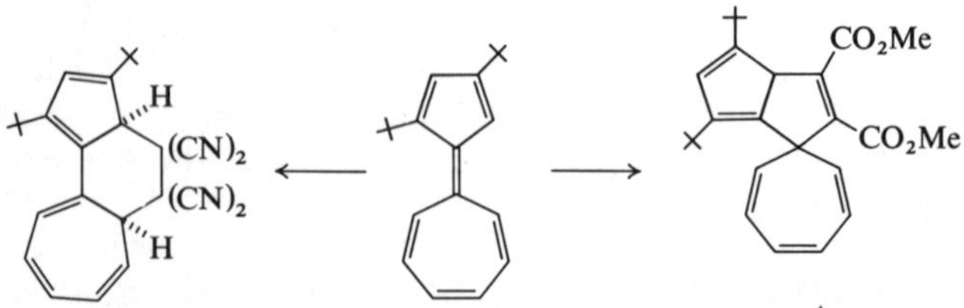

(99) (98) (100)

or of benzaldehyde tosylhydrazone gave heptafulvalenes (102) which are believed to be produced via (101)[99]. Recently, photolysis of 5-diazodibenzo[*a,d*]- (103) and 9-diazotribenzo[*a,c,e*]cycloheptene have also been reported[100].

(101) 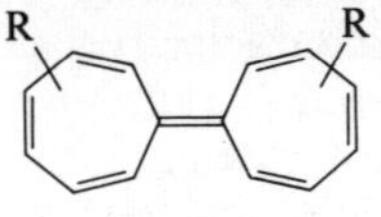 (102) 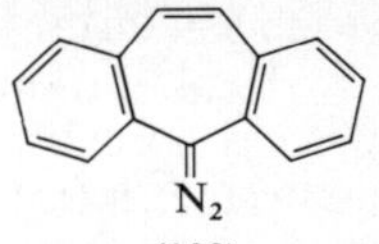 (103)

Treatment of perchlorotropylidene with n-butyl lithium at −78 °C lead to the formation of colourless crystals which transformed thermally into perchloroheptafulvalene (104). Upon heating to 340 °C, (104) converts in high yield to (105)[101]. In contrast to the instability of the parent hydrocarbon (102: R = H), the remarkable thermal stability of (104) indicates the stabilising effect of the complete chlorine substitution.

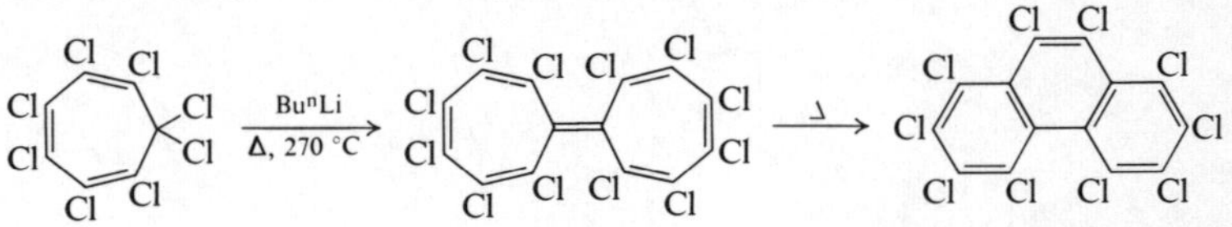

$\xrightarrow[\Delta,\ 270\ ^\circ C]{Bu^nLi}$ (104) $\xrightarrow{\Delta}$ (105)

Tetrabenzoheptafulvalene (106) may be obtained in two isomeric forms, *syn* or *anti*, with respect to the inter-ring double bond, by the action of copper with dibenzothiotropone or dibenzodichlorotropilidene[102].

Benzo-annelated derivatives of the hitherto unknown nonafulvalene, (107) and (108), have been prepared; both of the compounds appear to be non-planar, non-aromatic molecules[103].

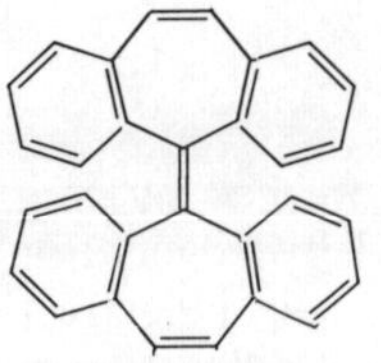

(106) 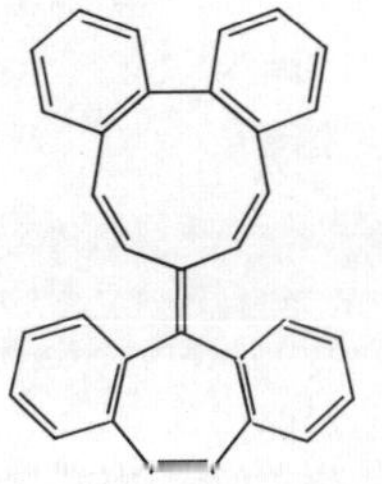 (107) 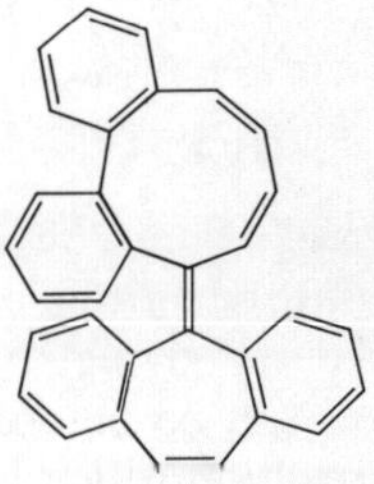 (108)

7.4 AZULENES

7.4.1 Synthesis

7.4.1.1 From troponoids and other seven-membered compounds

One of the most important type of azulene synthesis found by Nozoe and others[30d, 104] consists of the reaction of active methylene compounds with reactive troponoids (81: X = halogen, OMe or OSO_2Ar) or with 2*H*-cyclohepta[*b*]furan-2-ones (82) in the presence of a base to afford various polysubstituted azulenes (83) in one step. Examples illustrating the detailed preparation of some azulene derivatives were published recently and a reasonable mechanism for this characteristic transformation was presented[75, 105]. Heptafulvenes and cycloheptaphosphols can also be used instead of troponoids[30d]. It was found that hitherto difficulty obtainable 1,2-polymethyleneazulenes (109: R = CN, H; n = 3–5) could be made either from 8-cyanoheptafulvene (15: R = H)[106] or from 2*H*-cyclohepta[*b*] furan-2-one (82: R = H)[107] by enamine synthesis.

Pyrolysis of 8-(2,2-dicyanovinyl)heptafulvene (110: R = CN) at 450–500 °C affords 1,2-dicyanoazulene in good yield, while with the cyanomethoxycarbonyl derivative (110: R = CO_2Me) a mixture of three isomeric azulenes (112) is obtained[95]. This transformation could be explained as proceeding

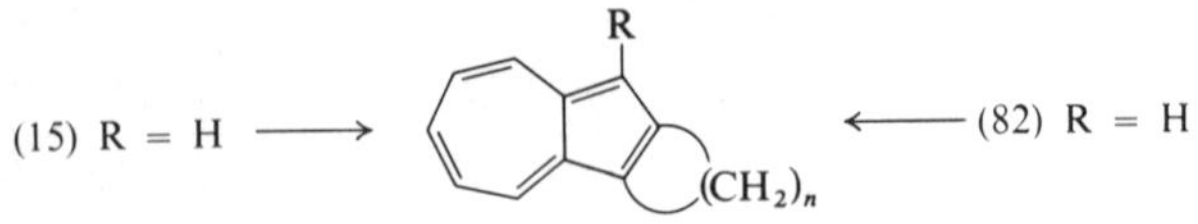

(15) R = H ⟶ (109) ⟵ (82) R = H

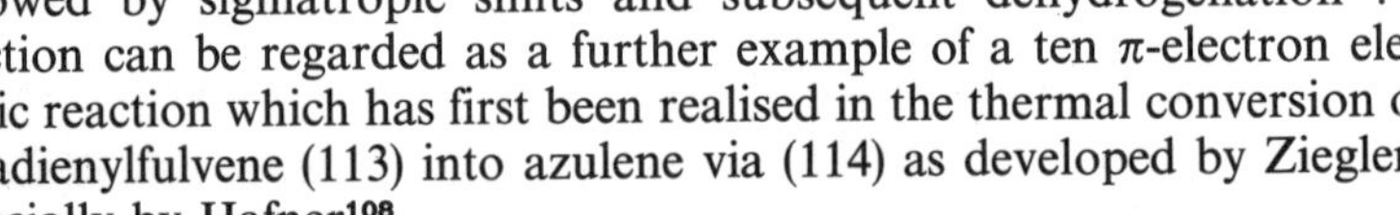

(110) ⟶ (111) ⟶ (112)

via an initial product (111: R = CN or CO_2Me) formed by ring closure followed by sigmatropic shifts and subsequent dehydrogenation[95]. This reaction can be regarded as a further example of a ten π-electron electrocyclic reaction which has first been realised in the thermal conversion of the butadienylfulvene (113) into azulene via (114) as developed by Ziegler and especially by Hafner[108].

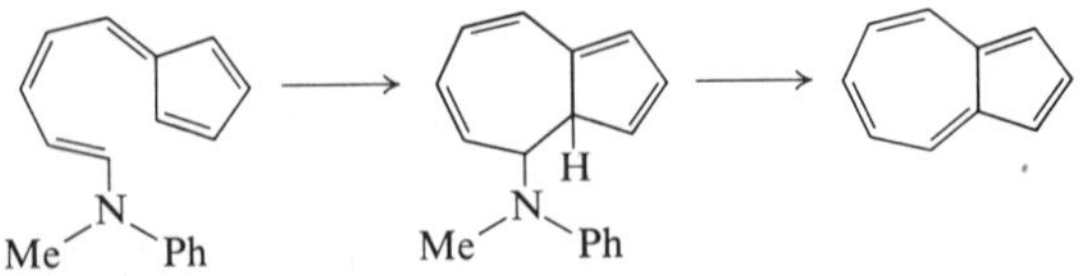

(113) ⟶ (114) ⟶

It is also interesting to note that the hydroazulene derivatives (117), which are easily transformed to azulenes, have been synthesised in good yield, by the catalysed [2+3] cyclocondensation of α,α′-dibromoketones (30) with the enamine (115) of cycloheptanone via (116)[27b].

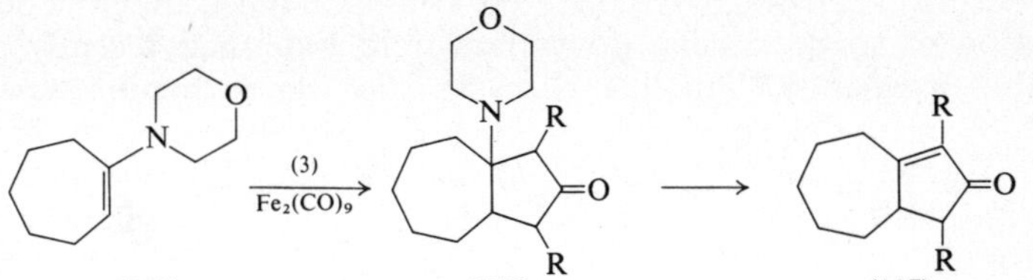

(3) $Fe_2(CO)_9$

(115) (116) (117)

7.4.1.2 *From fulvenes and other five-membered compounds*

A further azulene synthesis by the use of cyclo-addition of dimethylamino-vinylfulvene (118) and dimethyl acetylenedicarboxylate to give dimethyl azulenedicarboxylate (119) has been accomplished[109]. This reaction is assumed to proceed by a symmetry allowed $[\pi^8 s+\pi^2 s]$cyclo-addition, followed by loss of dimethylamine.

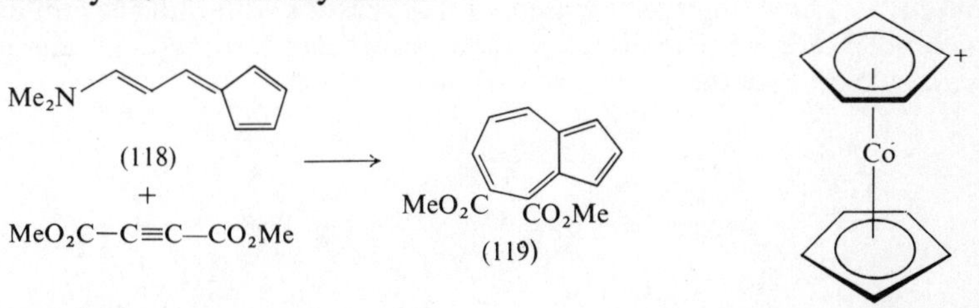

Me_2N (118) + $MeO_2C{-}C{\equiv}C{-}CO_2Me$ → MeO_2C CO_2Me (119)

Co (120)

The reaction of cobalticinium hexafluorophosphonate (120) with hydroxide or amide ions produces azulene in low yield[110]. The mechanism of this remarkable transformation was believed to involve a sequence of the nucleophilic attack of hydroxide ion followed by ring expansion, and finally a condensation reaction analogous to the Ziegler–Hafner azulene synthesis.

Another interesting example of azulene synthesis has appeared recently. Thus, bisdimethylaminopentalene (121) reacts with dimethyl acetylene-dicarboxylate to afford a 1,2-cyclo-adduct (122) which rapidly valence-isomerises to an azulene derivative (123)[111,112].

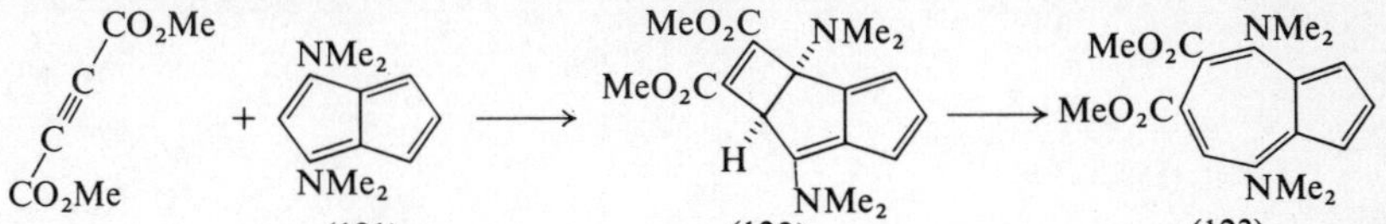

CO_2Me C C CO_2Me + NMe_2 NMe_2 (121) → MeO_2C MeO_2C NMe_2 H NMe_2 (122) → MeO_2C MeO_2C NMe_2 NMe_2 (123)

7.4.1.3 *Miscellaneous*

The reaction of the 'indenocyclone' (124) with cyclopropene affords a cycloheptatriene derivative (125), whose reduction and subsequent treatment with trityl perchlorate gives a number of substituted benzazulenes (126)[113].

Verdenoids (128), obtained by the photochemical dimerisation of *o*-bisethynylbenzenes (127: R = H, Ph), are azulenic hydrocarbons including

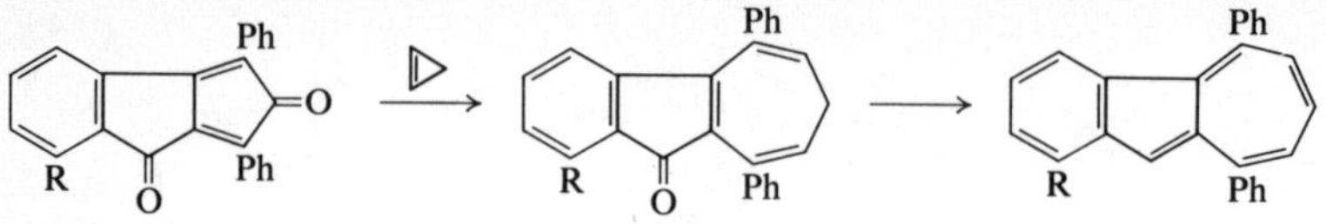

(124) (125) (126)

two triple bonds as substituents[114]. A 'ring closed' form, azuleno[3,3*a*,4-*a*,*b*] phenalene (129), was tentatively proposed for the verdenoids in the early stage of the research[115]. However, x-ray structure analysis on tetramethylverdene provides evidence for the 'open-azulene' structure (130)[116]. 1,2-Bis (phenylethynyl)cyclohexene (131) has also been converted into 'verdene' (132) by photochemical dimerisation and subsequent dehydrogenation with chloranil[117]. These findings together with the results with fluorine containing verdenoid[118] are all consistent with a molecular structure involving an 'open-azulene' structure with two acetylenic bonds.

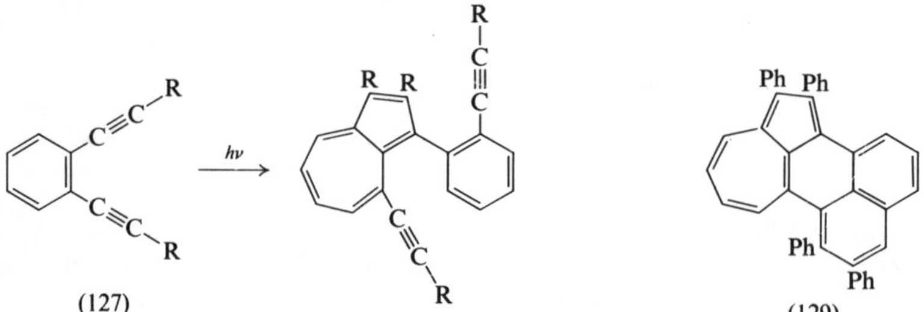

(127) (128) (129)

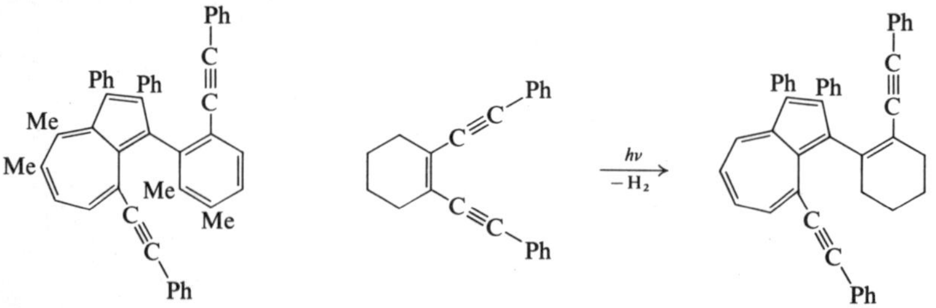

(130) (131) (132)

Similarly, irradiation of peri-bisethynylnaphthalene (133) affords polycyclic azulene derivative (134)[119].

Unlike azulenes, 5-aza-azulene (135) gives two kinds of cyclo-adducts (136) and (137) which on heating are easily converted into previously inaccessible azulenes, (138) and (139) respectively[120].

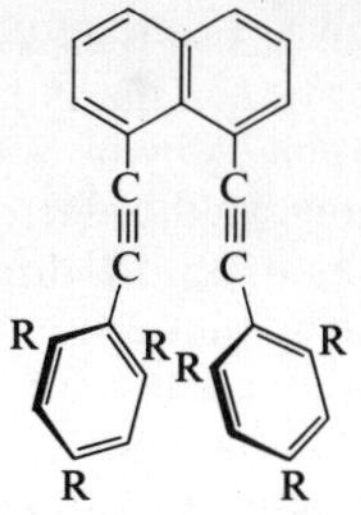

(133) R = Me

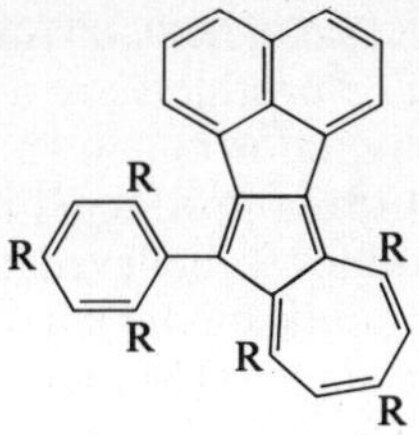

(134) R = Me

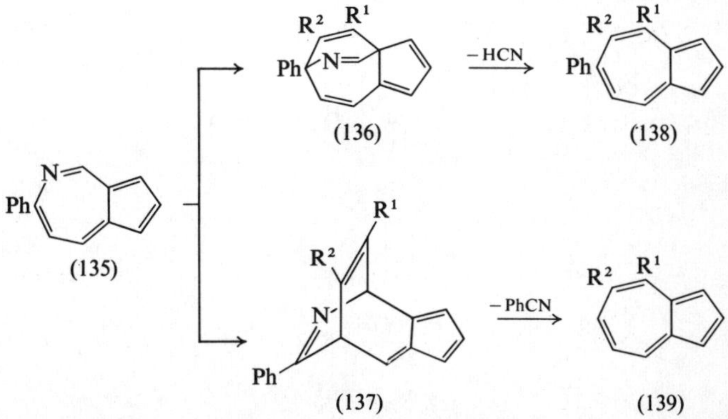

(135)

(136)

(137)

(138)

(139)

A new synthetically-useful route to azulene occurs by the transition metal induced rearrangement of a phenyl substituted bicyclo[1.1.0]butane (140). Thus, when (140) was treated with dicarbonyl rhodium chloride dimer in chloroform, the azulene (142) was formed along with other products. The proposed mechanism postulates the intermediacy of a metal-complexed carbene such as (141)[121].

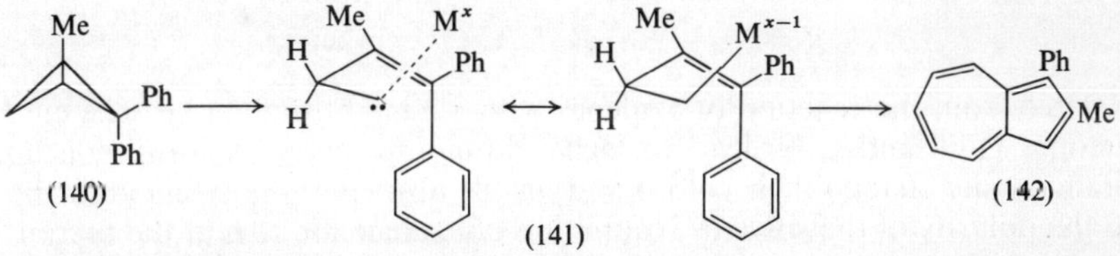

(140)

(141)

(142)

7.4.2 Reactions

7.4.2.1 *The ene reaction*

Unlike the troponoids, azulenes do not undergo cyclo-additions but readily undergo substitution reactions (ene-reactions) with typical enophiles such as maleic anhydride, diethyl acetylenedicarboxylate, diethyl azodicarboxylate as well as 1,4- and 1,2-quinones, wherein substitution takes place at the 1- and 3-positions[122]. Some of the reactions may be carried out at room temperature: the presence of heat, u.v.-irradiation or Lewis acids greatly facilitate these reactions. The products are usually formed in many tautomeric forms. For example, the reaction of azulene with maleic anhydride afforded a mono- (143) and disubstituted product (144). Judging by the electronic spectra, the disubstituted product is believed to exist as a yellow dipolar form (144a) in neutral and/or acidic media whereas the deep blue anion containing a true azulenic structure such as (144b) is suggested in alkaline medium[122]. Azulene and guaiazulene react easily with *p*-benzoquinone and 1,4-naphthoquinone in various solvents in either presence or absence of catalysts to give three types of isomers (145–147: R = 1-azulyl- or 3-guaiazulyl-). Two kinds of isomeric products (148, 149: R = 1-azulyl- or 3-guaiazulyl-) can also be

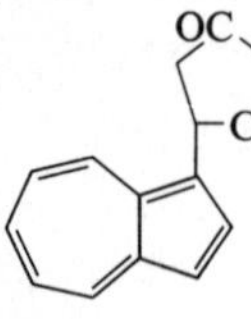

(143) 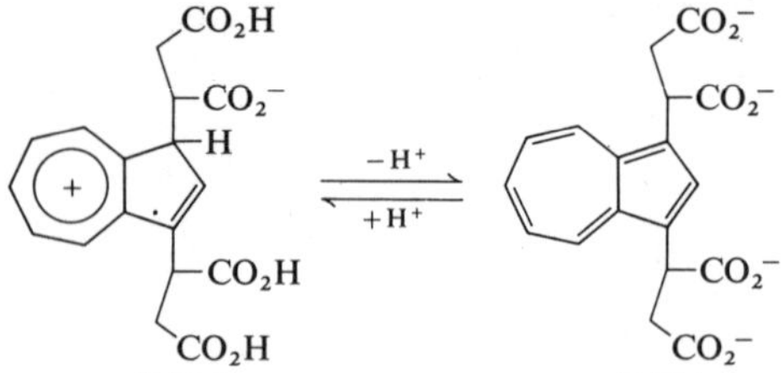 (144a) (144b)

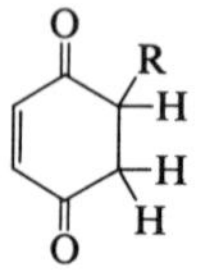

(145) 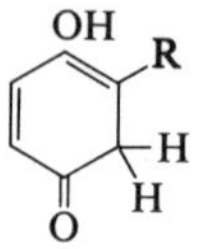 (146) 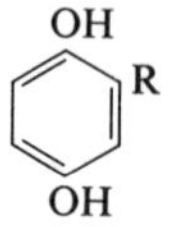 (147)

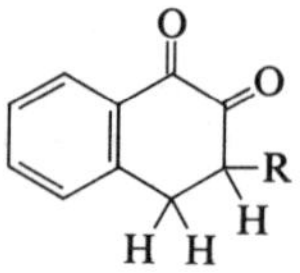

(148) 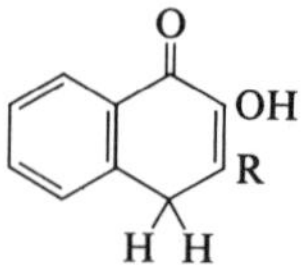 (149)

isolated from the reaction of azulenes with 1,2-naphthoquinone. The enol–ketones (148) and (149) lead to facile aluminium complex formation and dramatically change their colour and visible absorption maxima according to the polarity of the solvent. Irradiation of azulene for 10 h in the presence of *p*-benzoquinone in methanol gave (150). Although guaiazulene reacted with *p*-benzoquinone rapidly to give only resinous products, reaction of azulene with *p*-benzoquinone at room temperature afforded the single product (151)[123].

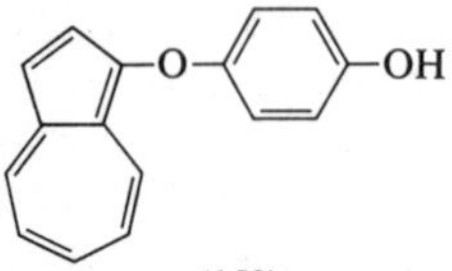

(150) 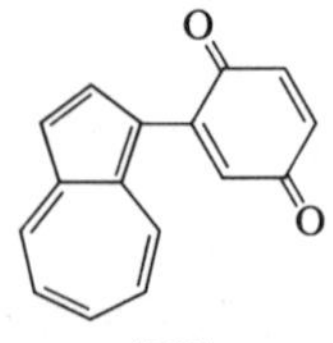 (151) 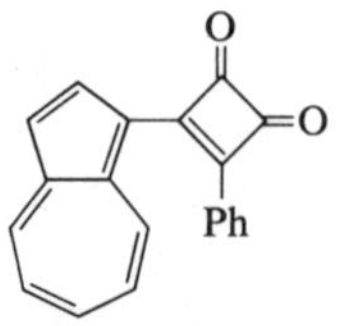 (152)

When azulene and 2-bromo-1-phenylcyclobutene-3,4-dione were heated in acetic acid, the substitution product (152) was obtained by the ene-reaction followed by dehydrobromination[124].

7.4.2.2 *Miscellaneous reactions*

Treatment of azulene with mercuric acetate and then calcium halide leads to the formation of 1-halogenomercuriazulene, which has quantitatively been converted to 1,3-dihaloazulene by the action of cupric halide[125]. Azulene reacted vigorously with sulphur dichloride in anhydrous ether even at −78 °C to give di-l-azulylsulphide, whereas di-l-azulylsulphoxide was obtained with thionyl chloride in acetonitrile at −45 °C[126].

Replacement of a nitro group by methoxyl or cyano groups has been observed when 1-nitroazulene was irradiated in the presence of sodium methoxide or sodium cyanide, respectively[127]. When 1-halogenoazulenes react with the copper salt of phenylacetylene in pyridine at elevated temperatures, a substitution reaction takes place to give 1-phenylethynylazulene or its 2-isomer[128].

Heating of 3-acetylguaiazulene (153) furnished a new isomer (155) in addition to the 2-acetyl derivative (154). The result indicates that acetyl and methyl groups in the five-membered ring moiety were thermally exchanged[129].

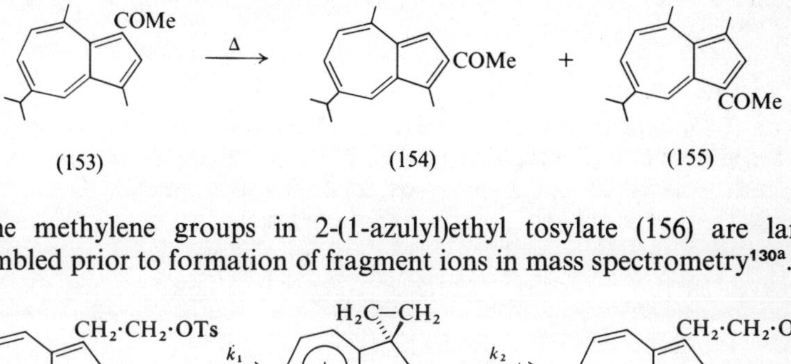

COMe Δ COMe + COMe

(153) (154) (155)

The methylene groups in 2-(1-azulyl)ethyl tosylate (156) are largely scrambled prior to formation of fragment ions in mass spectrometry[130a]. The

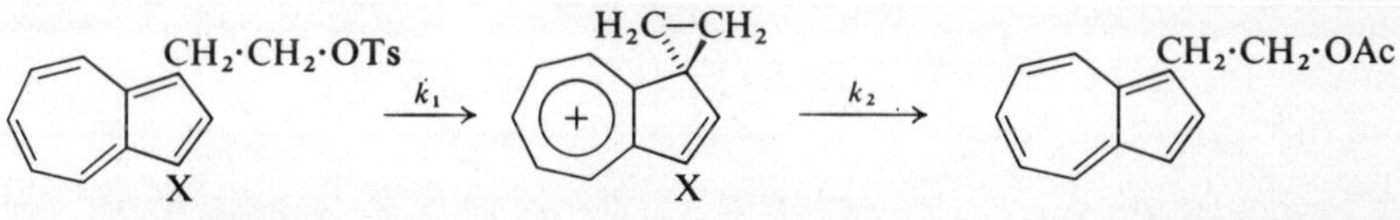

$CH_2 \cdot CH_2 \cdot OTs$ k_1 $H_2C—CH_2$ + X k_2 $CH_2 \cdot CH_2 \cdot OAc$ X

(156) (157)

O C Br + C O ⟷ Br + O C C O O= Ph =O Ph

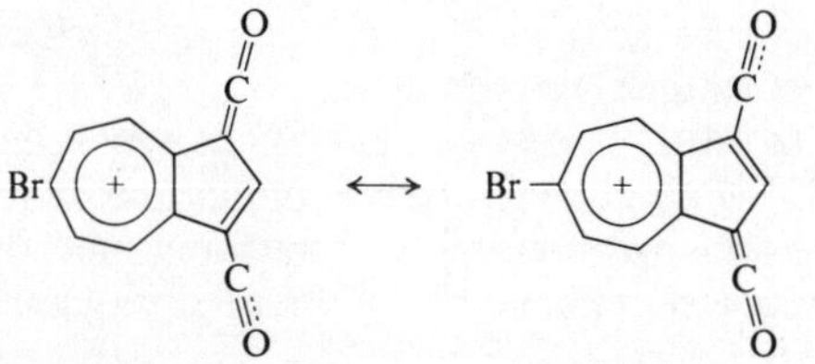

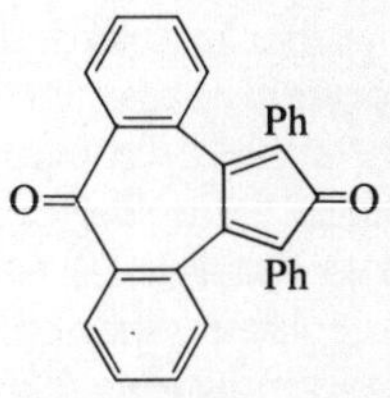

(158) (159)

kinetic studies on the acetolysis of (156) and certain derivatives indicate that the ionisation step, k_1, is rate determining, and the tremendous participating power of the azulene nucleus (157) contribute to the solvolysis[130b]. The stepwise formation of 6-bromoazulene-1,3-diacylium ion (158) is observed from the diconjugate acid of 6-bromoazulene-1,3-dicarboxylic acid or its diester via the monoacylium ion–monocarboxy–conjugate acid in SbF_5–FSO_3H–SO_2 system[131]. Various reactions of ketocyclone (159) which corresponds to an azulenoquinone have been studied[132].

7.5 TROPONOIDS WITH A CONDENSED HETEROAROMATIC NUCLEUS

7.5.1 Five-membered heterocyclic molecules

Acetone and its homologues react with heterocyclic dialdehydes (160) in the presence of a base to give the bicyclic tropone derivatives (161: X = S or NPh). From the reduction products of (161) a hydroxyl group may be abstracted by means of perchloric acid producing various 2-thionia-azulene or 2-azonia-azulene cations (162: X = S or NPh) having a heteroazulene skeleton[133]. Detailed studies on the reactions of these cations with nucleophiles[134, 135] and disproportionation reactions[136] have also been reported.

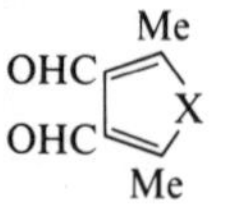

(160) 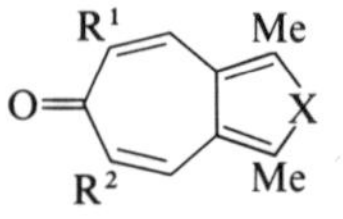 (161) 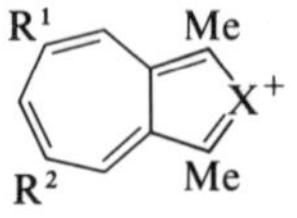 (162)

Cations (163) condense with aromatic aldehydes at the 6-methyl group to afford a variety of 6-styrylazoniaazulenes[137]. 6-Methoxy-derivatives of (162) have been isolated as deeply coloured crystalline compounds by the action of methyl iodide and $AgBF_4$ on the above-mentioned tropones (161)[138].

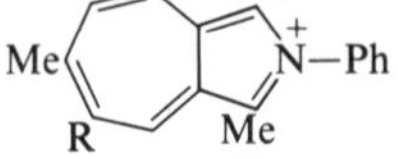

(163) 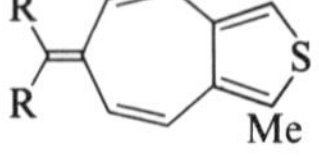 (164)

Condensation reactions of (161: $R^1 = R^2 =$ H, X = S) have been carried out to yield potentially 2π–10π and 6π–10π dipolar systems of the type (164) with molononitrile and tetrachlorocyclopentadiene, respectively[139].

Thiophenotropone (167) is obtained either from (165) or (166)[140], and the two isomeric dithiophenotropylium ions (168) and (169) are derived from dithiophenotropone (170) or its isomer, respectively[141].

The dimethylamino group in 6*H*-cycloheptoxazole (172), which has been prepared from the cation (171) with dimethylamine, undergoes rearrangement to the 4-(173) and 8-positions on standing in a solution, and the ratio of three isomers reached equilibrium presumably by the intermolecular shift of dimethylamino group via the cation (171)[142].

By irradiation of *o*-biphenylyl isothiocyanate (174) in cyclohexane at

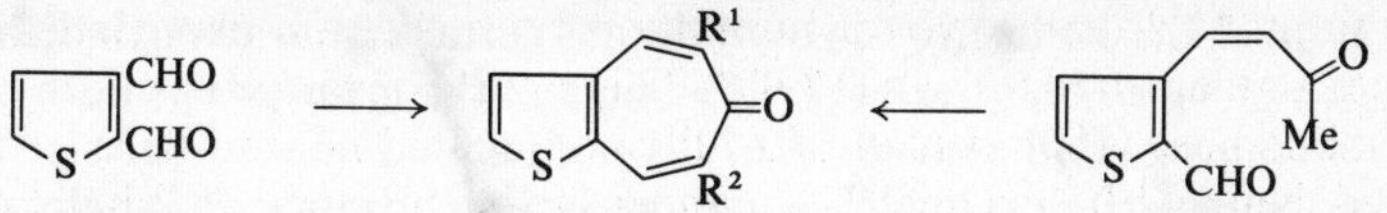

(165) (167) (166)

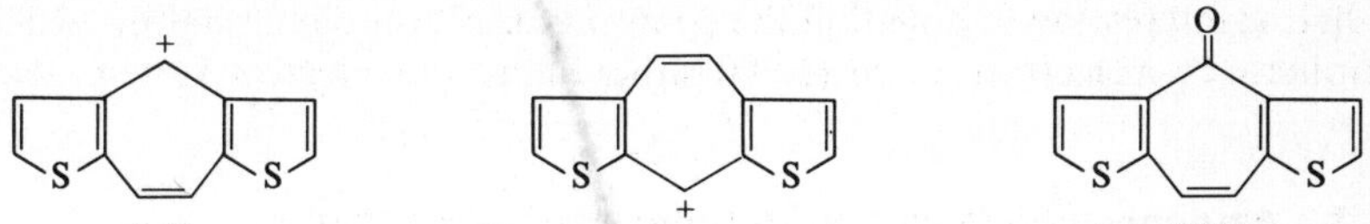

(168) (169) (170)

2537 Å, 1-azabenz[*b*]azulene (176) was obtained along with *o*-biphenylyl isocyanide (175) and sulphur[143]. In this new reaction is revealed the first

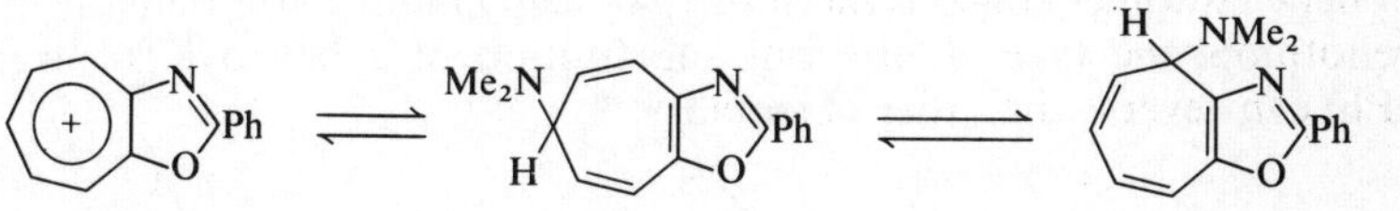

(171) (172) (173)

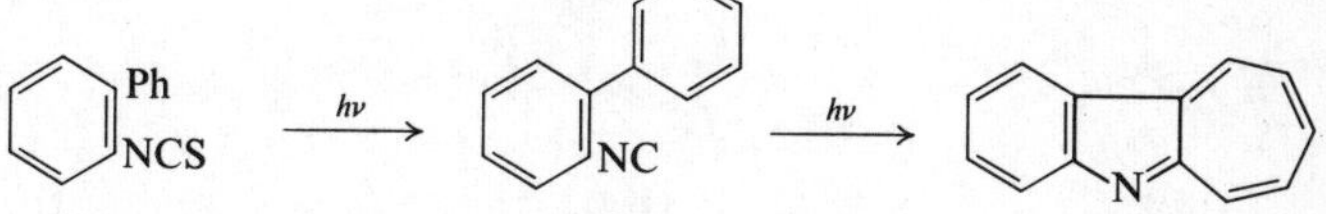

(174) (175) (176)

example of an aromatic ring expansion brought about by an insertion of an isocyanide carbon atom.

7.5.2 Six-membered heterocyclic molecules

Pyrido[3,2-*d*-]- and [2,3-*d*]tropones (177: R^1, R^2 = H and OH)[144], pyrido [3,4-*d*]tropone (178)[145] are obtained by the application of Thiele synthesis on the corresponding dialdehydes.

(177) (178)

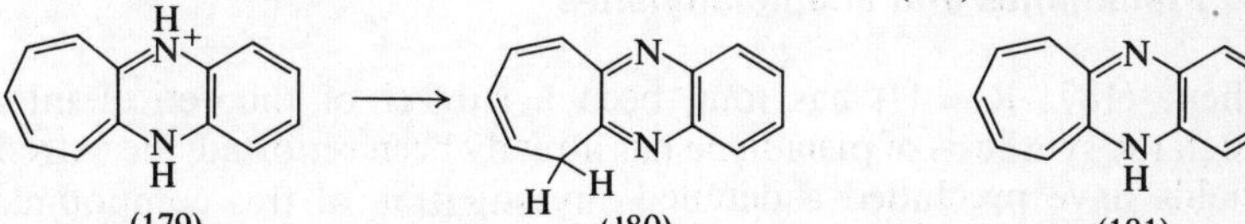

(179) (180) (181)

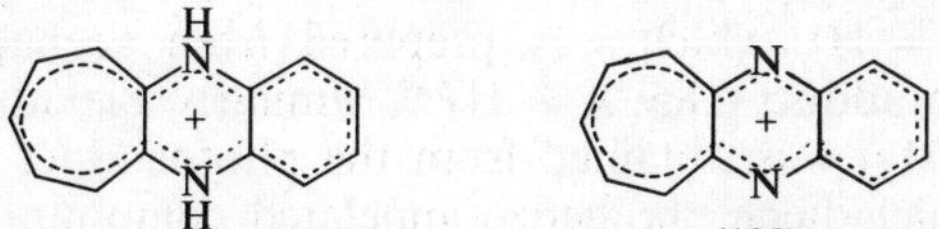

(179a) (182)

Reaction of 1,2-diethoxytropylium fluoroborate with *o*-phenylenediamine produces a greenish black salt (179)[146]. This salt is converted by triethylamine into a labile base (180) instead of a fully-conjugated neutral base (181). The so-called benzo[*b*]tropazine[147], a condensation product of 2-halogeno or 2-methoxytropone with *o*-phenylenediamine, is considered to be a dimer of (180). It is interesting to note that the compound (179) could be stabilised as a peripheral 16-π electron system (179a) rather than a 14-π electron system (182).

7.5.3 Azulenes condensed with heterocyclic molecules

An azulenothiophene derivative (184) was obtained from the thiocyanoazulyl ketone (183) and base, and it has been shown that the azulenodihydrothiophene tautomer (184a) is thermodynamically more stable than the heptafulvenothiophene type (184b); but substitution of a benzoyl group (R = COPh) can reverse the order of stability[148].

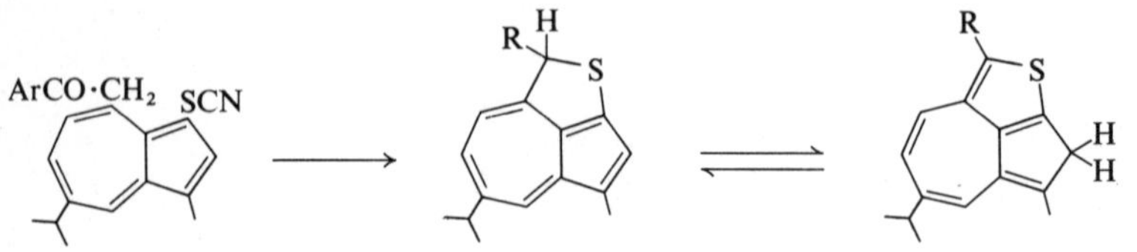

ArCO·CH_2 SCN

(183) (184a) (184b)

7.6 POLYCYCLIC TROPYLIUM COMPOUNDS

In the last two decades numerous monocyclic conjugated systems involving $(4n+2)$ and $(4n)$ π-electrons have been extensively studied concerned with the validity of the Hückel's rule. Especially, substantial knowledge has been accumulated on the non-benzenoid seven-membered monocyclic systems. In contrast to these systems pericondensed polycyclic systems having seven-membered ring contain the element of cross conjugation. Such conjugation should affect the properties of polycyclic systems and has become of interest with respect to both the theoretical and experimental points of view[179].

7.6.1 Pleiadienes and acepleiadylenes

Pleiadiene (187: R = H) has long been a subject of theoretical interest. Although the synthesis of pleiadiene has already been reported, the extremely low yields have precluded a detailed investigation of this compound. An easy synthesis of pleiadiene was carried out by the pyrolysing of the compound (186: R = H), which was prepared from acenaphthylene–maleic anhydride photo-adduct (185: X = H)[149]. Similarly, pleiadienedicarboxylate (187: R = CO_2Me) was obtained from the photoadduct (185: X = Br)[150]. In contrast to pleiadiene, the benzo-annelated compound, pleiadene (189) has never been directly observed. A stable solution of pleiadene in rigid

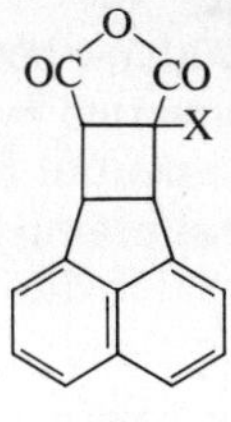

(185) 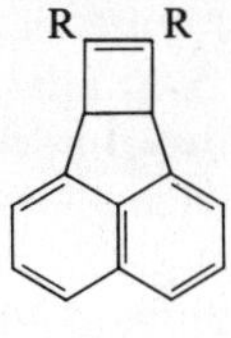 (186) 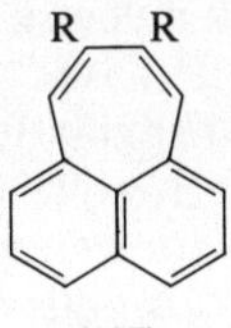 (187)

matrix at 77 K was prepared by a photochemically-induced valence isomerisation of the compound (188), the latter was obtained by addition of benzyne to acenaphthylene[151]. Pleiadene is green and its fine-structured electronic spectrum starts in the near-infrared region. These maxima are in good agreement with results of semi-empirical SCF–PPP calculations. When a rigid solution of (189) is allowed to warm up to 110 K, the green colour disappears in 19–15 s to produce a known dimer of pleiadene[151]. Experimental evidences predict that the photochemical conversion (188)→(189), occurs from one of the upper triplet states[151]. The photochemical conversion has also been applied to (186 → 187: R = H)[149]. Benzopleiadiene (190) was also prepared by the dehydrogenation of corresponding dihydro derivative with DDQ[152] and by the dehydration of the dihydro alcohol[153]. Unlike pleiadene, its isomer (190) is a stable hydrocarbon[152].

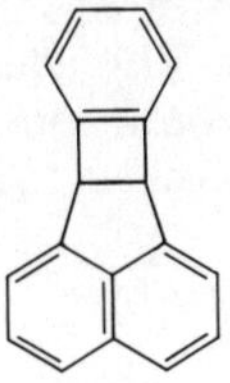

(188) 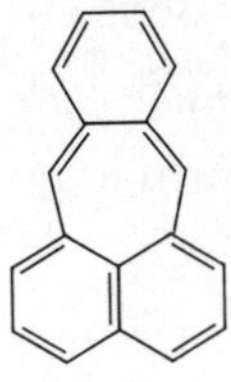 (189) 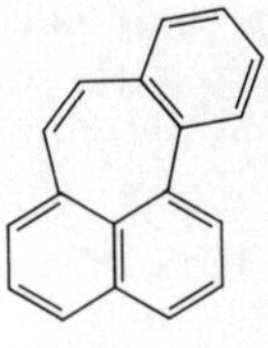 (190)

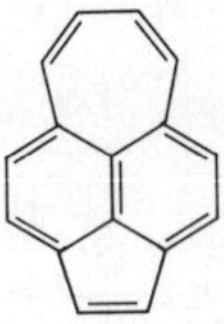

(191) 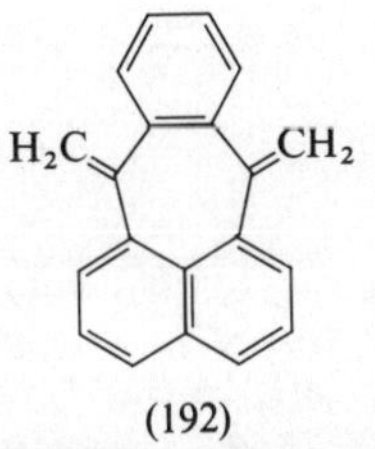 (192) 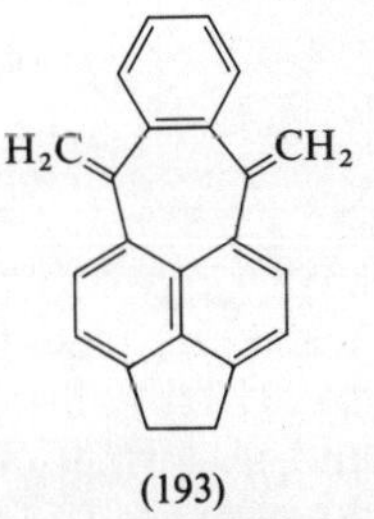 (193) 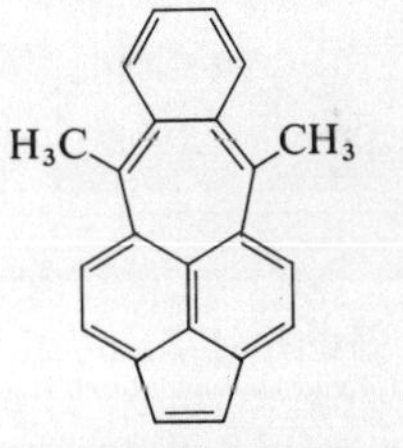 (194)

The conjugation effect in acepleiadylene[180] (191) was confirmed by several physical methods such as ^{13}C n.m.r.[154], diamagnetic susceptibility[155] and heat of hydrogenation[156]. These studies established conclusively that acepleiadylene has an aromaticity as a peripheral 14π-electron system, and provide strong support for Platt's concept that such molecules will have a stable peripheral π-electron shell separated from the inner π-electrons.

The seven-membered ring in quinodimethane analogues (192 and 193) of pleiadene and acepleiadene exist in a non-planar boat form[157]. When a dimethylsulphoxide solution of (193) was treated with potassium t-butoxide, a deep greenish blue colour appeared which showed the longest absorption maximum at 625 nm characteristic of the acepleiadylene moiety[157]. This was attributed to a base-catalysed tautomerisation of (193) to the 18-π peripheral aromatic 5,10-dimethylacepleiadylene (194). Compound (194) is stable only in dilute solution and upon attempted isolation yielded only polymers[157].

7.6.2 Pericondensed benzologues of azulene

Like acenaphthylene, the cyclopenta[*cd*]azulene (195) undergoes reaction with ethyl diazoacetate to yield a 1,2-cyclo-addition product which isomerises directly to 2*H*-benzo[cd]azulene (196). The heptafulvene system in (196) is protonated reversibly to the bridged benzotropylium ion (197) by 70% perchloric acid[158].

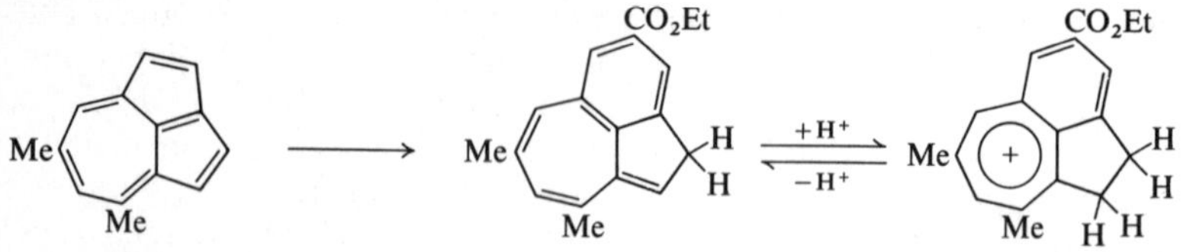

(195) (196) (197)

The reaction of azulene-1,3-dicarboxylate derivatives with phenylmagnesium bromide afford 4-phenylazulenes (198: X = Cl, H, Ph), which are converted to tetracyclic ketones (199)[159]. In strong acids a proton adds to the carbonyl group of the latter compound to give the hydroxy cation (200).

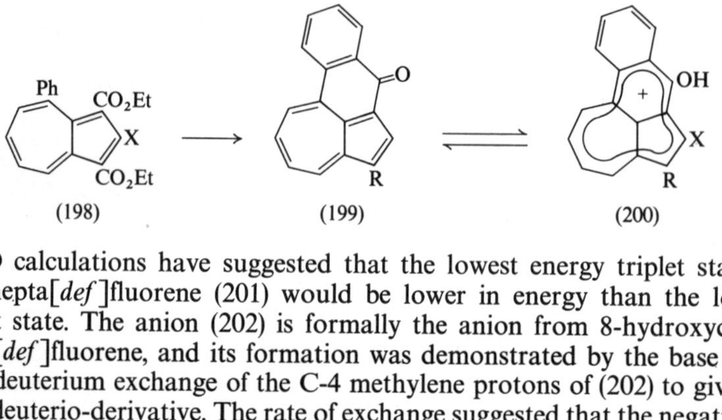

(198) (199) (200)

MO calculations have suggested that the lowest energy triplet state of cyclohepta[*def*]fluorene (201) would be lower in energy than the lowest singlet state. The anion (202) is formally the anion from 8-hydroxycyclohepta[*def*]fluorene, and its formation was demonstrated by the base catalysed deuterium exchange of the C-4 methylene protons of (202) to give the 4,4-dideuterio-derivative. The rate of exchange suggested that the negatively-charged system (202) may be regarded as an aromatic, delocalised anion[160].

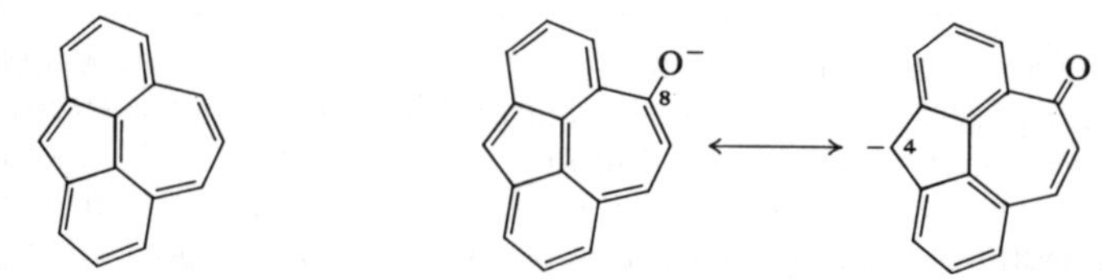

(201) (202)

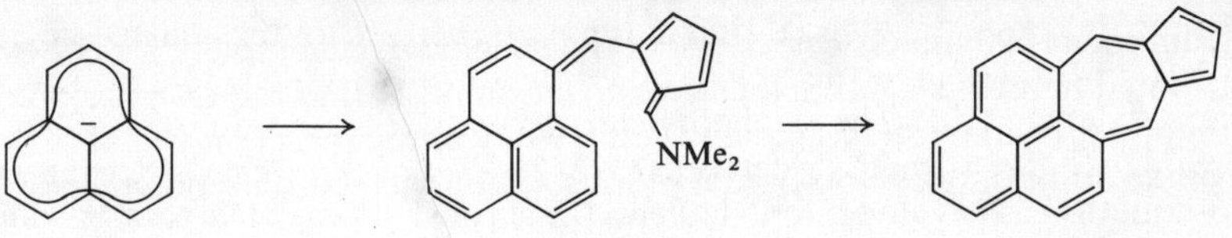

(203) (204) (205)

Phenalenide ion (203) reacts with 6-dimethylamino-2-dimethylimino-methylfulvene perchlorate easily to form a deep violet enamine (204) which can be converted to the green azuleno[5,6,7-*cd*]phenalene (205) by the thermal cyclisation with loss of dimethylamine[161]. Compound (205) is a stronger base than azulene and is extracted reversibly from benzene solution with 60% phosphoric acid.

7.6.3 Polycyclic systems with both five- and seven-membered rings

Systems composed from five- and seven-membered rings seem to be very interesting because these contain cross-conjugated systems as found in the well known penta- and hepta-fulvenes. The chemistry of this type of compound has already been compiled by Hafner[162]. Compound (195) reacts with dimethyl acetylenedicarboxylate at 25 °C at the 1,2- and/or 3,4-positions to give cyclobutene derivatives (206 and/or its isomer). Thermal valence isomerisation of these adducts leads to the aceheptylenes (207 and/or its

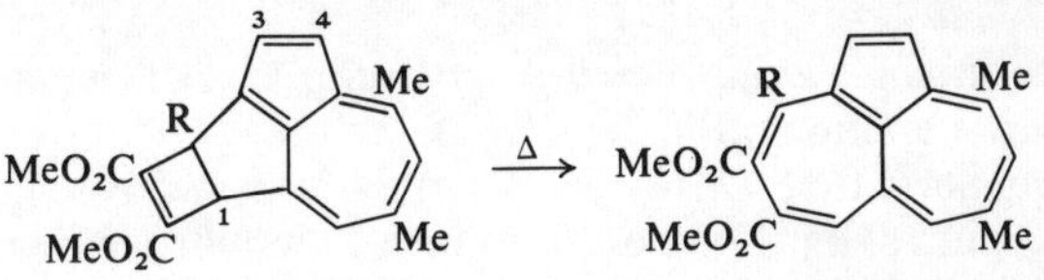

(206) (207)

isomer: R = H, Me, Ph) in good yields[163]. The parent hydrocarbon (210) could be prepared by an alternative route. Thus, the treatment of sodium azuleniate (208) with the immonium salt yields the dienamine (209) which

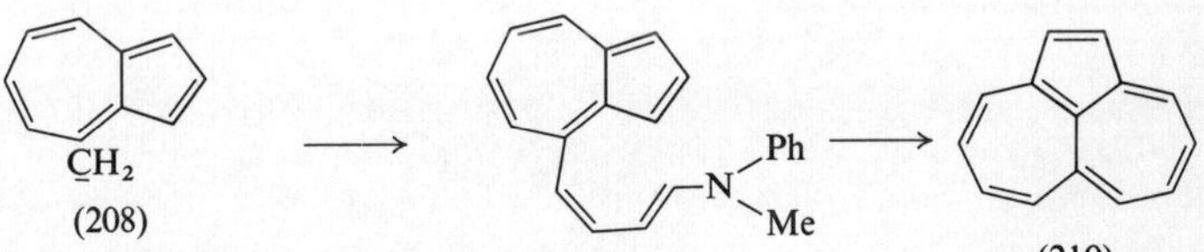

(208) (209) (210)

undergoes thermal intramolecular cyclisation to give aceheptylene (210)[164]. The concept that aceheptylenes can be regarded as a superposition of two azulenes and a heptalene unit is supported by x-ray analysis, n.m.r. studies and chemical reactivities[162].

One of the non-benzenoid isomers of pyrene is pentaleno[6,6a,1,2-*def*] heptalene (211: $R^1 = R^2 =$ H) which could be constructed by addition either of a second seven-membered ring in the peri-positions of the azulene portion of cyclopenta[*cd*]azulene (195) or of a second five-membered ring to aceheptylene (210). The aceheptylene derivative (212) can be converted directly to the trimethyl derivative (211: $R^1 = R^2 = Me$). Alternatively, 4,6-dimethyl-1,8-cyclopentenoazulene (213) can be subjected to a modified Vilsmeier reaction to yield the aldi-immonium salt (214) which on treatment with bases undergoes ring closure and on subsequent elimination of *N*-methylaniline yields the hydrocarbon (215) which is transformed to (211: $R^1 = Me$, $R^2 = H$) on treatment with chloranil. The ease of this dehydrogenation is indicative of the thermodynamic stability of the product (211).

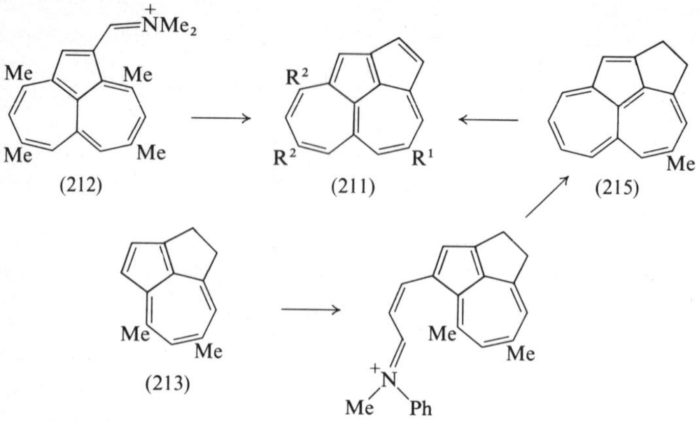

(212) (211) (215)

(213) (214)

The hydrocarbon (211) would seem to be composed of one azulene moiety conjugated with a triene unit[165].

The dihydro derivative (216) reacts with a Vilsmeier reagent to give a dihydro-immonium salt (217) and this in turn is dehydrogenated by chloranil to regenerate the aceheptylene structure (218). Base-catalysed ring closure yields the labile hydrocarbon, azuleno[8,8a,1,2-*def*]heptalene (219)[164]. The

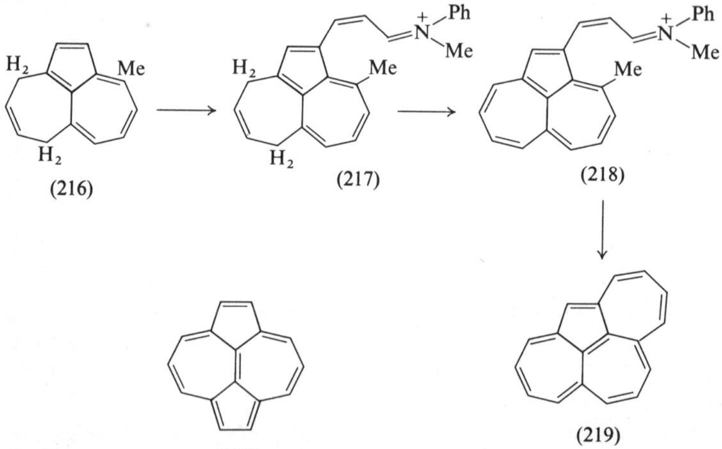

(216) (217) (218)

(220) (219)

azulenoheptalene (219) has a pronounced basicity and is soluble in 2N sulphuric acid from which it can be regenerated unchanged on dilution with water. The longest absorption of this compound in the electronic spectrum is found at 1075 nm with shoulders extending up to the region of 1500–1700 nm, a most unusual long-wave absorption for a hydrocarbon. The proton resonance signals of the seven-membered rings between τ, 5.6 and 8.4 indicate strong paramagnetic ring current in the three seven-membered rings.

Pyraceheptylene (220) and its benzo-derivative were prepared in alternative route by the application of the Ziegler–Hafner azulene synthesis[166].

7.7 PHYSICAL ASPECTS OF SOME MONOCYCLIC TROPONOIDS: CONCLUSION

In the early stage of troponoid chemistry, tropones (140) and heptafulvenes (15) were generally considered to be aromatic compounds having resonance contributing 6π dipolar structures (40a and 15a).

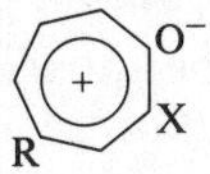

(40a) 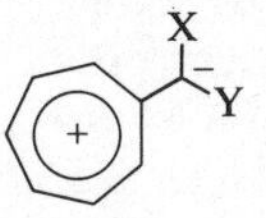 (15a)

Recent studies of tropone have demonstrated that it undergoes easy electrophilic addition and cyclo-addition reactions rather than substitutions with various reagents. Tropolone, on the other hand, has been assigned a more developed aromatic character from its thermal stability and ability to undergo ready electrophilic substitutions. From the detailed examination of the dipole moments (in conjunction with CNDO/2 calculations) and n.m.r. spectra, Bertelli *et al.*[167] suggested that the dipolar character of tropone, tropolone and heptafulvenes has hitherto been greatly overestimated and the ground states of these compounds exhibit a marked bond alternation characteristic of polyenones and polyenes. This conclusion is also supported by SCF–MO calculations[168].

The available data obtained from x-ray crystallographic analysis of 2-chlorotropone[169] and tropolone *p*-chlorobenzoate[170], and from electron diffraction study on tropone[171] have shown that these molecules should be formulated as a planar cycloheptatrienone (40) without contributions from the dipolar structure (40a).

An x-ray crystallographic study of free tropolone[172] indicated that the molecule had practically a planar heptagonal nucleus with intermolecular hydrogen bonds between two molecules in the form of dimer, and did not show marked bond alternation except for the C-1—C-2 bond. This is quite a similar situation in the case of recent x-ray studies of various chelates of tropolone and dithiotropolones[173].

It is also interesting to note that unsymmetrically substituted tropolones (40: X = OH) afford only one kind of acyloxytropones (almost quantitatively), which is unlike the case of *O*-alkyl ethers or *O*-arylsulphonate esters.

On the basis of the infrared and n.m.r. spectra of various acyloxytropones, it has been suggested that a highly mobile tautomerism exists between (222a) and (222b), presumably through a symmetrically-bridged intermediate (222c)[174].

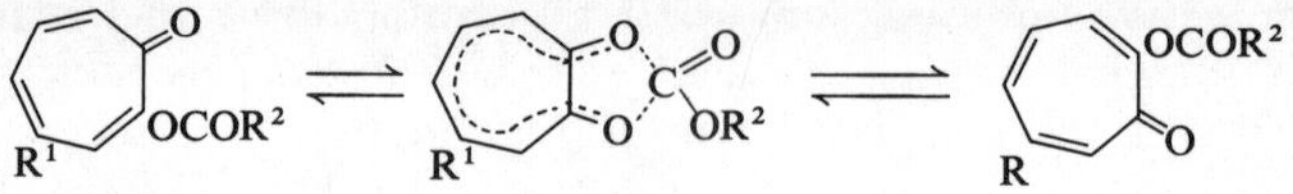

(222a) (222c) (222b)

The gas phase empirical resonance energies for tropone, 2-aminotropone and 2-aminotroponeimine (20: X = NMe, Y = NHMe) have been obtained from vapour-pressure and bomb-calorimetric measurements[175]. When the corrections for the intramolecular hydrogen bonding are applied for the empirical resonance energies, the stabilisation attributable to delocalisation is only *c.* 14 kcal mol^{-1} for the troponoid skeleton and is therefore not an appreciable amount considering their aromatic nature[176].

Although there are many arguments against the aromaticity[98, 176] of tropone and tropolone it seems reasonable to assume that these compounds could be regarded at least as potentially aromatic compounds because of the easy polarisation of the carbonyl group (low carbonyl-stretching vibrations, high boiling point and the hygroscopic nature of tropone and 2-methoxytropone to form hemihydrates, molecular complex formation such as picrates, etc.) so it would be convenient to classify all troponoids and heptafulvenes as members of non-benzenoid aromatics, especially belonging to the family of the truly aromatic tropylium ion.

Finally, it is interesting to note that tropones have now been synthetically connected with two types of bridged peripheral 10π aromatic systems, azulene on the one hand and methano[10]annulenes on the other[177].

Recently, Vogel and others[178] have developed the synthesis of various 10π analogues of tropylium ion (223), tropones (224) and tropolones (225) and a comparison of the chemical nature of troponoids and their 10π analogues are worthy of close attention.

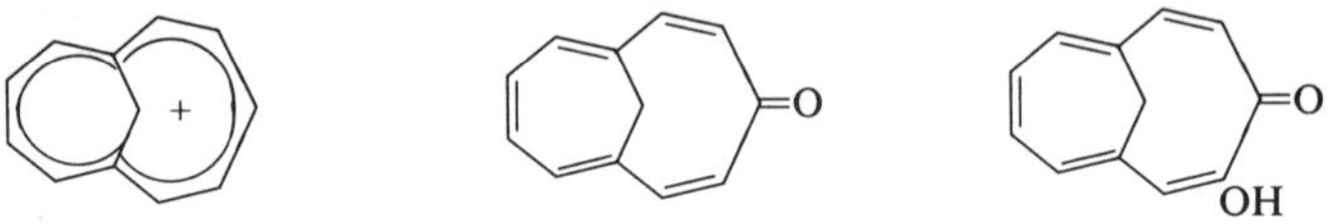

(223) (224) (225)

Acknowledgements

We are indebted to many chemists for providing us with their preprints and unpublished information during the time this manuscript was in progress. Much of the library work was done by one of the authors (T.N.) during his stay at the University of Alberta, Edmonton. We would like to thank Professor H. E. Gunning (Chairman of the Department of Chemistry) and to Professor S. Masamune for their helpful consideration and encouragement.

References

1. *International Symposium on the Chemistry of Nonbenzenoid Aromatic Compounds,* Sendai, August 24–28 (1970)
2. *The Jerusalem Symposia on Quantum Chemistry and Biochemistry, III,* Jerusalem, March 31–April 3 (1970)
3. Nozoe, T. (1971). *Pure Appl. Chem.,* **28,** 239
4. Watanabe, T. and Soma N. (1970). *Chem. Pharm. Bull.,* **18,** (a) 1595, (b) 1604
5. Sugimura, Y., Iino, K., Kawamoto, I. and Kishida, Y. To be published.
6. Itô, S. and Itoh, I. (1971). *Tetrahedron Letters,* 2969
7. Childs, R. F. and Taguchi, V. (1970). *Chem. Commun.,* 695
8. van Tamelen, E. E., Greeley, R. H. and Schumacher, H. (1971). *J. Amer. Chem. Soc.,* **93,** 6151
9. Akhrem, I. S., Dzhanbaeva, N. R. and Vol'pin, M. E. (1970). *Zhur. Org. Khim.,* **6,** 2489
10. Volz, H. and Volz-de Lecea, M. (1971). *Justus Liebigs Ann. Chem.,* **750,** 136
11. Oda, M., Tamate, K. and Kitahara, Y. (1971). *Chem. Commun.,* 347; Otomo, T., Oda, M. and Kitahara, Y. (1971). *Chem. Commun.,* 114
12. Kabuto, C., Oda, M. and Kitahara, Y. To be published.
13. Yagihara, M., Asao, T. and Kitahara, Y. To be published.
14. Ashkenazi, P., Lupan, S., Schwarz, A. and Cais, M. (1969). *Tetrahedron Letters,* 817
15. Harmon, K. M., Harmon, A. B. and MacDonald, A. A. (1969). *J. Amer. Chem. Soc.,* **91,** 323
16. (a) Haug, E. and Föhlisch, B. (1969). *Z. Naturforsch. B,* **24,** 1353; Föhlisch, B. and Haug, E. (1971). *Chem. Ber.,* **104,** 2324; Haug, E. and Föhlisch, B. (1971). *Chem. Ber.,* **104,** 2338; (b) Haug, E. and Föhlisch, B. (1971). *Chem. Ber.,* **104,** 2670
17. Hartmann, H. (1969). *Z. Chem.,* **9,** 146
18. Krebs, A. (1971). *Tetrahedron Letters,* 1901
19. Forbes, C. E. and Holm, R. H. (1970). *J. Amer. Chem. Soc.,* **92,** 2297
20. Kusuda, K., West, R. and Rao, V. N. M. (1971). *J. Amer. Chem. Soc.,* **93,** 3627
21. Föhlisch, B. (1969). *Justus Liebigs Ann. Chem.,* **721,** 48
22. Tochtermann, W. (1970). *Fortschr. Chem. Forsch.,* **15,** 378; Tochtermann, W. and Horstmann, H. -O. (1971). *Chem. Ber.,* **104,** 365
23. Garratt, P. J. and Vollhardt, K. P. C. (1971). *Chem. Commun.,* 1143
24. (a) Harmon, R. E., Suder, R. and Gupta, S. K. (1970). *Can. J. Chem.,* **48,** 195; (b) Harmon, R. E., Suder, R. and Gupta, S. K. (1969). *Chem. Commun.,* 1170
25. Collington, E. W. and Jones, G. (1969). *J. Chem. Soc. C,* 2656
26. Jones, G. (1970). *J. Chem. Soc. C,* 1230
27. (a) Noyori, R., Makino, S. and Takaya, H. (1971). *J. Amer. Chem. Soc.,* **93,** 1272; (b) Noyori, R., Yokoyama, K., Makino, S. and Hayakawa, Y. (1972). *J. Amer. Chem. Soc.,* in press
28. Stevens, H. C., Reich, D. A., Brandt, D. R., Fountain, K. R. and Gaughan, E. J. (1965). *J. Amer. Chem. Soc.,* **87,** 5257; Stevens, H. C., Rinehart, J. K., Lavanish, J. M. and Trenta, G. M. (1971). *J. Org. Chem.,* **36,** 2780; cf, Bartlett, P. D. (1969). *U.S. Patent,* 3 448 155
29. Bartlett, P. D. and Ando, T. (1970). *J. Amer. Chem. Soc.,* **92,** 7518
30. See ref. 3 (a) p. 271; (b) p. 250; (c) p. 254; (d) p. 264
31. (a) Shim, J. H. (1969). *Daehan Hwahak Hwoejee,* **13,** 75; 83; 89 [*Chem. Abstr.* (1970). **72,** 3194, 3195, 3196] (b) Tanaka, K. and Yoshikoshi, A. (1971). *Tetrahedron,* **27,** 4889
32. Turner, R. W. and Sedan, T. (1966). *Chem. Commun.,* 399
33. Brady, W. T. and Hieble, J. P. (1970). *Tetrahedron Letters,* 3205; For a review, see Brady, W. T. (1971). *Synthesis,* 415
34. Challand, B. D., Hikino, H., Kornis, G., Lange, G. and de Mayo, P. (1969). *J. Org. Chem.,* **34,** 794
35. Katritzky, A. R. and Takeuchi, Y. (1970). *J. Amer. Chem. Soc.,* **92,** 4134; (1971). *J. Chem. Soc. C,* 878
36. Saraf, S. D. (1971). *Synthesis,* 264
37. Saxena, M. K. and Bokadia, M. M. (1969). *J. Indian Chem.,* **66,** 855
38. (a) Sato, M., Kanemuro, T., Tsunetsugu, J. and Ebine, S. To be published: (b) Sato, M., Tsunetsugu, J. and Ebine, S. To be published

39. Rundel, W. and Kästner, P. (1970). *Justus Liebigs Ann. Chem.*, **737**, 87
40. Read, G. and Ruiz, V. M. (1969). *Chem. Commun.*, 735
41. Galantay, E. and Simpson, W. R. J. (1970). *Chem. Commun.*, 754
42. Bergmann, E. D. and Agranat, I. (1971). *J. Chem. Soc. C*, 3532
43. Föhlisch, B. and Widmann, E. (1969). *Z. Naturforsch.*, **24**, 464
44. Soyer, N. and Kerfanto, M. (1970). *Bull. Soc. Chim. Fr.*, 4512
45. Ravinovitz, M. and Bergman, H. (1969). *Israel J. Chem.*, **6**, 933
46. Moise, C. and Tirouflet, J. (1969). *Bull. Soc. Chim. Fr.*, 1182
47. Roberts, B. W. and Wissner, A. (1970). *J. Amer. Chem. Soc.*, **92**, 6382
48. Kinstle, T. H. and Carpenter, P. D. (1969). *Tetrahedron Letters*, 3943
49. Sasaki, T., Kanematsu, K. and Hayakawa, K. (1971). *J. Chem. Soc. C*, 2142
50. Kato, M., Okamoto, Y. and Miwa, T. (1971). *Tetrahedron*, **27**, 4013
51. Houk, K. N. and Woodward, R. B. (1970). *J. Amer. Chem. Soc.*, **92**, 4145
52. Houk, K. N., Luskus, L. J. and Bhacca, N. S. (1970). *J. Amer. Chem. Soc.*, **92**, 6392; Bhacca, N. S., Luskus, L. J. and Houk, K. N. (1971). *Chem. Commun.*, 109; Houk, K. N., Luskus, L. J. and Bhacca, N. S. Unpublished results.
53. Itô, S., Otani, H. and Narita, S. To be published
54. Vogel, E., Weyres, F., Lepper, H. and Rautenstrauch, V. (1966). *Angew. Chem.*, **78**, 754
55. Oda, M. and Kitahara, Y. (1971). *Synthesis*, 368
56. Houk, K. N. and Watts, C. R. (1970). *Tetrahedron Letters*, 4025
57. Franck-Neumann, M. (1970). *Tetrahedron Letters*, 2143
58. Luskus, L. J. and Houk, K. N. (1971). *Chem. Commun.*, in press.
59. Sugimura, Y., Soma, N. and Kishida, T. To be published
60. Cantrell, T. S. (1971). *J. Amer. Chem. Soc.*, **93**, 2540
61. Day, A. C. and Ledlie, M. A. (1970). *Chem. Commun.*, 1265
62. Baldwin, J. E. and Gano, J. E. (1969). *Tetrahedron Letters*, 1101
63. Hoffmann, R. W., Eicken, K. R., Luthardt, H. J. and Dittrich, B. (1969). *Tetrahedron Letters*, 3789; (1970). *Chem. Ber.*, **103**, 1547
64. Fukunaga, T., Mukai, T., Akasaki, Y. and Suzuki, R. (1970). *Tetrahedron Letters*, 2975
65. Jones, W. M., Hamon, B. N., Joines, R. C. and Ennis, C. L. (1969). *Tetrahedron Letters*, 3909
66. Tezuka, T., Yanagi, A. and Mukai, T. (1970). *Tetrahedron Letters*, 637
67. Hobson, J. D. and Malpass, J. R. (1969). *J. Chem. Soc. C*, 1499; (1970). 1935
68. Hoshino, M. and Ebine, S. (1970). *Bull. Chem. Soc. Jap.*, **43**, 1778
69. Harrison, R. M., Hobson, J. D. and Al Holly, M. M. (1971). *J. Chem. Soc. C*, 3084
70. Collier, P. D. (1969). *J. Chem. Soc. C*, 612; *J. Chem. Soc. B*, 637; For chemiluminescent oxidation, see: Salawinski, J. (1971). *Photochem. photobiol.*, 489
71. Sugimura, Y., Iino, K., Kuwano, M., Soma, N. and Kishida, Y. (1971). *Chem. Pharm. Bull.*, in press.
72. Machiguchi, T., Sumiya, M., Hoshino, M. and Ebine, S. Unpublished results
73. Föhlisch, B., Widmann, E. and Schupp, E. (1969). *Tetrahedron Letters*, 2355
74. Sasada, Y., Shimanouchi, H., Murata, I., Tajiri, A. and Kitahara, Y. (1969). *Tetrahedron Letters*, 1185
75. Nozoe, T., Takase, K., Kato, M. and Nogi, T. (1971). *Tetrahedron*, **27**, 6023
76. Higashi, H., Kurosawa, K. and Matsumura, H. (1970). *Bull. Chem. Soc. Jap.*, **43**, 3236; Nakama, S., Imafuku, K. and Matsumura, H. (1970). ibid **43**, 3265
77. Kagabu, S., Takahashi, K. and Takase, K. To be published
78. Barrow, M. J., Mills, O. S., Haque, F. and Pauson, P. L. (1971). *Chem. Commun.*, 1239
79. Ward, J. S. and Pettit, R. (1971). *J. Amer. Chem. Soc.*, **93**, 262
80. (a) Coxon, D. T., Holmes, A., Ollis, W. D. and Vora, V. C. (1970). *Tetrahedron Letters*, 5237; Coxon, D. T., Holmes, A. and Ollis, W. D. (1970). ibid., 5241; 5247; (b) Brown, A. G., Eyton, W. B., Holmes, A. and Ollis, W. D. (1969). *Nature (London)*, **221**, 742
81. Bryce, T., Collier, P. D., Fowlis, I., Thomas, P. E., Frost, D. J. and Wilkins, C. K. (1970). *Tetrahedron Letters*, 2789; Bryce, T., Collier, P. D., Mallows, R., Thomas, P. E., Frost, D. J. and Wilkins, C. K. (1971). ibid., in press; Korver, O. and Collier, P. To be published
82. Potěšilová, H., Alcaraz, C. and Šantavý, F. (1969). *Collect. Czech. Chem. Commun.*, **34**, 2128; Potěšilová, H., Wiedermannová, J. and Šantavý, F. (1969). ibid., **34**, 3642; Potěšilová, H., Šantavý, J., El-Hamidi, A. and Šantavý, F. (1969). ibid., **34**, 3540; Šantavý, F. (1970). ibid., **35**, 2857

83. Chommadov, B., Yusupov, M. K. and Sadykov, A. S. (1970). *Khim. Prir. Soedin,* **6,** 82; 275 [*Chem. Abstr.,* (1970). **73,** 45 635, 45 636]
84. (a) Canonica, L., Danieli, B., Manitto, P. and Russo, G. (1969). *Tetrahedron Letters,* 607 (b) Canonica, L., Danieli, B., Manitto, P., Russo, G., Bonati, A. and Bombardelli, E. (1969). *Gazz. Chim. Ital.,* **99,** 1059
85. Ramage, R. (1971). *Tetrahedron,* **27,** 1499
86. Šantavý, F., Sedmera, P., Snatzke, G. and Reichstein, T. (1971). *Helv. Chim. Acta.,* **54,** 1084; For a review see: Šantavý, F. (1968). *Planta Medica. Suppl.,* 46
87. Battersby, A. R., Ramage, R., Cameron, A. F., Hannaway, C. and Šantavý, F. (1971). *J. Chem. Soc. C,* 3514
88. Scott, A. I., Guilford, H. and Lee, E. (1971). *J. Amer. Chem. Soc.,* **93,** 3534
89. Takenaka, S. and Seto, S. (1971). *Agr. Biol. Chem. (Tokyo).* **35,** 862
90. McInnes, A. G., Smith, D. G. and Vining, L. C. (1971). *Chem. Commun.,* 325
91. Ehntholt, D. J., Emerson, G. F. and Kerber, R. C. (1969). *J. Amer. Chem. Soc.,* **91,** 7547; Ehntholt, D. J. and Kerber, R. C. (1970). *Chem. Commun.,* 1451; Rodeheaver, G. T., Farrant, G. C. and Hunt, D. F. (1971). *J. Organometallic. Chem.,* **30,** C22; Churchill, M. R. and Fennessey, J. P. (1970). *Chem. Commun.,* 1056
92. Johnson, B. F. G., Lewis, J., McArdle, P. and Randall, G. L. P. (1971). *Chem. Commun.,* 177
93. Combs, C. M., Lobeck, Jr. W. G. and Wu, Y. H. (1970). *J. Org. Chem.,* **35,** 275; Bergmann, E. D. and Solomonovici, A. (1970). *Synthesis,* 183; Nógrádi, M., Ollis, W. D. and Sutherland, I. O. (1970). *Chem. Commun.,* 158; Remy, D. C. and Van Saun, Jr. W. A. (1971). *Tetrahedron. Letters,* 2463
94. Oda, M. and Kitahara, Y. (1969). *Chem. and Ind. (London),* 920
95. Tsuruta, H., Sugimura, T. and Mukai, T. To be published
96. Kuroda, S., Oda, M. and Kitahara, Y. To be published
97. Prinzbach, H. and Knöfel, H. (1969). *Angew. Chem.,* **81,** 900; Prinzbach, H., Knöfel, H. and Woischnik, E. ref. 99, p. 269
98. Bergmann, E. D. and Pullman, B. (1971). *Aromaticity, Pseudo-Aromaticity, Anti-Aromaticity,* (Jerusalem: The Israel Academy of Sciences and Humanities)
99. Jones, W. M. and Ennis, C. L. (1969). *J. Amer. Chem. Soc.,* **91,** 6391; Joines, R. C., Turner, A. B. and Jones, W. M. (1969). ibid., **91,** 7754; Meyers, A., Joines, R. C. and Jones, W. M. (1970). ibid., **92,** 4740; Schissel, P., Kent, M. E., McAdoo, D. J. and Hedaya, E. (1970). ibid., **92,** 2147
100. Murahashi, S. -I., Moritani, I. and Nishino, M. (1971). *Tetrahedron,* **27,** 5131
101. Ishimori, M., West, R., Teo, B. K. and Dahl, L. F. (1971). *J. Amer. Chem. Soc.,* **93,** 7107
102. Schönberg, A., Sodtke, U. and Praefcke, K. (1969). *Chem. Ber.,* **102,** 1453
103. Garratt, P. J. and Knapp, K. A. (1971). *Chem. Commun.,* 1084
104. (a) Nozoe, T., Matsumura, S., Murase, Y. and Seto, S. (1955). *Chem. and Ind. (London),* 1257; Nozoe, T., Seto, S., Matsumura, S. and Asano, T. (1956). *Proc. Jap. Acad.,* **32,** 339; Nozoe, T., Seto, S., Takase, K., Matsumura, S. and Nakazawa, T. (1965). *Nippon Kagaku Zasshi,* **86,** 346; (b) For reviews see: Nozoe, T. (1957). *Croat. Chem. Acta.,* **29,** 207; Nozoe, T. and Itô, S. (1961). *Fortschr. Chem. Org. Naturst.,* **19,** 32
105. Nozoe, T., Takase, K. and Fukuda, S. (1971). *Bull. Chem. Soc. Jap.,* **44,** 2210; 2215; Nozoe, T., Takase, K., Nakazawa, T. and Fukuda, S. (1971). *Tetrahedron,* **27,** 3357
106. Oda, M. and Kitahara, Y. (1971). *Synthesis,* 367
107. Yang, P. W., Yasunami, M. and Takase, K. (1971). *Tetrahedron Letters,* 4275
108. Ziegler, K. and Hafner, K. (1955). *Angew. Chem.,* **67,** 301; Hafner, K. (1955). ibid., **67,** 301; Hafner, K. (1957). *Justus Liebigs Ann. Chem.,* **606,** 79; For a review see: Hafner, K. (1958). *Angew. Chem.,* **70,** 419
109. Alder, R. W. and Whittaker, G. (1971). *Chem. Commun.,* 776
110. Attridge, C. J., Baker, S. J. and Parkins, A. W. (1971). *Organometallic Chem. Syn.,* **1,** 183
111. Hafner, K. and Orfanos, V. Unpublished results
112. Hafner, K. (1971). *Pure Appl. Chem.* **28,** 156
113. Kapicak, L. A. and Battiste, M. A. (1971). *Synthesis,* 153
114. Müller, E., Sauerbier, M. and Zountsas, G. (1969). *Tetrahedron Letters,* 3003
115. Müller, E., Heiss, J. and Sauerbier, M. (1969). *Justus Liebigs Ann. Chem.,* **723,** 61; Müller, E., Sauerbier, M., Streichfuss, D., Thomas, R., Winter, W., Zountsas, G. and Heiss, J. (1970). ibid., **735,** 99

116. Brodherr, N., Narayanan, P., Zechmeister, K. and Hoppe, W. (1971). *Liebigs Ann. Chem.*, **750**, 53
117. Müller, E. and Zountsas, G. (1970). *Tetrahedron Letters*, 4531
118. Müller, E., Sauerbier, M., Streichfuss, D., Thomas, R., Winter, W., Zountsas, G. and Heiss, J. (1971). *Justus Liebigs Ann. Chem.*, **750**, 63
119. Staab, H. A. and Ipaktschi, J. (1971). *Chem. Ber.*, **104**, 1170
120. Hafner, K., Häring, J. and Jäkel, W. (1970). *Angew. Chem.*, **82**, 135
121. Gassmann, P. G. and Nakai, T. (1971). *J. Amer. Chem. Soc.*, **93**, 5897
122. Treibs, W. (1970). *Chimia*, **24**, 440; Treibs, W. and El Azem, K. (1970). *Justus Liebigs Ann. Chem.*, **740**, 112
123. Itô, S., Takeshita, H. and Makino, T. (1971). *Bull. Chem. Soc. Jap.*, **44**, 1982
124. Ried, W. and Kunstmann, W. (1969). *Chem. Ber.*, **102**, 1431
125. Nefedov, V. A. (1969). *Zhur. Obshch. Khim.*, **39**, 665
126. Replogle, L. L., Peters, G. C. and Maynard, J. R. (1969). *J. Org. Chem.*, **34**, 2022
127. Lok, C. M., Lugtenburg, J., Cornelisse, J. and Havinga, E. (1970). *Tetrahedron Letters*, 4701
128. Morita. T. and Takase, K. To be published
129. Kurokawa, S. (1969). *Tetrahedron Letters*, 3567
130. (a) Cooks, R. G., McDonald, R. N., Curtis, J. R. and Petty, H. E. (1971). *Org. Mass Spectrom.*, **5**, 785; (b) McDonald, R. N. and Curtis, J. R. (1971). *J. Amer. Chem. Soc.*, **93**, 2530
131. McDonald, R. N., Morris, D. L., Petty, H. E. and Hoskins, T. L. (1971). *Chem. Commun.*, 743
132. Ried, W. and Ehret, J. (1970). *Justus. Liebigs Ann. Chem.*, **737**, 61; Ried, W., Merkel, W. and Herrmann, H. J. (1971). ibid., **750**, 91
133. Ginesina, A. A., Kivokurtseva, L. N. and El'tsov, A. V. (1969). *Zhur. Org. Khim.*, **5**, 570
134. Kivokurtseva, L. N. and El'tsov, A. V. (1970). *Zhur. Org. Khim.*, **6**, 2126
135. El'tsov, A. V. and Ginesina, A. A. (1969). *Zhur. Org. Khim.*, **5**, 1135; El'tsov, A. V., Ginesina, A. A. and Kivokurtseva, L. N. (1969). ibid., **5**, 2072
136. Ginesina, A. A. and El'tsov, A. V. (1969). *Zhur. Obshch. Khim.*, **39**, 2601
137. El'tsov, A. V., Kul'bitskaya, O. V. and Ogol'tsova, N. V. (1969). *Zhur. Org. Khim.*, **5**, 2242
138. El'tsov, A. V., Ginesina, A. A. and Kivokurtseva, L. N. (1969). *Zhur. Org. Khim.*, **5**, 961
139. Seitz, G. and Monnighoff, H. (1970). *Angew. Chem.*, **82**, 938
140. Guilard, R. and Fournari, P. (1971). *Bull. Soc. Chim. Fr.*, 1437
141. Gronowitz, S. and Yomtov, B. (1970). *Z. Chem.*, **10**, 389
142. Nakazawa, J., Sugimura, Y. and Soma, N. (1970). *Chem. Pharm. Bull.*, **18**, 581
143. Boyer, J. H. and Jong, J. D. (1969). *J. Amer. Chem. Soc.*, **91**, 5929
144. Fugier, C., Quéguiner, G. and Pastour, P. (1969). *Compt. Rend. Acad. Sci., Paris, Ser. C*, **269**, 925
145. Quéguiner, G., Fugier, C. and Pastour, P. (1970). *Compt. Rend. Acad. Sci. Paris, Ser. C.* **270**, 551; *Bull. Soc. Chim. France*, 3636
146. Fukunaga, T. (1971). *23rd IUPAC Congres., Boston*, Abstract, 103
147. Nozoe, T., Kitahara, Y. and Sasaki, M. (1956). *Proc. Jap. Acad.*, **32**, 349
148. Ammon, H. L., Replogle, L. L., Watts, Jr. P. H., Katsumoto, K. and Stewart, J. M. (1971). *J. Amer. Chem. Soc.*, **93**, 2196
149. Meinwald, J., Samuelson, G. E. and Ikeda, M. (1970). *J. Amer. Chem. Soc.*, **92**, 7604
150. Shields, J. E., Gavrilovic, D. and Kopecký, J. (1971). *Tetrahedron Letters*, 271
151. Kolc, J. and Michl, J. (1970). *J. Amer. Chem. Soc.*, **92**, 4147; Michl, J. and Kolc, J. (1970). ibid., **92**, 4148; Michl, J. (1971). ibid., **93**, 523
152. Muller, J. F., Cagniant, D. and Cagniant, P. (1971). *Tetrahedron Letters*, 45
153. Craig, J. T., Tan, K. W. and Woolhouse, A. D. (1971). *Tetrahedron Letters*, 3209
154. Jones, A. J., Gardner, P. D., Grant, D. M., Litchman, W. M. and Boekelheide, V. (1970). *J. Amer. Chem. Soc.*, **92**, 2395
155. Dauben, Jr. H. J., Wilson, J. D. and Laity, J. L. (1969). *J. Amer. Chem. Soc.*, **91**, 1991
156. Turner, R. B., Lindsay, W. S. and Boekelheide, V. (1971). *Tetrahedron*, **27**, 3341
157. Lown, J. W. and Aido, A. S. K. (1971). *Can. J. Chem.*, **49**, 1848; 1861
158. Hafner, K. and Rieper, W. (1970). *Angew. Chem.*, **82**, 218
159. Abe, N., Morita, T. and Takase, K. To be published

160. Munday, R. and Sutherland, I. O. (1969). *J. Chem. Soc. C,* 1427
161. Jutz, C., Kirchlechner, R. and Seidel, H. -J. (1969). *Chem. Ber.,* **102,** 2301
162. Hafner, K. (1971). *Pure Appl. Chem.,* **28,** 168
163. Hafner, K. and Fleischer, R. (1970). *Angew. Chem.,* **82,** 217
164. Hafner, K., Hafner-Schneider, G. and Bauer, F. Unpublished results; cf. Ref. 162
165. Hafner, K., Friebe, W., Hafner-Schneider, G. and Bauer, F. Unpublished results p. 174
166. Jutz, C. and Schweiger, E. (1971). *Angew. Chem.,* **83,** 886; cf. ref. 162, p. 170 Anderson, Jr. A. G., MacDonald, A. A. and Montana, A. F. (1968). *J. Amer. Chem. Soc.,* **90,** 2993
167. Bertelli, D. J. and Andrews, Jr. T. G. (1969). *J. Amer. Chem. Soc.,* **91,** 5280; Bertelli, D. J., Andrews, Jr. T. G. and Crew, P. O. (1969). ibid., **91,** 5286
168. Dewar, M. J. S. and Trinajstić, N. (1970). *Croat. Chem. Acta.,* **42,** 1
169. Forbes, E. J., Gregory, M. J., Hamor, T. A. and Watkin, D. J. (1966). *Chem. Commun.,* 114; Watkin, D. J. and Hamor, T. A. (1971). *J. Chem. Soc. B,* 2167
170. Schaefer, J. P. and Reed, L. L. (1971). *J. Amer. Chem. Soc.,* **93,** 3902
171. Ogasawara, M. and Kimura, M. (1970). *International Symposium on the Chemistry of Non-benzenoid Aromatic Compounds, (Sendai),* Abstract 97
172. Schimanouchi, M. and Sasada, Y. (1970). *Tetrahedron Letters,* 2421
173. Hamor, T. A. and Watkin, D. J. (1969). *Chem. Commun.,* 440; Day, V. M. and Hoard, J. L. (1970). *J. Amer. Chem. Soc.,* **92,** 3626; Park, J. J., Collins, D. M. and Hoard, J. L. (1970). ibid., **92,** 3636; Khare, G. P., Schultz, A. J. and Eisenberg, R. (1971). ibid., **93,** 3579
174. Yasunami, M. and Takase, K. To be published
175. Jackson, W., Hung, T. S. and Hopkins, Jr. H. P. (1971). *J. Chem. Thermodyn.,* **3,** 347
176. Snyder, J. (1971). *Non-benzenoid Aromatics,* Vol. 2 (New York: Academic Press)
177. Vogel, E. and Roth, H. D. (1964). *Angew. Chem.,* **76,** 145; Vogel, E. and Böll, W. A. (1964). ibid., **76,** 784
178. Grimme, W., Reisdorff, J., Jünemann, W. and Vogel, E. (1970). *J. Amer. Chem. Soc.,* **92,** 6335; Reisdorff, J. and Vogel, E. (1971). *Angew. Chem.,* in press; For a review see: Vogel, E. (1971). *Pure Appl. Chem.,* **26,** Suppl. I, 275
179. Zahdradnik, R. (1971). ref. 176 p. 1
180. Flowerday, P., Perkins, M. J. and Arthur, A. R. J. (1970). *J. Chem. Soc. C,* 290

8
Annulenes and Related Compounds

P. SKRABAL
Swiss Federal Institute of Technology, Zurich

8.1 INTRODUCTION

8.1.1 Scope of the review

The large number of papers in the recent literature concerning annulenes and related compounds demonstrates the ever increasing interest in a field that embraces theoretical as well as synthetic organic chemistry. Reviews were written by Garratt and Sargent[1], Figeys[2], Jackman *et al.*[3] and Sondheimer[4, 5].

The vast amount of work that has been reported during 1970 and 1971 precludes a mention of every publication and only macrocyclic annulenes with more than 6π electrons will be discussed. For microcyclic compounds with 6π electrons and less, see Chapter 6 by Lloyd in this volume. For an introduction to the field of annulenes and to that of aromaticity see Refs. 6–8 and 164.

8.1.2 Criteria of aromaticity

For annulenes magnetic properties, particularly ring current effects (measured by chemical shifts) and coupling constants, have proved to be the most useful criterion[9]. The terms diatropic[5], paratropic[5] and atropic[5] will be used interchangeably with the terms aromatic, anti-aromatic and non-aromatic. Nevertheless, it should be pointed out that today a definition of these latter terms with respect to all available criteria is not possible[9].

8.2 ANNULENES AND DEHYDROANNULENES

Many annulenes and dehydroannulenes in the $(4n+2)$ and in the $4n$ series have been synthesised, chiefly by Sondheimer and his group[5]. When the chemical shifts of outer and inner perimeter protons are used as a criterion of aromaticity, these compounds obey 'Hückel's rule': $(4n+2)$ annulenes sustain a diamagnetic ring current and may therefore be called aromatic or diatropic, whereas $4n$ annulenes sustain a paramagnetic ring current and may be called anti-aromatic or paratropic. These ring currents, however, may be diminished or may vanish completely when non-planarity of the perimeter occurs. This non-planarity results in a smaller degree of π orbital overlap and hence a smaller degree of π electron delocalisation; this, in turn, results in bond alternation. Bond alternation has also been predicted for large ring annulenes with ring sizes above a certain limit. This limit was calculated to lie somewhere between the 22- and 26-membered compounds[10]. Some answers to these questions have been obtained recently.

8.2.1 (4n+2) Annulenes and dehydroannulenes

After a considerable amount of work[11], the isolation of pure crystalline [10]annulenes (2) and (3) has now been achieved[12] with the aid of chromatography at $-80\,°C$ after irradiation of *cis*-9,10-dihydronaphthalene (1) at $-60\,°C$. ^{1}H and ^{13}C n.m.r. spectra of (2) are independent of temperature in the range of -40 to $-160\,°C$, while those of (3) become independent of temperature at $-100\,°C$ but indicate a conformational isomerism at $-40\,°C$. The absorptions of both compounds lie in the olefinic region (τ 4.14–4.33), a fact which indicates the non-aromaticity of these [10]annulenes. In configurations (2) and (3), the considerable ring strain causes non-planarity of the molecules. Obviously the configuration (4) does not exist because of the severe overcrowding of two internal protons. Some discrepancies concerning the chemistry of (1), (2), (3) and the *trans*-isomer of (1) appear to have been resolved[13].

In 1.6-didehydro[10]annulene (5), a compound which has no internal protons, the repulsion of the in-plane π orbitals is important. A new approach

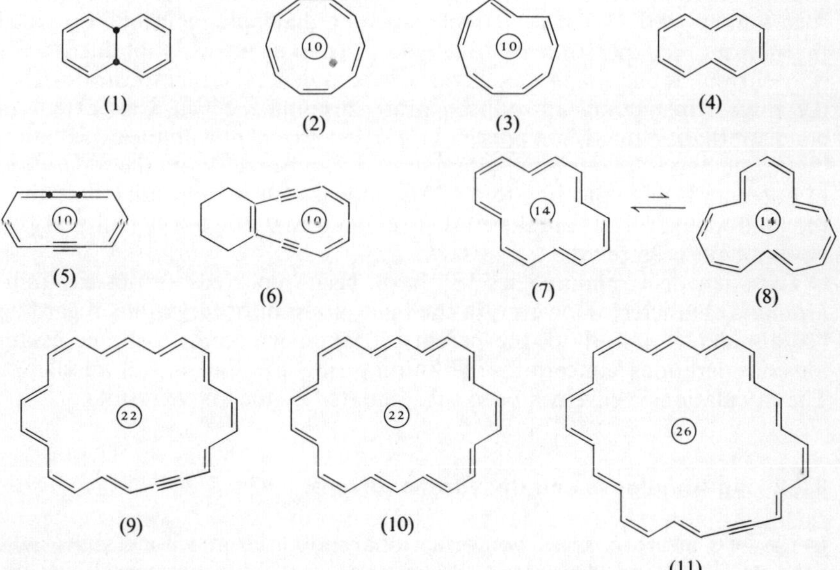

(1) (2) (3) (4) (5) (6) (7) (8) (9) (10) (11)

was used in the attempted preparation of (5), but was unsuccessful[14]. However, interesting preliminary results were reported[15] concerning the synthesis of the dialkylated 1,5-didehydro[10]annulene (6) and its subsequent isomerisation.

In the 14-membered series, the configurational isomerism of the [14]-annulenes (7) and (8) has now been proved by a detailed study of the ^{1}H n.m.r. spectra[5,16]. An interesting result of these studies is the observation that *configurational* isomerism occurs via *conformational* isomerism and *bond shift* when a planar transition state is possible. Without doubt the most important information that comes from these quantitative conformational mobility studies concerns the resonance energies of annulenes. The enthalpies of activation for the isodynamical conformational isomerisms in configurations (7) and (8) (10.2 and 6.0 kcal mol^{-1}) are upper limits for the resonance energies[16]. For [18]annulene, the conformational mobility studies led to the conclusion that its resonance energy cannot be greater than 19 kcal mol^{-1} [16,17]. These values, and those for [14]annulenes, are the first reasonable values for the resonance energies of annulenes (with the exception of benzene) to be estimated from experimental data.

An answer to the question of the limiting ring size of annulenes was given by the synthesis of monodehydro[22]annulene (9), the parent compound (10) and monodehydro[26]annulene (11). The ^{1}H n.m.r. spectrum at room temperature of (9)[18] is unchanged at $-60\,°C$ and shows multiplets at τ 1.55–3.75 and 6.55–9.30 ($\approx$13:7) for the outer and inner protons respectively, indicating a diatropic annulene. Compound (9) could be converted into the parent compound (10) by partial catalytic hydrogenation[19]. [22]Annulene (10) exhibits a temperature-dependent ^{1}H n.m.r. spectrum. The non-mobile spectrum is obtained at $-90\,°C$ (multiplets at τ 0.35–1.50 and 10.40–11.20) and shows clearly an induced diamagnetic ring current. In contrast to tridehydro[26]annulene[20], monodehydro[26]annulene[21] has a ^{1}H n.m.r. spectrum at $-90\,°C$ that is characteristic of a diatropic molecule. Provided the non-mobile spectrum of tridehydro[26]annulene was obtained[20], the ring current found in monodehydro[26]annulene (11) demonstrates that the three triple bonds introduced into [26]annulene cause an extent of bond alternation that is not present in (11). In agreement with the expectation that triple bonds favour bond alternation, it can be seen from the separation of outer and inner protons that [22]annulene (10) is more diatropic than the monodehydro[22]annulene (9). The conformations of (9), (10) and (11) have not yet been proved.

Two dehydro[30]annulenes[22,23] have been reported to possess non-aromatic character[5]. However, in the light of recent observations regarding [26]annulenes, a study of the parent [30]annulene seems to be necessary for considerations concerning the limiting ring size for ($4n+2$) annulenes. The calculations[10] have not been substantiated by the above results.

8.2.2 4n Annulenes and dehydroannulenes

In the $4n$ π electron series, conformational mobilities and bond shifts have been studied[16] for [8]annulene, [12]annulene and [16]annulene. From the enthalpies of activation for the bond shift and for the ring inversion, it is concluded that the resonance energy in a planar cyclo-octatetraene is *c.* -3.8 kcal mol^{-1}; this figure is in good agreement with the calculated values[10]. In [12]annulene, no bond shift could be detected up to -40 °C. Severe overcrowding of the internal protons prevents the formation of the planar transition state, that is necessary for such a shift to occur. However, conformational exchange was still observed at -170 °C[16]. For the [16]-annulenes (12) and (13), it may be concluded[16] from the values of the activation enthalpies that the resonance energy for a planar C-16 perimeter is close to zero (calculated[10]: $+0.99$ and -3.11 kcal mol^{-1}).

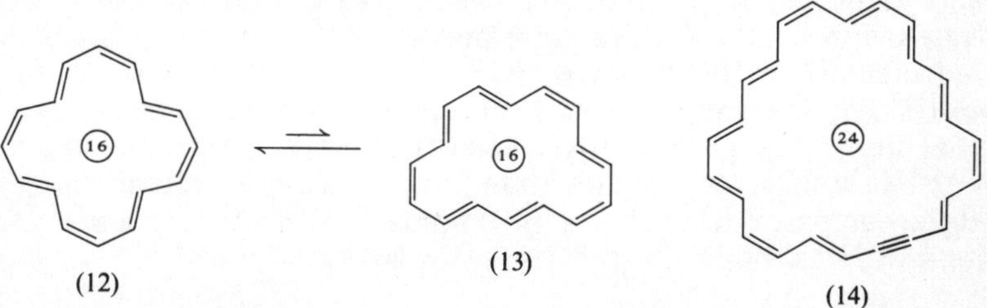

(12) (13) (14)

[20]Annulene, which completes the series of known $4n$ annulenes up to [24]annulene, and monodehydro[20]annulene have been synthesised[24]. As expected for annulenes having a 20 π electron system, they are paratropic annulenes. Monodehydro[20]annulene shows a non-mobile spectrum from -90 °C to room temperature, whereas the non-mobile spectrum of the parent compound is obtained at -105 °C.

Monodehydro[24]annulene (14)[25] was isolated as a by-product in the preparation of [24]annulene from 1,7,13,19-tetradehydro[24]annulene by catalytic hydrogenation. The separation (*c.* 6 p.p.m.) of bands in the ^{1}H n.m.r. spectrum corresponding to the outer and inner protons is indicative of a paramagnetic ring current. The integration for the outer and inner protons suggests the configuration (14).

8.2.3 Conclusion

The recent studies[21] of dehydro[26]annulene (11) demonstrate that bond alternation in higher annulenes is less important than predicted from calculations[10]. In contrast to the situation found for tridehydro[26]annulene, it has been confirmed that monodehydro[26]annulene sustains a diamagnetic ring current. Thus, the advantage of dehydroannulenes, namely the more readily available non-mobile ^{1}H n.m.r. spectra, is diminished by a possible bond alternation induced by the triple bonds; this extent of bond alternation is not necessarily present in the parent compounds. All known [24]annulenes (except (20), Section 8.3.2) are paratropic molecules. With regard to the limit of the ring size, the hitherto unknown $n = 7$ series of [28]annulenes and the parent [30]annulene will be of great interest. The most useful results from the theoretical point of view were obtained from a study of the temperature dependence of the ^{1}H n.m.r. spectra of the parent compounds [8]-, [14]-,

[16]- and [18]annulene[16, 17]; the resonance energies estimated from enthalpies of activation are in fairly good agreement with the calculated values[10].

8.3 CHARGED ANNULENES

8.3.1 Perimeters with an odd number of atoms

Cyclononatetraenyl anion (15), a 10π 9C annulene[26], is known to be a diatropic molecule[27]. The mono-*trans*-isomer of (15) which from symmetry rules is expected to be the precursor of (15) has also been reported[157]. Cyclononatetraenyl cation (16), the corresponding 8π 9C system, was proposed as an intermediate in the thermolysis of *anti*-9-chlorobicyclo[6.1.0]nona-2,4,6-triene[28]. The transient existence of (16) is now further supported by the finding that 9-d-g-chloro-*cis*[4]-cyclononatetraene (17), after initial ionisation at $-60\,°C$ in SO_2, undergoes rearrangement to (18) (see scheme (8.1)), and that this compound has an equal distribution of the deuterium label between its nine carbon centres[29]. Irrespective of the fact that thermal stability should

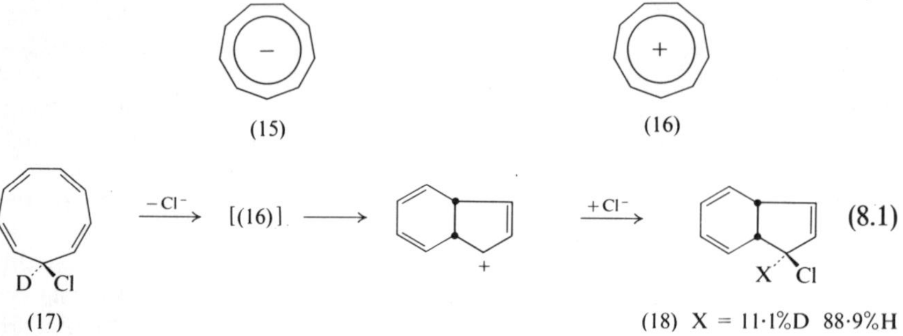

not be used as a criterion for aromaticity, the difference between the 10π system (15), which has been reported[30] to resist rearrangement on heating to $163\,°C$, and the 8π system (16) is obvious[29].

8.3.2 Perimeters with an even number of atoms

The two-electron reduction of $4n\pi$ to $(4n+2)\pi$ systems and of $(4n+2)\pi$ to $4n\pi$ systems offers the intriguing possibility to transform an anti-aromatic annulene into an aromatic one and vice versa.

The diatropic di-anion of cyclo-octatetraene has been known since 1960[31]. Since then several other annulenes have been reduced[32]. In every case the $(4n+2)\pi$ $4n$C systems sustain diamagnetic ring currents and the $4n\pi$ $(4n+2)$C systems sustain paramagnetic ring currents.

Recently, the di-anion of [12]annulene has been obtained[33]. Like the parent compound, it possesses the [12]-21-configuration[34] with three internal (τ 14.60) and nine external protons (τ 3.02 and 3.77). In contrast to [12]annulene, its di-anion does not undergo the conformational change between -90 and $+30\,°C$. From activation enthalpies one can conclude that its resonance energy is at least 8 kcal mol^{-1} greater than that of the neutral [14]-43-annulene (7)[33], despite the fact that the di-anion cannot achieve a planar geometry.

Similar results were obtained[35] in the study of the di-anion (14π 12C system) of 1,5-bisdehydro[12]annulene (see Section 8.4.3). The di-anion has a 'frozen' conformation at +30 °C, whereas the neutral 12π 12C system still undergoes conformational changes at −130 °C.

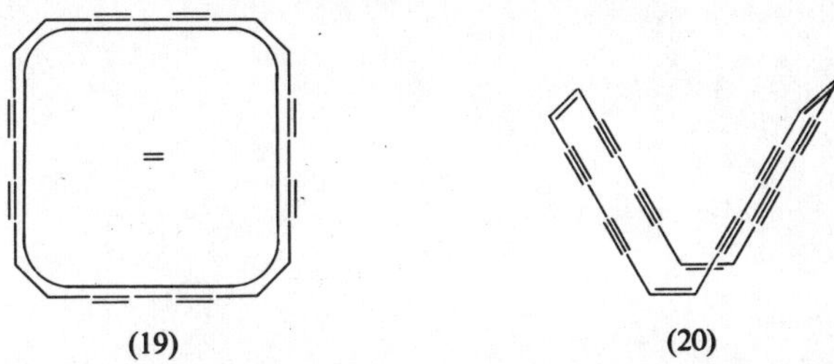

(19) (20)

In the light of the recent observation concerning the ring currents in the 26-membered series of annulenes (Section 8.2.1), the di-anion (19)[36] of the octadehydro[24]annulene (20) is of interest. Compound (20) is a non-planar, non-aromatic polyene (only outer protons at τ 3.85). The ^{1}H n.m.r. spectrum of the di-anion (19) exhibits a singlet at τ 3.88. If one roughly corrects[37] this value for two negative charges, one obtains a chemical shift for these protons of *c.* τ 3.05. This value implies that (19) is at least more planar than (20) and charge delocalisation and hence bond delocalisation is possible.

8.3.3 Conclusion

In the series of charged annulenes and dehydroannulenes the results obtained previously[32] have been further supported and indicate clearly that charge delocalisation is very important, particularly in the cases of annulenes with two formal charges. As the example of the di-anion of [12]annulene shows, the resonance energy in a charged species will usually exceed the resonance energy in the corresponding neutral π system.

8.4 BENZOANNULENES

In benzoannulenes or, more generally, fused annulenes, the fusion has sometimes from the synthetic point of view the advantage of holding polyene chains in a *cis* configuration and thus facilitating the ring closure reaction. In addition, one might expect a stabilisation of the molecule in question; but what are the effects of fusion on an annulene perimeter itself?

8.4.1 Benzo[8]-and benzo[9]annulenes

Calculations of the ring currents in cyclo-octatetraene (COT) and its benzo homologues show that successive additions of fused benzene rings decrease the paramagnetic ring current in the hypothetical planar 8π electron system[38, 39]. These calculations have now been corroborated by a low temperature ^{1}H n.m.r. study[39] of the tetraphenylene (21). A value of $\Delta G^{\ddagger}_{-116}$ = 5.7 ± 1 kcal mol^{-1} for the barrier of ring inversion was calculated; a very low value compared with that of the corresponding COT ($\Delta G^{\ddagger}_{-2}$ = 14.7 kcal mol^{-1})[40]. This difference can be rationalised by assuming a smaller destabilisation of the postulated planar transition state of ring inversion in the case of the tetraphenylene (21), i.e. planar (21) is less anti-aromatic than planar COT[39]. This, and the estimated resonance energy (Section 8.2.2), is the first evidence for the anti-aromaticity of planar [8]annulene.

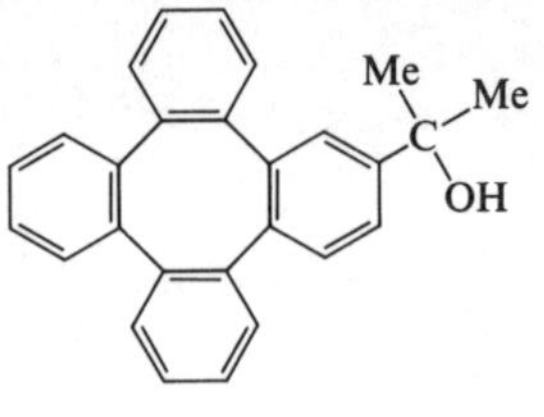

(21)

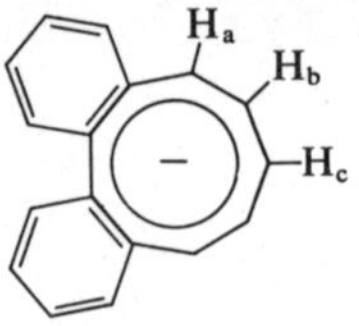

(22)

Recently, the synthesis and the ^{1}H n.m.r. spectrum of 3,4,5,6-dibenzocyclononatetraenyl anion (22) have been reported[41]. The chemical shifts for the protons H_a, H_b and H_c lie in the same region as those of the corresponding parent 10π 9C system (15). These data together with the size of the coupling constants, indicate a diatropic 10π system. Therefore, (22) should not suffer severe quenching of the ring current by the fused benzene nuclei, an observation already made for the di-anion of the sym-dibenzocyclooctatetraene[158]. From kinetic studies, a minimum value of 8.9 kcal mol^{-1} for the resonance energy of the 10π system (22) was calculated[41].

8.4.2 Benzo[10]annulenes

The synthesis of fused [10]annulenes has attracted considerable interest in the last few years because of the instability of the parent compound itself (see Section 8.2.1).

The first derivative that has been reported is the mono-*trans*-tribenzo[10]-annulene (23), a compound which is reasonably stable compared to the [10]annulenes (2) and (3); but its ^{1}H n.m.r. spectrum indicates a non-planar, non-aromatic 10π electron system[42].

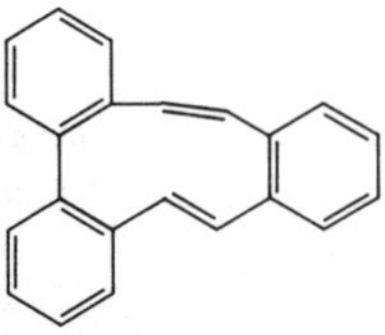

(23) 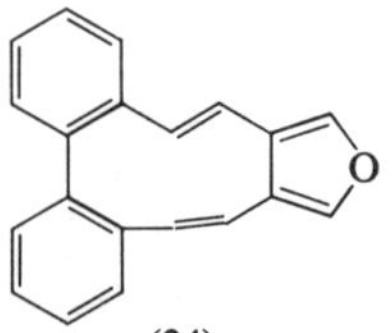 (24) 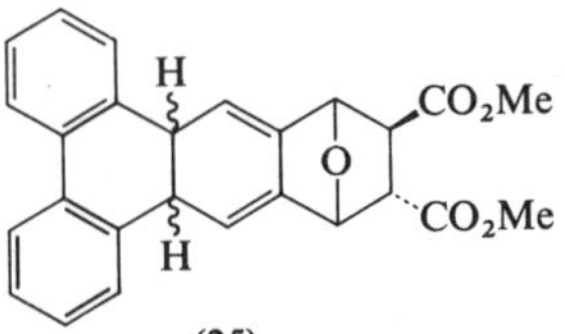 (25)

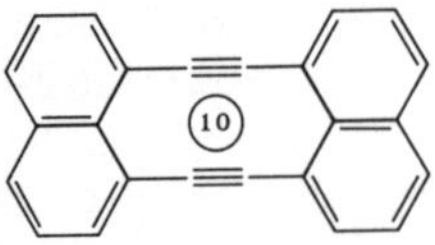

(26) 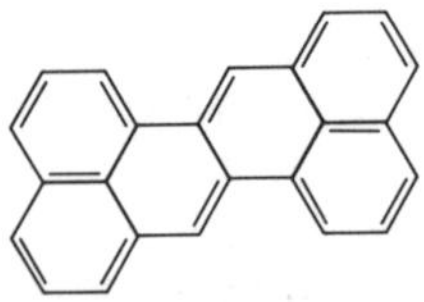 (27) 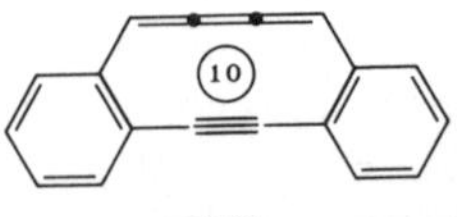 (28)

A recent re-examination[43] of compounds (24) and (25) has disproved structures which were previously postulated[44]. As in (23), the 10-membered perimeter in (24) is non-planar and atropic. The properties of these molecules demonstrate the expected stabilisation of [10]annulene by annelation of benzene rings besides reflecting the steric interaction of the internal protons and/or the bond-angle strain.

The importance of the interaction of in-plane π electrons is shown by the unsuccessful synthesis of the dinaphth-1,6-bisdehydro[10]annulene (26). This interaction results in the formation of (27)[45, 46]. Similar observations were made in the attempted synthesis[47] of the dibenzo-1,6-bisdehydro[10]-annulene (28) (see also (5) in Section 8.2.1).

8.4.3 Benzo[12]annulenes

Additional data on benzo[12]annulenes are available since Figeys' review of benzoannulenes was written[2].

In the light of the results obtained for [12]annulene[16], an extended ^{1}H n.m.r. investigation will be necessary in the case of the tribenzo[12]-annulene (29). There is no obvious reason why (29) should not undergo an isodynamical conformational isomerism. To date the ^{1}H n.m.r. spectrum has been investigated only down to $-60\,°C$ [48]. So far no conclusion can be drawn regarding the electronic structure of the 12-membered perimeter.

Recently, spectroscopic[49] and x-ray analysis[50] data of tribenzotridehydro-[12]annulene (30) have been reported. Compound (30) has a nearly planar structure, but shows no bond delocalisation in the annulene perimeter. In contrast to the paratropic parent tridehydro[12]annulene[51] no indication of a paramagnetic ring current in (30) has been obtained.

From the ^{1}H n.m.r. spectra of the benzo[12]annulenes (31) and (32), it was concluded[52] that these compounds are atropic molecules. Their *trans*-olefinic protons absorb as sharp singlets at τ 1.40 (unchanged down to $-60\,°C$) and τ 1.22 (room temperature). It has been pointed out[52] that, in a consideration of these low-field absorptions, the planar 12-membered perimeters cannot possess a perpendicular orientation of the trans double bonds; such an orientation would result in a real equivalence of the olefinic protons. The singlets must be caused by a conformational isomerism such as was found[53] for the parent 1,5-bisdehydro[12]annulene down to $-130\,°C$ [35]. The low-field shift of the *trans*-olefinic protons in (31) and (32) was explained[52] by a local anisotropy effect of the triple bonds. Thus the 12-membered perimeters of (31) and (32) are believed to be non-aromatic or, better, atropic. However, it should be emphasised that if the indication, by chemical shifts, of ring currents is accepted as a criterion for aromaticity one may conclude that (31) and (32) are anti-aromatic or paratropic annulenes for the following reasons:

The chemical shifts of the olefinic protons in model compounds[52] (33) and (34) compared with the chemical shifts of the corresponding protons in *trans*-stilbene (τ 2.90) demonstrate directly the deshielding by the triple bonds (down-field shift in (33): 0.65 p.p.m., in (34): 0.73 p.p.m.). If one adds the down-field shifts in (33) and (34) (1.38 p.p.m.), one can calculate the chemical shift for the internal proton in (31) to be τ 1.52, i.e. deshielded by 1.38 p.p.m. compared with stilbene. On multiplication of the down-field shift in (33) by two (1.30 p.p.m.), one obtains the chemical shift (τ 1.60) for the internal proton in (32), i.e. deshielded by 1.30 p.p.m. compared with stilbene. For both (31) and (32), the chemical shift of the external proton

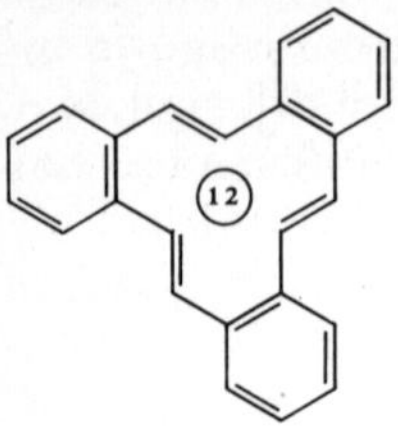

(29) 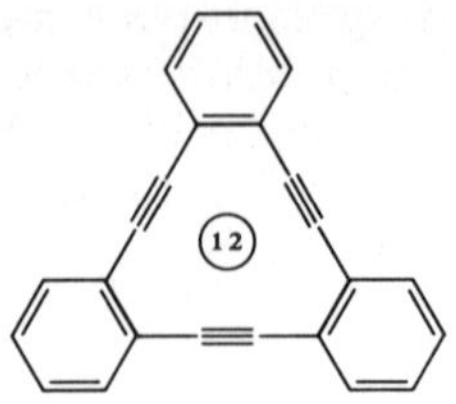 (30) 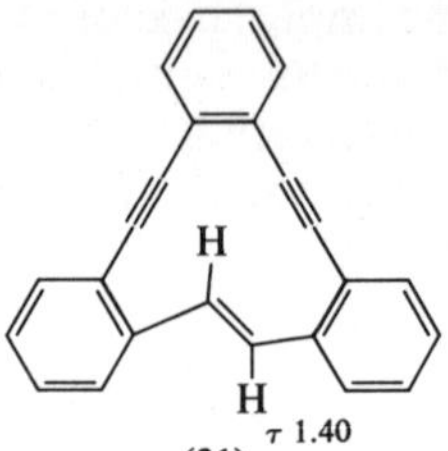 (31)

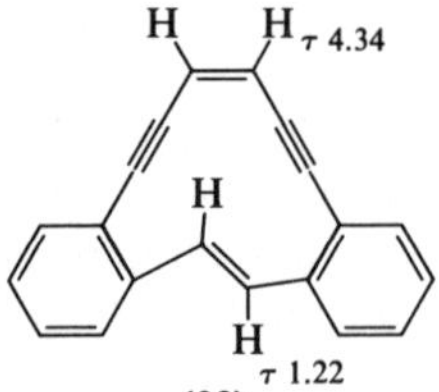

(32) 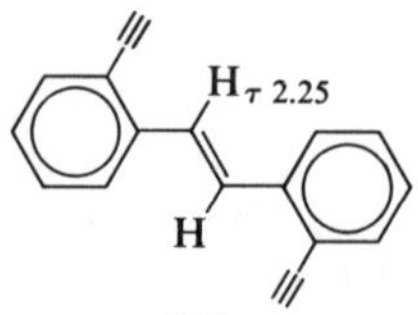 (33)

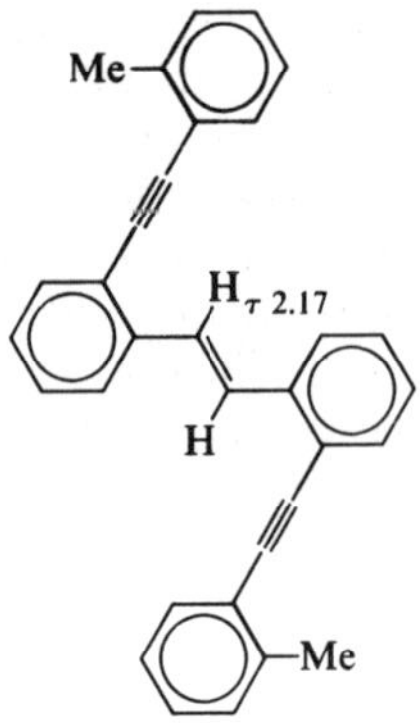

(34) 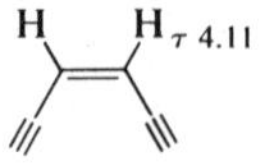 (35)

has to be taken as that of stilbene, τ 2.90. Thus one calculates chemical shifts of τ 2.21 and τ 2.25 for the exchanging protons in the atropic (31) and (32) respectively. However, the experimental values[52] are shifted down-field by 0.81 p.p.m. for (31) and 1.03 p.p.m. for (32). These down-field shifts may well be explained by the existence of paramagnetic ring currents. Thus (31) and (32) seem to be paratropic molecules. Of course, the extent of the paramagnetic ring current cannot be estimated because each pair of chemical shifts for the outer and inner protons can be chosen, provided they have equal distances from the observed average value. There is no doubt that fused benzene rings quench completely in some cases the ring currents of the parent annulenes; but this effect is dependent upon the type of perimeter (e.g. number of triple bonds) and the number of benzene rings. In qualitative agreement with this is the fact that the ring current intensity in (32) is higher than in (31), i.e. the exchanging *trans*-olefinic protons in (32) resonate at a lower field than those in (31). It is well known that internal protons are more affected by induced ring currents than external protons. For the parent compound 1,5-bisdehydro[12]annulene one can estimate the down-field shift of the internal proton ($\tau - 7.3$)[53] to be *c.* 10.2 p.p.m. (11.7 − 1.48 p.p.m. for deshielding by the triple bonds[54], referred to cyclohexene, τ 4.4), but the up-field shift of the outer protons is estimated to be only *c.* 0.7–1.7 p.p.m. (referred to (35)[51], τ 4.11). Therefore, the observed[52] chemical shift of the *cis*-olefinic protons in (32) (τ 4.34), referred to those in (35), does not prove the non-existence of a paramagnetic ring current, but indicates that it is diminished to some extent. Finally, the calculated down-field shift of 1.48 p.p.m.[54] for the internal proton in 1,5-bisdehydro[12]annulene caused by the triple bonds, is in very good agreement with the experimental values, obtained from models (33) and (34). A low-temperature ^{1}H n.m.r. study should verify this interpretation.

8.4.4 Higher-membered benzoannulenes

The syntheses and spectral properties of two fused 1,7,13-tridehydro[18]-annulenes have been reported recently[55]. The electronic spectra of (36) and (37) are very similar to that of the parent dehydro[18]annulene[56], but differ from that of (38) indicating cyclic π electron conjugation. However, in contrast to the ^{1}H n.m.r. spectrum of the diatropic parent compound, the ^{1}H n.m.r. spectra of (36) and (37) show no evidence for diamagnetic ring currents. This, of course, does not imply a contradiction, since the results of two different criteria, one of which depends also on the excited states of the molecules, are being compared with each other. In the cases of (36) and (37) the 'ring-current limiting' number of annelated benzene and naphthalene units respectively, and of triple bonds, is exceeded.

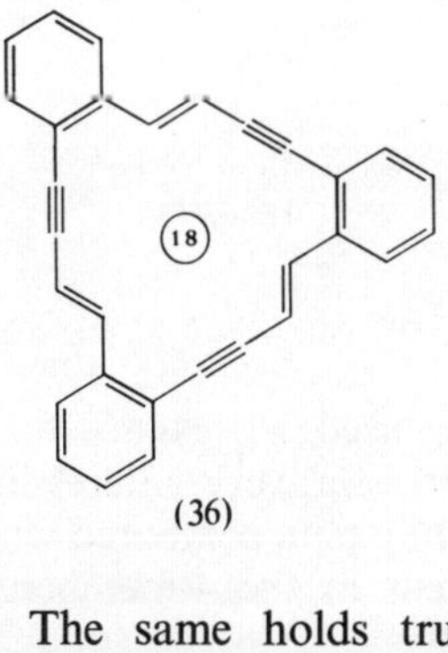

(18)

(36)

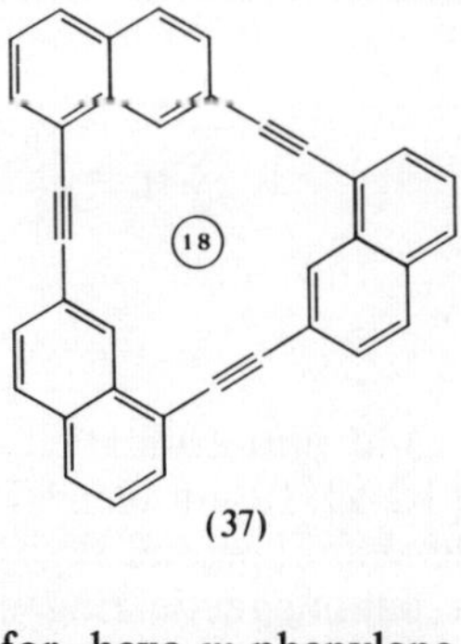

(18)

(37)

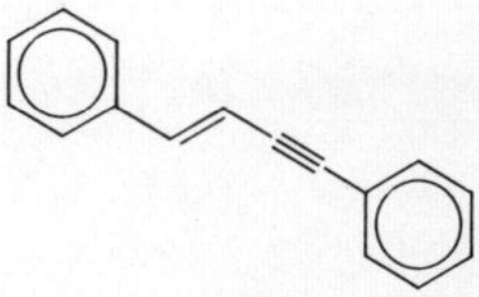

(38)

The same holds true for hexa-*m*-phenylene, a formal [18]annulene derivative for which x-ray analysis data are now available[57], and also for all fused derivatives of higher membered annulenes known to date[2].

8.4.5 Conclusion

There is experimental evidence for diatropic, paratropic and atropic fused annulenes. Annelation of benzenoid rings may nullify the intensity of the induced ring current or lower it to some extent. Delocalisation of π electrons may be restricted to fused 'sub-units'. Certainly, excess charges in the perimeter antagonise quenching of the ring current.

8.5 BRIDGED ANNULENES

Bridged annulenes will be discussed in this section in terms of the definition given by Jackman *et al.*[3]. Thus, so-called perturbed annulenes such as (39)[58] are not included since electronic interaction across the π system of the perimeter may occur.

8.5.1 [10]Annulenes

1,6-Bridged [10]annulenes are well known to be diatropic molecules[59]. Still one point of interest is the influence of bridge substituents on the hypothetical equilibrium[60] between 1,6-methano[10]annulene and bis-norcaradiene. From a series of 11-substituted compounds, the x-ray analysis of 11,11-difluoro-1,6-methano[10]annulene (40) has now been performed[61]. In accordance with predictions, (40) has an annulene perimeter. The study of diketo-derivatives[62] (41) and (42) and of their radical anions[63] throws an interesting light on to this problem.

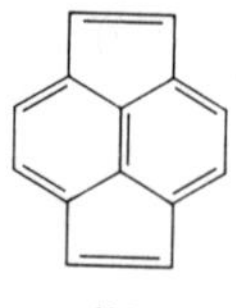

(39)

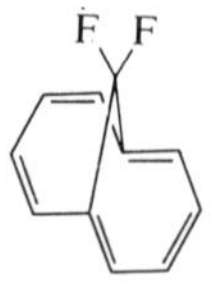

F F

(40)

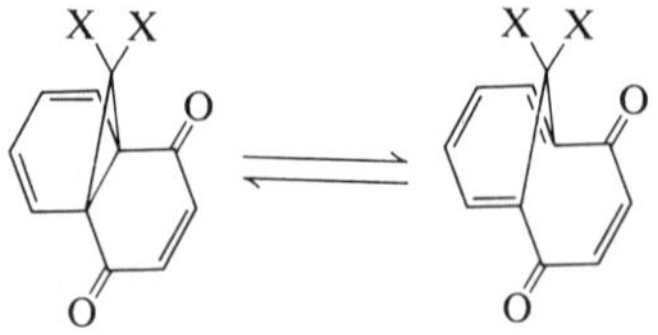

X X O O X X O O

(41) X = H, F

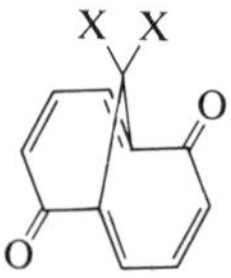

X X O O

(42) X = H, F

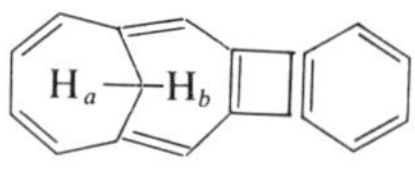

H_a — H_b

(43)

H_a : τ 9.71
H_b : τ 7.55

Regarding aromaticity and anti-aromaticity in fused π systems, 6,11-methano-cyclobutabenzo[1,2-a][10]annulene (43)[64] is a very interesting compound. The down-field shift of proton H_b by 2.16 p.p.m. compared with H_a was interpreted by a paramagnetic ring current in the 4-membered ring. It is one of the first $(4n_1+2)+4n_1+(4n_2+2)$ fused π systems ($n_1 = 1$, $n_2 > 1$; see also (93) and (94), Section 8.7.3) and obviously all three ring currents are operative[65].

8.5.2 [14]Annulenes

trans-15,16-Dimethyldihydropyrene turned out to be an interesting molecule with regard to symmetry conservation rules[66]. A useful alternative pathway for its synthesis has recently been found[67]. For steric reasons, the *cis* isomer (44) is less planar and less diatropic than the *trans* isomer[68].

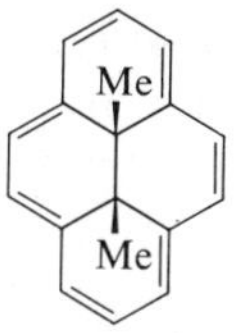

(44) 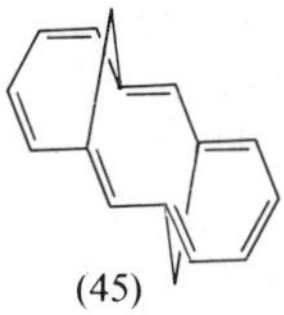 (45) 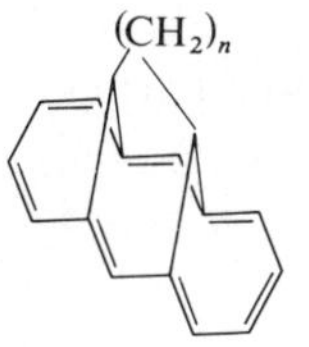 (46) $n = 1$ (47) $n = 2$ (48)

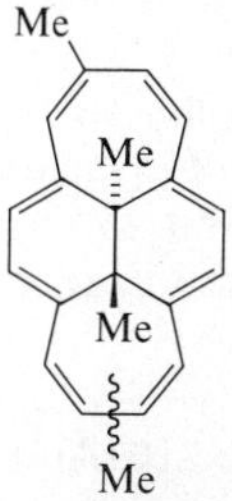

In the series of bridged [14]annulenes with anthracene-type perimeters, several x-ray analyses have been undertaken. The data verify the assumptions drawn from other criteria: The crystal structures of two derivatives[69, 70] of anti-1,6,8,13-bismethano[14]annulene (45) demonstrate clearly the bond alternation in this atropic molecule; the structure determinations of the diatropic [14]annulenes (46)[71] and (47)[72] show an increased bond alternation for (47) which is, however, far below that found for the above anti-bridged annulenes.

8.5.3 [16]Annulenes

Compound (48) seems to be the first bridged [16]annulene[73]. Its ^{1}H n.m.r. spectrum is highly complex and does not yet allow a definite interpretation. In contrast, its di-anion clearly sustains a diamagnetic ring current.

8.5.4 Conclusion

Bridged annulenes have proved to be excellent models for macrocyclic π systems provided the bridging groups do not cause a severe torsion of π orbitals, such as occurs in anti-bridged [14]annulenes.

8.6 ANNULENONES

Cyclopropenone, cyclopentadienone and cycloheptatrienone have been known for many years. These monocyclic, fully conjugated ketones (49) were named annulenones[74] and the first macrocycle was synthesised in 1968[74]. Recently, Sondheimer has reviewed the progress made in this field[4, 5].

The cyclic delocalisation of π electrons in annulenones is believed to depend on the extent of contribution of the mesomeric structure (50) to the

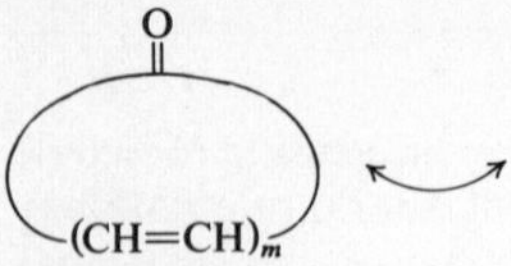

(49) 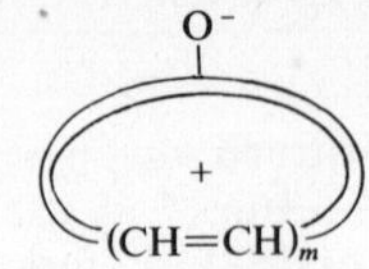 (50) 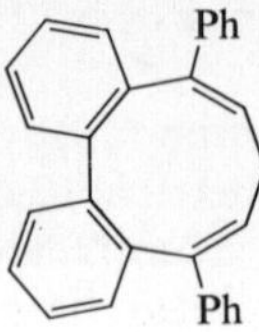 (51) 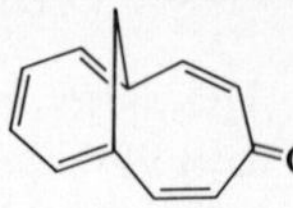 (52)

ground state of the molecule. In addition to the chemical shifts of peripheral protons, one expects the basicity of the carbonyl oxygen to be indicative of the contribution of structure (50). Protonation of the oxygen atom increases the importance of the mesomeric structure.

8.6.1 [9]Annulenones

The parent compound in the $(4n+1)$ π systems (49) ($m = 4$; $n = 2$) has not yet been synthesised, but the dibenzo derivative (51) has been reported recently[75]. The ^{1}H n.m.r. data (vinylic protons at τ 3.37) point to a non-planar atropic molecule. However, further information should be available from the ^{1}H n.m.r. spectrum of the protonated species.

8.6.2 [11]Annulenones

For 4,9-methano[11]annulenone (52)[76] an x-ray structure determination has been carried out[77]. The bond alternation is in a similar range to that found for the anti-bridged [14]annulenes (45) (Section 8.5.2). This indicates that (52) is best represented by the polyenone structure, a structure which has been concluded from ^{1}H n.m.r. spectroscopy[76].

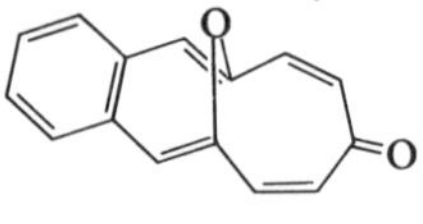

(53) 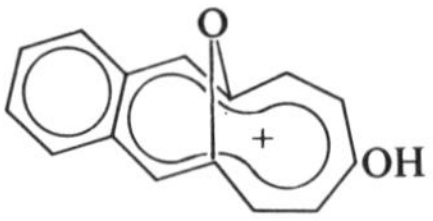 (54) 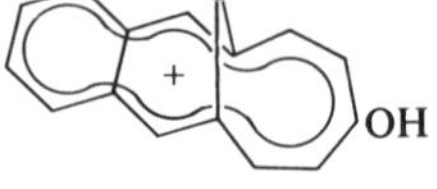 (55) 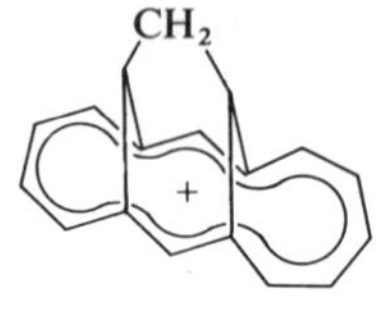 (56)

A very surprising observation has been made for the 6,7-benzo-4,9-oxido[11]annulenone (53)[78]. The ^{1}H n.m.r. spectra indicate the polyenone structure (53) in neutral solution and a diamagnetic ring current on protonation. However, in the electronic spectrum of protonated (53), the longest wavelength absorption exhibits a bathochromic shift of 165 nm compared with protonated (52). This behaviour was interpreted in terms of delocalisation in the C-14 perimeter (55) rather than in (54). Of course this conclusion needs additional support from a comparison with appropriate model compounds such as (56), which has recently been reported to be a diatropic molecule[79].

8.6.3 [13]Annulenones

The x-ray analysis data[80] for the first macrocyclic annulenone prepared, namely bis(tetramethylene)-2,6,8,12-tetradehydro[13]annulenone (57)[74,81] have been obtained. The extent of bond localisation is in accord with a very small paramagnetic up-field shift of the allylic protons[5,81], i.e. 0.35 p.p.m. compared to corresponding protons in the model compound (58) (Section 8.6.4).

8.6.4 [15]- and [17]Annulenones

The tertiary alcohol (58), obtained from 4,5-tetramethylene-2,6,8,14-tetradehydro[15]annulenone (59)[82] proved to be an excellent model, lacking a ring current, and therefore giving conclusive evidence for paramagnetic ring currents in (57) and in tetradehydro[17]annulenone[83], and for a diamagnetic ring current in (59).

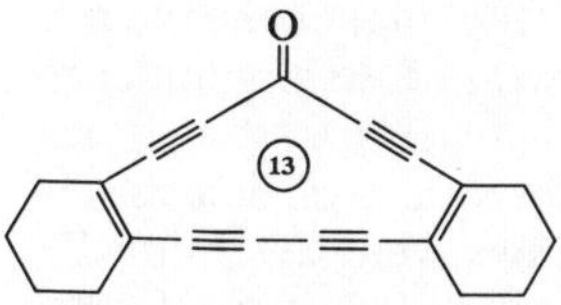

(57) 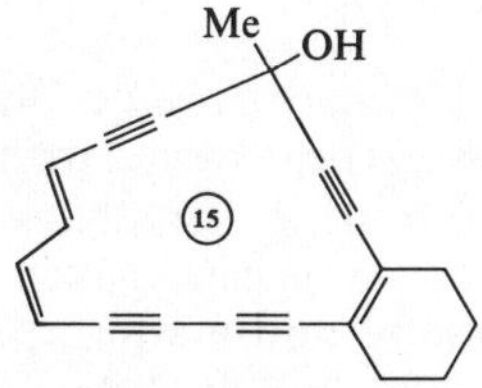 (58) 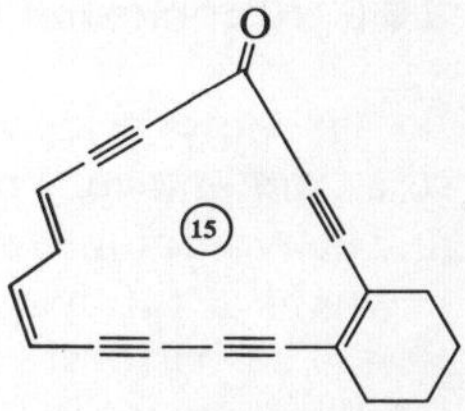 (59)

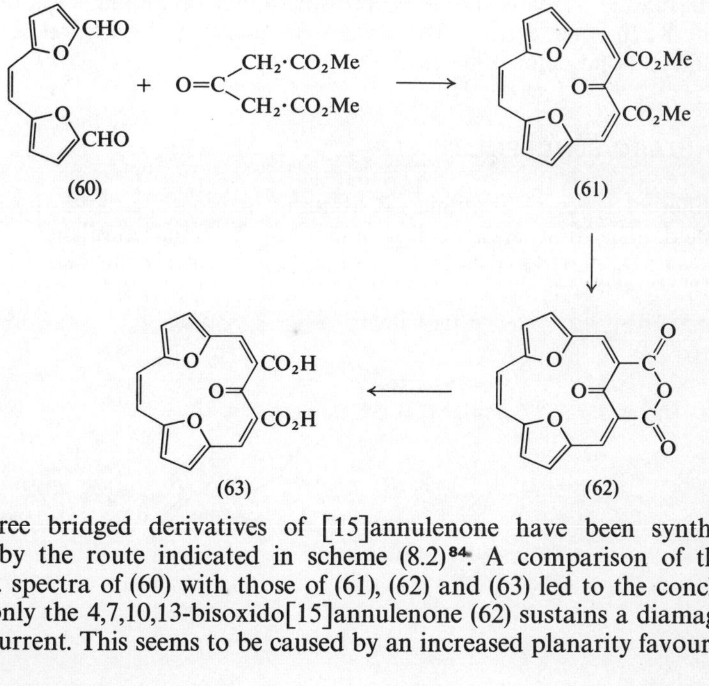

(60) (61) (8.2)

(63) (62)

Three bridged derivatives of [15]annulenone have been synthesised also by the route indicated in scheme (8.2)[84]. A comparison of the ^{1}H n.m.r. spectra of (60) with those of (61), (62) and (63) led to the conclusion that only the 4,7,10,13-bisoxido[15]annulenone (62) sustains a diamagnetic ring current. This seems to be caused by an increased planarity favoured by the anhydride substituent. An increase in ring-current intensity is found on protonation.

Recently the first oxygen-bridged [17]annulenone has been reported[85]. The trioxido[17]annulenone (64) exhibits up-field shifts of up to 0.85 p.p.m. for some protons compared with the linear ketone (65), and may therefore be quoted as a paratropic 16 π system.

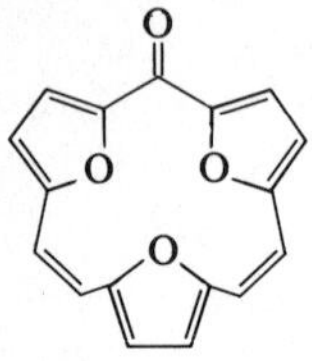

(64) 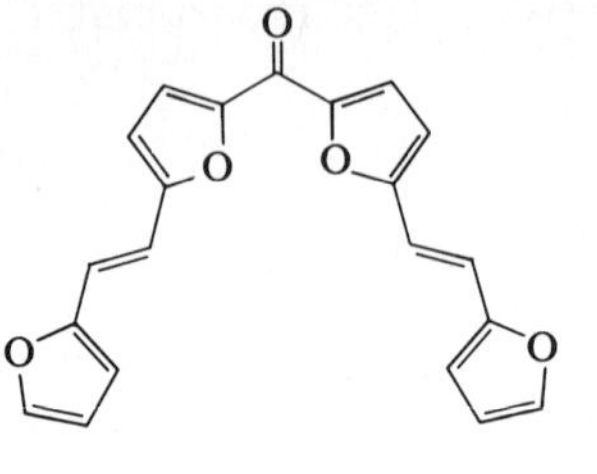 (65) 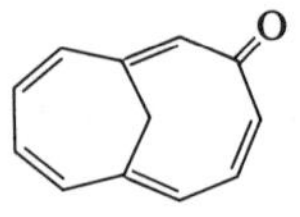 (66) 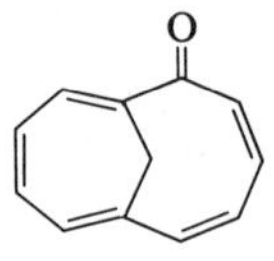 (67)

8.6.5 Conclusion

So far only bridged and dehydroannulenones in the higher-membered series are known. Their properties are in agreement with those of annulenes and dehydroannulenes, but the ring current intensities are less and sometimes negligible. Protonation of the oxygen atom increases the π electron delocalisation. The basicity gives useful information: pK_a values were determined[79] for the isomers (52), (66) and (67) as -0.6, $+0.5$ and -0.4, respectively, while those of ordinary carbonyl compounds are in the order of -6 to -7. Considering the observed influence of triple bonds on annulene perimeters (Section 8.2.3), one expects the parent annulenones to exhibit a more pronounced bond delocalisation.

8.7 HETEROANNULENES

Two different types of annulenes containing heteroatoms have to be distinguished. The first class, called π-excessive[86] heteroannulenes, covers heterocycles related to pyrrole, whereas the second class are the π-equivalent[86]

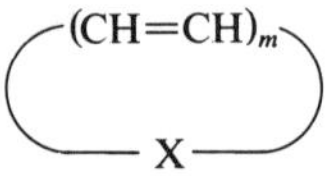

(68) X = NR, O, S 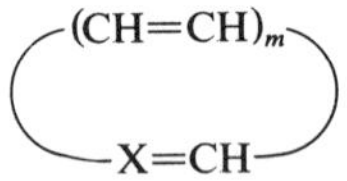 (69) X = N or O^+, etc. 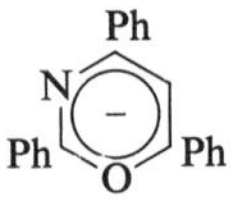 (70)  (71)

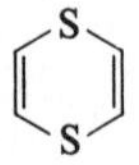

(72)  (73) X = NH, O, S 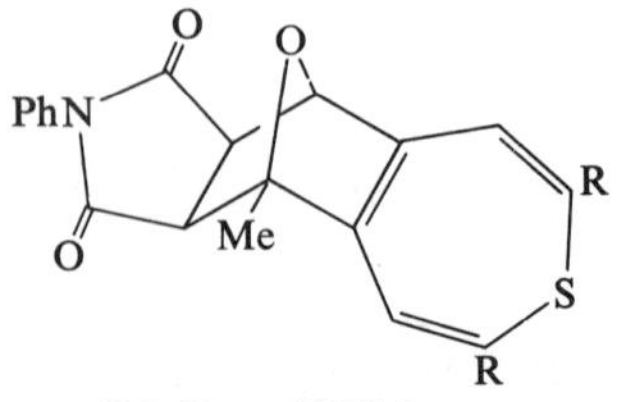 (74) R = $COHMe_2$ 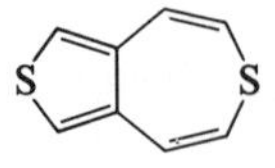 (75)

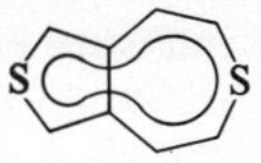

(76) 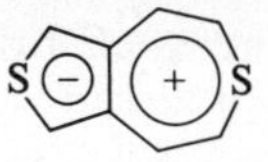 (77) 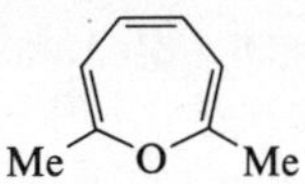 (78)

heteroannulenes related to pyridine. Also a combination of both types (68) and (69) may exist. Compounds for which m is an even number might be aromatic since they contain $(4n+2)$ π electrons. Conversely, when m is odd, the system possesses $4n$ π electrons and might be anti-aromatic.

The first attempted synthesis[87] of a heteroannulene was in 1957, but extensive investigations of heteroannulenes began only recently.

8.7.1 Hetero[7]annulenes

With the exceptions of intermediate (70), postulated very recently[88], the unstable 1,2-dithiins (71)[89] and the non-planar 1,4-dithiins (72)[90], the hetero-[7]annulenes (73) (X = NH,O) represent the smallest potential anti-aromatic 8 π systems[159]. However, they exhibit non-aromatic properties[91].

Thiepin (73) (X = S) is still unknown, but several derivatives have been reported. In accordance with calculations[92] for thiepin, the derivative (74) seems to have a paratropic 8 π system[93]. For the fused thiepin (75) a small positive resonance energy has been calculated[92]. The ^{1}H n.m.r. spectrum[94] and the x-ray analysis data[95] have been reported and the electronic structure has been described[94] in terms of the delocalised π system (77). Nevertheless, it seems doubtful whether conjugation of the type shown in the 12 π system (76) can be excluded.

The divergent properties of π-excessive and π-equivalent heteroannulenes have been demonstrated in an electron transfer reaction. 2,7-Dimethyloxepin (78), in contrast to azocines (see Section 8.7.2), does not undergo reduction to the 10 π electron oxepinyl di-anion[96].

8.7.2 Hetero[8]annulenes

The first heterocyclic analogue of COT, 2-methoxyazocine (79)[97], has been obtained[160]. Spectroscopic similarities with COT suggested that (79) is also a non-planar, non-aromatic polyene. Meanwhile, the properties of (79) and some derivatives have been studied extensively[98]. 2-Methoxyazocine, like COT, can be reduced to the diatropic 2-methoxyazocinyl di-anion. Polarographic studies demonstrate that (79) undergoes a two electron one-step reduction, whereas COT is reduced by two one-electron steps.

Diheterocines such as (80) may be formally regarded as 10 π systems. Synthesis of (81) and its thermolysis resulted in valence isomerisation to *N*,*N*-dimethylbenzo-1,6-dihydro-1,6-diazocine (82)[99]. The protons, attached to the 8-membered ring, show an absorption in the range typical of olefinic protons, indicating the atropic nature of the 10 π system. As expected from the higher electronegativity of oxygen, the dioxocine derivative (83) has also been shown to be atropic[100].

By analogy with furan, pyrrole and thiophene, one would expect cyclic delocalisation to be most important in the sulphur compound (80) (X = S).

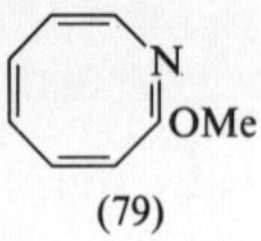

(79) 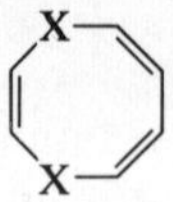 (80) X = NR, O, S 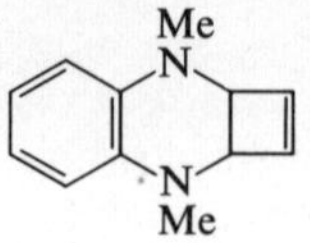 (81) 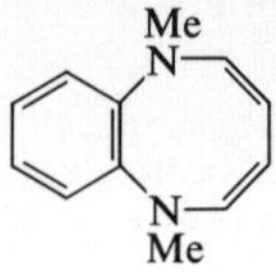 (82)

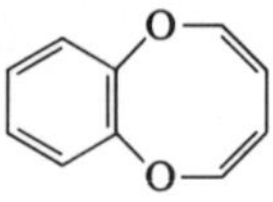

(83) 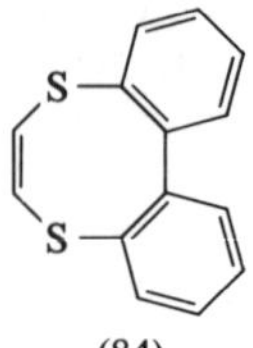 (84) 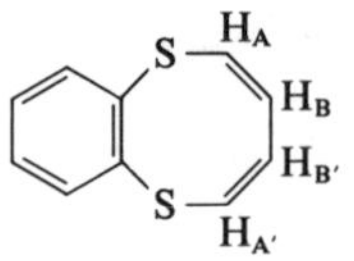 (85)

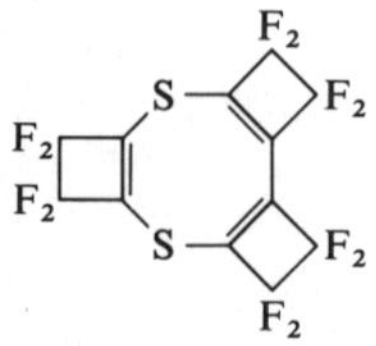

(86) 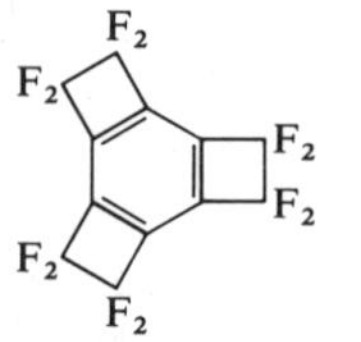 (87) 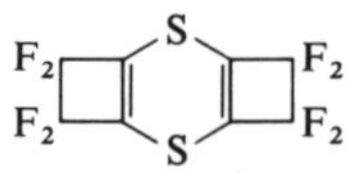 (88)

The first derivative of this sulphur heterocyclic compound reported[101] was the dibenzo-derivative (84). The chemical shift of τ 3.74 for the olefinic protons could be interpreted in terms of non-planarity of the heterocyclic ring, caused only by steric interaction of the benzene nuclei. The very similar resonance of respective protons in the recently described benzo derivative (85)[102] (τ 3.77), however, disproves this interpretation and suggests that these benzo-derivatives are non-aromatic molecules. The AA′BB′ multiplets of (85) have not yet been analysed and quite divergent coupling constants for (82) and (83) have been reported. Thus further information will be available from a detailed study of the coupling constants. Some evidence for a diamagnetic ring current in 1,4-dithiocine has been obtained now from the difluoromethylene derivative (86)[103]. The chemical shifts of the fluorine atoms are very similar to those in the benzene analogue (87), but deshielded by *c.* 12 p.p.m. compared to (88), which definitely is non-planar.

8.7.3 Hetero[9]annulenes

Hetero[9]annulenes (89) and (90) are of the π-excessive type (68) ($m = 4$) and possess a potentially aromatic 10 π system. The first two representatives prepared in this series were 4,5,6,7-dibenzoxonin (91) and 4,5,6,7-dibenzothionin (92)[104]. Their ^{1}H n.m.r. spectra are clearly indicative of atropic heterocyclic perimeters. The lack of ring current might have been a result of non-planarity caused by steric interaction and/or of quenching by the annelated benzene nuclei. Nevertheless, the study of oxonin (89, X = O)[105] and the thionin derivatives[106] (93) and (94) suggests that this presumption is probably incorrect. The ^{1}H n.m.r. spectra of all three indicate that their perimeters are non-planar and atropic.

The synthesis of 1-H azonin (90)[107] paved the way for the synthesis of a number of substituted azonins[108] via the potassium salt of (90) and an appropriate quencher.

(89) X = NR, O, S

(90)

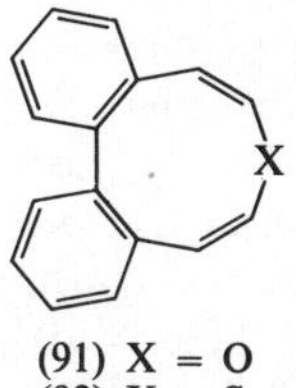

(91) X = O
(92) X = S

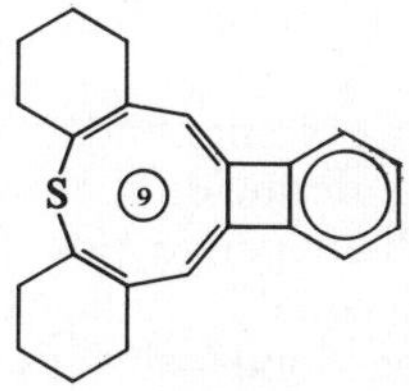

(93) 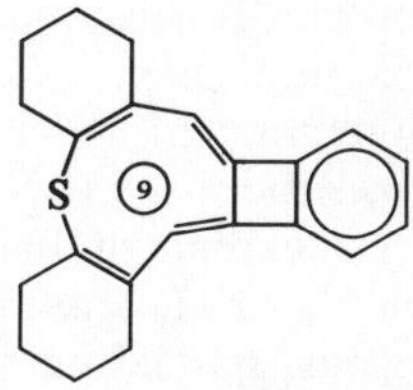 (94)

The electronic properties of these so-called heteronins (89) and (90) (only the parent sulphur heterocyclic compound is not yet known) have been discussed in terms of chemical shifts in the ^{1}H n.m.r. spectra and in terms of thermodynamic characteristics[108]. Considering the ^{1}H n.m.r. spectra, only 1-H azonin (90) is a diatropic 10 π system. The lack of deshielding, especially of α-hydrogens in *N*-alkyl substituted azonins (89, X = NR), is thought to be caused by distortions of these protons out of the molecular plane, arising from non-bonded interactions with substituents[108]. 1-H Azonin and its alkyl substituted derivatives (89, X = NR) were differentiated from oxonin and azonins substituted by electron withdrawing groups (such as CO_2Et[109]) not only by a consideration of the ^{1}H n.m.r. spectra, but also by a consideration of the thermodynamic properties[108]. No doubt there is some parallelism. Nevertheless, reactivity and thermodynamic properties are characteristics of the ground state *and* of the transition state. Therefore, all attempts to *classify* different species with the help of such properties are always arbitrary and only ground state properties are pertinent to aromaticity.

A recent communication[110] concerning the way in which ring anisotropies affect the chemical shift of a standard, has already prompted the investigation[111] of the heteronins (89, X = O,NMe) and (90) by this new technique. In agreement with the conclusion drawn from chemical-shift data, 1-H azonin is characterised by a large positive S term[112] ($S(0^0) = +1.35$), indicating strong diamagnetism, whereas oxonin has even a slightly negative S term ($S(0^0) = -0.07$), indicative of a small paramagnetism, perhaps resulting from homoconjugation. For (89, X = NMe), a small positive value ($S(0^0) = +0.34$) was found, pointing to a small diamagnetism. These values compare reasonably well with those of pyrrole, furan and benzene[110, 111]. The method appears to be promising[113].

8.7.4 Hetero[10]annulenes

The early synthesis[87] of the fused 1,6-diaza[10]annulene (95) was attempted in the hope that replacement of the two =CH— groups, bearing the two internal hydrogens in (4) (Section 8.2.1), by nitrogen would result in (95). In such a molecule, the repulsion of the nitrogen lone pairs might be expected to be less than the non-bonded interaction of the internal hydrogens in (4). However, the synthesis was unsuccessful.

Only two hetero[10]annulenes seem to be known to date. The tetra-alkylated 1,6-dithia[10]annulene (97) was obtained[114] by the route indicated in scheme (8.3). It is a π-excessive annulene, possessing a 12 π electron system which, however, is presumably non-planar. The ^{1}H n.m.r. spectrum is that of an atropic molecule.

The second 10-membered heterocyclic compound (98)[115] exhibits very interesting spectroscopic properties. The deshielding of the olefinic protons (τ 1.56–3.22) is best explained by a diamagnetic ring current, especially since either proton H_A or H_B shows two equal couplings of 7.5 Hz with two neighbouring protons, indicating strong bond delocalisation. Although a bicyclic delocalised structure (99) might be discussed, the macrocyclic delocalised annulenone perimeter (100) seems most likely. If so, (100) is the first example of a hetero[10]annulenone. An x-ray analysis should give the definite answer to this interesting question, which is directly connected with the question concerning the lone-pair repulsion in a *trans*-configuration such as in (95).

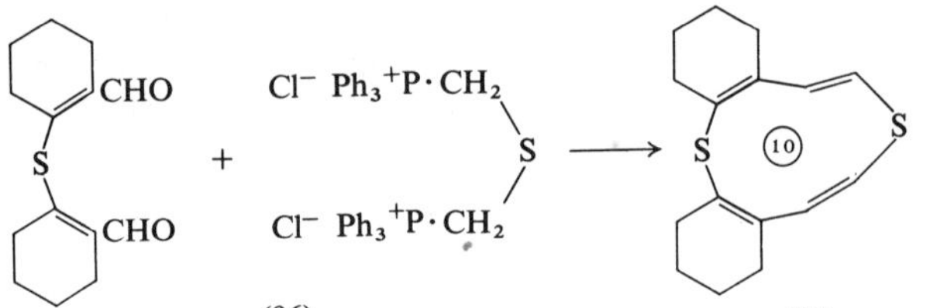

CHO S CHO + $Cl^- Ph_3{}^+P \cdot CH_2$ S $Cl^- Ph_3{}^+P \cdot CH_2$ ⟶ S (10) S (8.3)

(96) (97)

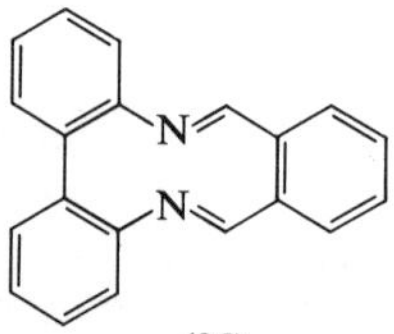

(95)

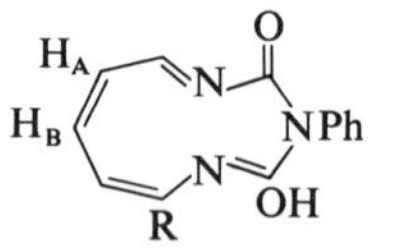

H_A H_B N O NPh N OH R

(98) R = $CH_2 \cdot CO_2Me$

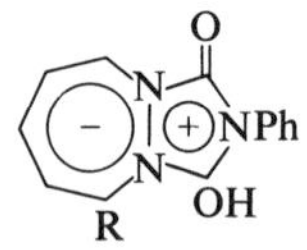

N O NPh N OH R

(99)

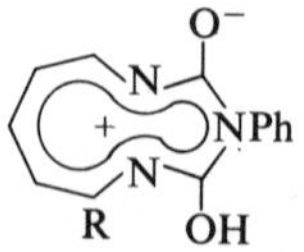

N O^- + NPh N R OH

(100)

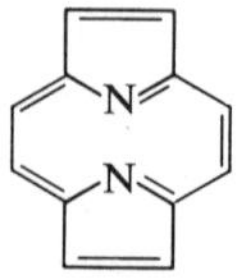

(101)

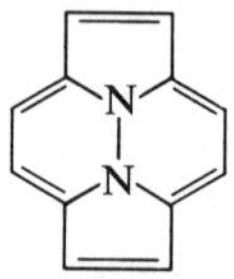

(102)

In this context, the extremely interesting heterocyclic compounds (101) and (102) should be mentioned. Since the synthesis of this π system has been reported[116], there is still no answer to the question, of whether the molecule is best represented as diaza[10]annulene (101), or as the *N*-bridged [12]annulene (102).

8.7.5 Hetero[11]-, hetero[12]-and hetero[13]annulenes

The dialdehyde (96) (equation (8.3)) proved useful for the synthesis of several interesting sulphur heterocyclic compounds. The reactions of (96) with the four Wittig reagents, shown in scheme (8.4), led to the thia-annulenes (103)[117], (104)[118], (105)[117] and (106)[117].

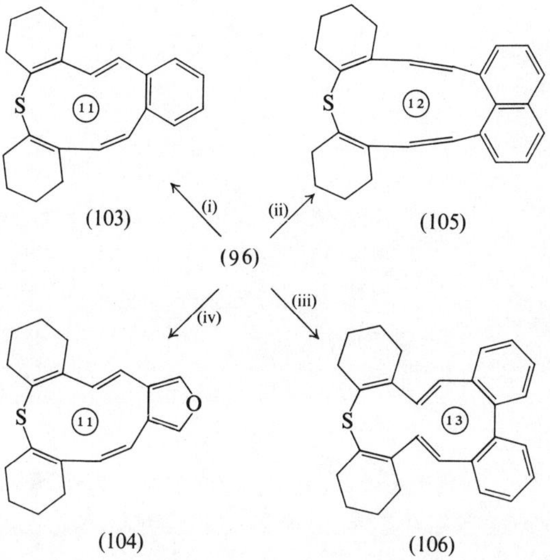

(103) (105) (96) (104) (106) (8.4)

Reagents

(i) $CH_2{\cdot}P^+Ph_3\ Br^-$ / $CH_2{\cdot}P^+Ph_3\ Br^-$

(ii) $CH_2{\cdot}P^+Ph_3\ Br^-$ / $CH_2{\cdot}P^+Ph_3\ Br^-$

(iii) $CH_2{\cdot}P^+Ph_3\ Br^-$ / $CH_2{\cdot}P^+Ph_3\ Br^-$

(iv) $CH_2{\cdot}P^+Ph_3\ Cl^-$ / $CH_2{\cdot}P^+Ph_3\ Cl^-$

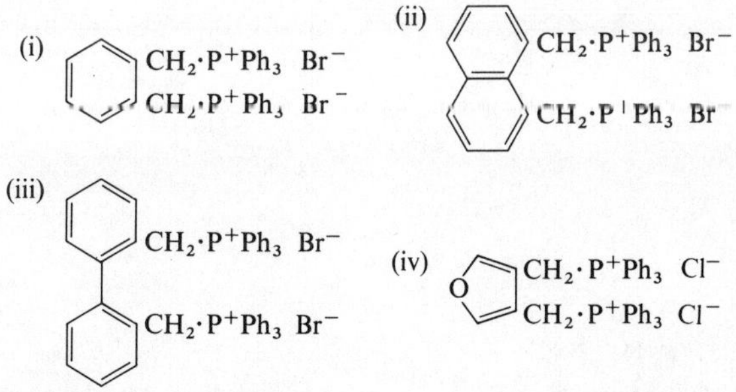

That the atropic character of the 12 π system in heterocyclic compound (103) is not caused by quenching of the ring current by the fused benzene nucleus, but rather by a non-planar perimeter, was demonstrated by the synthesis of (104) and conversion of this compound to the formally conjugated

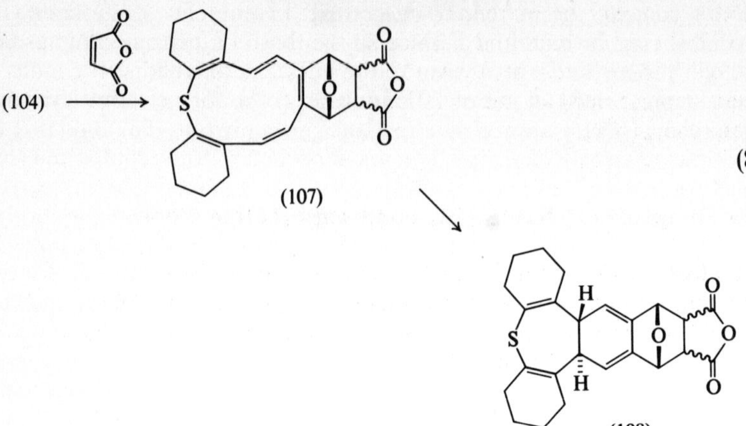

(104) → (107) (8.5)

(107) → (108)

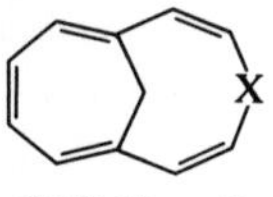

(109) X = O
(110) X = S

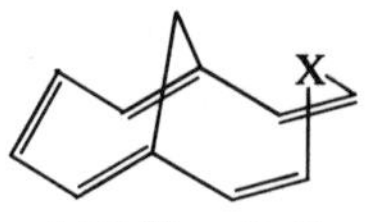

(111) X = O, S

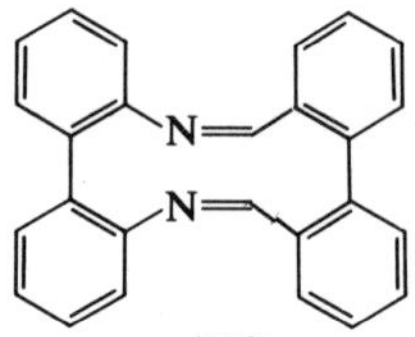

(112)

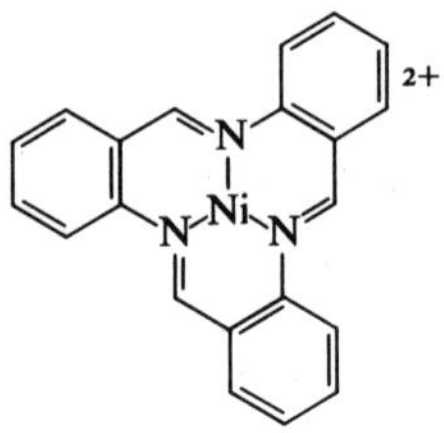

(113)

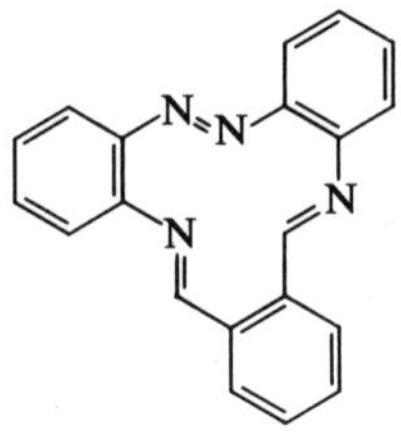

(114)

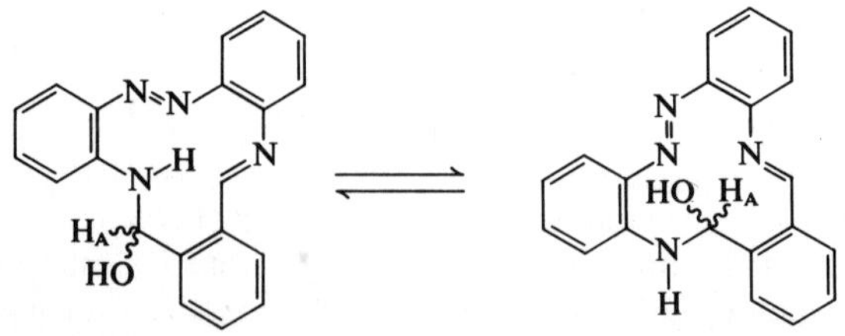

(115) ⇌ (116)

(107)[5, 118] (scheme (8.5)). Mono-*trans*-thia[11]annulene (107) shows no appreciable ring current, and because of the lack of a benzene ring, present in (103), it readily undergoes valence isomerisation to (108).

The chemical shifts of the olefinic protons in the 13 π electron system of (105) served as a very appropriate reference not only for (103) and (107) but also for the potentially diatropic 14 π system of heterocyclic compound (106). The ^{1}H n.m.r. spectrum of (106) does not indicate a perceptible ring current.

Similar observations were made recently for the first known bridged hetero[11]annulenes[79]: 4,9-Methano-oxa[11]annulene (109) and 4,9-methano-thia[11]annulene (110). From ^{1}H n.m.r. investigations, it was concluded that (109) and (110) hold the non-planar conformations shown in (111) and are therefore atropic molecules.

The first known metal-free nitrogen heterocyclic molecule in the 12-membered series was tetrabenzodiaza[12]annulene (112)[119]. Its electronic spectrum demonstrated that the 12-membered perimeter is definitely non-planar. The 12-membered nitrogen heterocyclic compound shown in (113) can be isolated only in the form of metal complexes from the metal template self-condensation of *o*-aminobenzaldehyde[120]. The x-ray analysis data for a Ni^{II} [121] and for a Co^{III} complex[122] demonstrate that the ligands have a localised 12 π system.

A tetra-aza[12]annulene, analogous to the ligand in (113), has been synthesised quite recently from 2,2′-diaminoazobenzene and phthalaldehyde. Surprisingly enough, only the carbinolamine (115) has been isolated[123] and not the fully conjugated annulene (114). Its ^{1}H n.m.r. spectrum is of the mobile type. At *c.* $-40\,°C$ both conformers (115) and (116) can be observed. The chemical shifts of proton H_A, which is an outer hydrogen in (115) but an inner one in (116) (the geometry of the molecule is very similar to that of [12]annulene[16]), indicate no cyclic delocalisation of the 12 π system, a process which would have to occur via homoconjugation (see Section 8.7.6).

8.7.6 Hetero[14]-, hetero[15]- and hetero[16]annulenes

The first aza[14]annulenes to be prepared are of the π-excessive type. Dibenzotetra-aza[14]annulene (117)[124] and several tetra-alkylated octa-aza-[14]annulenes of type (118) reported recently[125, 126] possess a potentially paratropic 16 π electron system; but so far no conclusions can be drawn from their spectral properties regarding the electronic structure.

More information has been obtained regarding the fused hexa-aza[14]-annulene (120). As in the case of tetra-aza[12]annulene (114), the mono-carbinolamine (120) has been isolated[127] and not the expected annulene (119). The fascinating question arises concerning the reasons for the stability of carbinolamines (115) and (120). Carbinolamines are usually unstable intermediates in the formation or the hydrolysis of Schiff bases. However, they or their ethers have been isolated from the azomethine (121)[128] and from metal complexes such as (122)[129] and (123)[130]. The stability of the carbinolamine ethers (124)[127, 131] is analogous to that of the corresponding derivatives of (121). These compounds bear NO_2 groups in a position

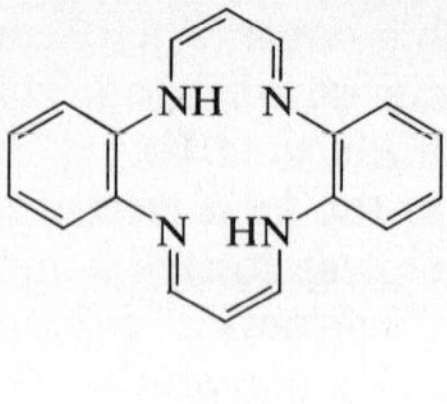

(117)

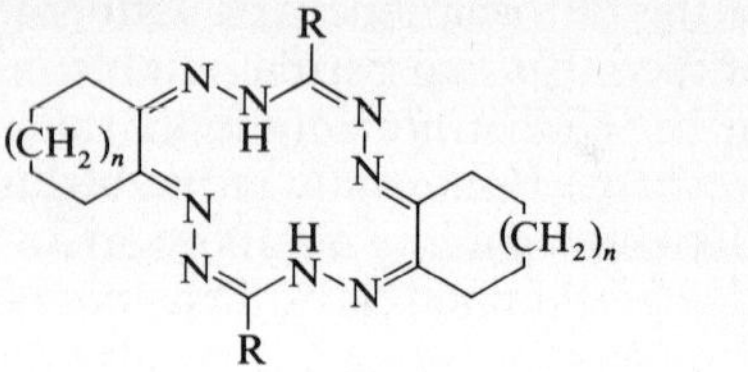

(118) n = 0–2, R = e.g.H

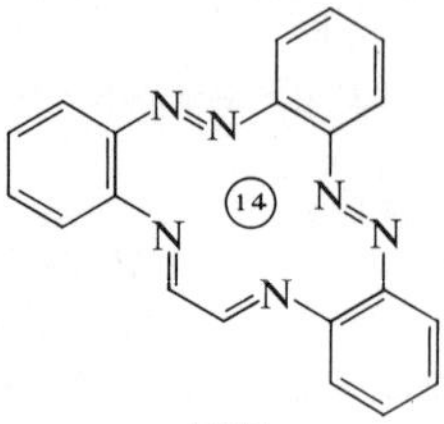

(119)

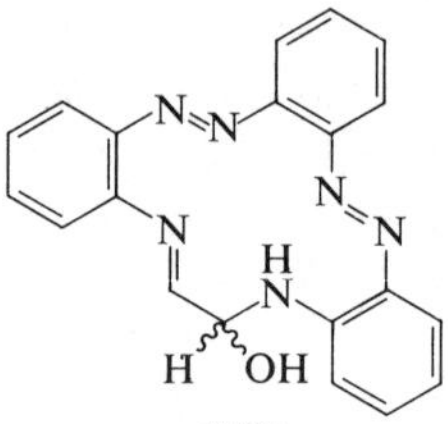

(120)

ortho or *para* to the aniline nitrogen, and therefore the increased electrophilicity of the —N=CH— carbon atoms is believed to be responsible for the stabilisation of their carbinolamines. Similarly, stabilisation of carbinolamines of (122) and (123) has been attributed to the electronic and steric requirement[4] of the metal ion.

Spectroscopic investigations indicate[127], that the stability of (120) is a result of the properties of the planar 14 π electron system. The carbinolamine CH proton in (120) is deshielded by *c.* 3.5 p.p.m. compared with the respective protons in conformers (115), (116) and (124). Since such severe *local* deshielding effects can be excluded from an inspection of molecular models, the deshielding suggests an induced diamagnetic ring current via homoconjugation

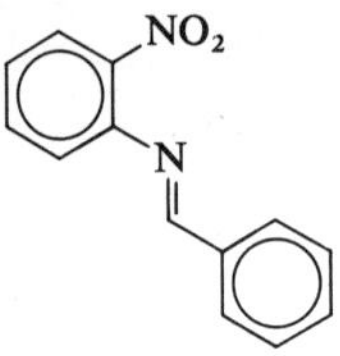

(121)

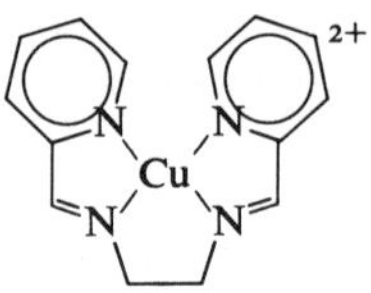

(122)

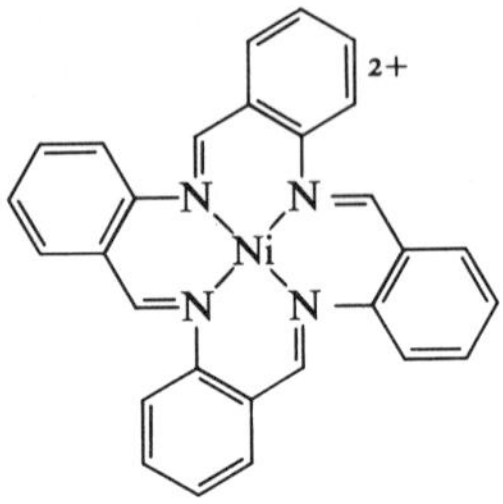

(123)

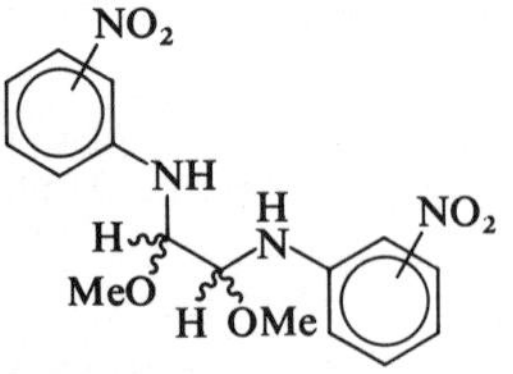

(124) $-NO_2$ = *o*, *p*

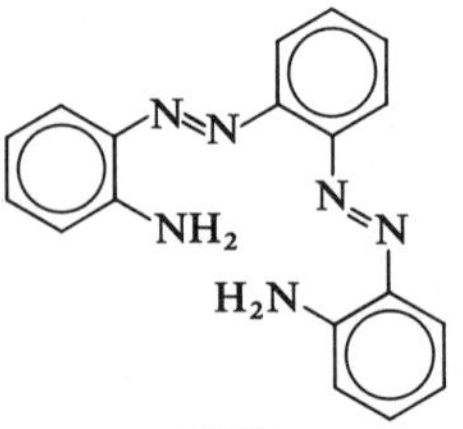

(125)

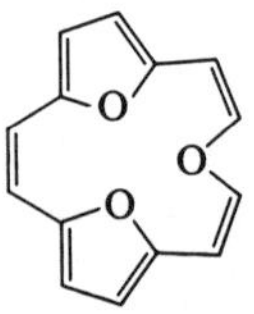

(126)

of the nitrogen lone pair with the C=N double bond. The chemical shifts of the other two perimeter protons support this interpretation; (120) seems to be the first *neutral* homoaromatic molecule detected by ^{1}H n.m.r. spectroscopy. Homoconjugation was further supported by a comparison of the electronic spectra of (120) and (125)[127]. From these results it appears that (120) is stabilised by cyclic delocalisation of the 14 π electron system and hence can be isolated as a fairly stable molecule.

At present it seems that the following reason may hold for the stabilisation of the carbinolamine (115). Namely that, compared to [12]annulene (or tribenzo[12]annulene (29)), the non-bonded interaction of three internal protons is replaced by the interaction of one internal proton with two nitrogen lone pairs in the hypothetical tetra-aza[12]annulene (114). Therefore, (114) could hold a more planar conformation than (29) resulting in a larger anti-aromatic destabilisation. This destabilisation can be avoided in the carbinolamine (115) since it holds a far less planar conformation which does not favour homoconjugation as the planar conformation of (120) does.

In the 15- and 16-membered series of heteroannulenes only 1-oxa-4,7,10,13-dioxido-cyclotetradecaheptaene (126)[161] and the above mentioned 16 π electron system (123) seem to be known. The ^{1}H n.m.r. spectrum of (126) indicates the existence of a paramagnetic ring current. Structure (123), like (113), is formed in the template self-condensation of *o*-aminobenzaldehyde and cannot be isolated without the metal ion[132].

8.7.7 Hetero[17]annulenes

Photolysis of the cyclo-octatetraene dimers (127) led to the hetero[17]-annulenes (128)[118] (see scheme (8.6))[133]. In spite of the potentially diatropic 18 π electron systems, their ^{1}H n.m.r. spectra indicate non-delocalised polyolefinic structures, an observation which parallels that made for the 9-membered series (Section 8.7.3), in which only 1-H azonin is diatropic. The configurations of (128) are not yet known.

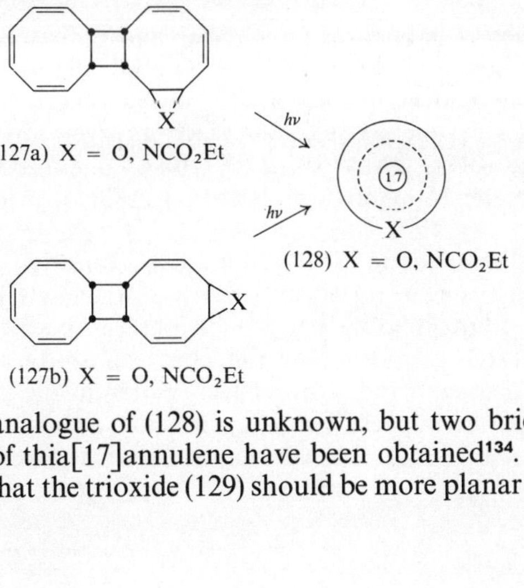

(127a) X = O, NCO_2Et

(128) X = O, NCO_2Et (8.6)

(127b) X = O, NCO_2Et

The sulphur analogue of (128) is unknown, but two bridged derivatives (129) and (130) of thia[17]annulene have been obtained[134]. Stuart–Briegleb models suggest that the trioxide (129) should be more planar than the dioxide

(130) with one thiophene unit. Nevertheless, ^{1}H n.m.r. spectra clearly demonstrate that both compounds possess non-delocalised, atropic 18 π electron systems. Obviously this bond localisation is caused only by the peripheral heteroatom, because the analogous bridged [18]annulenes (131) are well known to be diatropic annulenes[135].

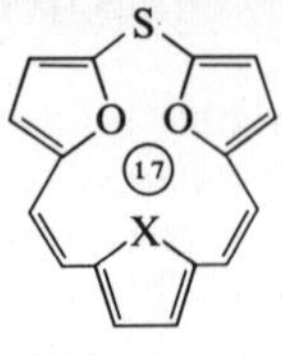

(129) X = O
(130) X = S

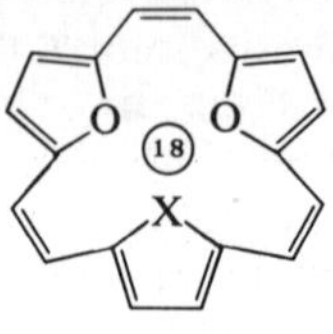

(131)

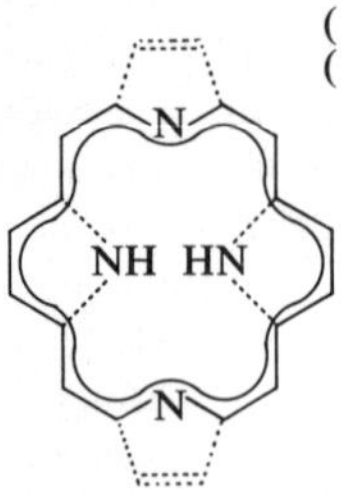

(132)

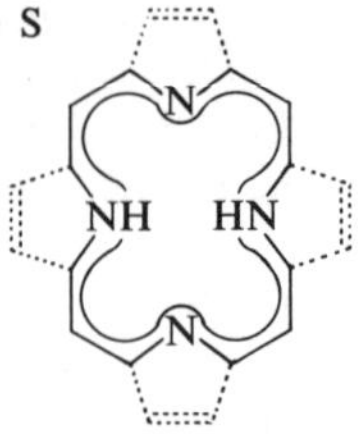

(133)

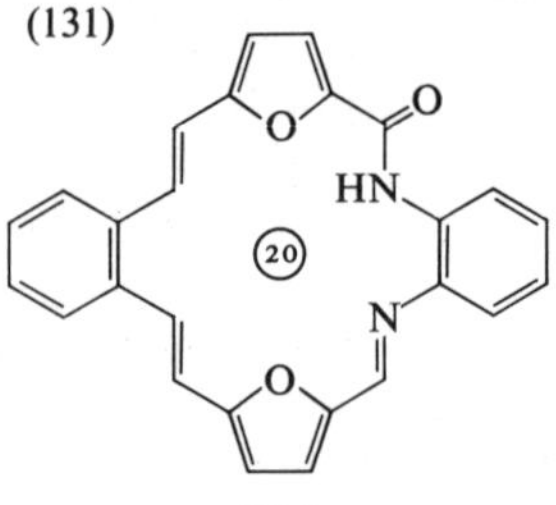

(134)

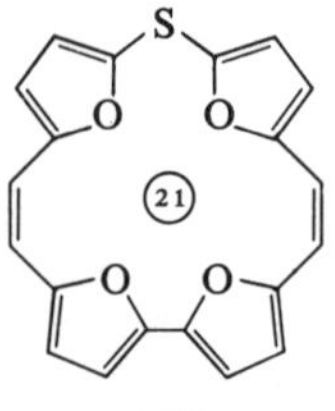

(135)

8.7.8 Higher-membered heteroannulenes

The large number of papers on porphyrins and phthalocyanins cannot be discussed in the scope of this chapter[162]. It might be mentioned, however, that recently porphyrins and their analogues have been considered[3] still as [18]annulene-like perimeters (132) with two nitrogen atoms in the path of conjugation. However, x-ray studies[136] have demonstrated that cyclic delocalisation in the 16-membered perimeter (133) (18 π 12C4N) is at least as important.

The diaza[20]annulenone (134) is the most interesting compound of the derivatives that have been reported[137] in the 20-membered series. From the ^{1}H n.m.r. data obtained so far no conclusions can be drawn regarding its electronic structure. According to the chemical shifts and the coupling constants, the 21-membered π-excessive thia-annulene (135)[134] definitely does not sustain a diamagnetic ring current; such a ring current is not to be expected, since it is not found even for the 17-membered analogues (129) and (130).

8.7.9 Conclusion

In contrast to the well known sequence of thiophene, pyrrole and furan (in the order of decreasing aromaticity), the data available at present demonstrate that there is no parallelism to be found in the higher membered series of heteroannulenes. Under the 9-membered heteronins, only 1-H azonin (90) sustains a perceptible ring current.

None of the reported thia-annulenes, neither in the $(4n+2)$ nor in the $4n$ series, shows a ^{1}H n.m.r. spectrum which indicates a ring current. The lack of ring current in the $4n$ π systems may be explained simply by the fact that they avoid the expected *destabilisation* by holding non-planar, strain-free conformations. However, what prevents the expected *stabilisation* in the $(4n+2)$ π systems? From ring current calculations[138], it is known that quenching of the ring current by bond alternation is particularly important in the higher annulenes; at a certain degree of bond alternation, increasing quenching occurs with increasing ring size. Obviously in thionin (89, X = S) and in higher membered thia-annulenes bond alternation, particularly in the C=C—S—C=C part of the perimeter, overcomes the effect of the smaller electronegativity of sulphur, which is the predominant factor in the 5-membered series. Thus it seems very likely that nitrogen plays an outstanding role as a heteroatom in macrocyclic perimeters; because of the reasonable C—N bond length and the smaller electronegativity of nitrogen compared with that of oxygen, the nitrogen lone pair can delocalise very readily*. Some preliminary evidence supporting these expectations is found in the 10 π system (100) and in the 14 π system (120). Good evidence comes from the studies of the electronic structure of porphyrins[136]. The cyclic delocalisation may become even more favoured by the presence of further =N— atoms in the perimeter, as is known for the 5-membered nitrogen heterocycles. The necessity of delocalisation of *N*-lone pairs obviously resembles the necessity of charge dispersal in carbocyclic annulenes.

Finally, in connection with triaza[10]annulenone (100), recent investigations[139] on 2-pyridone should be mentioned. From a consideration of tautomeric equilibria of 2-pyridone with 2-hydroxypyridine and a model compound, the aromatic resonance energy of 2-pyridone was estimated to be 25 kcal mol^{-1}. If one finds appropriate model compounds, the resonance energy of macrocyclic annulenes such as (100) or the 14-membered aza-annulenone (137)[140] might be determined also.

8.8 HOMOAROMATIC π SYSTEMS

Evidence for homoaromaticity has been adduced for a number of compounds[141]. Bishomoaromatic species became known only recently; the bicyclo[4.3.0]nonatrienyl cation (138)[142] and the 1,6-methano-bridged bis-

*In excellent agreement with these expectations are the properties of the methylene-bridged dehydroaza[17]annulenes (136), which have been synthesised very recently[163]. The diamagnetic ring current intensity decreases in the indicated order of substituents. The ring current found for (136) (R = CO_2Et) demonstrates that the atropic nature of the parent [17]annulene (128, X = NCO_2Et) must be caused by a non-planarity of its perimeter.

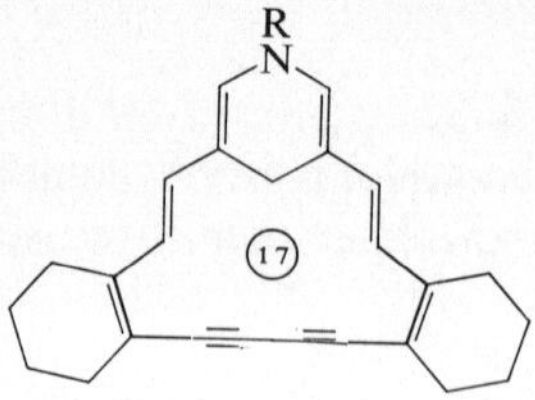

(136) R = Me, H, CO_2Et, COMe

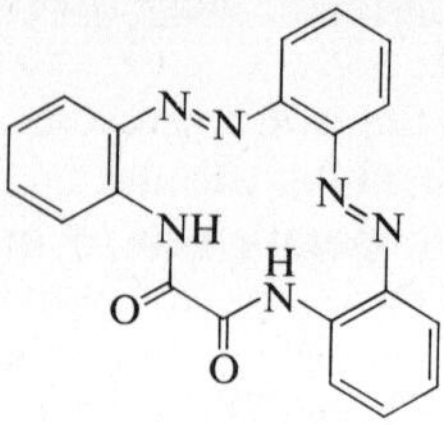

(137)

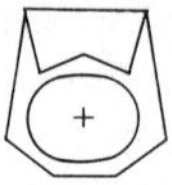

(138) 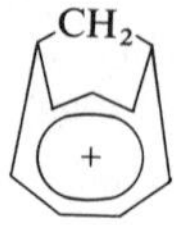 (139) 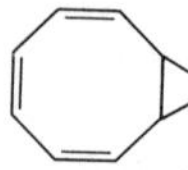 (140) 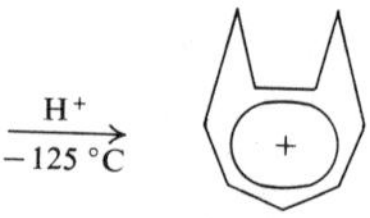 (141)

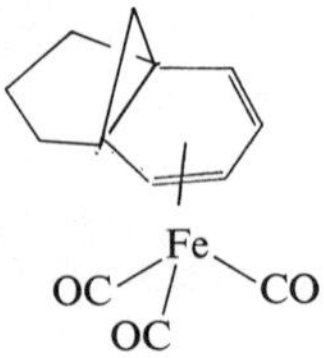

(142) 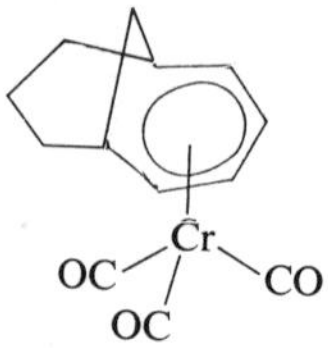 (143) 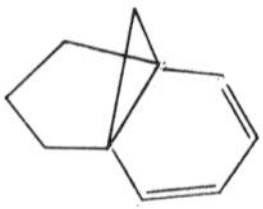 (144)  (145)

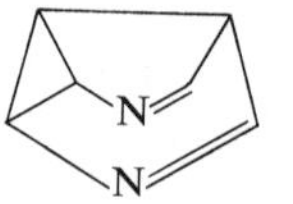

(146)  (147)

homotropylium ion (139)[143–145] are particularly good examples. The 1,3-bishomotropylium cation (141)[146] generated by protonation of (140) with FSO_3H/SO_2ClF, is the first known monocyclic bishomotropylium species.

Several metal carbonyl complexes of homoaromatic species have been reported[147]. The tricarbonyl-iron and -chromium complexes[148] (142) and (143) demonstrate the influence of complex formation on electronic structures. Both complexes were obtained from ligand (144). Coupling constants and chemical shifts in the ^1H n.m.r. spectra[148] and x-ray analysis data[149] indicate that in the complex formed with the dienophile $Fe(CO)_3$, the norcaradiene structure of the ligand (144) is retained, but the trienophile $Cr(CO)_3$ converts the ligand into a homoaromatic species.

Proton shifts are the most accessible data for the discussion of homoaromatic species. Recently, however, a criterion for homoconjugation has been tested which, because of its sensitivity, opens up new dimensions for the discussion of electronic structures. No evidence for homoconjugative interaction of double bonds in *cis-cis-cis*-1,4,7-cyclononatriene (145) has been obtained from ^1H n.m.r.[150] and x-ray analysis data[151] simply because of the small extent of this interaction. The analysis of the photo-electron spectrum[152] of (145) has now demonstrated clearly the small extent of these interactions. Unfortunately there are some limitations[152] to this otherwise useful method.

Finally, the interesting results of MINDO/2 calculations for the diazasemibullvalene (146) should be mentioned. These results[153] suggest that (146) has a bishomoaromatic ground state (147).

8.9 MOEBIUS TYPE PERIMETERS

The concept of Möbius-type arrays of π orbitals was introduced in 1964[154]. In contrast to Hückel-type perimeters, a cyclic delocalisation of $4n$ π electrons should result in stabilisation and that of $(4n+2)$ π electrons in destabilisation in such systems. This concept has been shown to be very useful for photochemical reactions[155]. Recently the temperature-dependence of the ^{1}H n.m.r. spectrum of the photolysis product from *cis*-bicyclodecatriene (148) was interpreted in terms of a rapid valence tautomerism between the cyclodecatetraene (149) and *trans*-bicyclodecatriene (150) via an 'aromatic' 8 π Möbius perimeter[156] (scheme (8.7)). However, the synthetic approach to a stable $4n$ π system is still missing.

hν

(148) (149) (150)

Acknowledgements

I wish to thank Prof. H. Zollinger to whom I owe my interest in this fascinating field of chemistry, and Professors A. G. Anastassiou, F. Sondheimer and E. Vogel for the communication of unpublished work. Professor Sondheimer was kind enough to read the manuscript and I appreciate his stimulating comments.

References

1. Garratt, P. J. and Sargent, M. V. (1969). *Advan. Org. Chem. M.*, **6,** 1
2. Figeys, H. P. (1969) in Lloyd, D., *Topics in Carbocyclic Chemistry,* I, 269, (London: Logos Press)
3. Haddon, R. C., Haddon, V. R. and Jackman, L. M. (1971). *Fortschr. Chem. Forsch.*, **16,** 103
4. Sondheimer, F. (1970). *International Symposium on the Chemistry of Nonbenzenoid Aromatic Compounds,* 24–2. (Sendai, Japan, August 24th–28th); (1971). *Pure Appl. Chem.*, **28,** 331
5. Sondheimer, F. (1972). *Accounts Chem. Res.*, **5,** 81
6. Lloyd, D. (1966). *Carbocyclic Non-Benzenoid Aromatic Compounds,* (Amsterdam: Elsevier Publishing Company); Badger, G. M. (1969). *Aromatic Character and Aromaticity,* (Cambridge: University Press)
7. Chemical Society (1967). *Aromaticity,* Special Publication No. 21, (London: The Chemical Society)
8. Snyder, J. P. (1969). *Nonbenzenoid Aromatics, I,* (New York: Academic Press) and (1971). Vol. 2

9. For a discussion of criteria of aromaticity, see Ref. 3, p.112; and *The Jerusalem Symposia on Quantum Chemistry and Biochemistry* (1971). Vol. 3, *Aromaticity, Pseudo-Aromaticity, Anti-Aromaticity,* (Jerusalem: Jerusalem Academic Press)
10. Dewar, M. J. S. and Gleicher, G. J. (1965). *J. Amer. Chem. Soc.,* **87,** 685
11. For a summary see Burkoth, T. L. and van Tamelen, E. E.; Ref. 8, Vol. 1, p.63
12. Masamune, S., Hojo, K., Hojo Kiyomi, Bigam, G. and Rabenstein, D. L. (1971). *J. Amer. Chem. Soc.,* **93,** 4966
13. Van Tamelen, E. E., with Pappas, B. C. T., Burkoth, T. L. and Greeley, R. H. (1971). *Chem. Commun.,* 601; *J. Amer. Chem. Soc.,* **93,** 6111, 6120
14. Alder, R. W. and Edley, D. T. (1971). *J. Chem. Soc. C,* 3485
15. Darby, N., Kim, C. U., Salaün, J. A., Shelton, K. W., Takada, S. and Masamune, S. (1971). *Chem. Commun.,* 1516
16. Oth, J. F. M. (1971). *Pure Appl. Chem.,* **25,** 573 and references therein
17. Gilles, J.-M., Oth, J. F. M., Sondheimer, F. and Woo, E. P. (1971). *J. Chem. Soc. B,* 2177
18. McQuilkin, R. M. and Sondheimer, F. (1970). *J. Amer. Chem. Soc.,* **92,** 6341
19. McQuilkin, R. M., Metcalf, B. W. and Sondheimer, F. (1971). *Chem. Commun.,* 338
20. Leznoff, C. C. and Sondheimer, F. (1967). *J. Amer. Chem. Soc.,* **89,** 4247
21. Metcalf, B. W. and Sondheimer, F. (1971). *J. Amer. Chem. Soc.,* **93,** 5271
22. Sondheimer, F. and Wolovsky, R. (1962). *J. Amer. Chem. Soc.,* **84,** 260
23. Sondheimer, F. and Gaoni, Y. (1962). *J. Amer. Chem. Soc.,* **84,** 3520
24. Metcalf, B. W. and Sondheimer, F. (1971). *J. Amer. Chem. Soc.,* **93,** 6675
25. Stöckel, K. and Sondheimer, F. (1972). *J. Chem. Soc. Perkin I,* 355
26. For nomenclature see Winstein, S., Ref. 7, p.5
27. Katz, T. J. and Garratt, P. J. (1963). *J. Amer. Chem. Soc.,* **85,** 2852; LaLancette, E. A. and Benson, R. E. (1963). ibid., **85,** 2853
28. Barborak, J. C., Su, T.-M., Schleyer, P.v.R., Boche, G. and Schneider, G. (1971). *J. Amer. Chem. Soc.,* **93,** 279
29. Anastassiou, A. G. and Yakali, E. (1972). *Chem. Commun.,* 92
30. LaLancette, E. A. and Benson, R. E. (1965). *J. Amer. Chem. Soc.,* **87,** 1941
31. Katz, T. J. (1960). *J. Amer. Chem. Soc.,* **82,** 3784, 3785
32. For a recent review see Garratt, P. J. and Sargent, M. V., Ref. 8, Vol. 2, p.207
33. Oth, J. F. M. and Schröder, G. (1971). *J. Chem. Soc. B,* 904
34. For notation see Ref. 16
35. Garratt, P. J., Rowland, N. E. and Sondheimer, F. (1971). *Tetrahedron,* **27,** 3157
36. McQuilkin, R. M., Garratt, P. J. and Sondheimer, F. (1970). *J. Amer. Chem. Soc.,* **92,** 6682
37. See the discussion in Ref. 3, p.190
38. Figeys, H. P. (1966). *Tetrahedron Letters,* 4625
39. Figeys, H. P. and Dralants, A. (1971). *Tetrahedron Letters,* 3901
40. Anet, F. A. L., Bourn, A. J. R. and Lin, Y. S. (1964). *J. Amer. Chem. Soc.,* **86,** 3576
41. Garratt, P. J. and Knapp, K. A. (1970). *Chem. Commun.,* 1215
42. Grohmann, K. and Sondheimer, F. (1967). *J. Amer. Chem. Soc.,* **89,** 7119
43. Bindra, A. P., Elix, J. A. and Sargent, M. V. (1971). *Aust. J. Chem.,* **24,** 1721
44. Bindra, A. P., Elix, J. A. and Sargent, M. V. (1969). *Aust. J. Chem.,* **22,** 1449 and (1968). *Tetrahedron Letters,* 4335
45. Mitchell, R. H. and Sondheimer, F. (1970). *Tetrahedron,* **26,** 2141
46. Staab, H. A., Ipaktschi, J. and Nissen, A. (1971). *Chem. Ber.,* **104,** 1182
47. Whitlock, H. W., Jr. and Reed, J. K. (1969). *J. Org. Chem.,* **34,** 874
48. Staab, H. A., Graf, F., Doerner, K. and Nissen, A. (1971). *Chem. Ber.,* **104,** 1159 and earlier communications
49. Staab, H. A. and Graf, F. (1970). *Chem. Ber.,* **103,** 1107
50. Irngartinger, H., Leiserowitz, L. and Schmidt, G. M. J. (1970). *Chem. Ber.,* **103,** 1119
51. Okamura, W. H. and Sondheimer, F. (1967). *J. Amer. Chem. Soc.,* **89,** 5991
52. Staab, H. A. and Bader, R. (1970). *Chem. Ber.,* **103,** 1157
53. Untch, K. G. and Wysocki, D. C. (1967). *J. Amer. Chem. Soc.,* **89,** 6386
54. Pople, J. A. and Untch, K. G. (1966). *J. Amer. Chem. Soc.,* **88,** 4811
55. Endo, K., Sakata, Y. and Misumi, S. (1970). *Tetrahedron Letters,* 2557; (1971). *Bull. Chem. Soc. Jap.,* **44,** 2465
56. Sondheimer, F. with Jackman, L. M., Amiel, Y., Ben-Efraim, D. A., Gaoni, Y., Wolovsky, R. and Bothner-By, A. A. (1962). *J. Amer. Chem. Soc.,* **84,** 260 and 4307

57. Irngartinger, H., Leiserowitz, L. and Schmidt, G. M. J. (1970). *Chem. Ber.*, **103,** 1132
58. Trost, B. M., Bright, G. M., Frihart, Ch. and Brittelli, D. (1971). *J. Amer. Chem. Soc.*, **93,** 737
59. Vogel, E., Ref. 7, p.113
60. Vogel, E. (1968). *Proc. Robert A. Welch Found. Conf. Chem. Res., XII. Organic Synthesis* (Houston, Texas, 1968), 215
61. Gramaccioli, C. M. and Simonetta, M. (1971). *Tetrahedron Letters,* 173
62. Vogel, E., with Lohmar, E., Böll, W. A., Söhngen, B., Müllen, K. and Günther, H. (1971). *Angew. Chem.*, **83,** 401, 403
63. Gerson, F., Müllen, K. and Vogel, E. (1971). *Helv. Chim. Acta,* **54,** 1046
64. Garratt, P. J. and Vollhardt, K. P. C. (1971). *Angew. Chem.*, **83,** 111
65. For a detailed discussion of biphenylenes see Ref. 2, p.319
66. Schmidt, W. (1971). *Helv. Chim. Acta,* **54,** 862
67. Blaschke, H., Ramey, C. E., Calder, I. and Boekelheide, V. (1970). *J. Amer. Chem. Soc.*, **92,** 3675
68. Mitchell, R. H. and Boekelheide, V. (1970). *Chem. Commun.*, 1555
69. Barrow, M. J. and Mills, O. S. (1971). *Chem. Commun.*, 220
70. Grammaccioli, C. M., Mimun, A., Mugnoli, A. and Simonetta, M. (1971). *Chem. Commun.*, 796
71. Casalone, G., Gavezzotti, A., Mugnoli, A. and Simonetta, M. (1970). *Angew. Chem.*, **82,** 516
72. Gramaccioli, C. M., Mugnoli, A., Pilati, T., Raimondi, M. and Simonetta, M. (1971). *Chem. Commun.*, 973
73. Mitchell, R. H. and Boekelheide, V. (1970). *Chem. Commun.*, 1557
74. Pilling, G. M. and Sondheimer, F. (1968). *J. Amer. Chem. Soc.*, **90,** 5610
75. Rabinovitz, M., Bergmann, E. D. and Gazit, A. (1971). *Tetrahedron Letters,* 2671
76. Grimme, W., Reisdorff, J., Jünemann, W. and Vogel, E. (1970). *J. Amer. Chem. Soc.*, **92,** 6335
77. Hudson, D. W. and Mills, O. S. (1971). *Chem. Commun.*, 153
78. Ogawa, H., Kato, H. and Yoshida, M. (1971). *Tetrahedron Letters,* 1793
79. Vogel, E. (1971). *XXIII. International Congress of Pure and Applied Chemistry* (Boston), Abstract of Papers p. 17
80. Duckworth, V. F., Hitchcock, P. B. and Mason, R. (1971). *Chem. Commun.*, 963
81. Pilling, G. M. and Sondheimer, F. (1971). *J. Amer. Chem. Soc.*, **93,** 1977
82. Cotterrell, G. P., Mitchell, G. H., Sondheimer, F. and Pilling, G. M. (1971). *J. Amer. Chem. Soc.*, **93,** 259
83. Brown, G. W. and Sondheimer, F. (1969). *J. Amer. Chem. Soc.*, **91,** 760
84. Ogawa, H., Shimojo, N. and Yoshida, M. (1971). *Tetrahedron Letters,* 2013. Two unsubstituted isomeres have been reported very recently by Ogawa, H., Yoshida, M. and Saikachi, H. (1972). *Tetrahedron Letters,* 153
85. Cresp, T. M. and Sargent, M. V. (1971). *Chem. Commun.*, 1457
86. Anderson, A. G., Jr., Harrison, W. F. and Anderson, R. G. (1963). *J. Amer. Chem. Soc.*, **85,** 3448 (footnote 9)
87. Blood, A. E. and Noller, C. R. (1957). *J. Org. Chem.*, **22,** 873
88. Schmidt, R. R. (1971). *Angew. Chem.*, **83,** 622
89. Schroth, W., Billig, F. and Reinhold, G. (1967). *Angew. Chem.*, **79,** 685
90. Parham, W. E., Wynberg, H., Hasek, W. R., Howell, P. A., Curtis, R. M. and Lipscomb, W. N. (1954). *J. Amer. Chem. Soc.*, **76,** 4957
91. For reviews see: Paquette, L. A., in Ref. 8, Vol. 1, p.249; Vogel, E. and Günther, H. (1967). *Angew. Chem.*, **79,** 429
92. Dewar, M. J. S. and Trinajstić, N. (1970). *J. Amer. Chem. Soc.*, **92,** 1453
93. Hoffman, J. M., Jr. and Schlessinger, R. H. (1970). *J. Amer. Chem. Soc.*, **92,** 5263
94. Schlessinger, R. H. and Ponticello, G. S. (1968). *Tetrahedron Letters,* 3017; see also (1969). ibid. 4361
95. Sakore, T. D., Schlessinger, R. H. and Sobell, H. M. (1969). *J. Amer. Chem. Soc.*, **91,** 3995
96. Paquette, L. A. and McCreadie, T. (1971). *J. Org. Chem.*, **36,** 1402
97. Paquette, L. A. and Kakihana, T. (1968). *J. Amer. Chem. Soc.*, **90,** 3897; Elix, J. A., Wilson, W. S. and Warrener, R. N. (1970). *Tetrahedron Letters,* 1837

98. Paquette, L. A., with Kakihana, T., Hansen, J. F., Philips, J. C. and Anderson, L. B. (1971). *J. Amer. Chem. Soc.,* **93,** 152, 161, 168, 174
99. Shue, H.-J. and Fowler, F. W. (1971). *Tetrahedron Letters,* 2437
100. Schroth, W. and Werner, B. (1967). *Angew. Chem.,* **79,** 684
101. Schroth, W., Billig, F. and Zschunke, A. (1969). *Z. Chem.,* **9,** 184
102. Coffen, D. L., Poon, Y. C. and Lee, M. L. (1971). *J. Amer. Chem. Soc.,* **93,** 4627
103. Riley, M. O. and Park, J. D. (1971). *Tetrahedron Letters,* 2871
104. Bindra, A. P., Elix, J. A., Garratt, P. J. and Mitchell, R. H. (1968). *J. Amer. Chem. Soc.,* **90,** 7372
105. Anastassiou, A. G. and Cellura, R. P. (1969). *Chem. Commun.,* 903; Masamune, S., Takada, S. and Seidner, R. T. (1969). *J. Amer. Chem. Soc.,* **91,** 7769; Anastassiou, A. G. and Cellura, R. P. (1969). *Chem. Commun.,* 1521; Holovka, J. M., Grabbe, R. R.. Gardner, P. D., Strow, C. B., Hill, M. L. and Van Auken, T. V. (1969). ibid., 1522
106. Garratt, P. J., Holmes, A. B., Sondheimer, F. and Vollhardt, K. P. C. (1970). *J. Amer. Chem. Soc.,* **92,** 4492
107. Anastassiou, A. G. and Gebrian, J. H. (1970). *Tetrahedron Letters,* 825
108. Anastassiou, A. G., Eachus, S. W., Cellura, R. P. and Gebrian, J. H. (1970). *Chem. Commun.,* 1133
109. Anastassiou, A. G. and Gebrian, J. H. (1969). *J. Amer. Chem. Soc.,* **91,** 4011; Masamune, S., Hojo, K. and Takada, S. (1969). *Chem. Commun.,* 1204
110. Anet, F. A. L. and Schenck, G. E. (1971). *J. Amer. Chem. Soc.,* **93,** 556
111. Anastassiou, A. G. and Yamamoto, H. (1972). *Chem. Commun.,* 286
112. For definition of the S term see Ref. 110
113. For a detailed discussion of total anisotropies and their several terms see Ref. 3, p.115 and references therein
114. Garratt, P. J., Holmes, A. B., Sondheimer, F. and Vollhardt, K. P. C. (1971). *Chem. Commun.,* 947
115. Sasaki, T., Kanematsu, K. and Hayakawa, K. (1971). *J. Chem. Soc. C,* 2142
116. Paudler, W. W. and Stephan, E. A. (1970). *J. Amer. Chem. Soc.,* **92,** 4468
117. Holmes, A. B. and Sondheimer, F. (1970). *J. Amer. Chem. Soc.,* **92,** 5284
118. Holmes, A. B. and Sondheimer, F. (1971). *Chem. Commun.,* 1434
119. Bindra, A. P. and Elix, J. A. (1970). *Tetrahedron,* **26,** 3749
120. Melson, G. A. and Busch, D. H. (1965). *J. Amer. Chem. Soc.,* **87,** 1706
121. Fleischer, E. B. and Klem, E. (1965). *Inorg. Chem.,* **4,** 637
122. Wing, R. M. and Eiss, R. (1970). *J. Amer. Chem. Soc.,* **92,** 1929
123. Blumer, M. (1972). *Ph.D. thesis* ETH, Zurich, in preparation
124. Hiller, H., Dimroth, P. and Pfitzner, H. (1968). *Justus Liebigs Ann. Chem.,* **717,** 137; Jäger, E.-G. (1969). *Z. Anorg. Allg. Chem.,* **364,** 177
125. Neunhoeffer, H. and Motitschke, L. (1970). *Tetrahedron Letters,* 655
126. Neunhoeffer, H., Stastny, J. and Motitschke, L. (1971). *Tetrahedron Letters,* 1601
127. Skrabal, P. and Zollinger, H. (1971). *Helv. Chim. Acta,* **54,** 1069
128. Smith, D. M., with Marshall, R. and Sears, D. J. (1970). *J. Chem. Soc. C,* 2144; (1971). ibid., 3510
129. Busch, D. H. and Bailar, J. C., Jr. (1956). *J. Amer. Chem. Soc.,* **78,** 1137; Cressey, M., McKenzie, E. D. and Yates, S. (1971). *J. Chem. Soc. A,* 2677
130. Taylor, L. T., Urbach, F. L. and Busch, D. H. (1969). *J. Amer. Chem. Soc.,* **91,** 1072
131. Skrabal, P., unpublished results
132. Cummings, S. C. and Busch, D. H. (1971). *Inorg. Chem.,* **10,** 1220
133. Schröder, G., Heil, G., Neuberg, R., Plinke, G. and Oth, J. F. M. (1971). *XXIII. International Congress of Pure and Applied Chemistry* (Boston), Abstract of Papers, p.37
134. Cresp, T. M. and Sargent, M. V. (1971). *Chem. Commun.,* 1458
135. Badger, G. M., with Elix, J. A., Lewis, G. E. and Singh, V. P. (1966). *Aust. J. Chem.,* **19,** 1221, 1461
136. Webb, L. E. and Fleischer, E. B. (1965). *J. Amer. Chem. Soc.,* **87,** 667; *J. Chem. Phys.,* **43,** 3100
137. Saikachi, H., Ogawa, H. and Sato, K. (1971). *Chem. Pharm. Bull.,* **19,** 97
138. For a detailed discussion see Ref. 3, p.119
139. Cook, M. J., Katritzky, A. R., Linda, P. and Tack, R. D. (1971). *Chem. Commun.,* 510
140. Heberlein, M. (1972). *Ph.D. thesis* ETH, Zurich, in preparation
141. For a review and the definition of homoaromaticity see Winstein, S. (1969). *Quart. Rev. Chem. Soc.,* **23,** 141; see also Ref. 3, p.201

142. Ahlberg, P., Harris, D. L. and Winstein, S. (1970). *J. Amer. Chem. Soc.*, **92,** 4454
143. Schröder, G., Prange, U., Bowman, N. S. and Oth, J. F. M. (1970). *Tetrahedron Letters,* 3251
144. Roberts, M., Hamberger, H. and Winstein, S. (1970). *J. Amer. Chem. Soc.*, **92,** 6346
145. Schröder, G., Prange, U., Putze, B., Thio, J. and Oth, J. F. M. (1971). *Chem. Ber.*, **104,** 3406
146. Warner, P. and Winstein, S. (1971). *J. Amer. Chem. Soc.*, **93,** 1284
147. See for example: Aumann, R. and Winstein, S. (1970). *Tetrahedron Letters,* 903
148. Bleck, W.-E., Grimme, W., Günther, H. and Vogel, E. (1970). *Angew. Chem.*, **82,** 292
149. Beddoes, R. L., Lindley, P. F. and Mills, O. S. (1970). *Angew. Chem.*, **82,** 293
150. Untch, K. G. and Kurland, R. J. (1964). *J. Molec. Spectrosc,* **14,** 156
151. Roth, W. R., Bang, W. B., Gœbel, P., Sass, R. L., Turner, R. B. and Yü, A. P. (1964). *J. Amer. Chem. Soc.*, **86,** 3178
152. Bischof, P., Gleiter, R. and Heilbronner, E. (1970). *Helv. Chim. Acta,* **53,** 1425
153. Dewar, M. J. S., Náhlovská, Z. and Náhlovský, B. D. (1971). *Chem. Commun.*, 1377
154. Heilbronner, E. (1964). *Tetrahedron Letters,* 1923
155. For a review see Zimmerman, H. E. (1969). *Angew. Chem.*, **81,** 45
156. Staley, S. W. and Henry, T. J. (1970). *J. Amer. Chem. Soc.*, **92,** 7612
157. Boche, G., Martens, D. and Danzer, W. (1969). *Angew. Chem.*, **81,** 1003
158. Katz, T. J., Yoshida, M. and Siew, L. C. (1965). *J. Amer. Chem. Soc.*, **87,** 4516
159. The synthesis of potentially aromatic 10 π systems has been reported also: Allinger, N. L. and Youngdale, G. A. (1962). *J. Amer. Chem. Soc.*, **84,** 1020
160. Very recently 1,2-diazacycloocta-2,4,6,8-tetraene has been reported to be a non-planar COT-analogue by Trost, B. M. and Cory, R. M. (1971). *J. Amer. Chem. Soc.*, **93,** 5573. See also the synthesis of a dibenzo[1,2]-diazocine and its dihydro derivative with 10 π electrons: Paudler, W. W. and Zeiler, A. G. (1969). *J. Org. Chem.*, **34,** 3237
161. Ogawa, H., Kubo, M. and Saikachi, H. (1971). *Tetrahedron Letters,* 4859
162. See Johnson, A. W. (1971). *Pure Appl. Chem.*, **28,** 195
163. Beeby, P. J. and Sondheimer, F. (1972). *J. Amer. Chem. Soc.*, **94,** 2128
164. Garratt, P. J. (1971). *Aromaticity* (Maidenhead: McGraw-Hill)

9
Quasi-Aromatic Compounds; Examples of Meneidic or Regenerative Systems

DOUGLAS LLOYD
University of St. Andrews, Scotland

9.1 GENERAL INTRODUCTION

The term quasi-aromatic appears to have been used in the first place to describe the *structures* of osazones. Fieser and Fieser[1] suggested that osazones could best be described as chelates, which might be further stabilised by resonance:

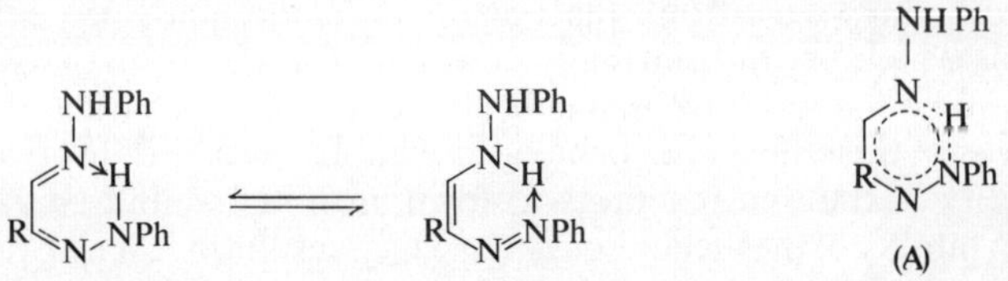

NHPh, N, H, R, N, NPh ⇌ NHPh, N, H, R, N, NPh; NH Ph, N, H, R, N, NPh (A)

This was studied in particular by Mester[2, 3] who concluded that 'the six-membered chelate ring containing conjugated double-bonds is to be regarded as a quasi-aromatic system',[2] alternatively represented[3] as in (A).

At the same time an alternative meaning of the term quasi-aromatic developed in connection with the metal chelates of β-diketones. The chromium(III), cobalt(III) and rhodium(III) complexes of acetylacetone had been shown to undergo a variety of electrophilic substitution reactions, as mentioned again later in this chapter, and they were thus said to demonstrate quasi-aromatic properties[4]; the reactions were described as quasi-aromatic substitution[5]. This usage of the term quasi-aromatic clearly refers to *chemical reactivity*. When n.m.r. spectroscopy became a tool in organic chemistry, the spectra of β-diketone metal complexes were studied to see if there were any indications of ring currents in the chelate rings (see references later in chapter). This added a further complication in that the term quasi-aromatic might now be applied to *magnetic*, as well as to structural or chemical properties. The situation was rapidly becoming as complex and as diffuse as that concerning aromaticity itself.

In an attempt to define the position Lloyd and Marshall proposed 'that molecules should be called quasi-aromatic only if they contain an acyclic conjugated π-electron system and show chemical properties typical of aromatic compounds, especially reaction by substitution with retention of type. A significant mesomeric stability is implied'[6]. This is the definition of quasi-aromatic which will be used in this chapter. This definition was in fact criticised, rightly, by Marchand[7], who said that there was a need to resolve the meaning of aromaticity in some suitable physical terms before introducing extra terms such as quasi-aromatic. Acting on this advice, Lloyd and Marshall have recently suggested that owing to the confusion about its precise meaning, the term aromaticity could well be discarded[8, 9]. They also felt that, since for more than a century 'aromatic' also embraced a characteristic sort of reactivity, namely the tendency to undergo substitution rather than addition reactions, it would be useful to have a specific term to mean, in Armit and Robinson's words[10], 'the tendency to retain the type', and suggested the terms *meneidic* or *regenerative*[8, 9]. The word 'meneidic' is of Greek origin, from *menein*, to stay, persist, remain; and *eidos*, type, form[11]; 'regenerative' is descriptive of the process which takes place—the original electronic system is regenerated after being temporarily lost in the transition state or in a steady state intermediate. It was intended that these terms should be restricted to substitution at sp^2-hybridised centres, and to what have been described[12] as S_A (for substitution through association) reactions, wherein a new bond is formed before the old bond breaks. In keeping with this suggestion, quasi-aromatic compounds are examples of compounds which are not normally considered to be aromatic compounds but which none the less have meneidic, or regenerative, character.

In all cases this meneidic behaviour can be ascribed to two factors, (a) the ability of the two reactants to form an initial stabilised σ-complex, and (b) the stability of the meneidic system which results in reaction proceeding from the σ-complex by loss of a group and an overall substitution reaction, rather than by a second addition step. In the following sections, some examples of compounds which show quasi-aromatic, or better, meneidic or regenerative character, will be discussed.

9.2 METAL CHELATES OF β-DIKETONES

It had been suggested by Calvin and Wilson in 1945 that the stability of these chelates could be accounted for by assuming a complete cyclic delocalisation in the chelate rings, akin to that in benzenoid compounds[13]. They expressed this in formulae as follows:

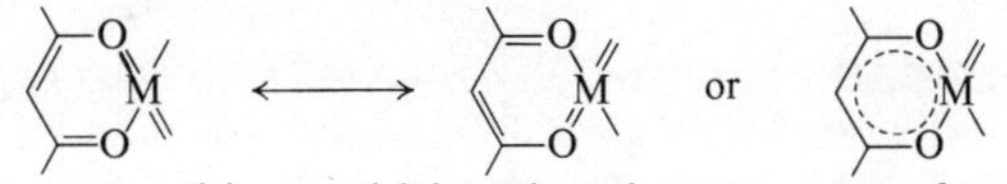

In contemporary terms this would involve the presence of an induced ring current when the n.m.r. spectrum was determined, whereas in an alternative formulation, involving only partial cyclic delocalisation, no ring current would be induced:

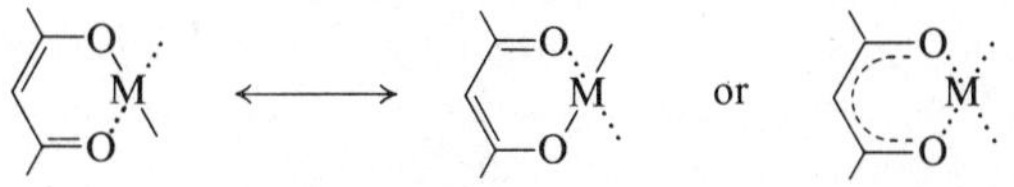

Some early investigators interpreted the chemical shifts observed in terms of ring-currents[14–16], but the n.m.r. spectra are currently interpreted in terms of structures lacking any induced ring-current[17–20]. The chelates therefore do not possess a completely cyclic delocalised system of π-electrons so that their meneidic properties are evidence of quasi-aromatic character as defined above[6].

The first examples of electrophilic substitution reactions in metal chelates were reported as far back as 1925[21]. Since then such reactions have been extensively investigated. The central carbon atom has been nitrated by copper nitrate in acetic anhydride[5, 23–27] and by dinitrogen tetroxide in benzene[22, 28], nitrosated[22, 28], halogenated by bromine[21, 23, 29], iodine monochloride[23, 29] or *N*-halogenosuccinimides[5, 23, 26, 27, 29–31], and thiocyanogenated[4, 5]. They are also formylated by the Vilsmeier method[24], acylated by means of acid chlorides and aluminium chloride[4, 5] or acetic anhydride and boron trifluoride etherate[23], and react with sulphur dichloride or arenesulphenyl chlorides without catalysts[4]. A mononitro-derivative has been reduced to the corresponding amine, which could be diazotised[32]. The resultant stable diazo-compound was in turn converted into a fluoro-compound under the conditions of the Balz–Schiemann reaction, and reduced with loss of nitrogen by ethanol[32]. These reactions, all reminiscent of those of aryl compounds, are described in more detail in a review article[33].

It is apposite to note that not only the metal chelates of β-diketones, but the enol forms of the ketones themselves may also show meneidic properties; for example they undergo halogenation. If a reaction such as bromination proceeds as shown below, substitution with retention of type ensues:

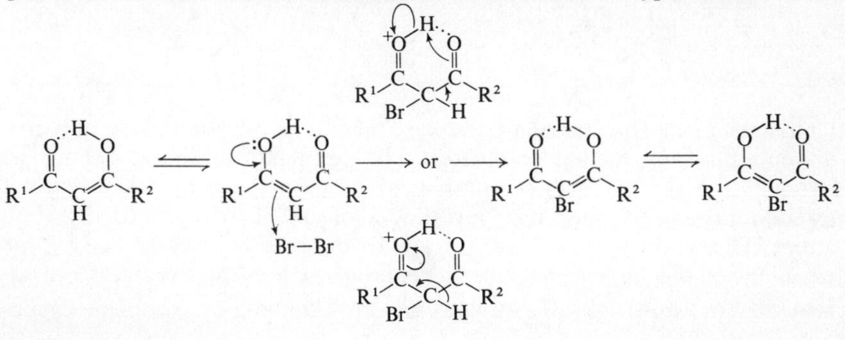

In the case of β-keto-enol ethers where such a mechanism cannot obtain, bromine reacts to give addition instead of substitution products[34].

9.3 TROPOLONES

Tropolones have been regarded as typically 'aromatic' compounds, and their ready electrophilic substitution has frequently been considered to derive from their 'aromatic' character. This is an unsatisfactory explanation since the 'aromatic' canonical form of tropolone with a delocalised sextet of electrons in the ring results in that ring bearing a positive charge; it is in fact a tropylium ring and tropylium rings are notably resistant to electrophilic attack. Tropolones are vinylogues of β-keto-enols and their meneidic character may be related to that of the keto-enols (cf. reference 35, pp. 120–122). Stabilised σ-complexes are formed in which the positive charge is delocalised over the hydroxyl group as well as the ring, while the electronic system of tropolone has sufficient stability to result in its regeneration by elimination of a proton rather than its destruction by addition of an anion to the intermediate cation:

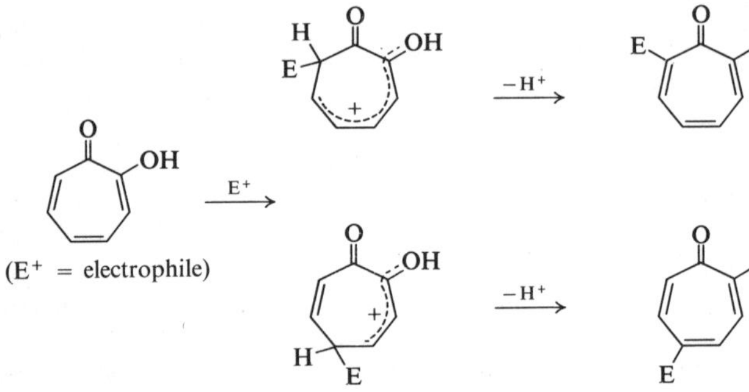

O OH E⁺ (E⁺ = electrophile)

H E O OH + −H⁺ E O OH

O OH + H E −H⁺ O OH E

Electrophilic substitution reactions of tropolones are listed in reference 35, pp. 143–145. 2-Aminotroponeimines are further examples of quasi-aromatic compounds which undergo electrophilic substitution in similar fashion[57, 58].

9.4 DIHYDRODIAZEPINIUM SALTS

2,3-Dihydro-1,4-diazepinium salts have the structure:

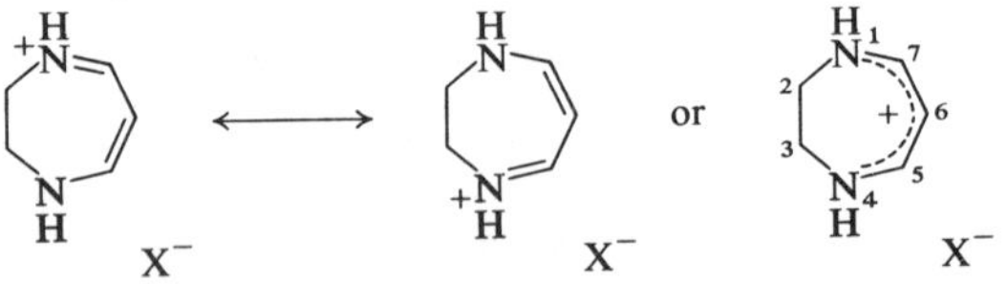

+H N N H X^- ⟷ H N +N H X^- or H 1 N 7 2 + 6 3 5 N 4 H X^-

Unlike the metal chelates of β-diketones, there is no question of the dihydrodiazepinium ring having complete cyclic conjugation, and it cannot be 'aromatic' in the classic structural sense because the conjugation cannot extend all round the ring since part of it is formed of sp^3-hybridised carbon atoms. Hence these salts have proved to be excellent substrates for the examination of quasi-aromatic or meneidic character. An extensive investigation of their properties has consequently been made.

Dihydrodiazepinium salts possess an extremely stable electronic system and are characterised by a great 'tendency to retain the type'[10]. They readily undergo the typical 'aromatic' electrophilic substitution reactions such as halogenation[36–40], protodehalogenation[39], nitration[41–44] and coupling with diazonium salts[45].

It is noteworthy that the latter reactions involve electrophilic attack by a cation on another cation. The reactions involve formation of a di-cation intermediate which loses a proton to regenerate the very stable dihydrodiazepinium system:

$$\text{(ring, 6-H)} \underset{}{\overset{E^+}{\rightleftharpoons}} \text{(di-cation, 6-H, 6-E)} \underset{}{\overset{-H^+}{\rightleftharpoons}} \text{(ring, 6-E)}$$

(E^+ = electrophile)

The 6-nitrodiazepinium salts can be reduced to 6-amino-derivatives which are converted into stable diazonium salts[44, 46]; the diazonium group can be replaced by chlorine[44, 46, 47] or by hydrogen[47]. No such stable di-cation intermediate can be drawn for electrophilic substitution at the 5,7-positions and comparative studies suggest a reactivity at the 6-position which is 10^9 or more greater than that at the 5,7-positions[40, 48]. The same structural feature is present in corrins, and nickel salts of tetradehydrocorrins similarly undergo halogenation and nitration[49, 59].

9.5 OPEN-CHAIN QUASI-AROMATIC COMPOUNDS

In the cases of the compounds discussed in Sections 9.2–9.4, the conjugated systems have formed part of ring structures and it is likely that this is a contributing factor to the character of these compounds, since they are thereby fixed in a favourably stable conformation. It none the less seems reasonable to expect some similar chemical behaviour to be shown by their open-chain analogues. Indeed, β-keto-enols mentioned in Section 9.2, might reasonably be regarded as open-chain compounds, although they are held in cyclic conformations by hydrogen bonding.

The open-chain analogues of dihydrodiazepinium ions are the α,ω-diazapolymethine cations. Such salts readily undergo quasi-aromatic electrophilic substitution at a β-methine group and have been brominated and nitrated[50]. A similar mechanistic pattern to that involved in dihydrodiazepinium salts can again be assumed, e.g.

$$[R_2N-CH=CH-CH=\overset{+}{N}R_2 \longleftrightarrow R_2\overset{+}{N}=CH-CH=CH-NR_2]$$

$$\downarrow E^+ \text{ (electrophile)}$$

$$R_2\overset{+}{N}=CH-CH\overset{+}{E}-CH=\overset{+}{N}R_2$$

$$\downarrow -H^+$$

$$[R_2N-CH=CE-CH=\overset{+}{N}R_2 \longleftrightarrow R_2\overset{+}{N}=CH-CE=CH-NR_2]$$

It may be noted that in the case of 1,3-diazapolymethine bases with secondary amino-groups, an alternative hydrogen-bonded structure involving complete cyclic delocalisation achieved by hydrogen bonding might be considered. Thus the formulation (1) might be interpreted in terms of complete (2) or

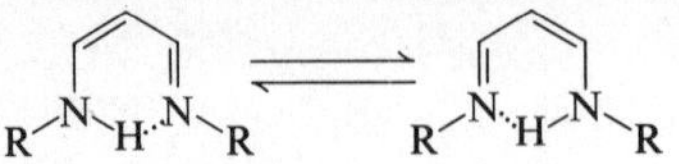

(1) 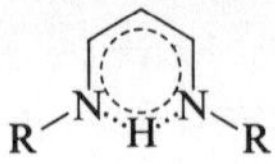 (2) 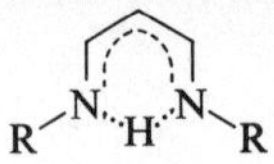 (3)

incomplete (3) cyclic delocalisation. It has been shown quite unequivocally by a detailed study of their u.v. and n.m.r. spectra that (2) is not a correct formulation for these compounds[51–53]. This is also the case for formazanes[54].

In similar fashion diazonium salts couple with the polymethine chains of cyanine dyestuffs[55]. Quasi-aromatic character has also been noted in α,ω-diphenylpolyenes which show meneidic properties in their methine chains. Thus α,ω-diphenyl-butadiene, -hexatriene and -octatetraene are all formylated at an α-methine group[56].

9.6 CODA

Some examples of unsaturated molecules which undergo substitution with retention of type rather than addition reactions have been given in the preceding sections. This behaviour has been defined as quasi-aromatic character[6]. Use of this term has been criticised, e.g. 'S_E reactions are no criterion of aromatic character, since many electrophilic substitution reactions are shown by polymethines and polyenes'[51, 53]; 'Use of the term quasi-aromatic is best avoided in connection with such systems since "non-aromatic" is not generally understood to be included under this heading'[19]. These comments rather beg the point since the term quasi-aromatic was intentionally defined to refer to compounds which, by their structure, no one considered to be aromatic, but which none the less underwent reactions which for about a century had been regarded as typical of 'aromatic' compounds. The criticism[7] that there was first a need to resolve the meaning of the term 'aromaticity' is more pertinent.

Since the tendency to undergo electrophilic substitution reactions is now regarded, at best, as marginal evidence of aromatic character, and since there is in any case considerable confusion as to the meaning of aromaticity, it does seem preferable to discontinue the use of the word quasi-aromatic and, as discussed in Section 9.1, to replace it by the terms meneidic or regenerative which are specifically defined[8, 9] without reference to the 'aromaticity' (whatever that may mean) of compounds.

References

1. Fieser, L. F. and Fieser, M. (1944). *Organic Chemistry,* 353. (Boston: D. C. Heath and Co.)
2. Mester, L. (1955). *J. Amer. Chem. Soc.,* **77,** 4301
3. Mester, L. (1965). *Angew. Chem.,* **77,** 580; *Angew. Chem. Internat. Ed.,* **4,** 574
4. Collman, J. P., Marshall, R. L., Young, W. L. and Sears, C. T. (1963). *J. Org. Chem.,* **28,** 1449

5. Collman, J. P., Blair, R. P., Marshall, R. L. and Slade, A. L. (1963). *J. Inorg. Chem.*, **2,** 576
6. Lloyd, D. and Marshall, D. R. (1964). *Chem. and Ind.*, 1760
7. Marchand, A. P. (1965). *Chem. and Ind.*, 161
8. Lloyd, D. and Marshall, D. R. (1971). *'Aromaticity, Pseudo-aromaticity, Anti-aromaticity'*, 85. (Jerusalem: The Israel Academy of Sciences and Humanities)
9. Lloyd, D. and Marshall, D. R. (1972). *Angew. Chem.*, **84**; *Angew. Chem. Internat. Ed.*, **11,** to be published
10. Armit, J. W. and Robinson, R. (1925). *J. Chem. Soc.*, **127,** 1604
11. This term is based on a suggestion kindly made by Professor K. J. Dover, Department of Greek, University of St. Andrews
12. Kosower, E. M. (1968). *An Introduction to Physical Organic Chemistry*, 63 *ff.* (New York: J. Wiley and Sons, Inc.)
13. Calvin, M. and Wilson, K. W. (1945). *J. Amer. Chem. Soc.*, **67,** 2003
14. Forman, A., Murrell, J. N. and Orgel, L. E. (1959). *J. Chem. Phys.*, **31,** 1129
15. Collman, J. P., Marshall, R. L. and Young, W. L. (1962). *Chem. and Ind.*, 1380
16. Hester, R. E. (1963). *Chem. and Ind.*, 1397
17. Fay, R. C. and Serpone, N. (1968). *J. Amer. Chem. Soc.*, **90,** 5701
18. Trestianu, A., Niculescu-Majewska, H., Bally, I., Barabás, A. and Balaban, A. T. (1968). *Tetrahedron*, **24,** 2499
19. Kuhr, M. and Musso, H. (1969). *Angew. Chem.*, **81,** 150; *Angew. Chem. Int. Edn.*, **8,** 147
20. Bock, B., Flatau, K., Junge, H., Kuhr, M. and Musso, H. (1971). *Angew. Chem.*, **83,** 239; *Angew. Chem. Int. Edn.*, **10,** 255
21. Reihlen, H., Illig, R. and Wittig, R. (1925). *Ber. Dtsch. Chem. Ges.*, **58,** 12
22. Djordjevič, C., Lewis, J. and Nyholm, R. S. (1959). *Chem. and Ind.*, 122
23. Collman, J. P., Moss, R. A., Goldby, S. D. and Trahanovsky, W. S. (1960). *Chem. and Ind.*, 1213
24. Collman, J. P., Marshall, R. L., Young, W. L. and Goldby, S. D. (1962). *Inorg. Chem.*, **1,** 704
25. Singh, P. R. and Sahai, R. (1968). *Inorg. Nucl. Chem. Lett.*, **4,** 513
26. Sasaki, T., Kanematsu, K. and Kinoshita, G. (1968). *J. Org. Chem.*, **33,** 680
27. Sasaki, T., Kanematsu, K. and Kinoshita, G. (1969). *J. Chem. Soc. C*, 951
28. Djordjevič, C., Lewis, J. and Nyholm, R. S. (1962). *J. Chem. Soc.*, 4778
29. Collman, J. P., Moss, R. A., Maltz, H. and Heindel, C. C. (1961). *J. Amer. Chem. Soc.*, **83,** 531
30. Collman, J. P. and Kittleman, E. T. (1961). *J. Amer. Chem. Soc.*, **83,** 3529
31. Collman, J. P. and Kittleman, E. T. (1962). *Inorg. Chem.*, **1,** 499
32. Collman, J. P. and Yamada, M. (1963). *J. Org. Chem.*, **28,** 3017
33. Collman, J. P. (1965). *Angew. Chem.*, **77,** 154; *Angew. Chem. Int. Edn.*, **4,** 132
34. Marshall, D. R. and Roberts, T. R. (1971). *J. Chem. Soc. B*, 797
35. Lloyd, D. (1966). *Carbocyclic Non-benzenoid Aromatic Compounds.* (Amsterdam: Elsevier)
36. Lloyd, D. and Marshall, D. R. (1958). *J. Chem. Soc.*, 118
37. Bell, R. P. and Marshall, D. R. (1964). *J. Chem. Soc.*, 2195
38. Barnett, C., Cleghorn, H. P., Cross, G. E., Lloyd, D. and Marshall, D. R. (1966). *J. Chem. Soc. C*, 93
39. Gorringe, A. M., Lloyd, D., Wasson, F. I., Marshall, D. R. and Duffield, P. A. (1969). *J. Chem. Soc. C*, 1449
40. Barnett, C., Marshall, D. R., Mulligan, L. A. and Lloyd, D. (1971). *J. Chem. Soc. B*, 1529
41. Barnett, C. (1967). *Chem. Commun.*, 637
42. Barnett, C. (1967). *J. Chem. Soc. B*, 2436
43. Gorringe, A. M., Lloyd, D., Marshall, D. R. and Mulligan, L. A. (1968). *Chem. and Ind.*, 130
44. Gorringe, A. M., Lloyd, D. and Marshall, D. R. (1970). *J. Chem. Soc. C*, 617
45. Grant, E. M., Lloyd, D. and Marshall, D. R. (1970). *Chem. Commun.*, 1320
46. Gorringe, A. M., Lloyd, D. and Marshall, D. R. (1968). *Chem. and Ind.*, 1160
47. Grant, E. M., Lloyd, D. and Marshall, D. R., unpublished work
48. Butler, A. R., Lloyd, D. and Marshall, D. R. (1971). *J. Chem. Soc. B*, 795
49. Grigg, R., Johnson, A. W. and Richardson, K. (1968). *Chem. Commun.*, 896
50. Kucera, J. and Arnold, Z. (1967). *Coll. Czech. Chem. Commun.*, **32,** 1704

51. Daltrozzo, E. and Feldmann, K. (1967). *Angew. Chem.*, **79,** 153; *Angew. Chem. Int. Edn.*, **6,** 182
52. Daltrozzo, E. and Feldmann, K. (1968). *Tetrahedron Lett.*, 4983
53. Daltrozzo, E. and Feldmann, K. (1968). *Ber. Bunsenges. Phys. Chem.*, **72,** 1140
54. Fischer, P. B., Kaul, B. L. and Zollinger, H. (1968). *Helv. Chim. Acta,* **51,** 1449
55. Treibs, A. and Zimmer-Galler, R. (1959). *Liebigs Ann.*, **627,** 166
56. Jutz, C. and Heinicke, R. (1969). *Chem. Ber.*, **102,** 623
57. Brasen, W. R., Holmquist, H. E. and Benson, R. E. (1960). *J. Amer. Chem. Soc.*, **82,** 995
58. Brasen, W. R., Holmquist, H. E. and Benson, R. E. (1961). *J. Amer. Chem. Soc.*, **83,** 3125
59. Hamilton, A. and Johnson, A. W. (1971). *J. Chem. Soc. C,* 3879

10
Organic Dyes and Optical Brightening Agents*

E. SIEGEL
Dyestuff Research Department, Farbenfabriken Bayer AG., Leverkusen, Federal Republic of Germany

* Translation of the original German manuscript by Dr. Laurence Gross, Department of Industrial and Engineering Chemistry, Swiss Federal Institute of Technology, (ETH), Zurich.

10.1 INTRODUCTION

The development of dyes and brighteners is by no means completed; rapid developments in the field of synthetic fibres, steady pressure for rationalisation of the dyeing processes, and lately, growing environmental demands create more and more problems for dye research.

Almost all the new and technically important developments originate from the laboratories of only a few big dye manufacturers in the Western world; recently, several Japanese firms have also come up with rather interesting contributions. Time and again it is also the worth while observations originating from the universities which lead toward industrial application.

10.2 AZO DYES

10.2.1 Metal-free azo dyes

10.2.1.1 Aryl-azo-enol dyes

Until recently, the primary importance of acetoacetic arylamide couplings was in the field of pigment dyes (Hansa yellows); at present they are also being successfully used in the reactive dyes in order to produce brilliant yellow colours with a strong greenish hue[1].

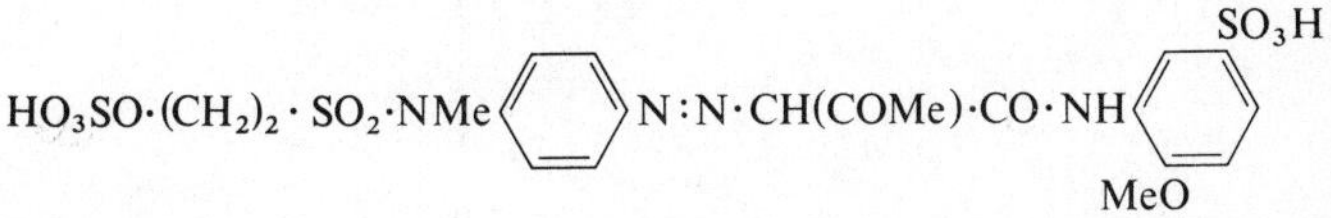

$$HO_3SO{\cdot}(CH_2)_2{\cdot}SO_2{\cdot}NMe{-}C_6H_4{-}N{:}N{\cdot}CH(COMe){\cdot}CO{\cdot}NH{-}C_6H_3(SO_3H)(MeO)$$

(1)

10.2.1.2 Aryl-azo-aryl dyes

(a) *Phenyl-azo-phenyl dyes*—The most technically important compounds of this type are the disperse dyes for synthetic fibres, especially polyester and polyamide fibres. Considerable industrial research has been concentrated on this group.

Worth while progress has been achieved in three different directions: (i) New type of substitution reactions on *o,o*-dihalogenoazo dyes, (ii) several special substitutions R^1–R^3 on nitrogen, or on the centre of components of aniline couplings and (iii) in the dimerisation of monoazo dye molecules,

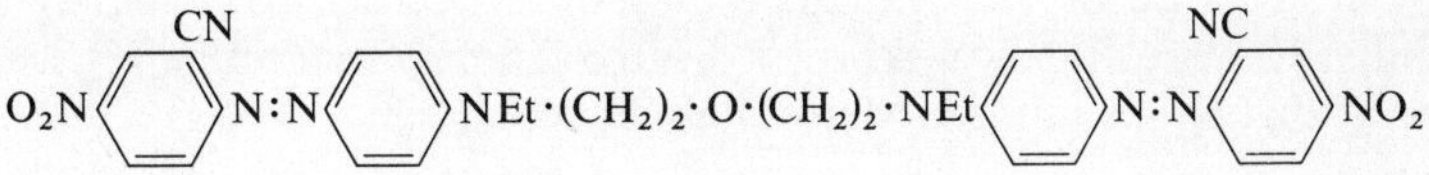

$$O_2N{-}C_6H_3(CN){-}N{:}N{-}C_6H_4{-}NEt{\cdot}(CH_2)_2{\cdot}O{\cdot}(CH_2)_2{\cdot}NEt{-}C_6H_4{-}N{:}N{-}C_6H_3(NC){-}NO_2$$

(2)

for example, over aliphatic bridge formation, (2) which leads to high sublimation-fast dyes[2].

As discovered by Gottschlich and Leverenz[3], the copper-catalysed nucleophilic exchange of a halogen atom in the *o*-position of the azo bridge (e.g. in (3)) occurs smoothly and often quantitatively at room temperature with CuCN in dimethyl formamide, nitrobenzene/pyridine or other polar solvents.

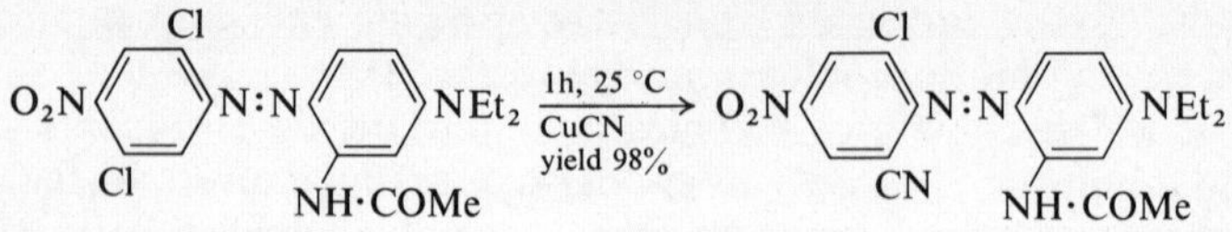

(3)

The new cyano exchange reaction is presumably based on a specific mechanism which is probably influenced by the complex formed by the azo bridge and CuCN. This opinion is strengthened by the catalytic effect of pyridine as well as the facilitation of the reaction through an acylamino group in the *ortho*-position to the azo bridge. The reaction is comparable to substitution in the presence of Cu^{II} salts of 2-halogeno-2′-hydroxyazo compounds through amines, alkoxides or sulphinates which take place also under very mild conditions[4].

Under somewhat more energetic conditions, *o*-halogenoazo compounds in the presence of copper (I) salts, can also be substituted through nitrite[5] to *o*-nitroazo and, through sulphinates[6], to *o*-sulphonylazo compounds. Such exchange reactions are of value in the synthesis of clear red and especially blue monoazo disperse dyes since these can be obtained through diazota-

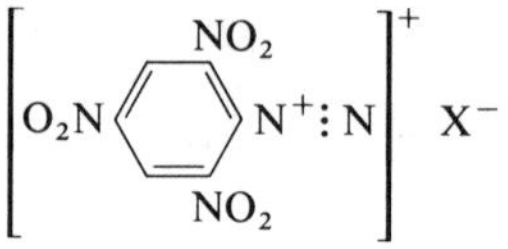

(4)

tion and coupling only in a very moderate yield because of the strongly negative substituted and, therefore, unstable diazo components (e.g. (4)).

The new blue polyester dyes of the light-fast and non-subliming types

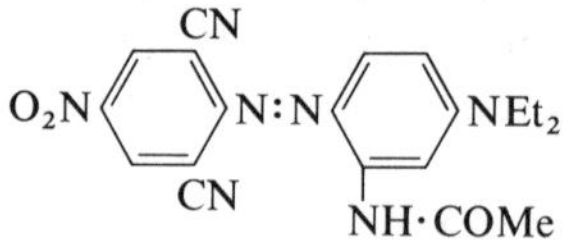

(5)

such as (5) are technically the most important[7]. The clearness of the hue of these compounds, hitherto not achieved for blue azo dyes, is attributed to the stretched structure of the cyanate group which causes co-planarity in the molecule. The previously unknown 4-nitro-2,6-dicyanoaniline (6) was also produced through a smooth Meisenheimer reaction[8]:

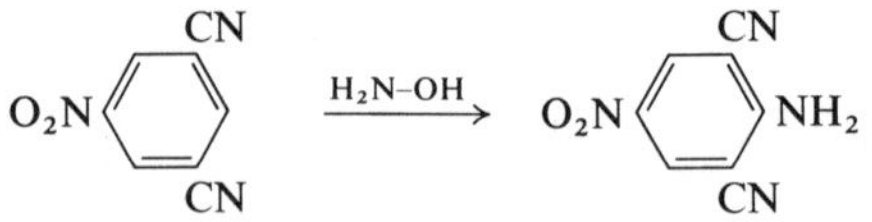

(6)

The 2-nitro-4,6-dicyanoaniline (7) as a diazo component for blueish reds and for blue disperse dyes is also new[9]; Roos and Wagner obtained the compound in the following way:

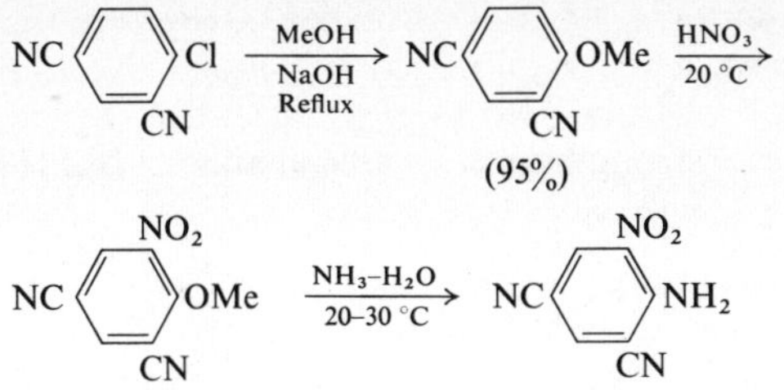

$$NC\text{-}C_6H_3(CN)\text{-}Cl \xrightarrow[\text{NaOH, Reflux}]{\text{MeOH}} NC\text{-}C_6H_3(CN)\text{-}OMe\ (95\%) \xrightarrow[20\,^{\circ}C]{HNO_3}$$

$$NC\text{-}C_6H_2(NO_2)(CN)\text{-}OMe \xrightarrow[20\text{–}30\,^{\circ}C]{NH_3\text{–}H_2O} NC\text{-}C_6H_2(NO_2)(CN)\text{-}NH_2$$

(7)

There are some new aniline coupling components which contain external cyclic imides of dicarboxylic acids, which form, for example, the orange dye

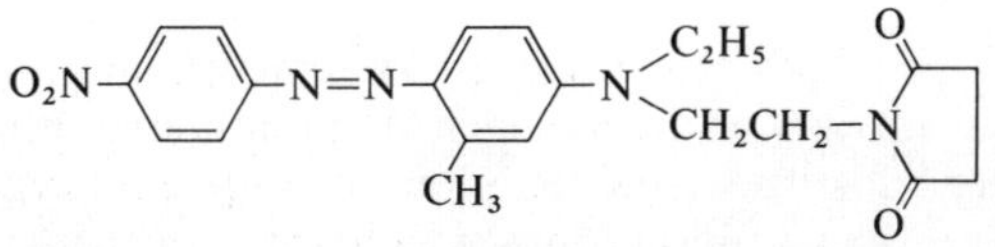

$$O_2N\text{-}C_6H_4\text{-}N{=}N\text{-}C_6H_3(CH_3)\text{-}N(C_2H_5)CH_2CH_2\text{-}N(CO\,CH_2CH_2\,CO)$$

(8)

(8). The synthesis is achieved from the corresponding aniline derivative with an external aliphatic amino group, by melting together with succinic anhydride. This type, which was intensely investigated by Eastman Kodak, resulted in a large number of variations[11]. Likewise, the phenoxyacetic acid has practical significance as the acylation agent for *N*-β-hydroxyethyl side chains on aniline derivatives[12], e.g. (9).

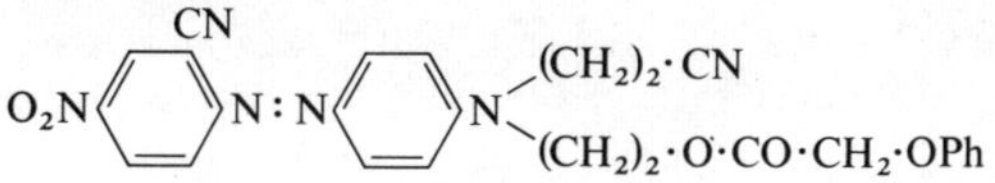

$$O_2N\text{-}C_6H_3(CN)\text{-}N{:}N\text{-}C_6H_4\text{-}N\begin{cases}(CH_2)_2\cdot CN\\(CH_2)_2\cdot O\cdot CO\cdot CH_2\cdot OPh\end{cases}$$

(9)

$$O_2N\text{-}C_6H_3(CN)\text{-}N{:}N\text{-}C_6H_4\text{-}N\begin{cases}(CH_2)_2\cdot CN\\(CH_2)_2\cdot Ph\end{cases}$$

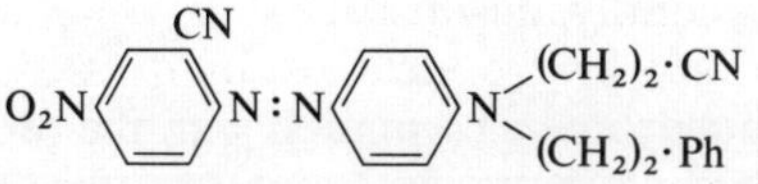

(10)

N-benzylaniline, *N*-β-phenylethylaniline and diphenyl amine, in certain combinations as coupling components, give high sublimation fastness and, in addition, good affinity and adequate light fastness. Dyes of this structure are not 'spoiled' in alkaline or acidic baths. In other words, the polar groups in the side chains do not hydrolyse[13]. (See e.g. the red dye (10)[14].)

There are a number of patent reports which claim polar substituents of such high acidity that water-insoluble disperse dyes under normal acidic application conditions dissociate in an alkaline environment, thereby becoming intermediately water soluble (11)[15].

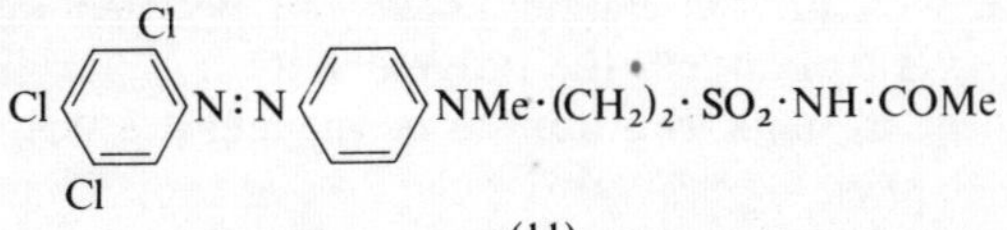

$$Cl\text{-}C_6H_2(Cl)_2\text{-}N{:}N\text{-}C_6H_4\text{-}NMe\cdot(CH_2)_2\cdot SO_2\cdot NH\cdot COMe$$

(11)

If a hydrogen atom is still present on the nitrogen of the aniline coupling component, then the necessary polar character of the dye can also be obtained through *ortho*-substituents. This then enables some hydrogen-bond formation which causes a strong hypsochromic change in hue (12)[16].

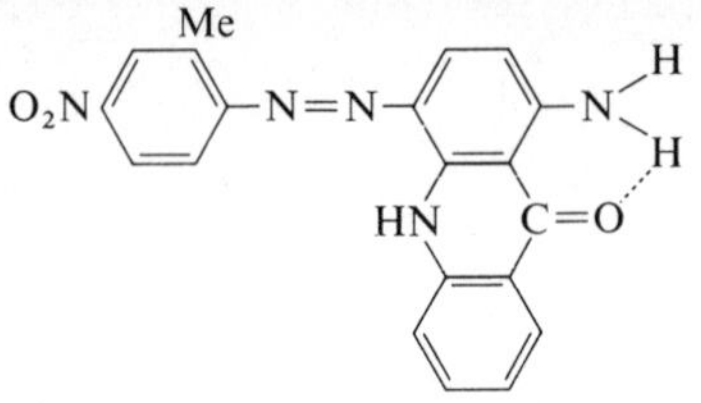

(12)

The polar groups which are necessary to achieve good non-sublimability can be introduced not only on the nitrogen of the aniline coupling components, but also on other substituents of the phenyl-azo-phenyl type dyes. Technically important are acyl amino groups which are in the *ortho*-position to the azo bridge in the coupling component in combination with polar substituents in the side chain (13)[17] as well as in unsubstituted *N,N*-dialkyl aniline derivatives or, even in aniline derivatives with a free amino group (14). Orange to red dyes of this type are useful for the colourists[18].

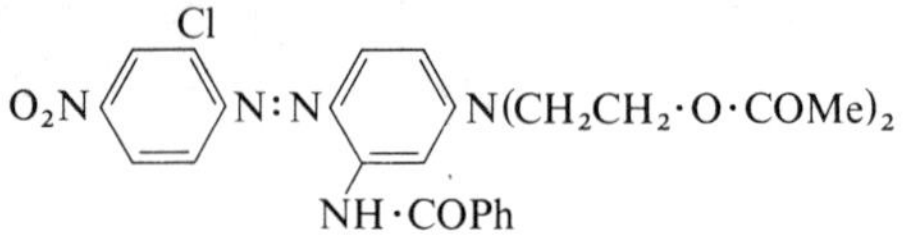

(13)

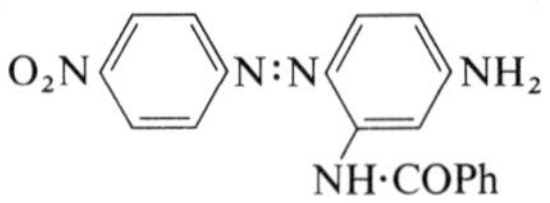

(14)

Coupling components of this type have two advantages: Firstly, they are easily prepared through acylation of 3-amino-dialkyl aniline and are accessible in a large variety and secondly, unsubstituted *N,N*-diethylamino groups in the *para*-position to the azo bridge yield strong dyes of especially clear hues which are shifted bathochrome if compared with the products obtained from negatively substituted *N,N*-dialkyl anilines (for example, *N,N*-dicyanoethylaniline).

In addition to the disperse dyes, copolymerisable dyes of the phenyl-azo-phenyl type have practical application, such as the yellow dye (15)[19].

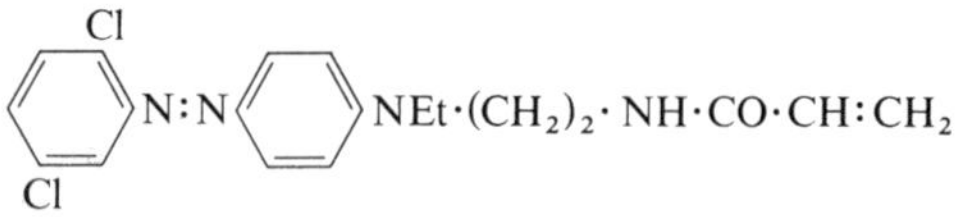

(15)

Coloured dispersions of polymers are obtained through emulsion copolymerisation with acrylic esters. These can be used in the finishing of leather. According to Lange[20], the polymerisation of acrylic esters, through these coloured monomers, is inhibited in varying degrees, according to the substituent.

(b) *Aryl-azo-naphthalene dyes (and other aryl-azo-aryl dyes)*—Schneider[21] discovered two water soluble acid dyes which, because of their unique coupling possibilities, are noteworthy:

(i) 8-Aminonaphth-1-ol-3,6-disulphonic acid (H-acid), coupled under alkaline conditions, can again be coupled (in good yield) in acetic acid in the form of their copper complexes (16).

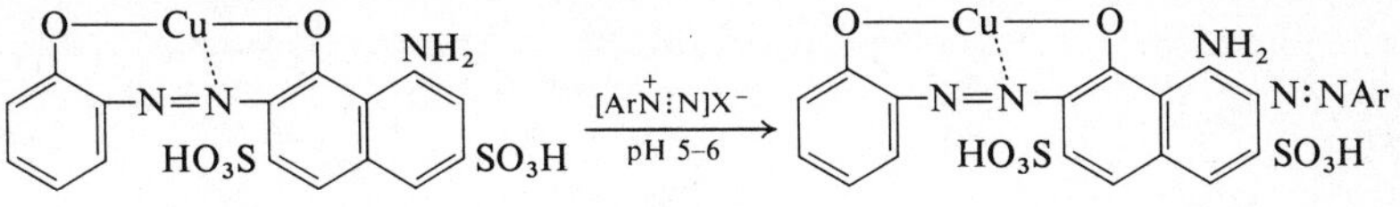

(16)

(ii) Disazo dyes can be obtained in 70% yield when 5-aminonaphth-1-ol-3-sulphonic acid (M-acid, 17), which is coupled to the hydroxyl group in the *peri*-position, is combined, in the presence of sodium carbonate, with diazotised 2-naphthylamine-4,6,8-trisulphonic acid.

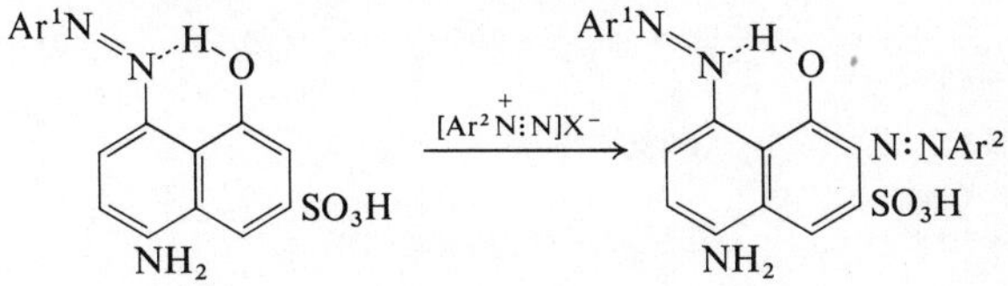

(17)

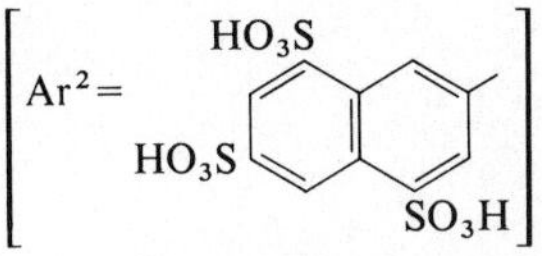

During the last few years 2-aminonaphth-8-ol-5-sulphonic acid, which is isomeric to γ-acid, has been successfully used as a component for the acid coupling in monoazo dyes as shown in formula (18)[22].

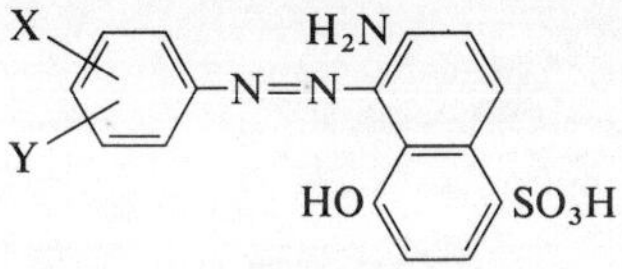

(18)

X = e.g. Cl, Me, SO_3H, NH–reactive group

Y = Substituent

The brilliant red acid dyes which are achieved in this manner are suitable for use in wool and polyamides and for the azo-silver bleaching process in light-fast colour pictures. It is possible for these dyes to contain reactive groups. In general, dyes built from 2-aminonaphth-8-ol-5-sulphonic acid are somewhat clearer, have a yellow hue and are more light-fast than the dyes derived from corresponding isomers of γ-acids.

Lithium salts of the yellow and orange substantive paper dyes in the stilbenazo or stilbenazoxy series show much better water solubility than the sodium salts, a fact that is important in the production of concentrated stock solutions[23].

Of considerable industrial value is the work which was done by Farbwerke Hoechst on fast pigment dyes of the phenyl-azo-naphthalene series which contain two carbonamide groups in the molecule and, along with their excellent light-fastness, are as migration-fast as disazo condensation pigments. Especially good results are obtained if one includes the carbonamide groups in the form of fused heterocyclic five- or six-membered rings (19)[24].

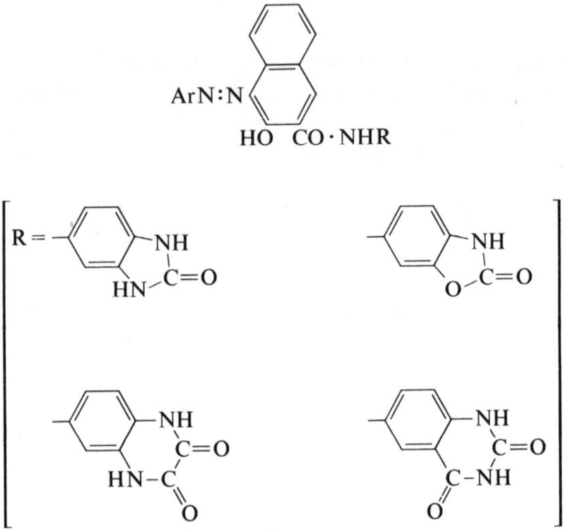

ArN:N HO CO·NHR

R = NH HN C=O; NH O C=O; NH C=O HN–C O; NH C=O C–NH O

(19)

For example, dyes like the red pigment (20) are technically important[25]. The production of the new coupling component is achieved through the

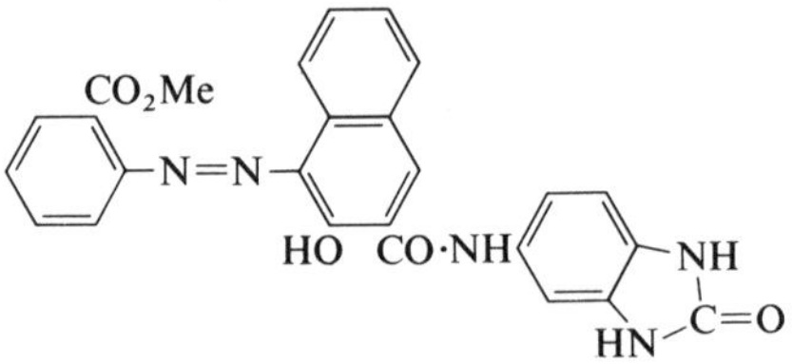

CO_2Me –N=N– HO CO·NH NH HN C=O

(20)

condensation of 2-hydroxynaphth-3-oic acid chloride with 5-aminobenzimidazolone.

Another new type of disazo condensation pigment which is, however, not as light-fast as are the Cromophtal dyes is noteworthy. In order to obtain these types of pigments, monoazo dyes from diazotised 3-aminophthalic acid and 2-hydroxynaphth-3-oic acid arylides are converted into their cyclic anhydride form in a high boiling organic solvent; these, then, 'dimerise' with diamines thereby creating two phthalimide rings (21)[26].

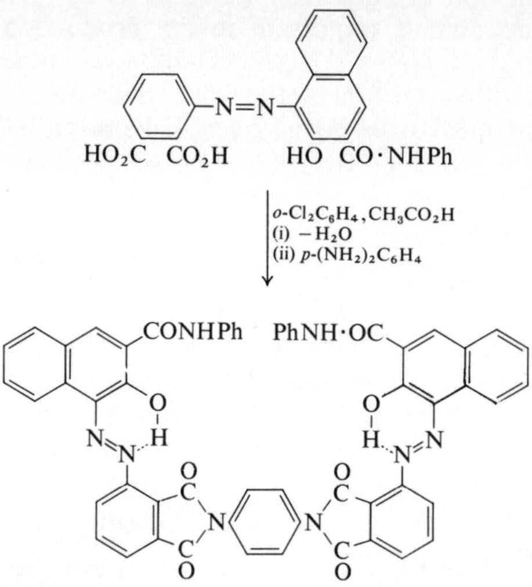

HO$_2$C CO$_2$H HO CO·NHPh

o-$Cl_2C_6H_4$, CH_3CO_2H
(i) $-H_2O$
(ii) p-$(NH_2)_2C_6H_4$

CONHPh PhNH·OC

(21)

10.2.1.3 Heteroaryl-azo-aryl dyes

These dyes are important mainly as disperse dyes; when they contain external cationic groups, they may also be of value in the dyeing of acrylic fibres and polyester or polyamide fibres which can be dyed with cationic dyes. In the last few years several new heterocyclic diazo components and coupling components were utilised.

(a) *Heterocyclic diazo components*—In this area especially noteworthy is the systematic work of BASF[27] on the properties of dyes derived from 2-aminobenzthiazole, 3-aminobenzisothiazole as well as from amines of the 1,2,4- or 1,3,4-thiadiazole series. 3-Amino-nitrobenzisothiazol (22) is accessible in the following way:

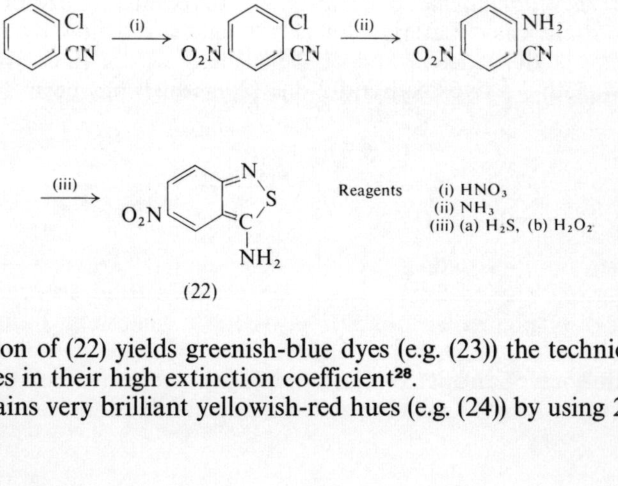

Reagents (i) HNO_3
(ii) NH_3
(iii) (a) H_2S, (b) H_2O_2

(22)

Diazotisation of (22) yields greenish-blue dyes (e.g. (23)) the technical value of which lies in their high extinction coefficient[28].

One obtains very brilliant yellowish-red hues (e.g. (24)) by using 2-amino-

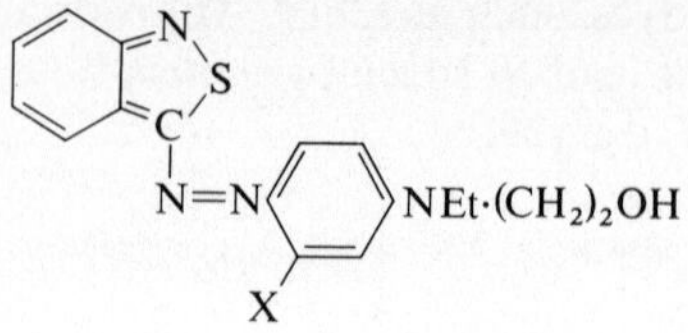

X = —H; —NH·COMe

(23)

5-phenyl-1,3,4-thiadiazole as the diazo component[29]. They are, however, sufficiently light-fast only on acetate fibres. 2-Amino-5-phenyl-1,3,4-thiadiazole is obtained from benzoylhydrazide with ammonium rhodanide in two steps[30].

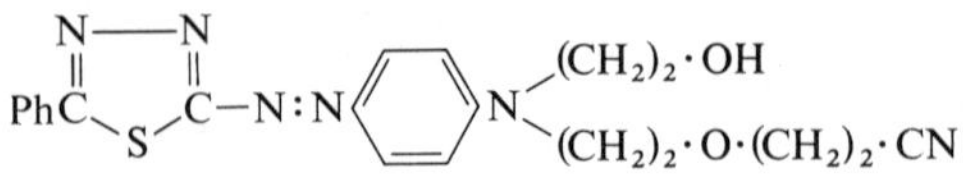

(24)

4-Aminobenzimidazole and 4-aminobenzitriazole, used as diazo components, create coupling products which, after alkylation and quaternation on the heterocyclic atoms, result in rather useful cationic dyes for polyacrylonitrile fibres, e.g. (25)[31].

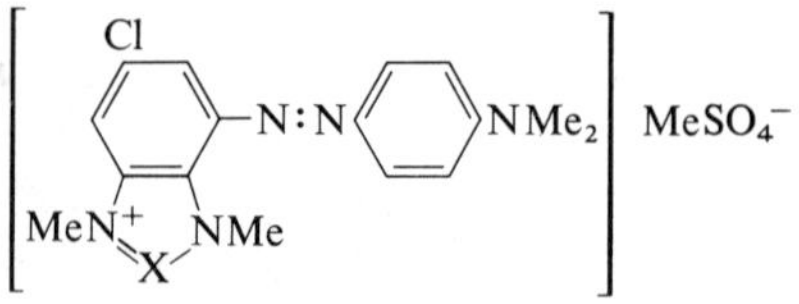

X = CH: scarlet
X = N: red
(25)

(b) *Heterocyclic coupling components*—Until recently pyrazolones were the most important basic materials for useful, light-fast yellow dyes, especially of the acid, of the reactive and of the disperse series. Recently, however, 3-cyano-4-methyl-6-hydroxypyrid-2-one (26) (which has been known for a

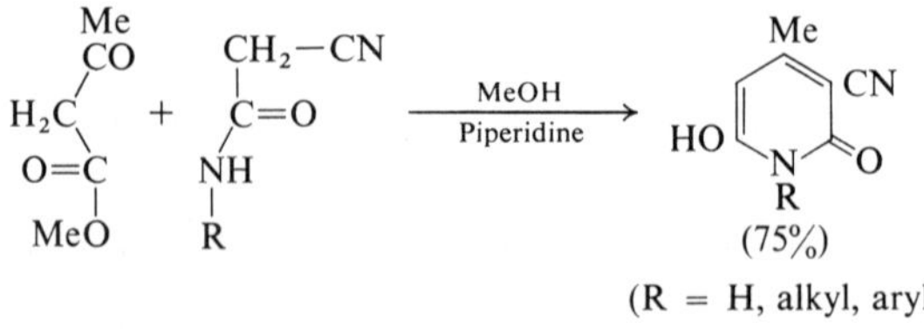

(75%)
(R = H, alkyl, aryl)
(26)

considerable time) and its *N*-derivatives have proven to be very valuable coupling components. One obtains clear disperse dyes, such as (27)[32] as well as interesting cationic dyes, such as (28)[33]. Appropriate positive substituents in the diazo component lead to special greenish hues, with light-fastness on polyester fibres up to 7, e.g. (29)[34].

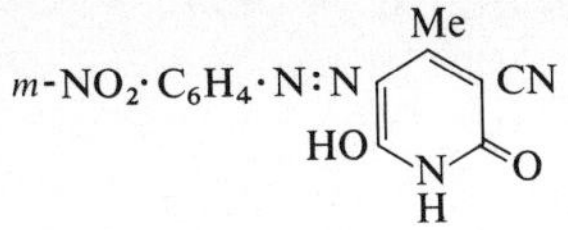

(27)

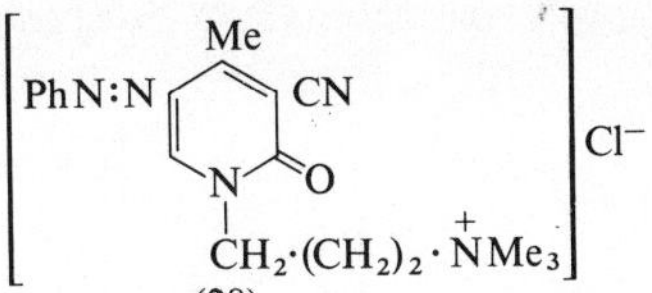

(28)

After the first patent report of Cassella, intensive work started in this new field; on the one hand, the new coupling components were built into acid and reactive dyes[35], into metal complex dyes[32, 36] and in pigment dyes[37].

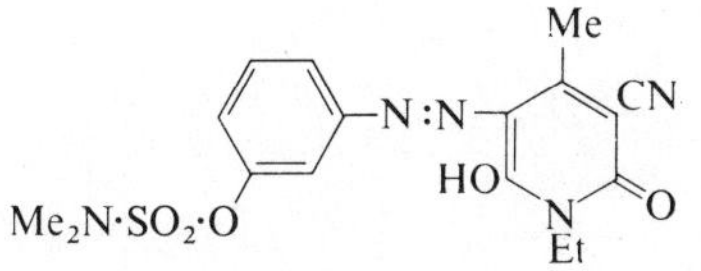

(29)

If one starts with 2-cyanomethylbenzimidazole instead of with cyano-acetamides, one obtains the corresponding condensed system (30)[38]. Besides

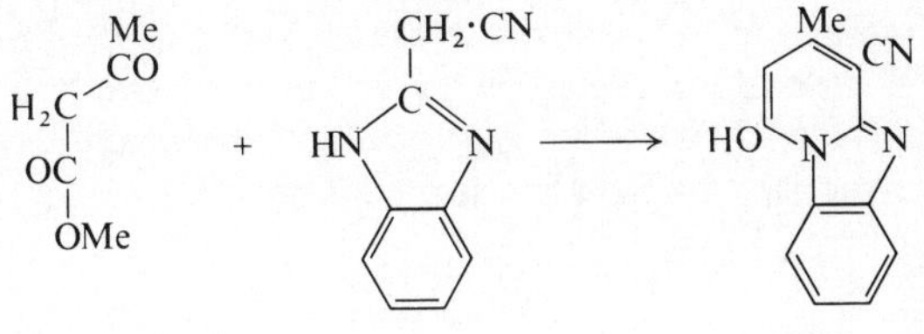

(30)

the 6-hydroxypyrid-2-one derivatives it should also be noted that the *N*-substituted barbituric acid derivatives were successfully applied in yellow disperse dyes by Geigy (31)[39].

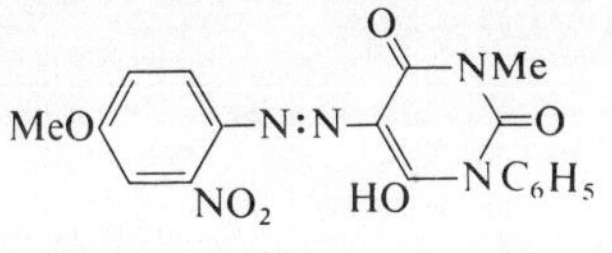

(31)

10.2.1.4 Cyclic azo dyes

Greenish yellow disperse dyes, e.g. (32), for polyester fibres are made through coupling of diazotised 3-aminoindiazole derivatives on triaminopyrimidine and cyclisation through the elimination of ammonia or amines[40].

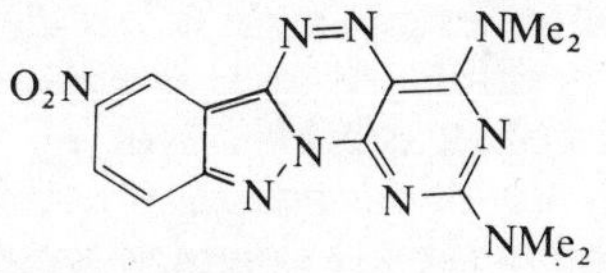

(32)

The strong tendency to cyclise observed in this series and in the case of the self-coupling of *N*-(1-naphthylamine-6-sulphonyl)-H-acid and *N*-(1-naphthylamine-7-sulphonyl)-H-acid was confirmed through further research on macrocyclic azo dyes (33), originally discovered by Allan and his co-workers[41].

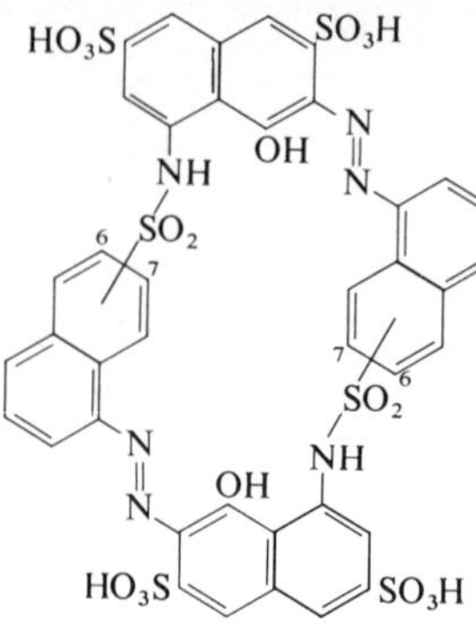

(33a) Sulphonamide groups in 6-positions
(33b) Sulphonamide groups in 7-positions

When the planar dye (33a) is in an alkaline medium, only the proton of the sulphonamide group dissociates; whereas, in the case of the non-planar macrocyclic molecule (33b), the hydroxy group is the most acidic function in the molecule. Recently, another type of macrocyclic azo compounds, e.g. (34) has been synthetised by Skrabal and Zollinger[42].

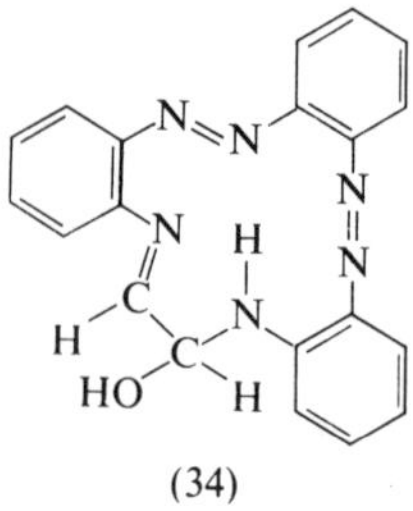

(34)

10.2.2 Metal complex forming azo dyes

10.2.2.1 Metal complex dyes

Hitherto only mixed 1:2 chrome complex dyes were known, as only the trivalent chrome delivers stable 1:1 complexes with *o,o'*-dihydroxyazo dyes the three coordination positions of which are saturated with water. Through conversion with a second (different) mole of *o,o'*-dihydroxyazo dye, the co-ordinately bound water is displaced and, as a result, the mixed 1:2 metal complex is formed. With cobalt complex dyes, on the other hand, because of the prevalent tendency for development of cobalt–nitrogen compounds, water which was the ligand in the 1:1 complex, would, under all preparative conditions, be immediately replaced by a second (identical) dye molecule:

in other words, there are always only symmetric cobalt 1:2-complex dyes available.

In a series of patent reports, ICI has now described for the first time stable cobalt 1:1 complexes in which the residual ligands are nitrogen-containing donors (35) (ammonia or an amine)[43].

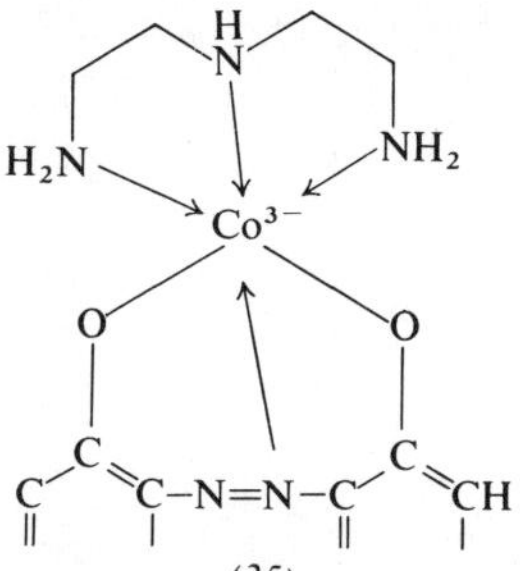

(35)

10.2.2.2 *Metal complex forming azo dyes for dyeing of nickel-stabilised polypropylene fibres*

Allied, Bayer, Hodogaya and other firms developed ranges of dyes, primarily of the monoazo dye type based on heterocyclic diazo components with a ring nitrogen atom next to the azo bridge and with a hydroxyl group in the *ortho*-position of the azo components (e.g. (36))[44]. A patent compilation of further monoazo and hydrazone dyes for nickel-stabilised polypropylene is given in Ref. 45. The nickel complex (37) (0.25–2% of the fibre weight) which is introduced into polypropylene protects the fibre against light and weather influences (photo-oxidation)[46].

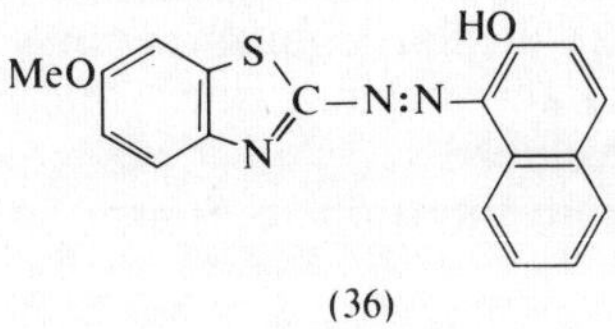

(36)

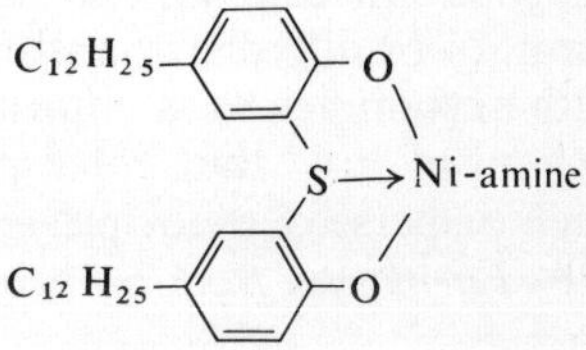

(37)

Dyes of type (36) form nickel complexes with (37). For example, the coupling product of diazotised 2-amino-6-alkoxybenzthiazole and naphth-2-ol (36) is dark red but the resulting nickel complex dye in the fibre is dark blue.

10.3 CARBONYL DYES

10.3.1 Anthraquinone derivatives

10.3.1.1 *New processes for the production of anthraquinone and α-aminoanthraquinone derivatives*

During the last few years industrial activities were concentrated on the development of new economically-advantageous production methods for anthraquinone and for the technically extremely important α-aminoanthraquinone derivatives. A new anthraquinone synthesis, discovered by and introduced into technology by BASF, is carried out in the following way: 1-Methyl-3-phenylindane (38) is first made in the form of a stereoisomeric mixture by an acid-catalysed dimerisation of styrene. This process is, incidentally, relatively easy and inexpensive. The 1-methyl-3-phenylindane is then oxidised in air with V_2O_5 as the catalyst to give *c.* 51% yield of anthraquinone[47].

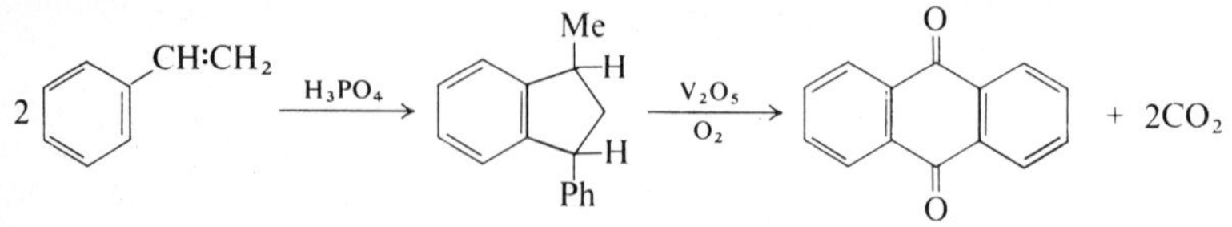

2 CH:CH$_2$ —H_3PO_4→ (38) [Me, H, H, Ph] —V_2O_5, O_2→ [anthraquinone] + $2CO_2$

(38)

10.3.1.2 1,4-Di- and 1,4,5,8-tetra-aminohydroxyanthraquinone derivatives

Practically all technically important red, violet and blue anthraquinone disperse dyes belong to this group. During the last few years a number of new types as well as new syntheses were described.

(a) *Formation over α-halogen derivatives*—Through the systematic work of Farbenfabriken Bayer, the 5,8-dichloro derivatives of the chinizarine[48], of the 1-aminoanthraquinone[49], and of the 1,4-diaminoanthraquinone[50] are readily accessible. The boric acid complexes of 1,4-diaminoanthraquinone and 1,4-diaminohydroxyanthraquinone are chlorinated at 50–80 °C in fuming sulphuric acid or in chlorosulphonic acid and in the presence of halogenation catalysts, as for example, J_2. Subsequently the chlorinated diamino- or dihydroxy-anthraquinone is again released through hydrolysis. With arylamines both chlorine atoms of the 5,8-dichlorochinizarins may be exchanged for the corresponding 5,8-diarylaminochinizarine, whereby green disperse dyes for polyester fibres are developed[51].

(b) *Formation from α-nitro derivatives*—Before nitration in 5-position of

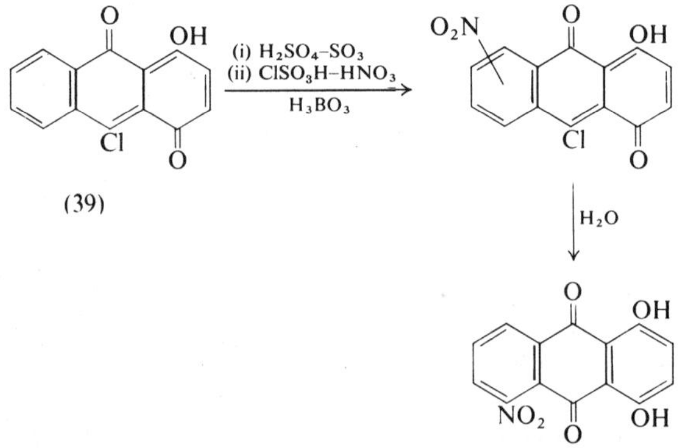

(39) [O, OH, Cl, O] —(i) H_2SO_4–SO_3 (ii) $ClSO_3H$–HNO_3, H_3BO_3→ [O_2N, O, OH, Cl, O] —H_2O→ [O, OH, NO_2, O, OH]

(39)

(40)

the chinizarines is possible, it is necessary to form the so-called Green's compound (39). Now the 2- and 3-positions will no longer interfere with the electrophilic substitution. The obtained mixture of the 5- and 8-nitro compound is hydrolysed to the chinizarine derivative (40)[52].

By reducing the 5-nitro to the 5-amino chinizarine, by brominating in the 6- and 8-position, and finally by stepwise exchange of the bromines in 8- and 6-position for amino and phenoxy (or mercapto) groups, respectively, interesting greenish-blue disperse dyes are formed.

(c) *β-Substitution on α-aminoanthraquinone or hydroxyanthraquinone derivatives*—1,4,5-Triamino-8-hydroxyanthraquinone brominated in the 2- or 3-position yields high affinity polyester dyes[53]. New research from ICI deals with the introduction[54] of a hydroxymethyl group in the 2-position of 1-aminoanthraquinone or hydroxyanthraquinone derivatives. The hydrooxymethyl group is then reacted with amines[55], alcohols, mercaptans or phenols[56], whereby mainly blue disperse dyes are obtained.

1,4-Diaminoanthraquinone-2-carboxylic acid chlorides (41) form, in the presence of amidoximes, their respective oxdiazoles, which are good disperse dyes for synthetic fibres[57].

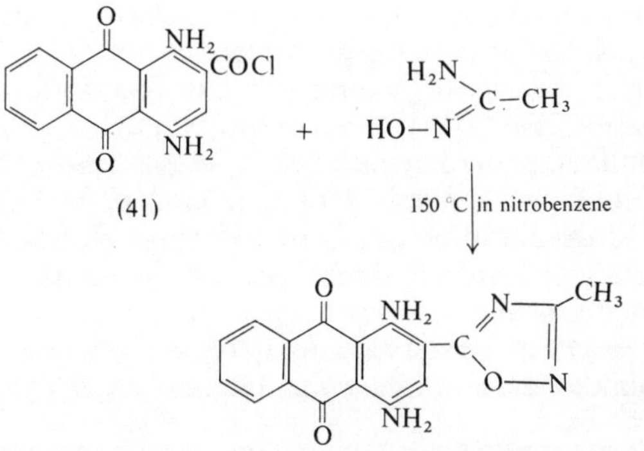

O, NH_2, COCl, O, NH_2 + H_2N, C—CH_3, HO—N

(41)

150 °C in nitrobenzene

O, NH_2, N, CH_3, C, O—N, O, NH_2

An improved synthesis of the technically important turquoise polyester-disperse dyes is obtained through the addition of alkaline cyanides onto 1-amino-4-nitroanthraquinone-2-carboxamides (42) in formamide, instead of the hitherto usual mixture of water and alcohol[58].

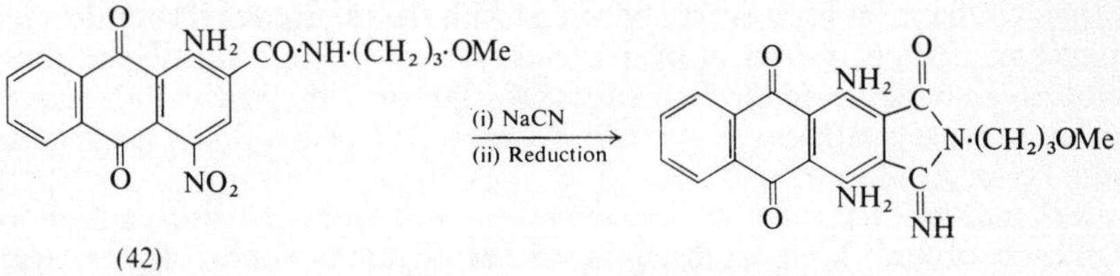

O, NH_2 $CO{\cdot}NH{\cdot}(CH_2)_3{\cdot}OMe$, O, NO_2

(i) NaCN
(ii) Reduction

O, NH_2, O, C, $N{\cdot}(CH_2)_3OMe$, C, NH_2, O, NH

(42)

The esterification of anthraquinone-β-sulphonyl acid chlorides with phenol results in disperse dyes with high fibre affinity[59]; with secondary amines, dialkylamides of β-sulphonyl acid are formed, and these too are suitable for the dyeing of synthetic fibres[60], e.g. (43).

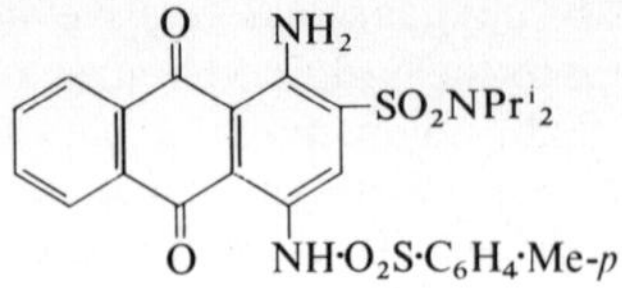

(43)

The reaction of 1-amino-4-hydroxyanthraquinone and 1,4-diamino-anthraquinone derivatives which contain an exchangeable substituent in the 2-position (halogen, sulpho- or phenoxy group) with some alcohols and phenols, is the subject of numerous patent applications. On these reactions one obtains the technically important bright pink and bright red types with improved sublimation fastness. Examples are tetrahydrofurfuryl alcohol[61], glycerol monoaryl ether[62], or *N*-β-hydroxyethylphthalimide[63], 4-hydroxy-diphenyl[64], β-naphth-2-ol[65], β-(3-hydroxyphenyl)-acrylic acid ester[66] or β-(4-hydroxyphenyl)-propionic acid ester[67].

(d) *External substitution on aminoanthraquinone and hydroxyanthraquinone derivatives*—Many of the groups dealt within Section 10.3.1.2(c) can be introduced not only in the core of the molecule, but also on already existing substituents such as in 2-phenoxy groups on anthraquinone. Examples from the literature include the sulphochlorination of β-phenoxy groups and subsequent esterification with phenol[68] or amidisation, for example, with dimethylamine[69] and the acylation of external hydroxy groups, for instance with arylsulphonic acid chlorides[70], dimethylaminosulphonyl chlorides[71] or esters of chlorocarbonic acid[72].

The addition of esters of acrylic acid to 1,4-diaminoanthraquinone-2,3-dicarboxylic acid imides leads to interesting blue polyester dyes like (44)[73].

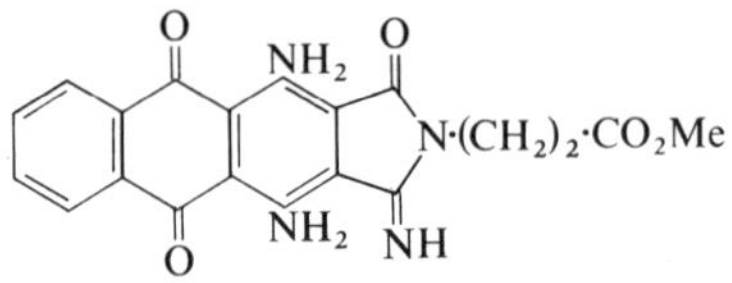

(44)

Provided the polar groups are connected with the external aryl core through a carbon–carbon instead of over a heteroatom, the introduction of these groups can be achieved through direct substitution with *N*-methylolamides[74] or with ketone synthesis, as a result of utilising a Friedel–Crafts mechanism with benzoyl chlorides and $AlCl_3$ [75].

(c) *Addition reactions on quinonimines*—Through dehydrogenation of 1,4-disubstituted aminoanthraquinone or hydroxyanthraquinone with MnO_2, quinonimines or diquinones are easy to make. They smoothly add nucleophilic compounds such as phenols, phenolethers[76], thiophenols[77], hydrochloric acid[78] or sulphinic acids[79]. By heating with aqueous $NaHSO_3$

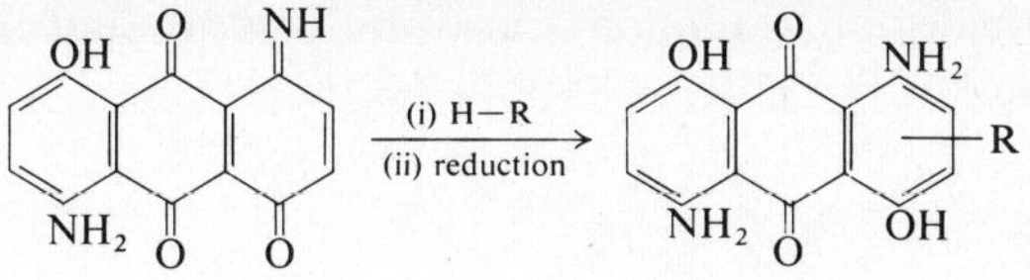

(45)

[R = $-C_6H_4 \cdot OMe$-*p* or $-SO_2Ph$]

solution, substituted amino- or hydroxy-anthraquinone derivatives (45) can be restored. In this way, technically-interesting disperse dyes can be easily produced.

10.3.1.3 1-Cyclohexylamino- and 1-isopropylamino-anthraquinones

The alkylation of the 1-amino group in acidic 1-amino-4-arylaminoanthraquinone dyes causes a desirable shift to a green hue but also a decrease in the light-fastness. As was elaborately investigated by Sandoz, considerable increase in light-fastness in the dyeing of wool and polyamide fibres is achieved if, in the 1-cyclohexylaminoanthraquinone derivatives[80], a secondary carbon atom is bound to the 1-amino group[46]. In contrast to the above-mentioned compounds, the likewise good light-fast isopropylamino dyes are better water soluble. While externally sulphonated dyes of this type were developed by Geigy[81], Bayer worked intensely on the 4-substituted derivatives of the 1-isopropylaminoanthraquinone-5-sulphonic acid[82]. Externally sulphonated 1-isopropylamino-4-arylaminoanthraquinone derivatives with a hydroxy group in the 5-position are greenish-blue to green dyes (47). These are especially valuable for wool and polyamides[83].

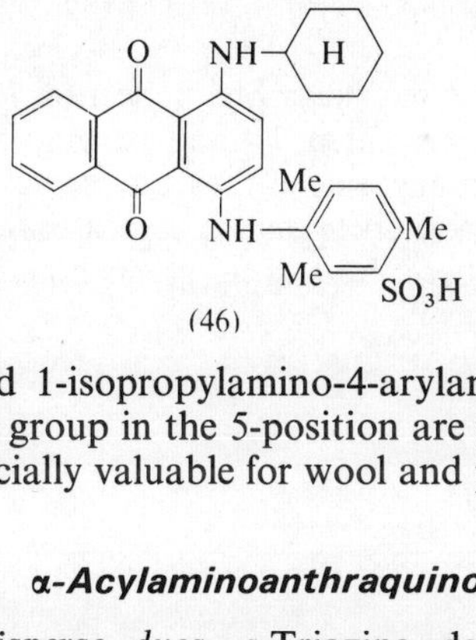

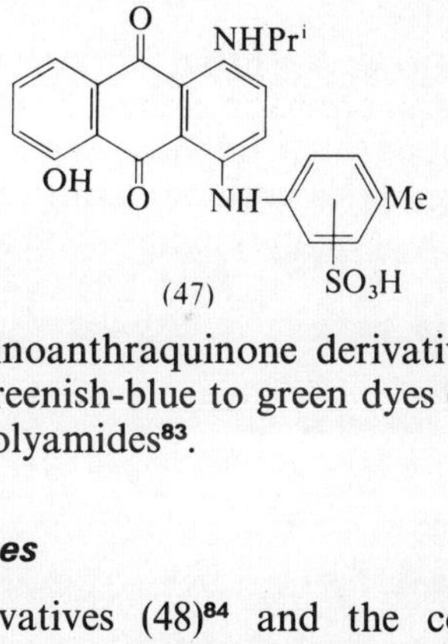

(46) (47)

10.3.1.4 α-Acylaminoanthraquinones

(a) *Disperse dyes*—*s*-Triazine derivatives (48)[84] and the comparably effective urea compounds (49)[84] are interesting derivatives of α-aminoanthraquinone. The condensation of α-aminoanthraquinone with dimethyl

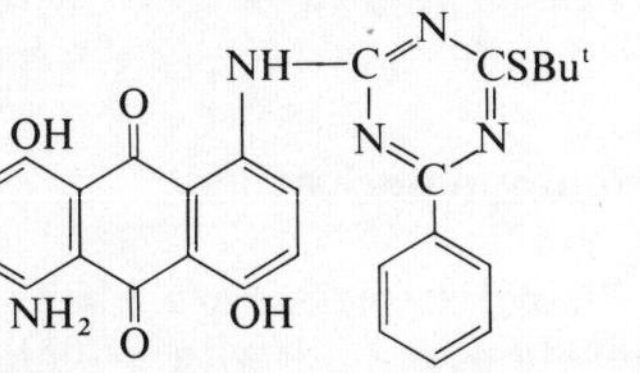

(48) 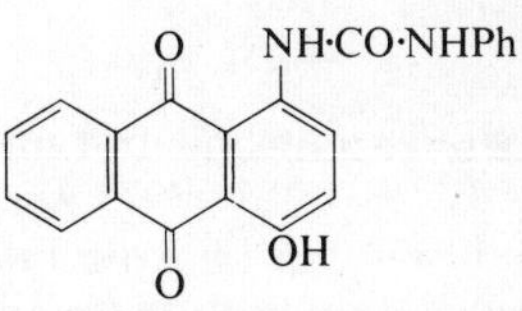 (49)

formamide–$POCl_3$ and cyanoacetic esters produces the yellow vinylogous acylation product (50)[85]. These compounds are so-called 'hard' disperse

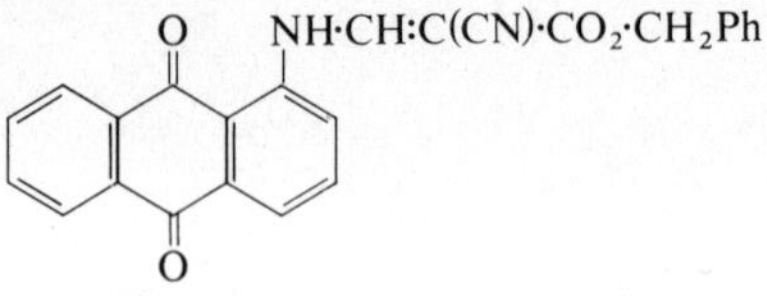

O NH·CH:C(CN)·CO_2·CH_2Ph

O

(50)

dyes. They are of special value in thermosol dyeing of polyester fibres in sublimation-fast hues. In addition, in the well-known yellow 5-acylamino-isothiazolanthrones and in the blue 1-acylaminoanthraquinone-3,4-(*N*)-benzacridones, the acyl groups were also varied[86].

(b) *Pigment dyes*—Yellow 1,5-diacylaminoanthraquinones, red 1,4-di-acylaminoanthraquinones and, depending upon the kind of acyl group, red to blue 1,5-diacylamino-4,8-dihydroxyanthraquinones, are obtained as pigment dyes by using special acylation agents. They show excellent light-fastness, good overlaquering qualities and excellent resistance against migration; for example (51a) is a blueish-violet (51b) a blue dye[87].

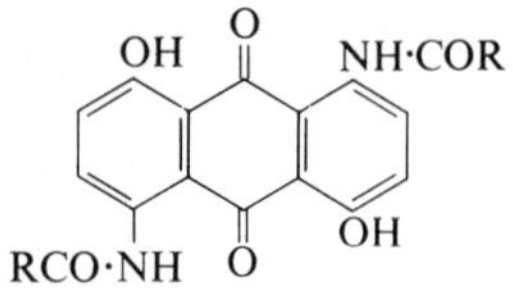

OH O NH·COR

RCO·NH O OH

(51a) R = (Cl, Cl, Cl, Cl-substituted phenyl)

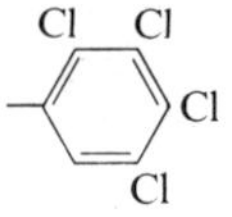

(51b) R = —C₆H₄—CO·NHPh

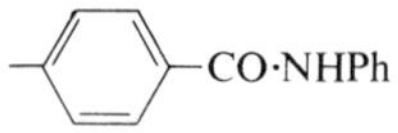

(c) *Reactive dyes*—From 5-nitrochinizarin, described in Section 10.2.2, are obtained, with aromatic amines in the *leuco* form, 1,4-diarylamino-5-amino-anthraquinone; after sulphonating both aryl nuclei they can be acylated on the amino group in position 5 with reactive components such as halogeno-5-triazines, halogenopyrimidines and others, into clear green reactive dyes, e.g. (52)[88].

SO_3H

O NH— —Me

ClC N C—NH O NH— —Me

N N

C

OMe SO_3H

(52)

10.3.2 Benzoquinone and napthoquinone derivatives

Polyester fibres can be dyed into excellent fast black hues when one treats them successively with 4-aminodiphenylamine and with *N*-chlorobenzo-quinone imines (53)[89].

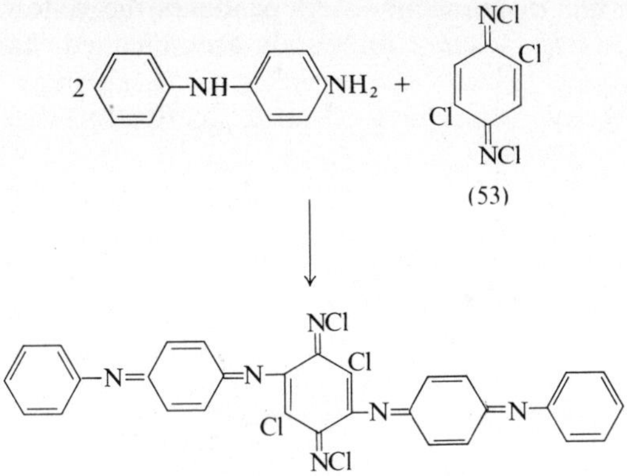

(53)

During the last few years this field was subject to intensive research and development. In the course of this work, amines were produced in the dye-bath through reductive cleavage of diazocyanine dyes[90] while *N*-chloroquinone imines were produced through oxidation of primary aromatic amines with *N,N*-dichloroisocyanourates[91]. Instead of the 4-aminodiphenylamine, hydroxydiphenylamines or hydroxyphenylnaphthylamines can also be inserted[92]. There was repeated research on phthaloylpyrrocolines[93] but they did not prove to be of practical value.

10.3.3 Derivatives of naphthalene-1,8-dicarboxylic acid and of perylenetetracarboxylic acid

In this area the fluorescent yellow, orange and red dyes (54) from Hoechst[94] which, on polyester fibres, result in bright hues with high light- and sublimation-fastness, represent the most interesting developments of the last few years.

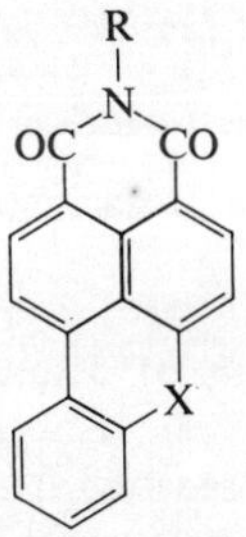

(X = S, O)

(54)

This new category of compounds is reminiscent of the familiar optical brighteners. To achieve the synthesis of the above mentioned xanthene and thioxanthene derivatives, one starts with 4-halogeno-5-nitronaphthalic acid anhydrides, condenses these with phenols or thiophenols, reduces the nitro group and after diazotisation cleaves the nitrogen with copper or copper salts to yield ring closure; finally, using suitable amines, imidisation of the 1,8-dicarboxylic acid anhydride ring occurs[95]. Through substitution with nitro and methoxy groups, the hue can be shifted to blueish red (55)[96].

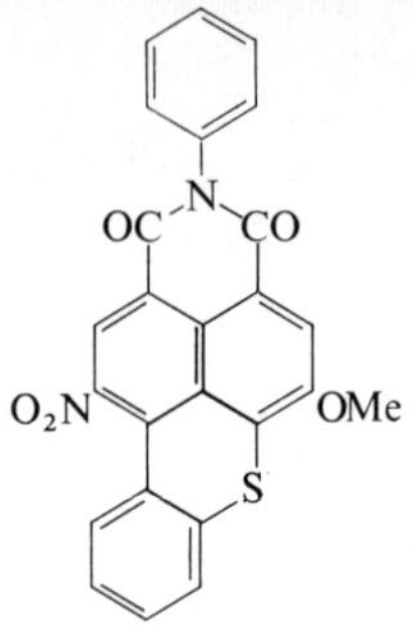

(55)

With *o*-phenylenediamine one obtains *N*-acylimidazoles from the *peri*-dicarboxylic anhydride grouping. The *N*-acylimidazoles have, however, when used as disperse dyes on polyester fibres, a far inferior affinity than the technically more important imides.

bis-Imides of perylenetetracarboxylic acid are important as high purity pigment dyes for lacquers and plastics. The clear *bis*-azobenzol derivative (56a)[97] and the *bis*-acylhydrazides (56b)[98] are new worth while types.

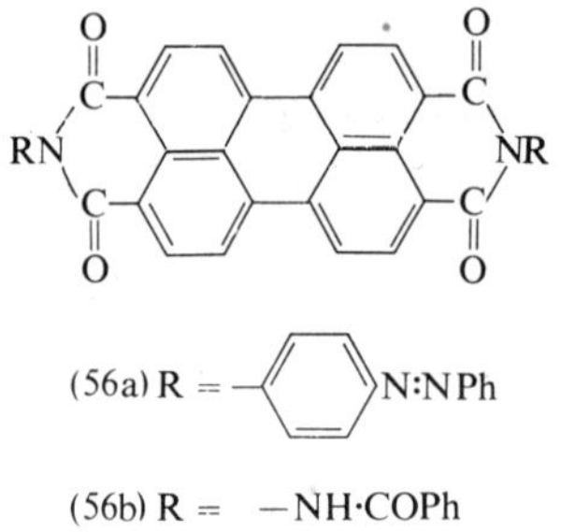

(56a) R = —C₆H₄—N:NPh

(56b) R = —NH·COPh

10.3.4 Cyclic metal complexes of open-chain dicarbonyl compounds

The 1:2 nickel complexes (57) obtained from nitrosated and oximated aryl-amides of acetoacetic acid are very good light-fast yellow to red pigment dyes[99].

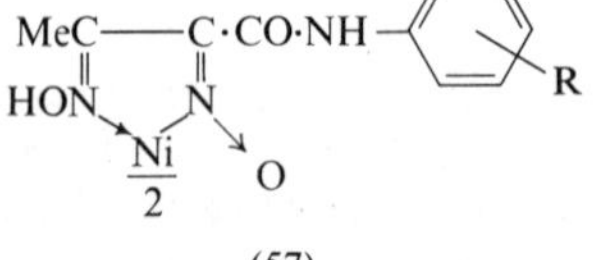

(57)

10.4 PHTHALOCYANINE DYES AND RELATED SYSTEMS

10.4.1 Phthalocyanine dyes

In sulphonated copper and nickel phthalocyanines the bridge links were widely varied between the chromophoric system and the reactive group by using aromatic, arylaliphatic and aliphatic diamines for condensation with phthalocyaninesulphonylchlorides. Thus, dyes can be built up containing the dissociated water soluble disulphimide bridge if the reactive component contains a sulphonic acid chloride group[100], e.g. (58).

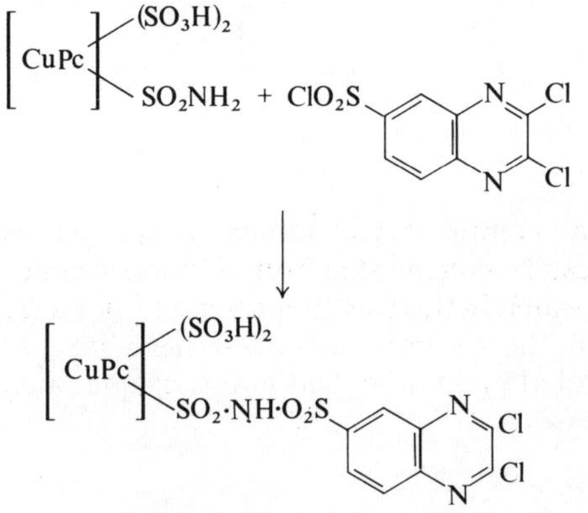

$$\left[\mathrm{CuPc}\right]\begin{matrix}(SO_3H)_2\\ SO_2NH_2\end{matrix} + ClO_2S\text{-(2,3-dichloroquinoxalin-6-yl)}$$

$$\downarrow$$

$$\left[\mathrm{CuPc}\right]\begin{matrix}(SO_3H)_2\\ SO_2\cdot NH\cdot O_2S\text{-(2,3-dichloroquinoxalin-6-yl)}\end{matrix}$$

(58)

The new developing dyes from the family of hexacoordinated phthalocyanine complexes having trivalent cobalt as the central atom are chemically and technically remarkable[101]. Their production proceeds either through exchange of the two chlorine atoms in dihalogenocobalt phthalocyanine complexes for amines or through simultaneous action of amines and oxidation agents on cobalt phthalocyanine[102].

By using diamines, for instance, 3-dimethylaminopropylamine as ligands, one obtains complexes which are soluble in dilute acids. After a brief heating to 80 °C the basic blue cobalt phthalocyanine (59) separates in the form of finely divided particles. These products are well suited for printing on cotton fabrics[103].

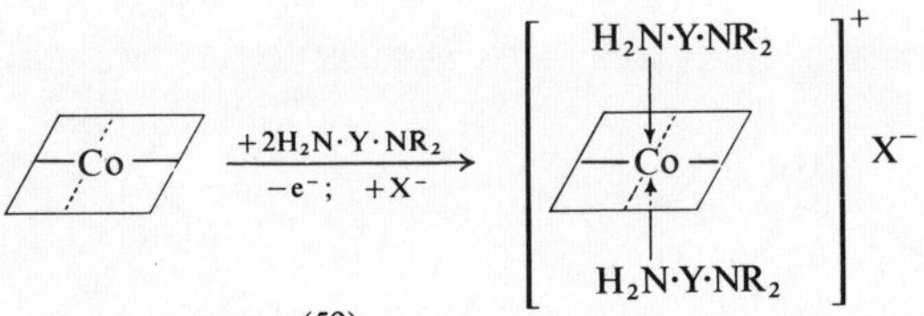

$$\mathrm{Co} \xrightarrow[-e^-;\ +X^-]{+2H_2N\cdot Y\cdot NR_2} \left[\begin{matrix}H_2N\cdot Y\cdot NR_2\\ \downarrow\\ \mathrm{Co}\\ \uparrow\\ H_2N\cdot Y\cdot NR_2\end{matrix}\right]^+ X^-$$

(59)

10.4.2 Aza-annulenes

Dihydrodibenzotetra-aza[14]annulene, a system related to phthalocyanine and a new type of macrocyclic in the form of small glimmering violet-red leaflets, is obtained through the reaction of propargyl aldehyde with *o*-phenylenediamine as transition metal chelates (60) directly from the reactants in the presence of, for instance, nickel salts[104].

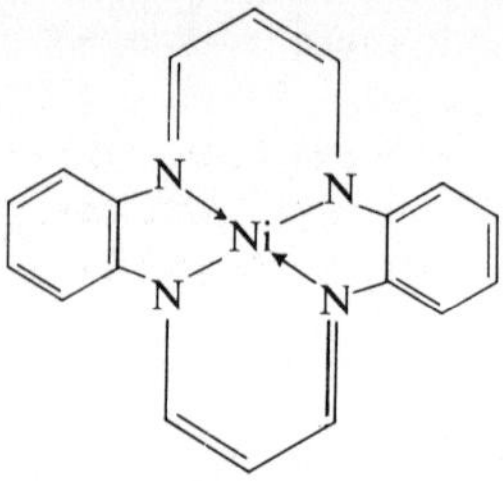

(60)

The transition metal complexes can sublime *in vacuo* above 300 °C without decomposing and can be precipitated from solution in concentrated sulphuric acid by addition of water. In this way they resemble the metal phthalocyanines. In this connection the recently described hexa-aza [14] annulene (34), synthesised by Skrabal[42] and described in this chapter already, is interesting for comparative reasons.

10.5 POLYMETHINE DYES

10.5.1 Neutrocyanines

10.5.1.1 Styryl dyes

With the exception of the fluorescent dyes, the light-fast, bright greenish yellows of the disperse series are derived from the styryl dyes of the type (61).

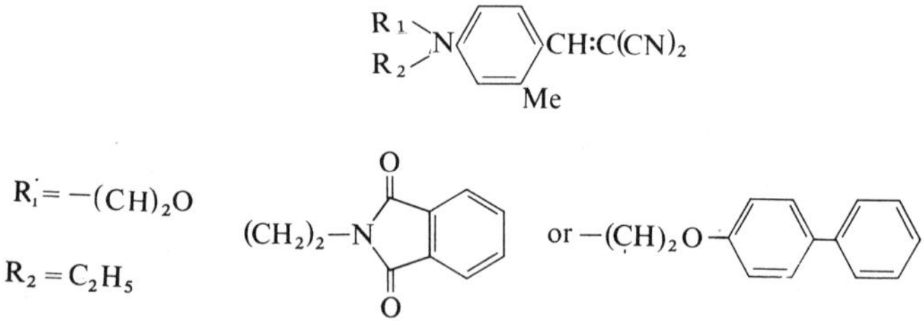

R_1 = $-(CH)_2O$ ... $(CH_2)_2-N$ ... or $-(CH)_2O-$

$R_2 = C_2H_5$

(61)

Similar to the azodisperse dyes, the sublimation-fastness and affinity in this series were improved through the introduction of appropriate substituents in the *N*-alkyl groups, through ring closure between R^1 and the *o*-position to the amino group, as well as through dimerisation of the chromophoric systems over non-conjugated bridges[105].

10.5.1.2 *Quinophthalone dyes*

As with the styryl dyes, this technically important class of yellow disperse dyes for polyester fibres yields, through polar substitutions, better sublimation-fast dyes (62)[106].

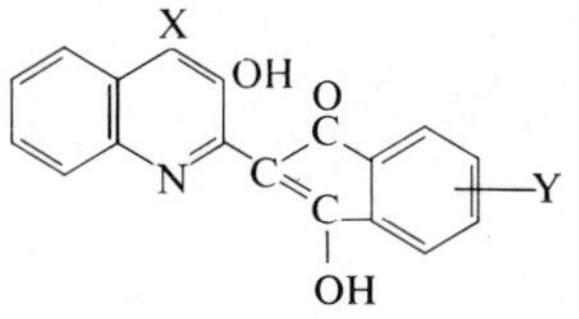

[X = H; Y = —NH·COMe; X = H; Y = —COPh,
X = $-SO_2Me$ or —OPh; Y = H]

(62)

10.5.1.3 *Other neutrocyanine dyes*

Through gradual exchange of both imino groups in amino-imino-isoindolenine first, for methylene active compounds and then, for a heterocyclic amine, bright fluorescent yellow to red dyes for polyester fibres (63) are obtained[107].

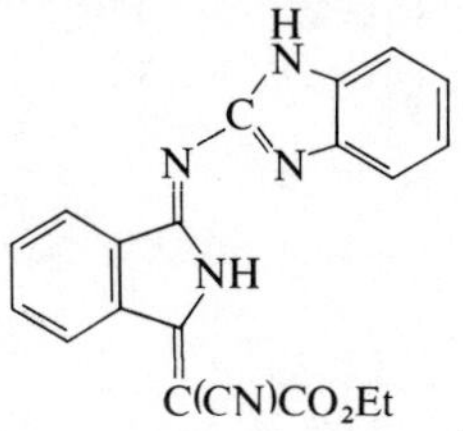

(63)

Yellow polyester dyes of formylmethylenephthalimides and methylene active compounds (64) are related to these compounds[108].

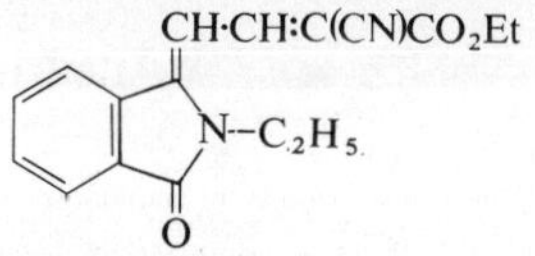

(64)

The azlactone dyes (65) from terephthalic dialdehydes and hippuric acid form an interesting new type of light- and sublimation-fast bright yellow dyes for polyesters[109].

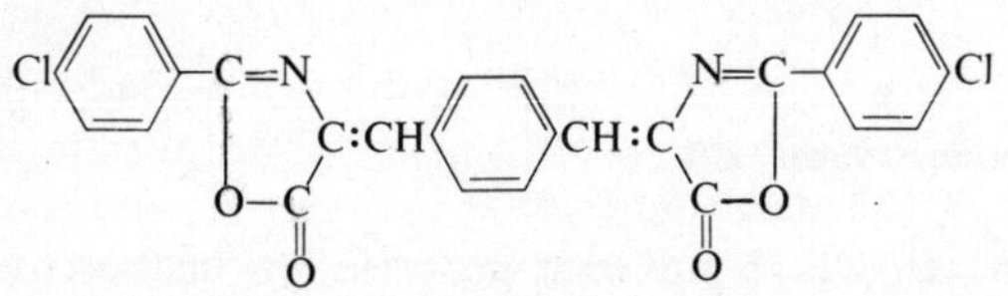

(65)

The technically important yellow fluorescent dyes for synthetic fibres belong to the 7-dialkylaminocoumarin series; they carry a benzimid-2-azolyl or a benzoxaz-2-olyl-group in the 3-position. Sumitomo developed the following variant (66)[110].

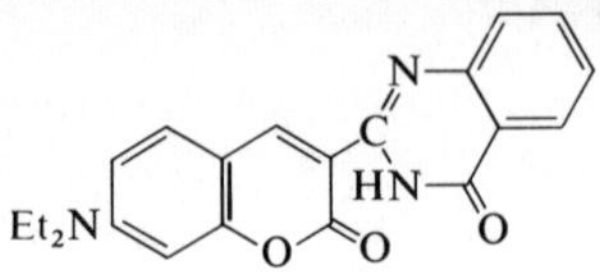

(66)

10.5.2 Polymethine, enamine and hydrazone dyes

The following new developments in this group, which is extremely important for the dyeing of acrylic fibres and acid modified polyester fibres, should be noted: (i) Hemicyanines containing carboxyl groups which equalise especially well, e.g. (67)[111] and (ii) Brilliant greenish yellow enamine dyes with

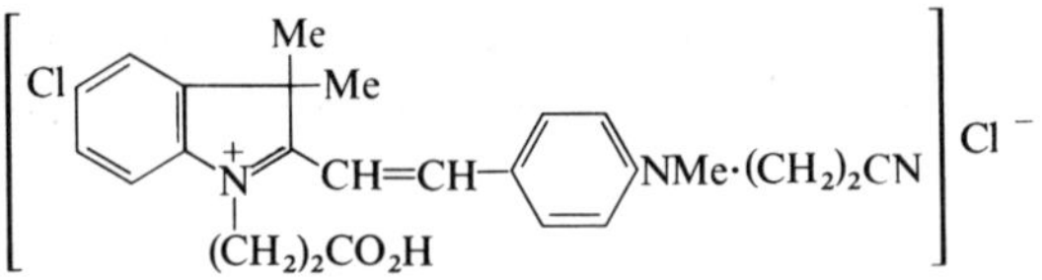

(67)

3-arylpyrazolines as the secondary amine component (68)[112]. Mitsubishi

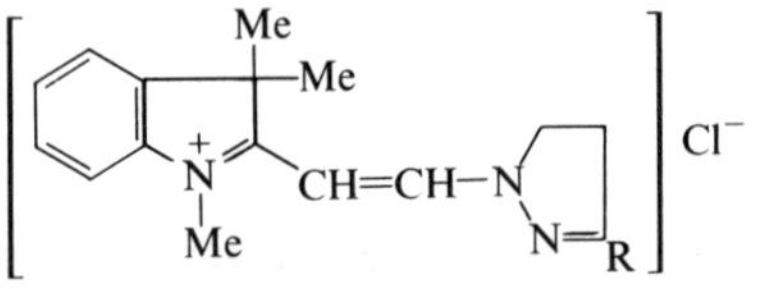

(68)

described hydrazone dyes which are alkylated on the methylene group of the Fischer-base (69)[113].

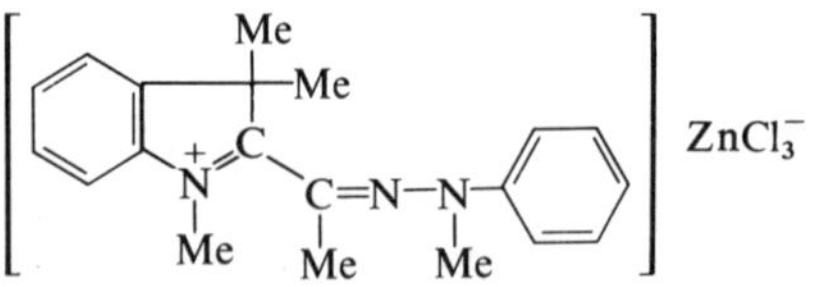

(69)

10.5.3 Diazacyanine dyes

During the last few years no significant progress has been made in this important field, except some new coupling, diazo and hydrazone components[114].

In addition, the dimerisation of dye molecules over non-conjugated bridge links of the nitrogen atom of the aniline coupling component is described[115].

10.5.4 Naphtholactam dyes

This group, which is important for cationic blue dyes, was enlarged through the use of β-cyanoethyl groups to quaternise the amino function (70)[116] as well as through the condensation of the *N*-alkylnaphtholactams with particular amines, such as diphenylamine, *N*-alkylcarbazoles (71) or *N*-alkyldihydrophenoxazines[117].

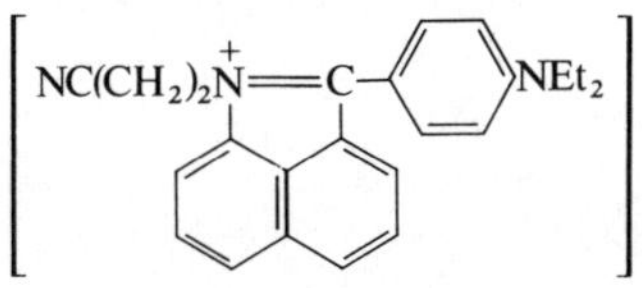

$[NC(CH_2)_2\overset{+}{N}{=}C{-}C_6H_4{-}NEt_2]\,Cl^-$

(70)

$[Et\overset{+}{N}{=}C{-}\ldots{-}NMe]\,Cl^-$

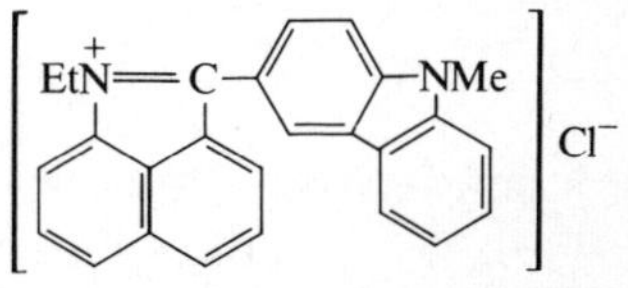

(71)

10.5.5 Triarylmethane and diarylheteroarylmethane dyes

10.5.5.1 Triarylmethane dyes

There are two new directions of development in this otherwise old field. Firstly, Mitsubishi and Sandoz discovered that types of Malachite green dyes with cyanoethylated amino groups (72) result in clear green hues on polyacrylonitrile fibres with very good light-fastness, especially if they carry another alkyl group in the neighbourhood of the central carbon atom[118].

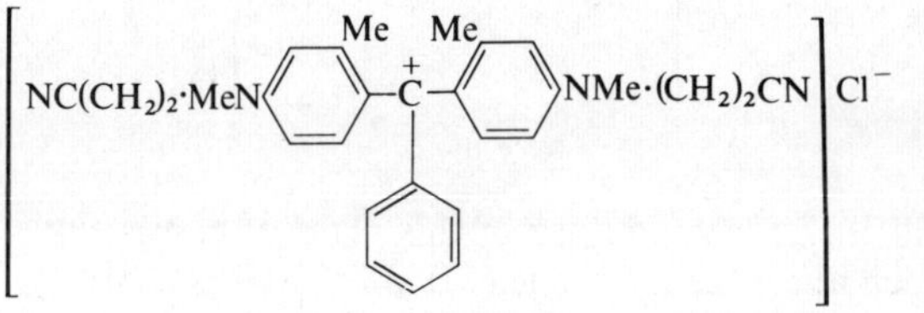

$[NC(CH_2)_2{\cdot}MeN{-}C_6H_3(Me){-}\overset{+}{C}(C_6H_5){-}C_6H_3(Me){-}NMe{\cdot}(CH_2)_2CN]\,Cl^-$

(72)

As the systematic work from Bayer[119] shows, one also obtains very good light-fast, clear scarlet and red hues with 4-alkoxy-4′-alkylaminotriphenylmethane dyes, for example (73). Secondly, Balli and Krüger[120] discovered

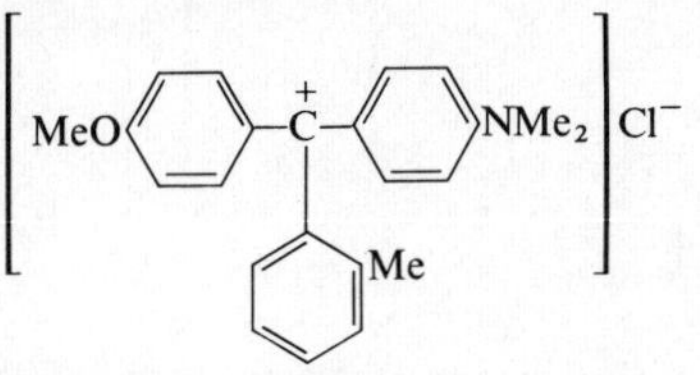

$[MeO{-}C_6H_4{-}\overset{+}{C}(C_6H_4Me){-}C_6H_4{-}NMe_2]\,Cl^-$

(73)

kryochromic systems in spiro compounds of the triphenylmethane series for the first time. These colourless compounds (74) were quickly changed into deeply coloured cations (75) upon cooling in a buffered solution of suitable pH value. The authors were able to formulate conditions for the kryochromic behaviour in which pH value, temperature and concentration of the ring-opening products are parameters.

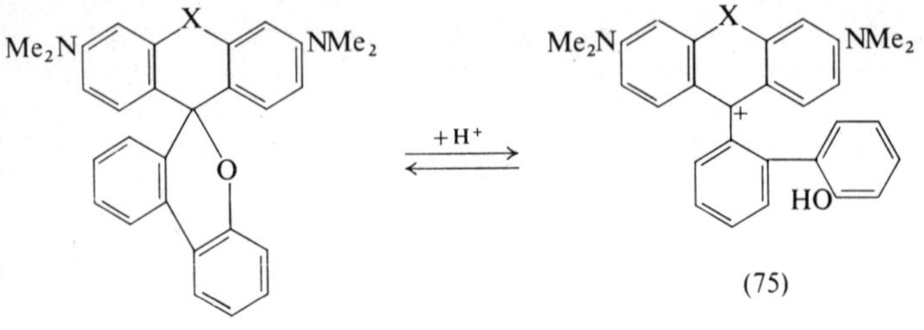

Me₂N X NMe₂ O $\xrightarrow{+H^+}$ Me₂N X NMe₂ + HO

(74) (X = O or S) (75)

10.5.5.2 *Diarylheteroarylmethane dyes*

Dyes of 4,4-diaminobenzophenone and indoline (76; X = NH_2) were described in 1900. They can be used for dyeing polyacrylonitrile fibres, but only with mediocre light-fastness. Surprisingly, the products with only one amino group (76; X = H) yield dyeings with excellent light-fastness.

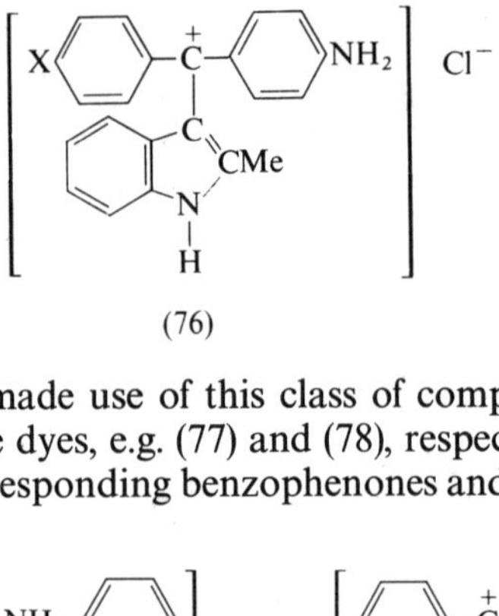

X C⁺ NH_2 Cl^- C CMe N H

(76)

BASF and Bayer[121] made use of this class of compounds and developed valuable green and blue dyes, e.g. (77) and (78), respectively. Their synthesis is possible from the corresponding benzophenones and indolines using acidic

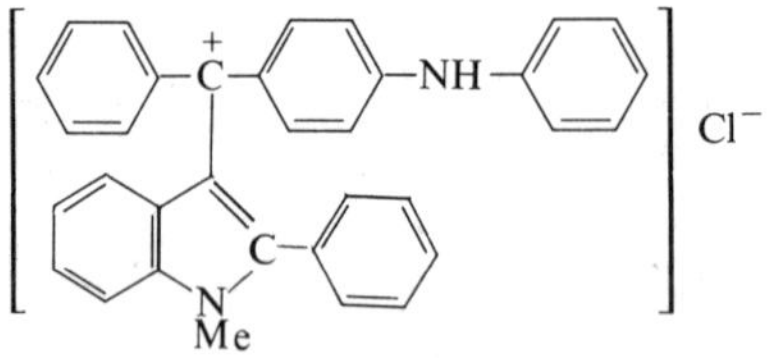

C⁺ NH Cl^- C N Me C⁺ NMe_2 Cl^- C N Me

(77) 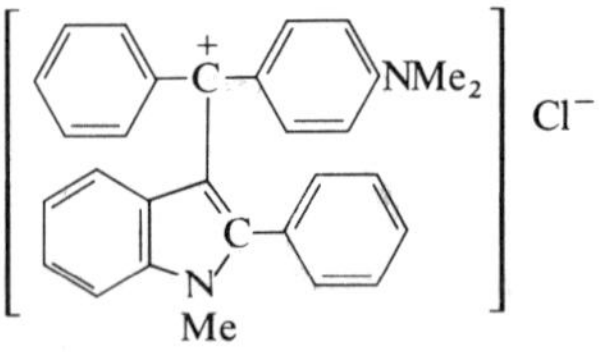 (78)

condensation reagents (for instance, $POCl_3$) as well as by dehydration of the *leuco* compounds.

10.6 OPTICAL BRIGHTENERS

During the last few years no fundamentally new and usable blue fluorescent systems have been discovered. However, new combinations and coupling methods for known fundamental structures have led to new technical developments.

10.6.1 Benzoxazol derivatives

Instead of the proven 1,2-bis-benzoxal-2-yl-ethylene derivatives (light-fast brighteners for synthetic fibres), Farbwerke Hoechst used 1,4-bis-benzoxazol-2-yl-naphthalene derivatives[122]. An interesting observation resulted from Hoechst's innovation: mixtures with α-styrylbenzoxazoles, such as (79) showed not only improved affinities, but also pronounced synergistic effects relative to the intensity of the fluorescence[123].

R^1–(benzoxazol-2-yl)C·CH:CH–C_6H_4–CO_2R^2

(R^1 and R^2 = H or alkyl)

(79)

10.6.2 *vic*-Triazol-2-yl derivatives

Brightening agents, which in a fluorophoric system, for example in stilbene, carry a 1,2,3-triazole group (80a) bound in the 2-position, are generally more light-fast and effective than the corresponding benzotriazole analogues (80b).

X–C_6H_3(Y)–CH:CH–C_6H_3(Y)–X

(80a) X = –N(N=CPh)(N=C–N(N=CPh)(N=CH))

(80b) X = –N(N)(N) (benzotriazol-2-yl)

The preferred synthetic method is the condensation of the appropriate hydrazines with oximinoketones and then dehydrating ring-closure[124]. This field, first investigated and developed by Bayer, was extended by Geigy by the use of a variation in the synthesis. The oximinohydrazone was oxidised to the *vic*-triazol-*N*-oxide (81). In the presence of HCl (or HBr), the ring is converted into the new 5-chloro-*vic*-triazole (82)[125].

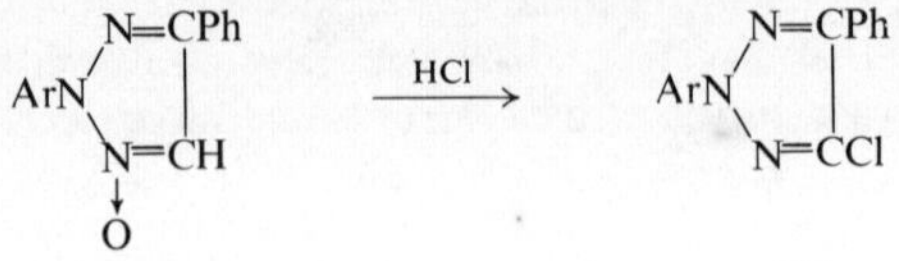

N=CPh / ArN / N=CH → O —HCl→ N=CPh / ArN / N=CCl

(81) (82)

Mitsui Toatsu[126] developed interesting *vic*-triazol brighteners (84) on the basis of the 2-phenyl-*vic*-triazol-4-aldehydes (83).

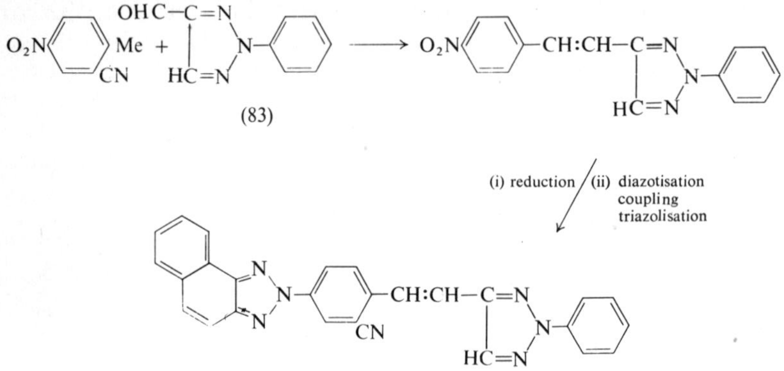

O_2N–C₆H₃(Me)(CN) + OHC–C=N / HC=N / N–Ph (83) ⟶ O_2N–C₆H₃(CN)–CH:CH–C=N / HC=N / N–Ph

(i) reduction / (ii) diazotisation coupling triazolisation

(84)

10.6.3 Coumarin derivatives

Practically all the research with this fluorophore was concentrated on variations of the substituents in the 3- and 7-positions (85). The best effects were achieved with electron-donating groups or hetero rings in the 7-position and with aromatic or heteroaromatic rings (if need be, quaternised) in the 3-position.

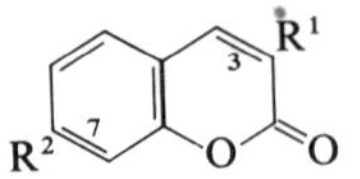

R^1, R^2, 3, 7, O, O

(85)

10.6.4 Inclusion of diphenyl groups in fluorescent systems

Ciba and Nippon Kayaku built diphenyl groups into a whole series of fluorescent systems which, in most cases, led to an increase in the efficiency and affinity for textile substrates in these products, for example (86)[127].

$[C_6H_5\text{–CH:CH–}C_6H_4\text{–}C_6H_4\text{–CH:CH–}C_6H_5]\text{–}(SO_3H)_n$

(n = 1, 2)

(86)

Asymmetrical types, such as (87)[128], are, according to Siegrist[129], easily accessible through the so-called 'anil synthesis'.

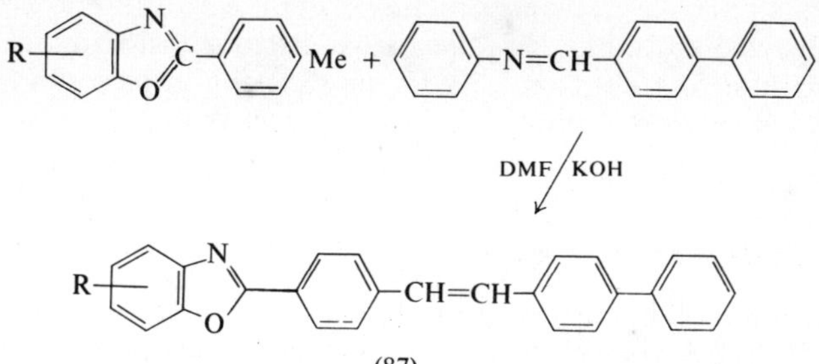

(87)

10.6.5 2-Styrylbenzotriazoles

2-Arylbenzotriazoles are easily accessible through diazotising arylamines, coupling of appropriately substituted aniline derivatives and oxidative ring-closure, to yield the triazole. The direct connection of benzotriazole over N-2 with an ethylene bridge is, however, preparatively more difficult. There is the possibility of alkylation of the isomers substituted at N-1 and condensation of the corresponding benzotriazolyl-2 acetic acid (88) with a benzaldehyde derivative[130].

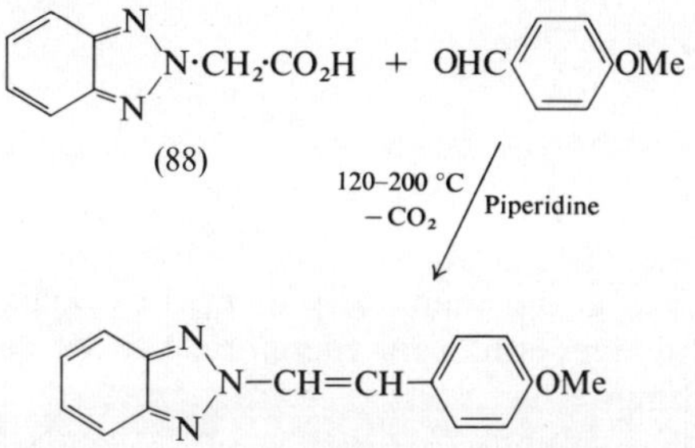

(88)

2-Styrylbenzotriazoles are chlorine-fast brighteners for man-made fibres whose fluorescence is greatly increased by adding another benzotriazole group, as in (89)[131].

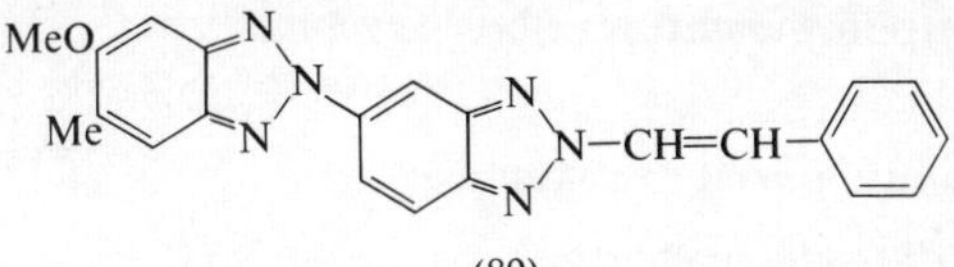

(89)

10.7 REACTIVE DYES (REACTIVE GROUPS)

During the last few years a number of technically-important developments were achieved, even though no basically new reactive groups were introduced. The developments were in optimising combinations based on the most well-known carrier systems and leaving groups. It should be emphasised the realisation that carrier systems (for example, pyrimidine, benzothiazole and isothiazole), which have only a slight activation potential and, consequently, lead to stable dye-fibre bonds, become fast-fixing if they are combined with strong electron withdrawing leaving groups (for example, fluoro- or alkyl-sulphonyl groups) and yield reactive groups which can be applied under long liquor ratio conditions and even at low temperatures.

10.7.1 Aliphatic carrier systems

Through variation of the leaving group X in β-position of the ethylsulphonyl group (90), Farbwerke Hoechst were able to achieve the optimum properties for their reactive dyes with β-substituted ethylsulphonyl groups with regard to solubility, stability in the padding liquor and in the printing paste, substantivity and suitability for the dyeing of wool.

$$DSO_2 \cdot (CH_2)_2X$$

(X = $-OSO_3H$, $-SSO_3H$, $-Cl$, $-O\cdot P(:O)(OH)_2$, $-NEt_2$, $-NMe(CH_2)_2SO_3H$)
(D = Chromophoric part of the dye molecule)

(90)

The first technically satisfactory wool reactive group was found by CIBA to be the α-bromoacrylamide (92) or its antecedent stage, α,β-dibromopropionylamino group (91)[132].

$$\underset{(91)}{DNH\cdot CO\cdot CHBr\cdot CH_2Br} \xrightarrow{-HBr} \underset{(92)}{D\cdot NH\cdot CO\cdot CBr:CH_2}$$

Contrary to cellulose, the chemically unfixed dye parts remain bound to wool fibres as acid dyes and can, therefore, not be completely removed through later washings.

As a result, later washings or other wet-treatments may cause the unfixed dye part to bleed slightly. In order to obtain a usable wet-fast wool reactive dye, at least 90% of the applied quantity of dye should chemically react with the wool protein. This requirement is fulfilled by β-bromoacrylamido dyes in which the fixation yields are 90% or more, with slight variations depending upon the chromophoric system[133].

10.7.2 Heterocyclic carrier systems

10.7.2.1 Pyrimidine derivatives

According to Bayer[134], fluoropyrimidines have proved to be valuable new reactive components. Especially 2,4,6-trifluoro-5-chloropyrimidine, which is easily accessible through substitution by fluorine of the three exchangeable chlorine atoms in tetrachloropyrimidine, is suitable for the syntheses of excellent wet-fast cellulose reactive dyes (93) which are for all application processes equally well appropriate and suitable for wool and polyamide dyes with fixation yields of *c.* 95%. The high acid stability of the system is thereby an advantage[135].

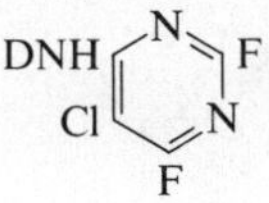

DNH, F, Cl, N, F

(93)

New technical methods of high temperature chlorination are described by H. Holtschmidt and co-workers[136] for the synthesis of tetrachloropyrimidine (94).

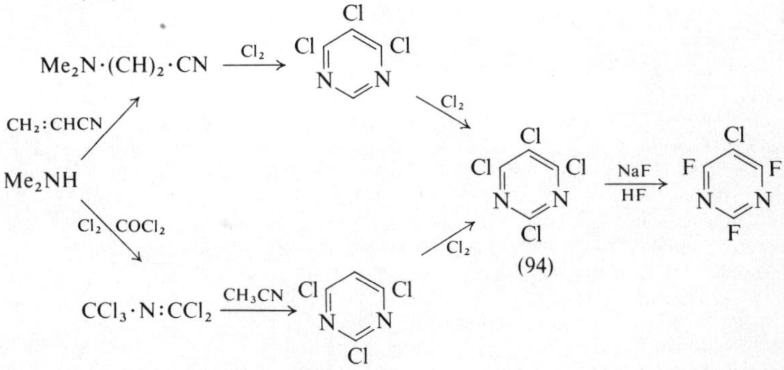

$Me_2N\cdot(CH)_2\cdot CN$ $\xrightarrow{Cl_2}$ trichloropyrimidine $\xrightarrow{Cl_2}$ (94)

$CH_2{:}CHCN$, Me_2NH, Cl_2 $COCl_2$

$CCl_3\cdot N{:}CCl_2$ $\xrightarrow{CH_3CN}$ trichloropyrimidine $\xrightarrow{Cl_2}$ (94)

(94) $\xrightarrow[HF]{NaF}$ trifluorochloropyrimidine

(94)

10.7.2.2 *s-Triazine derivatives*

Monochlorotriazinylamino dyes are considerably less active than the above derivatives and they are, therefore, not very suitable for fast fixation processes and for dyeing in long baths. Here, too, the substitution of fluorine for chlorine, for example (95) can create a considerable increase in the reactivity and a dye-fibre binding which is, of course, the same as that of the respective chlorotriazine dye[137].

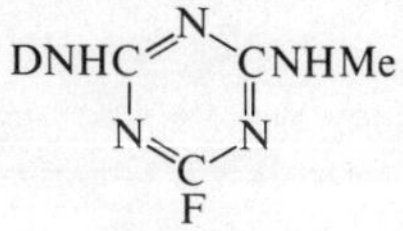

DNHC, CNHMe, N, N, C, F

(95)

Systematic research by ICI showed that the fixation yields of monochlorotriazinylamino dyes can be increased so much through the insertion of two reactive groups in the molecule, as in (96) that it is not necessary to remove a considerable part of the hydrolysate through washing. This is of considerable importance for the printing of textiles[138].

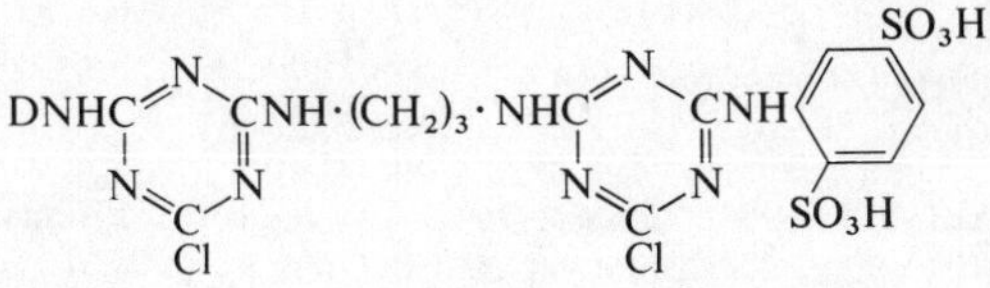

DNHC, CNH·$(CH_2)_3$·NHC, CNH, SO_3H, SO_3H, Cl, Cl

(96)

10.8 GENERAL BIBLIOGRAPHY

In 1970 and 1971 three additional volumes of Venkataraman's well-known standard handbook on dyes have been published[139]. They contain chapters on all classes of dyes, on intermediates, colours (electronic states of organic molecules) and brightening agents, written by 26 scientists from academic and industrial institutions.

A short treatise of the scientific fundamentals of the chemistry and the application of dyes was published by Rys and Zollinger[140].

References

1. Hoechst. (1967). *Belg. Pat.*, 712 351
2. Ciba. (1967). *German Offenlegungsschrift*, 1 619 406
3. Bayer. (1966). 1 290 915; (1966). *Belg. Pat.*, 694 264; (1967). *German Offenlegungsschrift*, 1 619 504; (1967). *Belg. Pat.*, 722 046; Compare: Cassella; (1966). *Belg. Pat.*, 699 118
4. Bayer. (1934). *Ger. Pat.*, 658 841; *Frdl.*, **24,** 679. Stepanov, B. J. (1962). *Chemistry of Natural and Synthetic Colouring Matters*, 451. (New York, London: Academic Press)
5. ICI. (1967). *Belg. Pat.*, 1 807 642; Cassella. (1968). *German Offenlegungsschrift*, 1 809 920
6. Sandoz. (1968). *Belg. Pat.*, 734 766; (1968). *German Offenlegungsschrift*, 1 946 459; Cassella. (1968). *German Offenlegungsschrift*, 1 809 921
7. ICI. (1967). *Brit. Pat.*, 1 220 448
8. Bayer. (1966). *German Offenlegungsschrift*, 1 543 620
9. Bayer. (1969). *German Offenlegungsschrift*, 1 917 165
10. Bayer. (1968). *German Offenlegungsschrift*, 1 811 717
11. Eastman Kodak. (1962). *U.S.A. Pat.*, 3 148 178: (1965). *Belg. Pat.*, 687 986; (1965). *U.S.A. Pat.*, 3 424 741; (1965). *Belg. Pat.*, 687 587; (1964). *Belg. Pat.*, 669 062; (1965). *Belg. Pat.*, 687 987; (1967). *U.S.A. Pat.*, 3 491 082; (1965). *U.S.A. Pat.*, 3 394 121
12. BASF. (1965). *French Pat.*, 1 466 675
13. Eastman Kodak. (1968). *German Offenlegungsschrift*, 1 923 592; (1968). *French Pat.*, 2 008 404; Sandoz. (1966). *Brit. Pat.*, 1 166 347; BASF. (1968)
14. Cassella. (1965). *Belg. Pat.*, 688 712; von Brachel, H. (1968). *Chimia Supplement*, 149
15. Hoechst. (1969). *German Offenlegungsschrift*, 1 917 691; ICI. (1966). *French Pat.*, 1 531 146
16. BASF. (1966). *French Pat.*, 1 533 677; (1967). *German Offenlegungsschrift*, 1 644 066
17. Du Pont. (1967). *U.S.A. Pat.*, 3 552 905
18. Sandoz. (1964). *Belg. Pat.*, 660 461; Ciba. (1967). *Belg. Pat.*, 716 197; Bayer. (1968). *Belg. Pat.*, 733 076
19. BASF. (1967). *Swiss Pat.*, 499 587
20. Lange, G. (1968). *Chimia Supplement*, 154
21. Schneider, L. (1968). *Helv. Chim. Acta*, **51,** 67; compare (1968). *Chimia Supplement*, 154
22. Bayer. (1965). *Brit. Pat.*, 1 094 343
23. Du Pont. (1965). *Brit. Pat.*, 1 114 944
24. Fuchs, O. (1968). *Chimia Supplement*, 159; Hoechst. (1960). *Ger. Pat.*, 1 188 299; *Ger. Pat.*, 1 235 467
25. Hoechst. (1961). *Ger. Pat.*, 1 217 008
26. Bayer. (1965). *Brit. Pat.*, 1 094 943; (1965). *Brit. Pat.*, 1 097 375; (1965). *Brit. Pat.*, 1 111 780; (1966). *Brit. Pat.*, 1 134 257
27. Wippel, H. G. (1969). *Melliand Textilberichte*, 1090
28. BASF. (1965). *Brit. Pat.*, 1 112 146; Bayer. (1966). *Brit. Pat.*, 1 134 579
29. BASF. (1959). *Ger. Pat.*, 1 117 232
30. Hoggarth, E. (1949). *J. Chem. Soc.*, 1163
31. Hoechst. (1963). *Belg. Pat.*, 657 118; (1965). *Belg. Pat.*, 667 795
32. Cassella. (1965). *French Pat.*, 1 489 492; Ciba. (1968). *German Offenlegungsschrift*, 1 956 142; compare ICI. (1968). *German Offenlegungsschrift*, 1 961 784
33. Sandoz. (1968). *Belg. Pat.*, 733 186; compare Ciba. (1968). *German Offenlegungsschrift*, 1 927 213
34. ICI. (1968). *German Offenlegungsschrift*, 1 932 806

35. Ciba. (1968). *German Offenlegungsschrift*, 1 930 491
36. Ciba. (1968). *German Offenlegungsschrift*, 1 930 492
37. ICI. (1968). *German Offenlegungsschrift*, 1 961 786; BASF. (1969). *German Offenlegungsschrift*, 1 910 475
38. BASF. (1970). *German Offenlegungsschrift*, 2 004 488
39. Geigy. (1965). *German Offenlegungsschrift*, 1 469 763
40. Geigy. (1968). *German Offenlegungsschrift*, 1 941 542
41. Allan, Z. I., Podstata, J. and Jarkowsky, J. (1969). *Collect. Czechoslov. Chem. Commun.*, **34**, 282; Vrba, Z. and Allan, Z. I. (1970). *Abstracts 4th Internat. Dye Symposium*, p.34
42. Skrabal, P. and Zollinger, H. (1971). *Helv. Chim. Acta*, **54**, 1069
43. ICI. (1965). *Ger. Pat.*, 1 094 746; (1965). *Ger. Pat.*, 1 139 503; (1965). *Ger. Pat.*, 1 138 833
44. I.C.I. (1961). *Belg. Pat.*, 614 566; Allied. (1962). *Belg. Pat.*, 632 652/3; Toyo Rayon. (1965). *Japan Pat.*, 8 954/68; Hodogaya. (1966). *German Offenlegungsschrift*, 1 719 086
45. Bayer. (1963). *German Offenlegungsschrift*, 1 444 240; (1963). *German Offenlegungsschrift*, 1 544 536; (1965). *German Offenlegungsschrift*, 1 544 531/2; (1968). *German Offenlegungsschrift*, 1 804 133; (1969). *German Offenlegungsschrift*, 1 958 707
46. Cyanamid Amer. (1966). *U.S.A. Pat.*, 3 454 350
47. BASF. (1969). *German Offenlegungsschrift*, 1 934 063
48. Bayer. (1963). *Ger. Pat.*, 1 199 279
49. Bayer. (1962). *Ger. Pat.*, 1 151 517; (1962). *Ger. Pat.*, 1 161 252
50. Bayer. (1962). *Ger. Pat.*, 1 173 907
51. Bayer. (1965). *Brit. Pat.*, 1 136 932/3
52. Bayer. (1966). *Brit. Pat.*, 1 130 499; (1965). *Brit. Pat.*, 1 090 190
53. Bayer. (1966). *Swiss Pat.*, 477 597
54. ICI. (1968). *French Pat.*, 2 009 682
55. ICI. (1968). *German Offenlegungsschrift*, 1 927 309
56. ICI. (1968). *German Offenlegungsschrift*, 1 927 312; (1969). *Belg. Pat.*, 733 588
57. BASF. (1969). *French Pat.*, 2 035 130
58. BASF. (1969). *Ger. Pat.*, 1 918 696
59. Bayer. (1965). *German Offenlegungsschrift*, 1 644 587
60. Eastman Kodak. (1967). *Belg. Pat.*, 713 407
61. Amer Aniline. (1964). *U.S.A. Pat.*, 3 428 411
62. Bayer. (1967). *Belg. Pat.*, 724 886
63. Eastman Kodak. (1965). *U.S.A. Pat.*, 3 394 144
64. Hoechst. (1964). *Ger. Pat.*, 1 289 214
65. Ciba-Geigy. (1969). *German Offenlegungsschrift*, 2 063 235
66. Hoechst. (1963). *Ger. Pat.*
67. Mitsubishi Chem. Ind, (1965). *Japan Pat.*, 2 678/70; (1965). *German Offenlegungsschrift*, 1 619 602
68. Geigy. (1963). *Belg. Pat.*, 650 735; Nippon Kayaku. (1964). *Japan Pat.*, 20 792/66; Bayer. (1964). *Belg. Pat.*, 666 632; Sumitomo Chem. (1965). *Japan Pat.*, 08 106/70; Geigy. (1963). *Belg. Pat.*, 650 733; Steinberg, Y. B., *Dyes Res. Inst.* (1966). *Rouman Pat.*, 1 978 838
69. BASF. (1965). *Brit. Pat.*, 1 139 310
70. Sumitomo Chem. (1965). *Japan Pat.*, 14 118/69
71. Geigy. (1968). *German Offenlegungsschrift*, 1 917 024
72. Nippon Kayaku. (1965). *Japan Pat.*, 24 909/69
73. BASF. (1967). *Belg. Pat.*, 714 010
74. ICI. (1966). *French Pat.*, 1 522 602; General Aniline. (1966). *French Pat.*, 1 546 344; Bayer. (1968). *Belg. Pat.*, 728 536
75. American Aniline. (1965). *U.S.A. Pat.*, 3 426 047
76. Bayer. (1963). *Ger. Pat.*, 1 228 734
77. Geigy. (1967). *Brit. Pat.*, 1 193 540
78. Bayer. (1963). *Ger. Pat.*, 1 266 424
79. Bayer. (1964). *Ger. Pat.*, 1 293 363
80. Sandoz. (1963). *Swiss Pat.*, 452 084; *Ger. Pat.*, 1 269 266; (1964). *French Pat.*, 1 455 722; (1963). *Ger. Pat.*, 1 267 359; (1967). *French Pat.*, 1 562 130; (1965). *Belg. Pat.*, 676 218; (1966). *German Offenlegungsschrift*, 1 644 659; (1967). *German Offenlegungsschrift*, 1 644 664; (1967). *Belg. Pat.*, 709 646; (1967). *French Pat.*, 1 517 006; (1967). *Belg. Pat.*, 712 362; (1967). 1 807 746; (1968). *Belg. Pat.*, 731 013; (1968). *Swiss Pat.*, 488 565

81. Geigy. (1965). *German Offenlegungsschrift*, 1 644 541
82. Bayer. (1968). *German Offenlegungsschrift*, 1 768 152; (1968). *German Offenlegungsschrift*, 1 644 626; (1969). *German Offenlegungsschrift*, 1 952 536
83. Bayer. (1969). *German Offenlegungsschrift*, 1 932 647; (1969). *German Offenlegungsschrift*, 1 932 646
84. Bayer. (1966). *Brit. Pat.*, 1 123 946; BASF. (1968). *German Offenlegungsschrift*, 1769 610; Bayer. (1965). *German Offenlegungsschrift*, 1 469 741
85. Bayer. (1964). *Brit. Pat.*, 1 044 499; (1966). *Brit. Pat.*, 1 131 005; (1966). *Brit. Pat.*, 1 119 429; (1968). *French Pat.*, 1 549 583
86. Sandoz. (1968). *German Offenlegungsschrift*, 1 900 305; Sumitomo Chem. (1965). *Japan Pat.*, 15 317/69; (1966). *Japan Pat.*, 21 431/69; Cassella. (1961). *Ger. Pat.*, 1 278 391; (1968). *German Offenlegungsschrift*, 1 769 209
87. Bayer. (1969). *German Offenlegungsschrift*, 1 907 783
88. Bayer. (1966). *German Offenlegungsschrift*, 1 644 611
89. Bayer. (1960). *Ger. Pat.*, 1 139 810; (1960). *Ger. Pat.*, 1 284 933; (1961). *Ger. Pat.*, 1 284 937
90. BASF. (1964). *Ger. Pat.*, 1 276 595
91. Bayer. (1965). *German Offenlegungsschrift*, 1 469 721; (1967). *German Offenlegungsschrift*, 1 619 519; (1967). *German Offenlegungsschrift*, 1 619 519
92. Toyo Kabushiki Kaisha. (1966). *Brit. Pat.*, 1 175 759
93. Allied Chem. (1965). *U.S.A. Pat.*, 3 413 302; Tilak, B. D. and Venkateswaran, P. R. (1968). *Chimia Supplement*, 95
94. Fuchs, O. and Tröster, H. (1970). *Abstracts 4th Internat. Dye Symposium*, p.66
95. Hoechst. (1967). *German Offenlegungsschrift*, 1 569 761, *Ger. Pat.*, 1 297 259
96. Hoechst. (1965). *German Offenlegungsschrift*, 1 569 737
97. BASF. (1962). *Ger. Pat.*, 1 257 096
98. Bayer. (1967). *German Offenlegungsschrift*, 1 569 757, 1 569 760
99. BASF. (1967). *Belg. Pat.*, 709 296
100. Bayer. (1964). *German Offenlegungsschrift*, 1 569 726
101. Vollmann, H. and Mertens, P. (1970). *Abstracts 4th Internat. Dye Symposium*, p. 40
102. Bayer. (1967). *German Offenlegungsschrift*, 1 569 725; (1967). *German Offenlegungsschrift*, 1 569 753; (1969). *German Offenlegungsschrift*, 1 940 180
103. Bayer. (1967). *German Offenlegungsschrift*, 1 619 546
104. Hiller, H. (1970). *Abstracts 4th Internat. Dye Symposium*, p.28; Hillier, H., Dimroth, P. and Pfitzner, H. (1968). *Liebigs Ann. Chem.*, **717**, 137
105. ICI. (1965). *U.S.A. Pat.*, 3 483 218; (1965). *Brit. Pat.*, 1 110 714; BASF. (1969). *German Offenlegungsschrift*, 1 959 706; BASF. (1967). *German Offenlegungsschrift*, 1 569 678
106. Bayer. (1967). *Belg. Pat.*, 725 664; BASF. (1967). *Austrian Pat.*, 1 569 672; Hoechst. (1968). *Belg. Pat.*, 733 604; (1968). *Belg. Pat.*, 733 605
107. Bayer. (1966). *French Pat.*, 1 537 299
108. BASF. (1968). *German Offenlegungsschrift*, 1 812 321
109. BASF. (1968). *Ger. Pat.*, 1 809 949
110. Sumitomo Chem. (1969). *German Offenlegungsschrift*, 2 005 933
111. Sandoz. (1969). *German Offenlegungsschrift*, 1 929 417
112. Bayer. (1969). *German Offenlegungsschrift*, 1 945 053
113. Mitsubishi. (1969). *German Offenlegungsschrift*, 2 043 192
114. Geigy. (1967). *Belg. Pat.*, 716 623; Etabl. Kuhlmann. (1964). *Brit. Pat.*, 1 104 391; Hoechst. (1965). *Brit. Pat.*, 1 094 309; Hodogaya. (1969). *German Offenlegungsschrift*, 2 022 624; (1969). *German Offenlegungsschrift*, 2 022 625
115. Mitsubishi. (1962). *Belg. Pat.*, 634 376
116. Yorkshire Dyeware. (1968). *Brit. Pat.*, 1 225 545; BASF. (1969). *German Offenlegungsschrift*, 1 944 797
117. BASF. (1966). *German Offenlegungsschrift*, 1 569 660; Bayer. (1970). *Belg. Pat.*, 763 853; (1967). *German Offenlegungsschrift*, 1 569 606
118. Mitsubishi. (1965). *French Pat.*, 1 470 451; Sandoz. (1966). *German Offenlegungsschrift*, 1 619 638; (1967). *German Offenlegungsschrift*, 1 644 662
119. Bayer. (1968). *German Offenlegungsschrift*, 1 811 336; 1 811 337; 1 811 338; 1 811 651; 1 811 652
120. Balli, H. and Krüger, U. (1970). *Abstracts 4th Internat. Dye Symposium*, p. 62
121. BASF. (1963). *Ger. Pat.*, 1 254 118; Bayer. (1966). *German Offenlegungsschrift*, 1 619 528;

(1967). *Belg. Pat.*, 714 130; (1968). *Brit. Pat.*, 1 195 004; (1967). *German Offenlegungsschrift*, 1 569 749; (1967). *German Offenlegungsschrift*, 1 569 750

122. Hoechst. (1964). *Ger. Pat.*, 1 282 592; (1964). 1 301 791; (1966). 1 302 052; compare. (1969). *German Offenlegungsschrift*, 1 955 310
123. Hoechst. (1967). *Belg. Pat.*, 726 255
124. Bayer. (1966). *Brit. Pat.*, 1 113 918
125. Geigy. (1969). *German Offenlegungsschrift*, 2 029 157; (1969). *German Offenlegungsschrift*, 2 029 142
126. Mitsui Toatsu. (1967). *German Offenlegungsschrift*, 1 805 371
127. Ciba. (1967). *Belg. Pat.*, 721 754; (1969). *German Offenlegungsschrift*, 2 009 156; Nippon Kayaku. (1966). *Japan Pat.*, 24 212/68; (1966). *Japan Pat.*, 23 025/69
128. Ciba. (1967). *German Offenlegungsschrift*, 1 768 204; compare. (1965). *German Offenlegungsschrift*, 1 594 822
129. Siegrist, A. E. (1967). *Helv. Chimica Acta*, 50, 906–957
130. Bayer. (1967). *German Offenlegungsschrift*, 1 670 841
131. Bayer. (1967). *German Offenlegungsschrift*, 1 670 852
132. Ciba. (1965). *Belg. Pat.*, 683 734
133. Zollinger, H. (1971). *Textilveredlung*, **6**, 57; Rouette, H. K., Wilshire, J. F. K., Yamase, I. and Zollinger, H. (1971). *Textile Res., J.*, **41**, 518
134. Bien, H. S. and Klauke, E. (1970). *Abstracts 4th Internat Dye Symposium*, p.46
135. Bayer. (1966). *Belg. Pat.*, 703 598; (1967). *Belg. Pat.*, 712 733
136. Holtschmidt, H. (1962). *Angew. Chemie*, **74**, 849; Bayer. (1967). *Belg. Pat.*, 706 761; *Belg. Pat.*, 707 203
137. Bayer. (1967). *Belg. Pat.*, 713 937; (1967). *Belg. Pat.*, 716 013; (1967). *Belg. Pat.*, 716 014
138. ICI. (1968). *German Offenlegungsschrift*, 1 902 109
139. Venkataramen, K. (ed.,). (1970, 1971). *The Chemistry of Synthetic Dyes*, vol. 3–5 (New York: Academic Press)
140. Rys, P. and Zollinger, H. (1970). *Leitfaden der Farbstoffchemie*. (Weinheim: Verlag Chemie); (1972). *Fundamentals of the Chemistry and the Application of Dyes* (London: John Wiley)

11
Aromatic Intermediates

N. N. VOROZHTSOV, JR.
Institute of Organic Chemistry, Novosibirsk, U.S.S.R

11.1 SULPHONATION

11.1.1 Synthesis of sulphonic acids and their derivatives

11.1.1.1 Benzenesulphonic acids

Benzene vapours can be sulphonated with sulphuric acid in a counterflow in five towers. The first tower is sprayed at 120–140 °C with concentrated sulphuric acid, while each of the remaining four is sprayed with the product from the respective preceding tower (the major part of water having been distilled off from the product). The temperature increases to 175 °C in the last tower. The reaction mixture emerging from the last tower contains up to 96% benzenesulphonic acid, *c.* 3% sulphuric acid and less than 1% water[1].

The *para*isomer (86%) is formed in the sulphonation of toluene with 10% oleum at 160 °C under pressure (1.16–1.20 mole of H_2SO_4 per mole of toluene). A maximum percentage (45%) of toluene-*m*-sulphonic acid is obtained with 3 mol of 94% H_2SO_4 at 180 °C[2].

Pure sodium salts of benzene homologue sulphonic acids can be obtained by treating the sulphonation mass with lime and passing the solution through an acidic form of an ion-exchange resin (such as Dowex 50) and then neutralising the solution of sulphonic acids with sodium hydroxide[3].

Nitro compounds react with sulphuric anhydride (or with its SO_2 solution) at 90–100 °C rapidly forming first pyrosulphonic acids ($ArSO_3SO_3H$)[4] and sulphonic acid anhydrides[5]. The reaction is completed at 110–113 °C (having preliminarily distilled off SO_2 in case it is used). The reaction proceeds smoothly (the yield of sulphones is 0.2–0.5%) and presents no explosion hazards[6].

The nitro compound can be treated with gaseous sulphuric anhydride below 40 °C. The resulting adduct SO_3-$ArNO_2$ rearranges after heating to 130 °C[7].

The sulphonation of *m*-chlorobenzenesulphonic acid with oleum gives mainly chlorobenzene-3,5- but also chlorobenzene-2,5-disulphonic acid. The content of the 2,5-isomer diminishes with an increase in the reaction temperature (170–250 °C) and duration[8].

1,3-bis-(Heptaisopropyl)benzene-5-sulphonic acid produced with SO_3 at

60–86 °C, is used as an insecticide and its chloride makes cotton water-repellent[9].

2-Sulphoterephthalic acid and 5-sulphoisophthalic acid are produced from the corresponding benzenedicarboxylic acids by treating them with oleum in the presence of 1–2% HgO[10].

Acetanilide can be sulphonated and hydrolysed to sulphanilic acid with 70% H_2SO_4. The relative rates of the consecutive reactions were investigated as a function of the acid concentration[11].

11.1.1.2 Naphthalenesulphonic acids and sulphonic acids of other polycyclic hydrocarbons

Naphthalene-1-sulphonic acid can be produced by passing gaseous SO_3 (including a mixture of gases obtained by contact oxidation of SO_2) into a solution of naphthalene in tetrachloroethylene at 0–30 °C with a yield of 88%[12].

Naphthalene-1-sulphonic acid with a small amount of 2-isomer can be obtained with chlorosulphonic acid in dichloroethane solution at −35 °C[13].

1-Aminonaphthalene-3-sulphonic acid and its derivatives can be sulphonated in the 4-position by reacting with sodium bisulphite and an oxidant (MnO_2)[14].

1-Naphthol-3,6-disulphonic acid is produced smoothly from 1-naphthol-3,6,8-trisulphonic acid by the reducing elimination of the sulpho group through the action of cast iron chips in a 40–45% aqueous solution of sodium hydroxide by boiling[15].

Anthraquinone-1,2,3,4-tetrasulphonic acid (the first representative of aromatic vicinal tetrasulphonic acids) has been produced from 1,2,3,4-tetrafluoroanthraquinone or 1,2,3,4-tetrachloroanthraquinone by treating them with sodium sulphite in a boiling aqueous solution of dioxane[16].

The methods of producing anthraquinonesulphonic acids were reviewed[17].

11.1.1.3 Sulphochlorides and esters of sulphonic acids

Sulphochlorides from the reaction mass with excess chlorosulphonic acid can be isolated by adding HCl(>15%) at −10 to −15 °C[18]. It is suggested that sulphochlorides are produced by reacting sulphofluorides with $AlCl_3$[19] or by reacting sulphamides with thionyl chloride and then with H_2SO_4[20]. A mixture of sulphochlorides and sulphones, for example from chlorobenzene, can be obtained by treating it with SO_3 and excess thionyl chloride. Sulphur chlorides added in small quantities enhance the yield of the reaction products[21].

m-Benzenedisulphonic acid and a number of its substituted compounds and homologues can be converted to corresponding dichlorides by treating the former with excess CCl_4 in the presence of sulphuric, phosphoric or polyphosphoric acid. The reaction proceeds under heating to 170 °C in accordance with the equation[22]:

$$XC_6H_3(SO_3H)_2 + 2CCl_4 \xrightarrow{[H^+]} XC_6H_3(SO_2Cl)_2 + 2COCl_2 + 2HCl$$

m-Sulphochloride is an intermediate for reactive dyes; it is obtained by treating 2-chloroethylphenyl sulphone first with chlorosulphonic acid and then with thionyl chloride[23].

Aromatic sulphonic acids can be converted to methyl and ethyl esters by heating with excess of a corresponding boric acid ester[24].

The esterification of sulphonyl chlorides are carried out in the presence of acetone[25].

11.1.2 Analysis of sulphonic acids

Interesting results were obtained by mass-spectroscopic methods. For the analysis, the sample is converted to a solution of free acid by passing it through a column with Amberlite IR-120-H. The resulting sulphonic acid solution is converted to a tetramethyl ammonium salt (by neutralisation with a 10% $(CH_3)_4N^+OH^-$ solution and evaporation) or to a methyl ether with CH_2N_2. The tetramethyl ammonium salt dissociates in the mass spectrometer, giving $ArSO_2OMe + Me_3N$.

Consideration of the mass spectrum allows the determination of the series of aromatic compounds to which the sulphonic acids belong (benzene, naphthalene or anthraquinone), the number of sulpho groups and the character of other substituents present in the sulphonic acid (as well as the presence of admixtures of other sulphonic acids). The positions of substituents in the nucleus are determined from the n.m.r. spectra[26].

The application of g.l.c. is suggested for the analysis of sulphonic acids in the form of their chlorides or methyl ethers[27], or fluorides[28]. A practically quantitative conversion of sulphonic acids to corresponding hydroxy compounds and sulphite when the acids are fused with excess potassium hydroxide makes it possible in some cases to determine the character and quantity of the sulphonic acid by the g.l.c. analysis of the hydroxy compounds and quantitative determination of the sulphite[29]. This method, however, is of little use for the analysis of amino- and hydroxy-sulphonic acids.

In a number of cases the application of pyrolysis–g.l.c. may prove successful for the analysis of sulphonic acids. With the help of a pyrolysis microreactor[30] it is possible to carry out the analysis by using 10–50 μl of a 0.1% solution of sulphonic acid salt. When a number of sulphonic acids of the benzene and naphthalene series are heated to a temperature above 700 °C with the addition of SiO_2, SO_2 is formed quantitatively, so that the content of sulphonic acids can be determined with an accuracy of $\pm 2\%$. In this procedure a non-sulphonated compound corresponding to the sulphonic acid is formed simultaneously, but with yields, as a rule, far from being quantitative (usually of the order of 50%). The addition of carbohydrazide somewhat increases the yield of the non-sulphonated compound[31].

Spectral determination of individual sulphonic acids can be effected after the separation of their mixture with the use of ion-exchange chromatography[32].

The use of the n.m.r. techniques for the quantitative determination of the content of different sulphonic acids of benzene and its homologues is discussed in reference 33.

11.2 NITRATION

11.2.1 Synthesis of nitro derivatives of benzene series

Solutions of $Pd(MeCO_2)_2$ in acetic acid catalyse the nitration of benzene by $NaNO_2$ or N_2O_4. Phenyl acetate is formed as a by-product (up to 50% of the nitrobenzene)[34]. Silver nitrate, on silicic acid, nitrates aromatic compounds at room temperature in carbon tetrachloride[35]. Substitution of the carboxyl group by the nitro group can take place when carboxylic acids (e.g. *o*- and *p*-nitrobenzoic acids) are heated with 20–50% nitric acid to 215–250 °C[36].

Nitration by means of a mixture of anhydrous nitric acid and SO_3 in excess liquid SO_2 in a number of cases gives interesting results, e.g. from benzoic acid a mixture containing 93% of *m*-isomer is obtained. The change in the orientation is, possibly, due not to the peculiarity of the medium, but to the low reaction temperature[37].

When nitrating alkyl benzenes, the orientation of the nitro group into the *p*-position to the alkyl is enhanced by using mixtures of nitric acid and phosphoric or polyphosphoric acid. A mononitro product is thus obtained from toluene with a ratio $p{:}o = 1.11–1.52$ (instead of the usual 0.63). In mononitroethyl benzenes $p{:}o = 1.37$, in mononitrocumenes, 3.03[38].

From *o*-dialkyl benzenes and vicinal trialkyl benzene nitro derivatives of polyalkyl biphenyls can be produced by nitration with concentrated nitric acid at −40 – +30 °C. From *o*-xylene, for example, 2-nitro-3′,4,4′,5-tetramethylbiphenyl and 2,2′-dinitro-4,4′,5,5′-tetramethyl biphenyl are obtained[39].

The predominant entrance of the nitro group into the *p*-position to the hydroxy group when nitrating phenol and *m*-cresol is attained by the use of a mixture of nitric, sulphuric and nitrous acids (introduced in the form of N_2O_4)[40].

The nitration of *p*-cresol into the *meta*-position to the hydroxyl and into the *ortho*-position to methyl can be effected by introducing its monosulphonic acid ether into the reaction[41].

The nitration of sodium benzenesulphonate with a nitrating mixture containing 3% of free SO_3 results in an increased content (84%) of *m*-nitrobenzenesulphonic acid if the reaction temperature is lowered to −50 °C[42].

11.2.2 Synthesis of polycyclic nitro derivatives

It is possible to nitrate naphthalene with an excess of 85–90% nitric acid and then isolate 1,5-dinitronaphthalene by cooling[43]. The nitration of 1,5-nitronaphthalenesulphonic acid by KNO_3 in oleum (preferably 20%) gives 1,7-dinitronaphthalene-5-sulphonic acid[44].

The nitration of naphtho-(1,4)-quinone in a medium of 98% hydrogen fluoride at 10 °C gave 5-nitronaphtho-(1,4)-quinone[45] identical to that synthesised earlier from 1,8-dinitronaphthalene[46].

The nitration of 1,2,3,4-tetrafluoronaphthalene by nitric acid or a nitrating mixture gives the 5- and the 6-nitro derivative in a 9:1 ratio[47].

A survey of the production and application of nitroanthraquinone derivatives is given in Ref. 48.

The nitration of 1-acylaminoanthraquinone-2-sulphonic acid gave 1-acylamino-4-nitroanthraquinone-2-sulphonic acid which is the starting product for 2,3-dicyan-1,4-diaminoanthraquinone[49].

11.2.3 Analysis in nitration

Rapid determination of nitric acid in oleum is possible with a nitrate ion selective electrode[50]. Thermal analysis of the system SO_3–N_2O_5–H_2O is given in Ref. 51. Gas–liquid chromatography allows quantitative determination of all the toluene nitration products (nitrotoluenes, phenylnitromethane, *p*-tolylacetate)[52].

The specific character of the mass spectra of all ten isomers of dinitronaphthalene makes mass spectroscopy applicable for the analysis of mixtures of isomers[53]. The separation, quantitative determination and identification of nitrobenzene and a number of its homologues (in cigarette smoke) in quantity of the order of 5–10 ng have been carried out with the help of gas–liquid chromatography and mass spectroscopy[54].

11.3 HALOGENATION

11.3.1 Synthesis of fluoro derivatives

An increase in the interest shown in the technical application of fluoro derivatives[55] has been observed. Some of such fluoro compounds are expected to be used in the synthesis of intermediates by replacing the considerably mobile fluorine atom by other substituents[56]. Other fluoro derivatives (such as *o*-fluorobenzoic acid, fluoro derivatives of anthraquinone, etc.) are employed in the synthesis of dyes[57].

Aromatic fluoro derivatives are synthesised, mostly, by indirect methods, namely, by the replacement of chlorine atoms or the amino group (through the Schiemann reaction) by fluorine. The first-mentioned method in many cases gives good results when producing polyfluoroaromatic compounds (a survey is given in Ref. 58).

Contrary to the rather widespread opinion, aromatic fluoro derivatives have recently been shown to be obtainable with a satisfactory yield also by a direct replacement of hydrogen atoms by fluorine atoms. Thus, the reaction between gaseous fluorine (diluted with nitrogen or helium) and a 6% acetonitrile solution of benzene at −35 °C gave a mixture of *m*-, *o*- and *p*-difluorobenzene and monofluorobenzene, approximately in a 1:4:5:60 ratio. The fluorination of toluene at −70 °C (without a solvent) gave a mixture consisting of 62% of toluene and 38% of a mixture of monofluorotoluene isomers ($m{:}o{:}p = 1{:}5{:}4$). Monofluoronitrobenzenes with a predominance of the *meta*-isomer were produced in a similar way from nitrobenzene (in acetonitrile), and a mixture of fluorochlorobenzenes (with a predominance of *para*-isomer) was produced from chlorobenzene. From naphthalene (in acetonitrile), 1- and 2-fluoronaphthalenes were produced in a ratio of 3:1. For the reaction to be successful, low temperatures (of −20 to −70 °C) and a deficiency of the fluorine are essential. The use of excess fluorine results in the formation of polymeric products, evidently, of polyfluoro derivatives of polycyclohexene[59]. The replacement of the hydrogen atom by fluorine is effected by anodic fluorination of aromatic compounds in anhydrous organic solvents at potentials lower than the fluorine anion oxidation potential.

Thus, with the electrolysis of naphthalene in a solution of $Et_4N^+F^-$ in acetonitrile at +1.8 V (with a Pt anode), 1,4-difluoronaphthalene is obtained with a yield of 70%. When the electrolysis is carried out in the presence of $Et_4N^+F^- \cdot 3HF$, 1-fluoronaphthalene is the main product. From benzotrifluoride *m*-fluorobenzotrifluoride and 3,5-difluorobenzotrifluoride are obtained. Fluoro derivatives have been obtained also from benzene and its other derivatives[60].

Fluorination can be effected also with the use of certain fluoro compounds, e.g. xenon fluorides. Benzene and a number of its derivatives have been converted into monofluoro compounds with a 50–80% yield for the reacted compound, the maximum yields taking place by fluorination of benzotrifluoride and nitrobenzene, by reacting XeF_2 with their CCl_4 or CH_2Cl_2 solutions at room temperature. The initiation of the reaction requires the presence of hydrogen fluoride. By-products of the reaction are high-boiling condensation products (but not those of fluorine addition). In the case of benzene, the formation of biphenyl and its monofluoro derivatives is proved to take place. No substitution into the side chain of toluene occurs[61]. The substitution in benzene and its derivatives is also effected by the action of XeF_2 in the gaseous phase at a temperature of 100–167 °C. The by-products are those of addition and polymerisation[62]. For the production of aromatic fluoro compounds XeF_4 may also be used. The yields of the products are low[63].

The fluorination of benzene to monofluorobenzene (with a 65% yield) has been performed by the action of CF_3OF at −78 °C under intensive irradiation. 10% of trifluoromethoxybenzene $PhOCF_3$ are formed as a by-product. Under the irradiation, CF_3OF dissociates with the formation of $F^{\cdot}$ and $CF_3O^{\cdot}$, one of which functions as the chain fluorination initiator. Under these conditions toluene gives *o*-fluorotoluene (34%) and benzyl fluoride (25%), while anisole gives mainly *o*-fluoroanisole[64].

Practical utilisation of direct fluorination is associated with difficulties arising from the fact that one has to work with incomplete conversion of the compound being fluorinated. The proximity of the boiling points of the mono- and polyfluoro-substituted compounds and of the starting compound makes the processing of the reaction mass extremely complicated. The finding of effective methods for the separation of such mixtures will, probably, be decisive for their further development.

Methods have recently been proposed for preparing mono- and difluoro derivatives of aromatic compounds from 1,3-dienes (aliphatic or cyclic) and fluoro-substituted methane or ethylene.

The pyrolysis of, e.g. cyclopentadiene together with difluorochloromethane or tetrafluoroethylene at 400–600 °C gives monofluorobenzene with a 70% yield; with methylcyclopentadiene and CF_2HCl, a mixture of fluorotoluenes (47%) is obtained; indene and CF_2HCl at 600 °C give 2-fluoronaphthalene (51%)[65].

p-Fluoroaniline is produced from nitrobenzene by heating the latter with triphenyl phosphine[66] and other reducing agents[67] in anhydrous HF.

11.3.2 Synthesis of chloro and bromo derivatives

11.3.2.1 Chlorination of benzene and its derivatives

The vapour-phase chlorination of benzene, chlorobenzene and benzonitrile with addition of perchlorodimethyl sulphone (CCl_3–SO_2–CCl_3) has been investigated. The chlorination into the nucleus (without the formation of considerable quantities of addition products or biaryls) proceeds at 180 °C. Perchlorodimethyl sulphone, evidently, initiates the radical chain reaction, losing SO_2 and forming $CCl_3^{\cdot}$ radicals. Benzene reacts with a rate *c.* seven times as high as chlorobenzene which has a rate approximately twice that of benzonitrile. In the absence of an initiator the chlorination takes place only at a temperature above 260 °C [68]. The continuous chlorination of various aromatic compounds is investigated intensively[69].

p-Phenolsulphonic acid is suggested as a starting product for producing *o*-chlorophenol. The monochloro derivative produced by chlorination in an aqueous medium at 90 °C is converted into *o*-chlorophenol by hydrolysis in aqueous sulphuric acid at 200 °C [70].

It is suggested that *p*-chlorophenol and *p*-chloro-*o*-cresol should be produced by chlorinating corresponding triaryl phosphates in the presence of conventional chlorination catalysts ($FeCl_3$, $AlCl_3$, I_2, etc.) with subsequent hydrolysis with the help of acids or alkalies. The ratio of *para*- and *ortho*-isomers in the phenol chlorination product is 92:8 (Ref. 71).

2,6-Dichlorophenol is suggested to be produced by chlorinating *o*-chlorophenol in a non-polar solvent (CCl_4, benzene, hexane) at its boiling point till 2,4,6-trichlorophenol appears in the reaction mass. The content of 2,6-dichlorophenol reaches 82% (17% of 2,4-isomer). Pure product can be isolated from the mixture by rectification or, better, by conversion into phenolate and counter-current extraction in a column by using a solution of acetic acid in tetrachloroethylene. The aqueous solution emerging from the top of the column contains 99% 2,6-dichlorophenol[72].

The analysis of phenol chlorination products containing 1–3 chlorine atoms in the nucleus is effected by the gas–liquid chromatography method[73].

The pentachlorination of benzoic acid and phenol have been investigated[74].

p-Chlorotoluene is chlorinated with a predominant formation of 2,4-dichlorotoluene when the reaction is carried out at 10–30 °C in the presence of $ZrCl_4$ with addition of iron. By the distillation of the reaction mass dichlorotoluene is produced containing more than 80% of the 2,4-isomer (3,4-dichlorotoluene being the balance)[75].

2,6-Dichlorotoluene can be produced from toluene substituted in the 3- and 5-positions of the nucleus by tertiary alkyl groups, for example, from

3,5-di-t-butyl toluene by chlorinating it with SO_2Cl_2 in the presence of $AlCl_3$. 2,6-Dichloro-3,5-di-t-butyl toluene (t-butyl groups sterically hindering the entrance of chlorine into the 4-position) is debutylated by passing in a stream of N_2 over a catalyst K-106 (at 350 °C) to 2,6-dichlorotoluene[76].

n-Propylbenzene in the dark at −60 °C and in the absence of catalysts is chlorinated only in the nucleus. The *o*:*p* ratio is 46:54 (Ref. 77).

2,3,5-Trichloro-4-methylbenzenesulphonic acid, 2,3,5,6-tetrachloro-4-methylbenzenesulphonic acid and mixtures of these two sulphonic acids are suggested to be produced by chlorination of toluene-*p*-sulphonic acid (or its salts) in sulphuric acid in the presence of iodine or iodides. The hydrolysis of tetrachlorosulphonic acid in sulphuric acid gave 2,3,5,6-tetrachlorotoluene[78].

It is reasonable that 2,6-dichloro-4-nitrotoluene should be produced from *p*-nitrotoluene by chlorinating it in the presence of $SbCl_3$ onto a mixture of 2,6-dichloro derivative with tri- and tetrachloro derivatives. Polychloro derivatives are converted into 2,6-dichloro-4-nitrotoluene by the substitution of the chlorine atom which is in the *ortho*-position to the nitro group by hydrogen; this is attained through the action of powdered copper in the medium of benzoic acid or in a mixture of chlorobenzene with acetic acid[79].

11.3.2.2 Chlorination and bromination of polycyclic compounds

Naphthalene adds chlorine in CCl_4 at *c.* 40 °C and under illumination. The resulting 1,2,3,4-tetrachlorotetralin, being boiled with water (with the addition of a surface-active agent), is converted into 2,3-dichloro-1,4-dioxytetralin. By oxidation with nitric acid, dichlorohydrin is converted into 2-chloronaphtho-1,4-quinone. The chlorination of the monochloro derivative to 2,3-dichloronaphtho-1,4-quinone is conducted in acetic acid in the presence of iodine[80].

1,4,5-Trichloroanthraquinone is suggested to be produced from anthraquinone-1-sulphonic acid by chlorination first in 20–25% oleum (with KJ as a catalyst), and then after dilution with water to $d = 1.1$ and addition of hydrochloric acid and ammonium chloride (for the initiation of radical chlorination)[81].

Halogenation of aminoanthraquinones can be successful with the use of mixed media. Thus, 2,4-dibromo-1-aminoanthraquinone can be produced from 1-aminoanthraquinone in the presence of hydrochloric acid and chlorobenzene. Prior to bromination, amine is mixed with this medium for a long time so that its particles swell. Then bromine and hypochlorite are added. The yield is 97%[82].

A U.S. patent recommends media for the halogenation of aminoanthraquinones, consisting of a mixture of an alcohol (preferably methyl alcohol) with mono- or poly-nitrobenzene or halobenzene (e.g. $C_6H_5NO_2$, *o*-$C_6H_4Cl_2$). Thus, in the production of 4-bromo-1-methylaminoanthraquinone the bromination of 1-methylaminoanthraquinone is started in a suspension in methyl alcohol. After the introduction of 0.2 mol bromine, nitrobenzene is added and then the rest of the bromine. The yield is 94.5% (without the nitrobenzene, 70.5%)[83].

1-Acylamino-4-bromoanthraquinone is produced from 1-acylamino-anthraquinone by bromination in a nitrobenzene solution with the addition of water and an alkali metal bromate[84]. At the same time 1-amino-4-bromo-anthraquinone-2-carboxylic acid is suggested to be produced from 1-amino-anthraquinone-2-carboxylic acid by bromination in 90–100% sulphuric acid[85] and 1-amino-2,4-dichloroanthraquinone and 1,5-diamino-2,4,6,8-tetrachloroanthraquinone are suggested to be produced by chlorinating corresponding amines in dimethyl formamide[86].

The chlorination of quinizarine in acetic acid gave its 2,3-dichloro derivative, and the same process with 2-methylpurpurine (1,4-dihydroxy-2-methoxyanthraquinone) gave its 3-chloro derivative[87].

11.3.2.3 Chlorination without the use of elementary chlorine

The use of noble metals, particularly Pd, as catalysts for the oxidative chlorination of aromatic compounds is of current interest. Thus, for instance, chlorobenzene has been produced by heating benzene with sodium chloride, nitric and acetic acids or with HCl gas in the presence of palladium salts[88].

It is assumed that the reaction of aromatic compounds with salts of bivalent palladium in the presence of oxidants and nucleophiles, e.g. halide ions, proceeds either with the formation of salts of tetravalent palladium, which react with the aromatic compound with the formation of benzene halide, or via Ar—PdX compounds. These compounds react with the nucleophile and oxidant, regenerating the palladium salt with the formation of a substituted aromatic compound[89].

The reaction of chlorine with a complex aryl-palladium compound formed from $PdCl_2$ and azobenzene (with the elimination of hydrogen chloride) gives mono- and poly-chloro derivatives of azobenzene containing chlorine atoms only in the *o*-position to the azo-group. The $PdCl_2$ eliminated during the chlorination again enters into the reaction with the azo-compound. Thus in this case $PdCl_2$ is a catalyst of the homogeneous-catalytic chlorination which proceeds through the formation of an arylpalladium compound. The complexation of the Pd atom with the azo-group leads to palladium and chlorine entering only into the *ortho*-position of the aromatic nucleus[90].

The chlorination of a number of compounds has been effected by the action of cupric chloride. Thus, mesitylene, being heated with $CuCl_2$ to boiling point becomes converted into a 2-chloro derivative (90% yield). From *p*-xylene the yield of the chloro derivative is 20%. The reaction, probably, proceeds through a cation–radical stage by the following scheme[91]:

$$\mathrm{HArMe} \xrightarrow[-\mathrm{Cl}^-,\ -\mathrm{CuCl}]{+\mathrm{CuCl_2}} [\mathrm{HArMe}]^{+\cdot} \xrightarrow[-\mathrm{CuCl},\ -\mathrm{H}^+]{+\mathrm{CuCl_2}} \mathrm{ClArMe}$$

A good yield of 4-chloro-3,5-dimethyl phenol is attained by heating a solution of sym-xylenol in toluene with cupric chloride. Phenol and its homologues can be chlorinated by passing oxygen and hydrogen chloride through their solution in 5 mol l^{-1} hydrochloric acid containing cupric chloride in a quantity equimolar to phenol. The main resulting product is *p*-chlorophenol.

For the chlorination of aromatic amines it is reasonable to use 2 mol of $CuCl_2$ per mole of the amine and run the reaction at maximum possible concentration of hydrogen chloride. In this way *p*-chloroamines have been produced from aniline, *N,N*-dimethylaniline and *o*-toluidine. The 2,4-dichloro derivative has been produced from 1-naphthylamine[92].

The chlorination of benzene, acetanilide, *N*-methylacetanilide and *p*-acetotoluidide is effected by the action of equimolar quantities of $TiCl_4$ and trifluoroperacetic (or *m*-chloroperbenzoic) acid in a solution of dichloromethane at 0 °C. Into the *ortho*-position to the acetamino group the chlorine enters only *after* the substitution of the *para*-atom of hydrogen. The addition of $Et_4N^+Cl^-$ in the case of acetanilide leads to quantitative formation of 2,4-dichloroacetanilide[93].

A survey on the halogenation by the use of NCl_3 and other *N*-chloro- and *N*-bromo-amines is given in Ref. 94.

2-Chloro- and 2,7-dichlorofluorenes and fluorenones, 5-chloroacenaphthene (and also the corresponding bromo derivatives) are produced by the halogenation of fluorene, fluorenone and acenaphthene with *N*-chloro- and *N*-bromo-succinimide in the medium of propylene carbonate, acetic acid or aqueous sulphuric acid[95].

The chlorination of benzene, other aromatic hydrocarbons and a number of substituted benzenes can be effected with the help of chloroisocyanuric acid. With the use of $FeCl_3$ and 50% H_2SO_4 and H_3PO_4 as catalysts, the substitution into the nucleus takes place[96].

11.3.2.4 Chlorination into the side chain

It is suggested that benzyl chloride should be produced from toluene, the chlorination being initiated by the diacetyl derivative of $\alpha\alpha$-dihydroxyperoxide of chloral at 60–70 °C. The yield is 90%. Unlike benzoyl peroxide, the initiator is insensitive to shock and friction[97].

As initiators of the chlorination of toluene, *p*-xylene, etc. into the side chain, nitrates of mono- and polyatomic alcohols are proposed, such as methyl nitrate, tetranitroerythritol, etc.[98].

Benzyl chloride is produced with a moderate yield from toluene by reacting toluene with trichloroisocyanuric acid in the presence of benzoyl peroxide[96].

The investigation of the regularities of the chlorination of n-propyl benzene has shown that under sufficient illumination at 25 °C only hydrogen atoms in the side chain are substituted, predominantly in the α-position (then $\beta > \gamma$). The addition of carbon disulphide sharply enhances the selectivity (81.5% into the α-position). At 160 °C and under illumination the substitution into the β-position becomes predominant (48.5%)[77].

11.3.2.5 Methods of producing halogen derivatives, other than by substitution of hydrogen

Halogen derivatives can be produced smoothly by the decarbonylation of acid halides by heating them with 1–5% chlorotris(triphenylphosphine) rhodium, $RhCl(PPh_3)_3$ or chlorocarbonylbis(triphenylphosphine)rhodium $RhClCO(PPh_3)_2$ to a temperature of 200 °C or higher. The yields are mostly of the order of 80–90% of the theoretical quantity[99].

The first of these catalysts ensured high yield of chlorobenzene and *p*-dichlorobenzene from benzene and *p*-chlorobenzene sulphochloride, respectively, and also from a number of other sulphochlorides of the aromatic series (to 6-chrysenesulphonyl chloride inclusive). From sulphobromides bromo derivatives are produced. From sulphofluorides only traces of fluoro derivatives are obtained[100].

The substitution of the sulpho group is suggested to be effected by treating sulphonic acids, their salts or sulphonyl chlorides by $COCl_2$ or CCl_4 (or their mixture) at a high temperature. In this way it is possible to produce chlorobenzene, polychloro derivatives of benzene (e.g. 1,3,5-trichlorobenzene), chloro-substituted of polycyclic aromatic hydrocarbons[101].

By reacting 1,3,5-trichloro-2,4,6-trinitrobenzene with LiCl in acetone or isopropanol 1,2,3,5-tetrachlorodinitrobenzene is produced. The facility with which the nitro group is replaced by the chlorine atom can be explained in this case by the fact that in the σ-complex which forms with the addition of the chloride ion two nitro groups are nearer to the plane of the ring and therefore its formation is energetically advantageous[102].

Finally, it is possible to produce aromatic chloro derivatives from non-aromatic chloro compounds. Thus, scarcely available 1,2,3,4-tetrachlorobenzene can be produced by pyrolytic elimination of two hydrogen bromide molecules from 1,6-dibromo-2,3,4,5-tetrachlorohexa-2,4-diene (at 250–280 °C)[103].

2-Chloronaphthalene is produced by the pyrolysis of indene together with $CHCl_3$ or tetrachloroethylene[104].

11.4 SYNTHESIS OF AMINES

11.4.1 Reduction of nitro compounds

11.4.1.1 Catalytic reduction with hydrogen

The reduction of nitro compounds with hydrogen in the presence of various catalysts seems to be most promising at present. The industrial application of this method becomes all the more extensive and, accordingly, most attention is paid to it by research workers.

It is shown that aniline can be produced by *homogeneous* catalytic reduction of nitrobenzene (and azobenzene) with hydrogen in dimethyl formamide in the presence of the soluble complex catalyst $(Py)_3RhCl_3$–$NaBH_4$ [105].

The homogeneous catalytic reduction of nitrobenzene with the help of hydrogen and carbon monoxide can be effected in the presence of ruthenium carbonyls $Ru_3(CO)_{12}$, $Ru(CO)_5$. Aniline and Diphenyl urea is also formed[106].

Heterogeneous catalytic hydrogenation of aromatic nitro compounds (nitrobenzene, its homologues, nitrophenols) to corresponding amines is effected in the presence of Pt-catalysts prepared by reducing chloroplatinic acid or platinum dioxide with triethyl silane[107], or by applying Pt (as well as Pd) onto ε-alumina[108].

The hydrogenation of *o*-nitrophenol has been effected in an alcoholic medium on a colloidal platinum catalyst stabilised with polyvinyl alcohol[109].

m-Nitroacetophenone can be reduced to the corresponding amino ketones with hydrogen in the presence of $Pd/CaCO_3$ at 20–50 °C and under 1–3 abs. atm[110].

The kinetics has been studied of the hydrogenation of technical dinitrotoluene (containing 74–76% of 2,4-, 19–21% of 2,6-, 2.4–2.6% of 3,4-dinitrotoluene and smaller quantities of other dinitrotoluenes) and also of *m*-dinitrobenzene in alcohols (methyl and isopropyl ones) in toluene and benzene in the presence of Pd–C and Pt-catalysts under a pressure of up to 40 gauge atmospheres. The yields of the corresponding diamines are close to quantitative ones. The slowest stage which determines the speed of the whole process is the diffusion of hydrogen in the liquid to the surface of the catalyst (at 80–100 °C)[111].

2-Amino-3-hydroxynaphtho-1,4-quinone is produced by the reduction of 2-nitro-3-hydroxynaphtho-1,4-quinone (with subsequent oxidation by air of the resulting primary 2-amino-1,3,4-trihydroxynaphthalene) in the presence of PtO_2 [112].

Partial reduction of 2,4-dinitro-*N*-alkylanilines to 4-amino-2-nitro derivatives can be effected in the presence of Pt- or Pd-catalyst in polar organic solvents containing an organic acid[113].

Techniques are suggested for precluding the elimination of chlorine in hydrogenation of nitrochlorobenzene derivatives[114].

To preclude the formation of hydrazo compounds, the catalytic reduction of chloronitro compounds by hydrogen is suggested to be carried out in the medium of a corresponding amine (Pt and Pd as catalysts, addition of morpholine for diminishing the dechlorination)[115].

The reduction of isomeric nitrobenzenesulphonic acids, *p*-nitrotoluenesulphonic acid and *p*-nitroanilinesulphonic acid has been studied for the case of an aqueous solution (maximum speed at pH = 5) with nickel on kieselguhr as a catalyst. At 15 atm and 50 °C the hydrogenation of 0.4 $mol\,l^{-1}$ solutions is completed during 10–20 min[116].

Raney nickel is suggested as a catalyst for reducing various compounds, e.g. nitrobenzoyl- and nitrophenyl-sulphonylsulphamic acids in an aqueous medium[117], 3,3′-dinitro-4,4′-diaminodiphenyl sulphoxide in an alcoholic medium[118].

To preclude the hydrogenation of the nucleus during the reduction of aromatic nitro compounds, it is recommended that 0.05–15% sulphur dioxide be added to the hydrogen[119].

Dyestuffs Division I.C.I. is reported to have been awarded a prize for the invention and development of a novel hydrogenation process for the production of aniline[120].

Catalytic reduction can also be effected by hydrogen which is eliminated from the nitro compound. Thus, aniline and its homologues can be produced from nitrocyclohexane (and its homologues, respectively) by passing their vapours in a stream of nitrogen over dehydrogenating catalysts (Pd or Pt on Al_2O_3)[121].

11.4.1.2 Reduction by agents other than hydrogen

The reduction of various aromatic nitro compounds to corresponding primary amines can be effected by the action of SO_2 (under a pressure of up to 10 atm) in sulphuric acid medium with addition of small quantities of hydroiodic acid and compounds of metals belonging to Groups IB, IIB, IV, VA, VIA and VIII of the Periodic Table as catalysts. This method is of interest, mainly, for the reduction of nitrosulphonic acids, since they can be introduced into the reaction without separation from the reaction mass[122].

Aromatic bis-*o*-nitroamines (e.g. derivatives of diphenyl-methane, and propane, biphenyl) are reduced to corresponding tetramines by hydrazine in the presence of Raney nickel as catalyst[123].

The reduction of nitro compounds to derivatives of amines (isothiocyanates) has been realised by the action of carbon disulphide and an alkoholate or phenolate of an alkali metal[124].

Isocyanates are produced by the action of carbon monoxide on nitro compounds in the presence of oxides of iron (and other metals, e.g. Mo, Ge, Pd, Rh) and of a compound with a nitrogen-containing heterocyclic molecule[125]. The same process is suggested to be carried out in the presence of $PdCl_2$ in the medium of saturated nitrile (p(CO) = 50–900 gauge atm, temperature 150–225 °C). From polynitro compounds nitroisocyanates are produced, e.g. from *m*-dinitrobenzene, 3-nitro-phenylisocyanate (72.3 %) and *m*-phenylenedi-isocyanate (14.2 %)[126].

The reduction by means of carbon oxide of 2,4-dinitrotoluene in the medium of *o*-dichlorobenzene (60 gauge atm 200 °C) in the presence of oxides of Cu and Mn leads, as a result of the reduction of one of the nitro groups, to the formation of a mixture of 3-nitro-*p*-toluidine and 5-nitro-*o*-toluidine in a ratio of 2:1. The degree of conversion of the dinitro compound and the yield of amines are not high[127].

The reduction of one of the nitro groups in 2,4-dinitroaniline (with the formation, mainly of 4-nitro-1,2-diaminobenzene) and in 4,4′-dinitrostilbene-2,2′-disulphonic acid is realised by the action of an equimolecular quantity of resorcinol in a sodium bicarbonate solution. Some nitroazo-dyes are thus reduced to aminoazo-dyes[128].

A ethanol solution of sodium hydrosulphide is the best reducing agent for producing nitroamines by partial reduction of biphenyl dinitro derivatives[129].

11.4.1.3 Reduction to hydrazo compounds

Aromatic hydrazo compounds are suggested to be produced by the reduction of nitro compounds in an alkaline medium with the use of aliphatic or cyclo-aliphatic glycols containing 2–12 carbon atoms in the presence of naphthoquinones or their derivatives. Glycols are converted into acids: ethylene glycol, for example, is converted into oxalic and glycolic acid. The yields of the hydrazo compounds are high[130].

A 1000 A electrolytic cell is described for the production of hydrazobenzene by electrochemical reduction of nitrobenzene. The catholyte is a 8 % solution of NaOH containing PbO and the anolyte is 30 % NaOH[131].

The reduction of aromatic nitro compounds by sodium borohydride in dimethyl sulphoxide or sulpholane in certain cases allows the production of azoxy compounds with a good yield. Further reduction gives a mixture of azobenzene and aniline[132].

The importance of hydrazo compounds as intermediates is now diminishing, since the benzidine and its substituted compounds are carcinogenic. The necessity of adhering to safety regulations is emphasised when working with these and some other aromatic amines (e.g. with 4-aminostilbene) even under laboratory conditions[133].

11.4.2 Replacement of other substituents by an amino group

11.4.2.1 Replacement of hydroxy group

Until recently the main field where the replacement of the hydroxy group by the amino group was practically employed was the chemistry of naphthalene and anthraquinone. However, a growing demand for aniline (mainly for the production of chemicals for rubber and polymeric isocyanates from diphenylmethanedi-isocyanate) and improvements in the production of phenol (cumene process) led to the development of practical and interesting methods of producing aniline from phenol.

In a recent U.S. patent (related to the improvements in the method disclosed earlier[134]) to prepare aniline from phenol is suggested to be produced by mixing vapours of phenol and ammonia that are produced by their individual evaporation under a pressure of *c.* 16 abs. atm (phenol at 320 °C; ammonia at 45 °C), the mixture of the vapours being then directed onto an aluminosilicate catalyst. The reaction temperature is 385 °C, pressure is 15 abs. atm. With the 95 % conversion of phenol, the yield is 93 %[135].

As initial materials for the production of aniline it is also suggested that phenol hydrogenation products—a mixture of cyclohexanone and cyclohexanol (prepared by oxidising cyclohexane with air) should be used. The passing of this mixture with excess ammonia over nickel on kieselguhr at 315 °C results in a smooth formation of aniline[136].

The heretofore unknown 1,7-diaminonaphthalene-4-sulphonic acid is suggested to be produced from 1-amino-7-hydroxynaphthalene-4-sulphonic acid, 1,7-dihydroxynaphthalene-4-sulphonic acid and, especially, from 1,7-dihydroxy-2-carboxynaphthalene-4-sulphonic acid by heating with ammonia and ammonium bisulphite to a temperature of *c.* 150 °C. By using primary and secondary amines instead of ammonia, corresponding *N*-alkyl derivatives are produced[137].

1-Amino-2,4-dihydroxynaphthalene can be produced by the action of ammonia on 2-hydroxy-1,4-naphthoquinone and subsequent reduction[138].

The production of 1-hydroxy-4-arylaminoanthraquinones by the action of aromatic amines on quinizarin or on leucoquinizarin is recommended to be carried out in an aqueous medium in the presence of surface-active agents[139].

11.4.2.2 Replacement of halogen atoms

N,*N*-Dialkyl derivatives of 2-nitro-4-hydroxyaniline are produced by reacting secondary amines (at 15–26 °C in the medium of dimethyl formamide, acetonitrile or sulpholane) with ethers of sulphonic acids of 4-chloro-3-nitrophenol with subsequent elimination of the sulphonic acid residue by reaction with alkalies[140].

N,*N*-Dialkyl derivatives of 2-amino-4-trifluoromethyl-5-sulphamidobenzoic acid can be produced from 2-halogeno-4-trifluoromethyl-5-sulphamidobenzoic acid by reaction with a secondary amine in ethylene glycol–dimethyl ether at 160 °C[141].

N-Arylated 1-aminoanthraquinones are produced by the action of chlorobenzene and its derivatives on 1-aminoanthraquinone in the presence of sodium bicarbonate and copper and copper acetate in nitrobenzene at 180–210 °C[142].

1-Cyclohexylamino-5-chloroanthraquinone is produced from 1,5-dichloroanthraquinone with cyclohexylamine in nitrobenzene, with addition of pyridine, at 160 °C [143].

2,3-Difluoro-substituted 1,4-diaminoanthraquinones are produced by the reaction of ammonia or amines on 1-amino-2,3,4-trifluoroanthraquinone derivatives in a non-polar solvent at room temperature. In polar solvents, however, 1,3-diamines or their mixture with 1,4-isomers are formed[144].

1,4,5,8-Tetra-aminoanthraquinone can be produced from 1,4,5,8-tetrachloroanthroquinone with aqueous ammonia at 150–190 °C in the presence of sodium dithionite[145].

A survey on the reactions of halogenoanthraquinones with amines is given in [146]. The use of fluoroderivatives for the synthesis of amines is discussed in Section 11.3.1.

11.4.2.3 Replacement of sulpho group

1,8- and 1,5-Diaminoanthraquinone can be produced from salts of the corresponding sulphonic acids by reaction with an aqueous solution of ammonia at 250–400 °C by the addition of conventional oxidants. The high rate makes the reaction in a tubular reactor suitable for a continuous process[147].

The reaction of 1,5-anthraquinonedisulphonic acid with an aqueous solution of 2-propylamine gave 1-isopropylaminoanthraquinone-5-sulphonic acid which by the action of ammonia was converted into 1-amino-5-isopropylaminoanthraquinone[143].

11.4.3 Production of amines by direct substitution of hydrogen atom by an amino group

Direct substitution of the hydrogen atom by an amino group is of interest since it opens up the simplest way for the synthesis of amines.

One of the long known reactions of a direct introduction of the amino group – by the action of hydroxylamine in an alkaline medium – is suggested to be employed for producing 4-nitro-2,6-dicyanoaniline from 3,5-dicyanonitrobenzene[148].

By the same methods but with moderate yields 9-amino-10-nitrophenanthrene (from 9-nitrophenanthrene) and 2-amino-3-nitrobiphenylene (from 2-nitrobiphenylene) have been synthesised[149].

Interesting results are reported about the formation of small quantities of aromatic amines by reactions with elementary nitrogen in the presence of $TiCl_4$ and Na or $TiCl_4$ and Li or with lithium alone. 1-Naphthylamine, its 5,8-dihydro- and 5,6,7,8-tetrahydro-derivatives and 2-naphthylamine are produced from naphthalene in this way. Biphenyl gives *o*- and *p*- aminobiphenyl[150].

11.5 SYNTHESIS OF PHENOLS

11.5.1 From chloro derivatives

It is suggested that *o*- and *p*-nitrophenol (should be produced from the respective nitro chloro benzenes with sodium hydroxide in dimethyl sulphoxide) while passing air through it (or under air pressure)[151]. The use of oxidant (hydrogen peroxide) is also recommended when producing 2,4-dinitrophenol from dinitrochlorobenzene with an aqueous solution of sodium hydroxide[152].

4-(4′-Chlorophenylsulphonyl)phenol is produced from bis-(4-chlorophenyl)sulphone by treating it with 2 mol of an aqueous solution of NaOH or KOH with addition of dimethyl sulphoxide[153].

Pyrocatechol is suggested to be produced from *o*-chlorophenol by treating it with excess of a 60–85% solution of sodium hydroxide at 180–220 °C in the presence of copper or its compounds, preferably Cu_2O [154].

4,6-Dichloro-2-nitro-3-phenoxyaniline can be synthesised by adding a 50% aqueous solution of KOH to a mixture of phenol with 2-nitro-3,4,6-trichloroaniline at 130 °C [155].

When naphthol is produced by treating chloronaphthalene with sodium hydroxide at 275 °C in the presence of copper, cuprous oxide and cupric oxide or a mixture of CuO–Cu_2O, the content of 2-isomer in naphthol is higher than in the starting material in the presence of the mixture Cu–Cu_2O lower than in the initial chloronaphthalene[156].

11.5.2 From amines

1,7-Diaminonaphthalene-5-sulphonic acid (produced by the reduction of 1,7-dinitronaphthalene-5-sulphonic acid by conventional methods) is converted into 1-naphthol-7-amino-5-sulphonic acid when treated with an aqueous solution of sodium bisulphite[44]. The *leuco* compound of 1,4,5,8-tetrahydroxyanthraquinone is formed from 1,8-dihydroxy-4,5-dinitroanthraquinone by treating it with an alkaline solution of sodium dithionite[157], or by treating a mixture of 1,8-dihydroxy-4,5-diamino- and 1,5-dihydroxy-4,8-diaminoanthraquinone with the same reagent[158].

11.5.3 Replacement of hydrogen atom by hydroxy group

11.5.3.1 With oxygen

The problem of direct introduction of the hydroxy group into the aromatic nucleus has long been of interest to chemists, but no practically-applicable solution of this problem has yet been found. In recent publications palladium is often suggested as a catalyst.

It is suggested that phenol should be produced from benzene by treating it at room temperature with a mixture of oxygen and hydrogen diluted with an inert gas in the presence of a dilute solution of hydrochloric and sulphuric acids, containing small quantities of titanous sulphate and palladium on silica gel[159].

It is suggested that in the presence of palladium on silica gel, phenol from benzene should be produced by oxidising the benzene with air in formic acid as a solvent (the latter being oxidised also)[160].

The direct oxidation of benzene to phenyl acetate in acetic acid in the presence of palladium catalysts has been studied. In the liquid phase the best results are obtained with Pd–Bi catalysts on silica gel. In the gaseous phase the best results were obtained with Pd–Au catalysts under addition of cadmium acetate (also on silica gel). From chlorobenzene and anisole the main products obtainable are *o*-substituted phenyl acetates, while naphthalene gives only 1-naphthyl acetate. Phenyl acetate can be converted into phenol either by hydrolysis or by thermolysis (with a concurrent formation of ketene)[161].

The acetoxylation of anisole and chlorobenzene in the presence of Pd (O_2CMe) in acetic acid in a stream of oxygen gives *m*-substituted phenyl acetates, almost without exception. Evidently, in this case the reaction mechanism differs from that when metallic Pd is used[162].

It is suggested that phenol should be produced from benzene with oxygen by irradiation of the mixture of its vapours in the presence of cyclohexane as a promotor[163]. A survey on the oxidation of benzene to phenol is given by Denisov[164].

11.5.3.2 With peroxides

A mixture of phenol, pyrocatechol and hydroquinone has been obtained in good yield by treating excess benzene with hydrogen peroxide at 0 °C in liquid hydrogen fluoride. With this method, hydroxy compounds are produced from a number of benzene homologues[165].

Phenols are formed by reacting hydrocarbons with organic hydroperoxides (such as $PhCH_2O_2H$) in the presence of boric acid and its derivatives. When using a tertiary hydroperoxide, a small quantity of an amine (e.g. of pyridine) is added[166].

The oxidation of phenol with performic or peracetic acid is recommended to be carried out at 50–150 °C in the presence of a small quantity of phosphoric acid. The yield of the pyrocatechol and hydroquinone mixture comes to 79 %[167].

The kinetics and orientation have been studied for the case when phenol is oxidised by $K_2S_2O_8$ to the mixture of pyrocatechol and hydroquinone. Phenol, evidently, enters into the reaction as a phenolate ion[168].

Phenol esters are produced by treating aromatic hydrocarbons with corresponding diacyl peroxides and cupric chloride. With the use of di-isopropyl peroxydicarbonate, 85 % of tolylisopropylcarbonates were produced from toluene (benzoyl peroxide gave only 40 % of tolylbenzoates)[169].

Benzoates of a number of phenols were also produced from aromatic hydrocarbons by treating them with benzoyl peroxide and iodine at 90 °C. The yields are mostly *c.* 60 %[170].

Aryl esters of *o*- and *m*-nitrobenzenesulphonic acids are produced by reacting solutions of alkyl benzenes in ethylacetate with nitrobenzenesulphonyl peroxides[171].

In these reactions with peroxides, the orientation of the hydroxy or acylhydroxy group shows under all the above-stated conditions that the reaction takes place with an electrophilic reagent.

References

1. Chem. Werk Lowi. (1971). *Ger. Pat.*, 1 418 841
2. Spryskov, A. A. and Kozlov, V. A. (1969). *Izv. Vysshykh Uchebn. Zavedenii, Khim. i Khim. Tekhnol.*, **12**, 900
3. Union Oil Co. (1970). *U.S. Pat.*, 3 496 224
4. Koeberg-Telder, A. and Cerfontain, H. (1971). *Rec. Trav. Chim.*, **90**, 193
5. Guyer, P., Fleury, R. and Reich, H. U. (1968). *Chimia (Aarau)*, **22**, 40
6. Geigy. (1969). *Brit. Pat.*, 1 164 752
7. Allied Chem. Corp. (1970). *U.S. Pat.*, 3 492 342
8. Kharitonov, V. V., Spryskov, A. A. and Leshchev, V. P. (1971). *Izv. Vysshykh Uchebn. Zavedenii, Khim. i Khim. Tekhnol.*, **14**, 238
9. Allied Chem. Corp. (1970). *U.S. Pat.*, 3 501 522
10. Chem. Fabr. Witten. (1967). *U.S.S.R. Pat.*, 201 991
11. Barnett, J. W. and O'Connor, C. J. (1970). *Chem. Ind. (London)*, 1172
12. Sandoz. (1968). *U.S. Pat.*, 3 410 895
13. Gnedin, B. G., Spryskov, A. A. and Lokshina, E. G. (1967). *Izv. Vysshykh Uchebn. Zavedenii, Khim. i Khim. Tekhnol.*, **10**, 744; (1966). *U.S.S.R. Pat.*, 186 499
14. Ugine-Kuhlmann. (1968). *Fr. Pat.*, 92057
15. Karandasheva, N. N. and Serjakova, G. S. (1969). *U.S.S.R. Pat.*, 242 912
16. Gaeva, L. A., Dokunikhin, N. S. and Zemskova, I. I. (1971). *Zh. Vses. Khim. Obshchestva im. D.I. Mendeleeva*, **16**, 114
17. Gaeva, L. A. and Dokunikhin, N. S. (1969). *Anthraquinonesulphonic Acids* in *Chemistry of Anthraquinone*, 57, (Moscow, Publishing House 'Chemistry')
18. Strongin, G. M. *et al.* (1968). *U.S.S.R. Pat.*, 228 677
19. Sokolskii, G. A. and Ragulin, L. I. (1968). *U.S.S.R. Pat.*, 230 137
20. Kozlov, V. A. and Spryskov, A. A. (1968). *U.S.S.R. Pat.*, 214 533
21. Union Carbide Corp. (1970). *Brit. Pat.*, 1 190 433
22. Chevron Research Co. (1970). *Brit. Pat.*, 1 198 065
23. Crompton & Knowles Corp. (1969). *Brit. Pat.*, 1 151 545
24. Gertsev, V. V. (1969). *U.S.S.R. Pat.*, 234 398
25. Shiriaev, M. M. and Kljuev, V. N. (1969). *U.S.S.R. Pat.*, 257 499
26. Hegwood, A., Mathias, A., Williams, A. E. (1970). *Analyt. Chem.*, **42**, 11, 1272
27. Kirkland, J. J. (1960). *Analyt. Chem.*, **32**, 1388; Kozlov, V. A. and Spryskov, A. A. (1970). *Izv. Vysshykh Uchebn. Zavedenii, Khim. i Khim. Tekhnol.*, **13**, 1752
28. Parsons, J. S. (1967). *J. Gas Chromatogr.*, **5**, 254; Spryskov, A. A., Kozlov, V. A. (1968). *Izv. Vysshykh. Uchebn. Zavedenii, Khim. i Khim. Tekhnol.*, **11**, 785; (1970) **13**, 1752; Kharitonov, V. V., Spryskov, A. A., Leshchev, V. P. (1971). *Izv. Vysshykh Uchebn. Zavedenii, Khim. i Khim. Tekhnol.*, **14**, 238
29. Siggia, S., Whitlock, L. R., Tao, J. C. (1969). *Analyt. Chem.*, **41**, 1387
30. Ettre, K. and Varadi, P. F. (1963). *Analyt. Chem.*, **35**, 69
31. Siggia, S., Whitlock, L. R. (1970). *Analyt. Chem.*, **42**, 1719
32. Stehl, R. H. (1970). *Analyt. Chem.*, **42**, 1802
33. Prinsen, A. J., Loeberg-Telder, A., and Cerfontain, H. (1970). *Rec. Trav. Chim.*, **26**, 1953
34. Tisue, T. and Downs, W. J. (1969). *Chem. Commun.*, 410

35. Gordon, J. E. (1970). *J. Org. Chem.*, **35,** 2722
36. Bofors, A. G. (1970). *U.S. Pat.*, 3 507 924; *Brit. Pat.*, 1 208 502
37. Stamicarbon Chemical Co. N.V. (1967). *Belg. Pat.*, 693 500
38. Bofors, A. G. (1970). *U.S. Pat.*, 3 488 397
39. Standard Oil Co. (1970). *U.S. Pat.*, 3 492 361
40. American Cyanamid Co. (1968). *Brit. Pat.*, 1 098 717
41. L'Oréal (1969). *Brit. Pat.*, 1 143 589
42. Barvinskaya, I. K. and Spryskov, A. A. (1970). *Izv. Vysshykh Uchebn. Zavedenii, Khim. i Khim. Tekhnol.*, **13,** 802
43. BASF. (1970). *Ger. Pat. Aus.*, 1 618 109; *U.S. Pat.*, 3 506 725
44. Bayer. (1970). *Brit. Pat.*, 1 198 575
45. Institut Przemyslu Organicznego. (1968). *Brit. Pat.*, 1 135 083
46. Vorozhtsov, jr., N. N., Gerasimova, T. N., Karpova, E. N. and Lisenkova, G. S. (1960). *Zh. Vses. Khim. Obshchestva im. D.I. Mendeleeva,* **5,** 474
47. Coe, P. L., Pearl, G. M. and Tatlow, J. C. (1971). *J. Chem. Soc., C,* 604
48. Moiseeva, Z. Z. (1969). *Chemistry of Anthraquinone,* 70. (Moscow: Publishing House 'Chemistry')
49. CIBA. (1969). *Belg. Pat.*, 728 613
50. Ridden, J. M. C., Barefoot, R. R. and Roy, J. G. (1971). *Analyt. Chem.*, **43,** 1109
51. Vitse, P. and Potier, A. (1971). *Bull Soc. Chim. Fr.*, 1266
52. Hartshorn, S. R., Moddie, R. B. and Schofield, K. (1971). *J. Chem. Soc. B,* 1256
53. Britain, E. F. K., Wells, C. H. J., Paisley, H. M. and Stickley, D. J. (1970). *J. Chem. Soc. B,* 1714
54. Hoffmann, D. and Rathkamp, G. (1970). *Analyt. Chem.*, **42,** 1643
55. Schiemann, G. and Cornils, B. (1969), *Chemie und Technologie cyclischer Fluorverbindungen.* (Stuttgart: F. Enke)
56. Bil, M. (1969). *Chem. Ind. (London)*, 1466; (1970) 1080. (1971) *Chem. Ind. (London)*, 656; Clairol Ltd. (1968). *Belg. Pat.*, 718 734, 719 109
57. Dokunikhin, N. S. and Salov, B. V. (1969). *Fluoroderivatives of Anthraquinone and related Dyes* in *Chemistry of Anthraquinone,* 80. (Moscow: Publishing House 'Chemistry'); CIBA. (1969). *Brit. Pat.*, 1 170 899
58. Vorozhtsov-jr., N. N. (1970). *Zh. Vses. Khim. Obshchestva im. D.I. Mendeleeva,* 52
59. Grauskas, V. (1970). *J. Org. Chem.*, **35,** 723
60. Rozhkov, I. N., Bukhtiarov, A. V. and Knunjanz, I. L. (1970). *Dokl. Akad. Nauk SSSR,* **193,** 1322. Rozhkov, I. N., Bukhtiarov, A. V., Galpern, E. G. and Knunjanz, I. L. (1971). *Dokl. Akad. Nauk SSSR,* **199,** 369. Knunjanz, I. L., Rozhkov, I. N. and Bukhtiarov, A. V. (1971). *Izv. Akad. Nauk SSSR,* Ser. Khim. No. 6, 1369
61. Shaw, M. J., Hyman, H. H. and Filler, R. (1969). *J. Amer. Chem. Soc.*, **91,** 1563; (1970). **92,** 6498
62. Mackenzie, D. R. and Fajer, J. (1970). *J. Amer. Chem. Soc.*, **92,** 4994
63. Shien, T. C., Feit, E. D., Chernick, C. L. and Yang, N. C. (1970). *J. Org. Chem.*, **35,** 4020
64. Kollonitsch, J., Barash, L. and Doldouras, G. A. (1970). *J. Amer. Chem. Soc.*, **92,** 7495
65. Nefedov, O. M. and Ivashenko, A. A. (1968). *U.S.S.R. Pat.*, 214 530; (1970), 277 759, 279 608. *Izv. Akad. Nauk SSSR,* Ser. Khim., 446, 1417
66. Scott, P. H., Smith, C. P., Rober, E. and Churchill, J. W. (1970). *Tetrahedron Letters,* 1153
67. Olin Mathieson Chemical Corp. (1971). *U.S. Pat.*, 3 558 707, 3 580 951
68. Dorrepaal, W. and Louw, R. (1971). *Rec. Trav. Chim.*, **90,** 700
69. Universal Oil Products Co., (1967). *U.S. Pat.*, 3 347 942; Chemolimpex Magyar Vegyiaru Kulkeres Kedelmi Vallalat; (1967). *Brit. Pat.*, 1 195 716
70. Simonov, V. D. *et al.* (1970). *U.S.S.R. Pat.*, 273 184
71. Sumimoto Chemical Co. (1970). *U.S.S.R. Pat.*, 285 634
72. Zubarev, S. V. *et al.* (1970). *U.S.S.R. Pat.*, 252 319
73. Nosal, A., Pasternak, A. and Wilek, S. (1969). *Chem. Anal. (Warsaw)*, **14,** 1115
74. Monsanto Co. (1969). *Brit Pat.*, 1 213 090; Weidenbruch, M. and Böke, S. (1970). *Chem. Ber.*, **103,** 510
75. Tenneco Chemicals Inc. (1968). *U.S. Pat.*, 3 366 698
76. Shell International Research Mij N.V. (1968). *Brit. Pat.*, 1 110 029
77. Partchamazad, I. (1968). Contribution á l'étude de l'action photochimique du Cl_2 sur les hydrocarbures alkylaromatiques. *Thesis* (Universite d'Aix-Marseille)

78. Ishihara Sangyo Kaisha Ltd. (1969). *Ger. Pat.*, 1 279 672
79. Merck & Co. (1969). *U.S. Pat.*, 3 423 475
80. Koppers Co. (1968). *Brit. Pat.*, 1 128 115
81. Poznjakevich *et al.* (1970). *U.S.S.R. Pat.*, 255 232
82. Rashevskaja, S. T. *et al.* (1967). *U.S.S.R. Pat.*, 196 763
83. Allied Chem. Corp. (1970). *U.S. Pat.*, 3 530 151
84. CIBA. (1968). *Belg. Pat.*, 714 864
85. BASF. (1969). *Belg. Pat.*, 728 545
86. CIBA. (1967). *Brit. Pat.*, 1 094 503; (1966). *Fr. Pat.*, 1 448 505; (1965). *Belg. Pat.*, 670 204
87. Sargent, M. V. and Smith, D. O'N. (1970). *Tetrahedron Lett.*, 2065
88. Gulf Research & Development Co. (1969). *Brit. Pat.*, 1 163 236. National Distillers and Chem. Corp. (1970). *Brit. Pat.*, 1 179 373
89. Henry, P. M. (1971). *J. Org. Chem.*, **36**, 1886
90. Fahey, D. R. (1970). *Chem. Commun.*, 417
91. Cummings, C. A. and Milner, D. J. (1971). *J. Chem. Soc. C*, 1571
92. Crocker, H. P. and Walser, R. (1970). *J. Chem. Soc. C*, 1982
93. Grossert, J. S. and Chip, G. K. (1970). *Tetrahedron Lett.*, 2611
94. Kovacic, P., Lowery, M. K. and Field, K. W. (1970). *Chem. Rev.*, **70**, 659
95. Dewhurst, F., Shah, P. K. J. (1970). *J. Chem. Soc. C*, 1737
96. Juenge, E. C., Beal, D. A. and Duncan, W. P. (1970). *J. Org. Chem.*, **35**, 719
97. Kozlov, E. M. *et al.* (1968). *U.S.S.R. Pat.*, 213 783
98. Andrukhov, N. A. *et al.* (1970). *U.S.S.R. Pat.*, 273 186
99. Tsuji, J. and Ohno, K. (1969). *Synthesis*, No. 4, 157
100. Blum, J. and Scharf, G. (1970). *J. Org. Chem.*, **35**, 1895
101. du Pont. (1970). *Brit. Pat.*, 1 209 403
102. Gore, P. H., Hammond, S. D. and Morris, D. F. C. (1970). *Tetrahedron Lett.*, 2747
103. Akopjan, A. N. *et al.* (1969). *U.S.S.R. Pat.*, 257 485
104. Nefedov, O. M. and Ivashenko, A. A. (1968). *U.S.S.R. Pat.*, 209 434
105. Jardine, I. and Mc Quillin, F. J. (1970). *Chem. Commun.*, 626
106. L'Eplattenier, F., Matthys, P. and Calderazzo, F. (1970). *Inorg. Chem.*, **9**, 342
107. Eaborn, C., Pant, B. C., Peeling, E. R. A. and Taylor, S. C. (1969). *J. Chem. Soc. C*, 2823
108. W. R. Grace & Co. (1969). *U.S. Pat.*, 3 472 897
109. Tjurenkova, O. A. (1969). *Zh. Fiz. Khim.*, **43**, 135
110. Engelhard Industries Inc. (1969). *U.S. Pat.*, 3 423 462
111. Dovell, F. S., Ferguson, W. E. and Greenfield, H. (1970). *Ind. Eng. Chem., Prod. Res. Develop.*, **9**, 224
112. Podrebarac, E. G. and Cheng, C. C. (1970). *J. Org. Chem.*, **35**, 281
113. Bristol-Myers Co. Ltd. (1968). *Ger. Pat.*, 1 264 446
114. Geigy. (1968). *Fr. Pat.*, 1 534 361; (1969). *U.S. Pat.*, 3 474 144; du Pont. (1970). *U.S. Pat.*, 3 546 297; Uniroyal Incorporated. (1968). *U.S. Pat.*, 3 350 450; (1967). *Belg. Pat.*, 702 709
115. du Pont. (1968). *U.S. Pat.*, 3 361 819; (1970). *U.S. Pat.*, 3 499 034
116. Lager, P. (1968). Dissertationsschrift Zürich, Eidgenössische Technische Hochschule
117. Beech, W. F. (1970). *J. Chem. Soc. C*, 515
118. Ivanov, A. V. *et al.* (1969). *U.S.S.R. Pat.*, 247 294
119. ICI. (1967). *Brit. Pat.*, 1 098 236; (1966). *Belg. Pat.*, 690 846
120. *Chem. Brit.* (1969). **5**, 246
121. Atlas Chem. Ind. (1969). *U.S. Pat.*, 3 427 355
122. Bayer. (1968). *Brit. Pat.*, 1 122 597. Bayer. (1969). *Brit. Pat.*, 1 145 659
123. Korshak, V. V. *et al.* (1968). *U.S.S.R. Pat.*, 218 898
124. Otmann, G. and Kober, E. (1969). *Angew. Chem.*, **81**, 782
125. (1970). *Neth. Pat. Appl.*, 6913; (1970). *Ind. Eng. Chem.*, **62**, 415
126. du Pont. (1970). *Brit. Pat.*, 1 189 556
127. Olin Corp. (1971). *U.S. Pat.*, 3 578 713
128. Kvapil, J., Vrba, Z. and Allan, Z. J. (1970). *Collect. Czech. Chem. Commun.*, **35**, 366
129. Idoux, J. P. (1970). *J. Chem. Soc. C*, 435
130. Montecatini Edison Sp. A. (1970). *U.S. Pat.*, 3 527 805
131. Balakrishnan, T. D., Udupa, K. S., Subramanian, G. S. and Udupa, H. V. K. (1969). *Chem. Ing. Tech.*, **41**, 776
132. Hutchins, R. O., Lamson, D. W., Rua, L., Milewski, C. and Maryanoff, B. (1971). *J. Org. Chem.*, **36**, 803

133. Searle, C. E. (1970). *Chem. Brit.*, **6,** 5
134. Halcon International Inc. (1966). *U.S. Pat.*, 3 272 865
135. Halcon International Inc. (1971). *U.S. Pat.*, 3 578 714; Chem. Eng. News. (1969). **47,** No. 14, 44
136. Witco Chemical Co. (1971). *U.S. Pat.*, 3 553 268
137. Bayer. (1969). *Brit. Pat.*, 1 130 707
138. Lantz, R. (1971). *Bull. Soc. Chim. Fr.*, 249, 253
139. Plakidin, V. L. *et al.* (1968). *U.S.S.R. Pat.*, 210 184
140. Bayer. (1968). *Ger. Pat.*, 1 260 468
141. CIBA. (1968). *U.S.S.R. Pat.*, 222 274
142. Gudzenko, V. I. and Krasnoselkaja, M. I. (1970). *U.S.S.R. Pat.*, 261 392
143. Havlickova, L. and Arient, J. (1971). *Collect. Czech. Chem. Commun.*, **36,** 2074
144. Fokin, E. P. *et al.* (1969). *U.S.S.R. Pat.*, 259 854
145. Karpukhin, P. P. *et al.* (1968). *U.S.S.R. Pat.*, 232 283
146. Kurdjumova, T. N. (1969). *Chemistry of Anthraquinone,* 94 (Moscow: Publishing House 'Chemistry')
147. BASF. (1969). *Brit. Pat.*, 1 141 655
148. Bayer. (1968). *Brit. Pat.*, 1 127 085
149. Barton, J. W., Grinham, A. R. and Whitaker, K. E. (1971). *J. Chem. Soc. C,* 1384
150. Volpin, M. E., Belyi, A. A., Shur, V. B., Katkov, N. A., Nekaeva, I. M. and Kudryavtsev, R. V. (1971). *Chem. Commun.*, 246
151. F.M.C. Corp. (1969). *Brit. Pat.*, 1 143 170
152. Ovechkina, L. F. *et al.* (1969). *U.S.S.R. Pat.*, 245 798
153. ICI. (1970). *U.S.S.R. Pat.*, 252 219
154. Dow Chem. Co. (1970). *U.S. Pat.*, 3 536 767
155. Agripat S.A. (1969). *Brit. Pat.*, 1 170 098
156. Frontier Chem. Co. (1968). *U.S. Pat.*, 3 413 357
157. Belkin, I. D. *et al.* (1968). *U.S.S.R. Pat.*, 230 188
158. Karpukhin, P. P. *et al.* (1969). *U.S.S.R. Pat.*, 237 162
159. ICI. (1968). *Brit. Pat.*, 1 125 971
160. ICI. (1968). *Brit. Pat.*, 1 108 256
161. Arpe, H.-J. and Hörnig, L. (1970). *Erdöl Kohle, Erdgas, Petrochem.*, **23,** 79
162. Eberson, L. and Gomez-Gonzales, L. (1971). *Chem. Commun.*, 263
163. Timofeev, V. D. *et al.* (1969). *U.S.S.R. Pat.*, 245 128
164. Denisov, E. T. and Metelitza, D. I. (1968). *Usp. Khim.*, **37,** 1547
165. Vesely, J. A. and Schmerling, L. (1970). *J. Org. Chem.*, **35,** 4028; (1969). *Brit. Pat.*, 1 158 173; Universal Oil Products Co. (1969). *U.S. Pat.*, 3 453 332
166. Rhone-Poulenc. (1967). *U.S.S.R. Pat.*, 196 659
167. Rhone-Poulenc. (1968). *U.S.S.R. Pat.*, 223 678
168. Ogata, Y. and Akada, T. (1970). *Tetrahedron,* **26,** 5945
169. Kovacic, P., Reid, C. G. and Kurz, M. E. (1969). *J. Org. Chem.*, **34,** 3302; Kurz, M. E. and Kovacic, P. (1968). ibid., **33,** 1950
170. Kovacic, P., Reid, C. G. and Brittain, M. J. (1970). *J. Org. Chem.*, **35,** 2152
171. Dannley, R. L., Gagen, J. E. and Stewart, O. J. (1970). *J. Org. Chem.*, **35,** 3076